Catalysis of
Organic Reactions

CHEMICAL INDUSTRIES

A Series of Reference Books and Textbooks

Founding Editor

HEINZ HEINEMANN
Berkeley, California

Series Editor

JAMES G. SPEIGHT
Laramie, Wyoming

1. *Fluid Catalytic Cracking with Zeolite Catalysts,* Paul B. Venuto and E. Thomas Habib, Jr.
2. *Ethylene: Keystone to the Petrochemical Industry,* Ludwig Kniel, Olaf Winter, and Karl Stork
3. *The Chemistry and Technology of Petroleum,* James G. Speight
4. *The Desulfurization of Heavy Oils and Residua,* James G. Speight
5. *Catalysis of Organic Reactions,* edited by William R. Moser
6. *Acetylene-Based Chemicals from Coal and Other Natural Resources,* Robert J. Tedeschi
7. *Chemically Resistant Masonry,* Walter Lee Sheppard, Jr.
8. *Compressors and Expanders: Selection and Application for the Process Industry,* Heinz P. Bloch, Joseph A. Cameron, Frank M. Danowski, Jr., Ralph James, Jr., Judson S. Swearingen, and Marilyn E. Weightman
9. *Metering Pumps: Selection and Application,* James P. Poynton
10. *Hydrocarbons from Methanol,* Clarence D. Chang
11. *Form Flotation: Theory and Applications,* Ann N. Clarke and David J. Wilson
12. *The Chemistry and Technology of Coal,* James G. Speight
13. *Pneumatic and Hydraulic Conveying of Solids,* O. A. Williams

Catalysis of Organic Reactions

Edited by

Stephen R. Schmidt

W. R. Grace & Co.
Columbia, Maryland

CRC Press
Taylor & Francis Group
Boca Raton   London   New York

CRC Press is an imprint of the
Taylor & Francis Group, an **informa** business

CRC Press
Taylor & Francis Group
6000 Broken Sound Parkway NW, Suite 300
Boca Raton, FL 33487-2742

First issued in paperback 2019

© 2007 by Taylor & Francis Group, LLC
CRC Press is an imprint of Taylor & Francis Group, an Informa business

No claim to original U.S. Government works

ISBN-13: 978-0-8493-7557-6 (hbk)
ISBN-13: 978-0-367-38977-2 (pbk)

Library of Congress Cataloging-in-Publication Data

Catalysis of organic reactions / [edited by] Stephen R. Schmidt.
    p. cm.
  Includes bibliographical references and indexes.
  ISBN 0-8493-7557-6 (acid-free paper)
  1. Organic compounds--Synthesis--Congresses. 2. Catalysis--Congresses. I.
Schmidt, Stephen R. (Stephen Raymond), 1956- II. Title.

QD262.C35535 2006
547'.215--dc22   2006050677

Visit the Taylor & Francis Web site at
http://www.taylorandfrancis.com

and the CRC Press Web site at
http://www.crcpress.com

Contents

Laboratory of Industrial Chemistry, Process Chemistry Centre, Åbo Akademi University, Biskopsgatan 8, FIN-20500 Turku, Finland

xviii

Board of Editors

Chronology of Organic Reactions
Catalysis Society Conferences

Conf	Year	Chair	Location	Proceedings Publisher
1	1967	Joseph O'Connor	New York City	NYAS
2	1969	Joseph O'Connor	New York City	NYAS
3	1970	Mel Rebenstrot	New York City	NYAS
4	1973	Paul Rylander	New York City	NYAS
5	1974	Paul Rylander & Harold Greenfield	Boston	Academic Press
6	1976	Gerry Smith	Boston	Academic Press
7	1978	Bill Jones	Chicago	Academic Press
8	1980	Bill Moses	New Orleans	Academic Press
9	1982	John Kosak	South Charleston	Dekker
10	1984	Bob Augustine	Williamsburg	Dekker
11	1986	Paul Rylander & Harold Greenfield	Savannah	Dekker
12	1988	Dale Blackburn	San Antonio	Dekker
13	1990	Bill Pascoe	Boca Raton	Dekker
14	1992	Tom Johnson & John Kosak	Albuquerque	Dekker
15	1994	Mike Scaros & Mike Prunier	Phoenix	Dekker
16	1996	Russ Malz	Atlanta	Dekker
17	1998	Frank Herkes	New Orleans	Dekker
18	2000	Mike Ford	Charleston	Dekker
19	2002	Dennis Morrell	San Antonio	Dekker
20	2004	John Sowa	Hilton Head	CRC Press
21	2006	Steve Schmidt	Orlando	CRC Press

Preface

This volume of *Catalysis of Organic Reactions* compiles 57 papers presented at the 21st conference organized by the Organic Reactions Catalysis Society, (ORCS) (www.orcs.org). The conference occurred on April 2-6, 2006 in Orlando, Florida, where these papers reported on significant recent developments in catalysis as applied to production of chemicals.

Each of the papers documenting these developments and published here was edited by ORCS members (drawn from both academia and industry) and was peer-reviewed by experts in related fields of study.

The volume is organized into the following sections reflecting symposia in the conference program (including papers presented as posters):

I- Catalysis in Organic Synthesis
II- Catalytic Oxidation
III-Catalytic Hydrogenation
IV- Novel Methods in Catalysis of Organic Reactions
V- Acid and Base Catalysis
VI-"Green" Catalysis
VII-Other Topics in Catalysis

The Catalytic Hydrogenation section includes the 2006 Murray Raney Award Lecture by Professor Isamu Yamauchi, Osaka University, Japan. Similarly the Novel Methods section features an invited lecture by Gadi Rothenberg, University of Amsterdam, the 2006 Paul N. Rylander Award winner. 2005 Rylander Award winner Jean-Marie Basset is also acknowledged for presenting a lecture in the symposium on Catalysis in Organic Synthesis.

A number of recent emerging themes in catalysis appeared repeatedly in the various symposia. Examples include novel homogeneous and immobilized organometallic catalysts, semi-empirical calculation methods for catalyst selection, and synthetic processes based on "renewable", agriculturally-derived feedstocks (e.g. oils and sugars). There is a remarkable diversity of topics, some of which defied our simple attempts at sorting into a small number of categories. This reflects the current degree of specialization within our broad subject area.

Support from our sponsors greatly contributed to a successful conference. Specifically we thank these organizations: *W. R. Grace (Davison Catalysts), Parr Instrument, Degussa, North American Catalysis Society, ACS-PRF, Air Products and Chemicals, CRI Catalysts, Engelhard,, Eli Lilly & Co., Merck & Co., Süd Chemie, Umicore, Amgen and Nova Molecular Technologies, Inc.*

As Chairman I greatly appreciate the dedication, teamwork and perseverance exhibited by our ORCS Board members Alan Allgeier, Mike Ford, Anne Gaffney, Kathy Hayes, Steve Jacobson, Yongkui Sun, John Super, and Angelo Vaccari, and Executive Committee members Mike Prunier, Helene Shea, and John Sowa. And finally I am most thankful to my wife Zsuzsanna and son Tim for their unending support, tolerance, and love during my tenure as chairman and editor-in-chief.

To Zsuzsanna and Tim

I. Symposium on Catalysis in Organic Synthesis

1. On the Use of Immobilized Metal Complex Catalysts in Organic Synthesis

Christopher W. Jones,[1,2] Michael Holbach,[2] John Richardson,[1] William Sommer,[2] Marcus Weck,[2] Kunquan Yu,[1] and Xiaolai Zheng[1,2]

[1]*Georgia Institute of Technology, School of Chemical & Biomolecular Engineering,* [2]*School of Chemistry and Biochemistry, Atlanta, GA 30332 USA*

cjones@chbe.gatech.edu

Abstract

Heterogenization of homogeneous metal complex catalysts represents one way to improve the total turnover number for expensive or toxic catalysts. Two case studies in catalyst immobilization are presented here. Immobilization of Pd(II) SCS and PCP pincer complexes for use in Heck coupling reactions does not lead to stable, recyclable catalysts, as all catalysis is shown to be associated with leached palladium species. In contrast, when immobilizing Co(II) salen complexes for kinetic resolutions of epoxides, immobilization can lead to enhanced catalytic properties, including improved reaction rates while still obtaining excellent enantioselectivity and catalyst recyclability.

Introduction

Supported metal complexes (1) have been studied for many years due to their potential for combining the best attributes of both homogeneous and heterogeneous catalysis – high reaction rates and selectivities coupled with easy catalyst recovery. Unfortunately, in many cases, immobilized metal complex catalysts display the disadvantages of each class of catalysts, poor recyclability due to catalyst leaching, low reaction rates due to diffusional limitations, and poor selectivities due to the presence of multiple types of active sites. Indeed, although hundreds of different metal complexes have been immobilized on virtually every type of known catalyst support, supported metal complex catalysts still are relatively poorly understood compared to the more typical homogeneous (e.g. soluble metal complexes) and heterogeneous (e.g. supported metals) catalysts that dominate commercial processes.

To this end, we have undertaken a detailed, long-term investigation of two families of supported metal complex catalysts, supported Pd pincer complexes for use in C-C couplings such as Heck and Suzuki reactions and supported Co salen complexes for epoxide ring-opening reactions. These two catalyst systems represent interesting targets for detailed study. Pd(II) pincer palladacycles have been proposed in the literature to be well-defined, stable Pd(II) catalysts that are active in Heck or other coupling reactions (2-7), potentially via a controversial Pd(II)-Pd(IV) catalytic cycle (8). Here we summarize our studies of supported Pd(II) PCP and SCS pincer complexes on both insoluble (porous silica) and soluble supports (polymers). Co(III) salen complexes represent powerful enantioselective catalysts for epoxide ring-opening reactions (9). In this case, the design of the proper support is of paramount importance, as the transition state of the reaction involves two Co salen centers, and hence the supported system must be able to accommodate such a transition state. Here we explore our use of different soluble polymeric supports of differing backbone and side-chain structure and evaluate the role of the support on the catalytic properties in the hydrolytic kinetic resolution of *rac*-epichlorohydrin.

Results and Discussion

A variety of Pd(II) palladacycle complexes have been reported over the past decade for applications in Heck, Suzuki and other coupling reactions (10). These precatalysts appeared quite stable under reaction conditions and little evidence was observed for the formation of Pd(0), the usual form of active palladium in these coupling reactions. For this reason, a new catalytic Heck cycle was hypothesized to account for the catalytic activity observed when using these precatalysts a Pd(II)-Pd(IV) cycle, rather than the usual Pd(0)-Pd(II) cycle. Over the last 5-7 years, it has been systematically shown that bidentate palladacycles based on SC, NC and PC ligands (Figure 1) decompose to liberate soluble, ligand-free Pd(0) that is the active catalyst (11-17). However, up until 2004, tridentate palladacycle such as Pd(II) pincer complexes of NCN, SCS, or PCP ligation (Figure 1) had still been thought to potentially be stable complexes that catalyze the Heck coupling by a Pd(II)-Pd(IV) cycle (18-19). Here we share our recent results that conclusively show that SCS (20-21) and PCP (22) Pd(II) pincer complexes decompose to liberate soluble catalytic Pd(0) species, and that solid-supported Pd(II) pincers simply represent reusable soluble precatalyst sources, rather than the previously hypothesized stable recyclable catalysts (20-23).

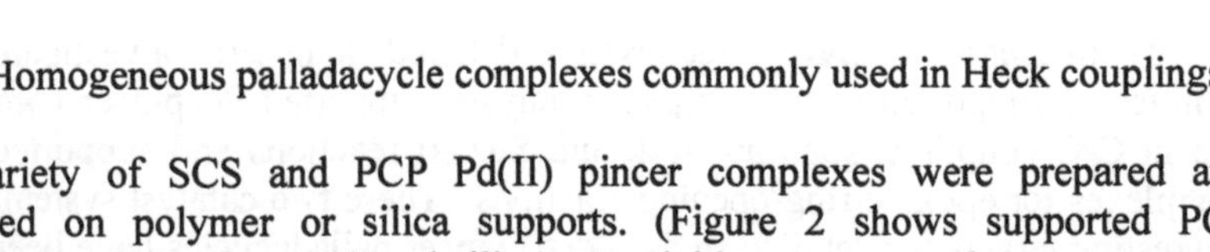

Figure 1 Homogeneous palladacycle complexes commonly used in Heck couplings.

A variety of SCS and PCP Pd(II) pincer complexes were prepared and immobilized on polymer or silica supports. (Figure 2 shows supported PCP complexes on poly(norbornene) and silica). Insoluble supports such as mesoporous silica and Merrifield resins along with soluble supports such as poly(norbornene) allowed for generalization of our observations, as all immobilized catalysts behaved similarly. The application of poisoning tests, kinetics studies, filtration tests, and three-phase tests in the coupling of iodobenzene and n-butyl acrylate showed that only Pd(0) species that leached from the supports were active. In particular, the use of poly(vinylpyridine) as a unique poison for only soluble, leached species was used to conclusively show for the first time that there was no catalytic activity associated with intact, immobilized Pd(II) pincer species (20-22). Mechanistic studies of the precatalyst decomposition pathway combining both experimental and computational efforts outlined one potential pathway for liberation of free palladium species (22).

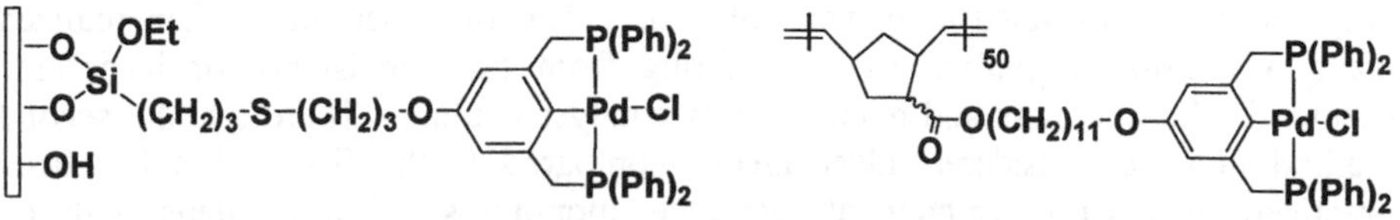

Figure 2 Immobilized PCP Pd(II) pincer complexes used in our Heck couplings. Other SCS and PCP pincers were also studied.

Our studies, as well as those of other authors, rule out the possibility of effectively using immobilized pincer complexes as recoverable and recyclable catalysts in a variety of coupling reactions that include a Pd(0) intermediate including Heck (20-25) and Stille (26) couplings. However, immobilized pincer species still find utility in reactions where a M(0) species is not required, for example, in the work of van Koten (27) and others (28). This conclusion is important because these complexes represented almost the last of the Pd(II) complexes that were thought to be truly stable under reaction conditions (N-heterocyclic carbene based CNC pincer complexes still have not been shown to decompose (29)), and thus the concept of using immobilized metal-ligand complexes of this type as stable, recyclable catalysts may not be generally feasible using currently known, well-defined metal-ligand complexes. The knowledge that all known precatalysts, where the mechanism has been clearly elucidated, operate by a Pd(0)-Pd(II) cycle now suggests that immobilization of discrete metal-ligand complexes may hinge on the development of ligands that effectively bind and stabilize Pd(0). Indeed, this suggests that other routes for obtaining high turnover numbers may be more attractive commercially, including application of homeopathic palladium (30) or supported Pd(0) metal particles (31-32). For difficult to activate reactants, Pd(II) homogeneous complexes will also continue to be important (31). Nonetheless, supported forms of precatalysts like SCS and PCP pincers may still provide utility in the production of fine chemicals if they are applied as recoverable, recyclable materials that slowly release soluble, active Pd(0) species.

Knowledge that the above catalysts operate by a Pd(0)-Pd(II) mechanism might dissuade the catalyst designer from immobilizing Pd(II) precatalysts on solid supports in the hope of achieving stable, solid catalysts. However, mechanistic knowledge in other cases can encourage the designer to create new, supported forms of catalysts. In the case of Co(III) salen complexes, it has been shown the transition state of the reaction involves two Co salen centers (9). Given this knowledge, Jacobsen has shown that the design of homogeneous complexes based on Co salen dimers (34), dendrimer-immobilized Co salens (35), and cyclic oligomeric Co salens (36) give enhanced catalytic rates as a consequence of facilitating the bimolecular transition state. Thus, proper design of immobilized Co salen catalysts could allow for not only enhanced activity compared to the homogeneous complex but also increased turn-over numbers if the catalyst can be recycled. Polymer-supported Co salens have been reported, having been synthesized via two strategies, (i) grafting reactions of salen ligands onto insoluble supports such as resins (37-38), or (ii) polymerizations of salen monomers (39-43). The first method is often realized using a multi-step grafting route, suffering from the coexistence of ill-defined species in the polymers and relatively low catalyst loading. Therefore, the second method may be considered advantageous compared to the first, yet it has been practiced only with symmetrical salens as monomers. While salens with C_2 symmetry are readily available from a synthetic point of view, polymerization or copolymerization of such monomers introduces the salen cores along the main-chain or as a crosslink of the polymer matrix, respectively, which undesirably hinders the accessibility and flexibility of the catalytic sites. Therefore, in comparison with their homogeneous counterparts, the polymer-bound salen complexes often exhibit poor enantio-control and reduced reactivity. We have derived two soluble polymer supported Co salen catalysts for the hydrolytic kinetic resolution (HKR) of racemic epichlorohydrin that contain side-chains functionalized with pendant Co(III) salen complexes.

The first system is based on a poly(norbornene)-supported Co salen and the second system is based on a poly(styrene) backbone. For the poly(norbornene) system, a homopolymers and several different copolymers were prepared (44), with varying fractions of Co salen side-chains and spacer side-chains (Figure 3, copolymer 1a-c and homopolymer 1d). For the poly(styrene) system, both homopolymers of salen-containing monomers and copolymers with styrene (45,46) were prepared (Figure 3, copolymer 2a-c and homopolymer 2d).

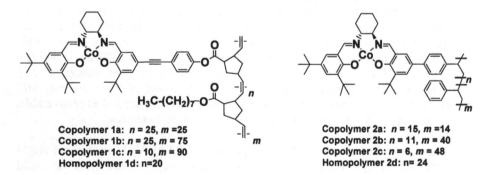

Copolymer 1a: $n = 25$, $m = 25$
Copolymer 1b: $n = 25$, $m = 75$
Copolymer 1c: $n = 10$, $m = 90$
Homopolymer 1d: n=20

Copolymer 2a: $n = 15$, $m = 14$
Copolymer 2b: $n = 11$, $m = 40$
Copolymer 2c: $n = 6$, $m = 48$
Homopolymer 2d: n= 24

Figure 3 Immobilized Co(II) salen complexes are oxidized to Co(III) and utilized in the hydrolytic kinetic resolution of *rac*-epichlorohydrin.

In the HKR of *rac*-epichlorohydrin, the Co(III) norbornene polymers **1a-d** were dissolved in a mixture of methylene chloride, reactant and chlorobenzene as an internal standard, followed by the addition of 0.7 equivalents of water to start the resolution. The addition of some methylene chloride as a solvent was necessary because the copolymers were not fully soluble in the epoxide. The reaction kinetics were studied via chiral GC-analysis. Using either the homopolymer **1d** or the two copolymers **1a** and **1b**, the (R) epoxide was fully converted after five hours to its corresponding diol, leaving pure (S) epoxide in the reaction mixture in above 99% enantiomeric excess. After this time period, 55% of the racemic epoxide was converted, meaning that all of the unwanted (R) enantiomer was consumed while only 5% of the desired epoxide was converted. This selectivity is similar to the original Jacobsen Co(III)OAc salen catalyst (53% conversion, >99% ee under solvent-free conditions). The epoxidation rates with **1a** and **1b** were slightly higher than the ones using **1d**. This observation may seem to go against the hypothesis that the reaction involves a bimetallic mechanism for the HKR, however, we suggest that the result may be a consequence of the higher backbone flexibility of the copolymers in comparison to the sterically more congested homopolymer. It is clear, however, that the density of the salen complexes in the copolymers is also important, as further dilution of the salen-moieties along the polymer backbone via use of **1c** resulted in a dramatic drop of the activity, with only 43% conversion and 80% ee after five hours). This result suggests that the extreme dilution of the catalytic moieties along the polymer backbone results in decreased reaction rates due to the low probability of two complexes being in close proximity, a prerequisite for the bimetallic catalytic pathway.

While the poly(norbornene) system represents an easy to manipulate polymeric support as it can be prepared via a living polymerization and the polymerization process is tolerant to a wide array of functional groups, it also represents a relatively expensive way of immobilizing a catalyst. To this end, we pursued poly(styrene) systems as well. The conversions and enantiomeric excesses (ee) of the substrate in the HKR of *rac*-epichlorohydrin using the styrene polymers **2b, 2d** and the homogeneous Co salen complex were again monitored by GC analysis. The catalytic data are presented in a kinetic plot of ee vs. reaction time in Figure 4. All the poly(styrene)-supported catalysts are highly reactive and enantioselective for the HKR of epichlorohydrin. As shown in Table 2, the copolymer-supported catalysts **2b** and **2c** showed the most desired catalytic performances. The remaining epichlorohydrin was determined to have enantiomeric excesses higher than 99% within one hour with a conversion of 54%. In comparison, the homogeneous Co(III) salen catalyst gave 93% ee in 49% conversion under the same reaction conditions and it took 1.5 h for it to reach >99% ee. It is noteworthy that the copolymer-supported catalysts **2a-c** in general exhibited improved reactivity and enantioselectivity compared with the homopolymer **2d**. We attributed this observation to the greater complex mobility in the copolymer-bound salen catalysts. Dilution of the salen moieties in the poly(styrene) main-chain might make the catalytic sites more accessible to the substrate. In addition, the copolymers might have more flexible polymer backbones that would increase the possibility of intramolecular cooperation between cobalt catalytic sites. Clarification of the role of polymer structure on the observed (improved) catalytic properties continues to be a subject of ongoing investigation.

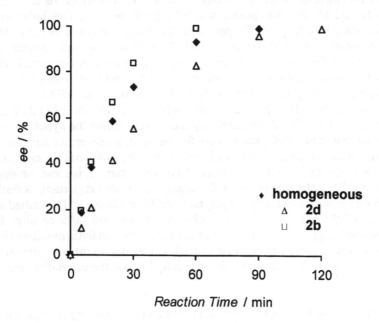

Figure 4 Kinetic plot of the HKR of *rac*-epichlorohydrin using the homogeneous and poly(styrene) supported Co(III) salen catalysts.

A key motivation for developing immobilized metal complexes lies in their potential for facile recovery and reuse in subsequent reactions. The recycling of the copolymer-bound Co(salen) complex **2b** (Table 2) was studied by precipitation of the catalyst after the HKR of epichlorohydrin by the addition of diethyl ether. The precipitated catalyst was reactivated with acetic acid and then reused under identical conditions to the first run. Whereas the enantioselectivity of the reused catalyst essentially did not change after four cycles, the catalytic reactivity decreased gradually. To evaluate whether the deactivation was due to leaching of catalyst, more racemic epichlorohydrin and water were charged into the pale yellowish organic phase from the workup of the first catalytic run of precatalyst **2b**. About 4% additional epichlorohydrin was consumed in sixty minutes. Further control experiments showed that no background reaction was detected in the absence of the catalyst or in the presence of the non-metallated salen polymer. These results indicated that, at least in part, the loss of catalysts during workup was responsible for the observed deactivation on recycle (45). It is not clear what the role of other factors such as potential morphological changes of the polymers have on the long term performance characteristics of the catalysts. This represents an area of continuing study.

Table 1 Catalytic data for homogeneous and poly(styrene) supported Co(III) salen complexes in the HKR of *rac*-epichlorohydrin.

Entry	Catalyst	Time [h]	Conv. [%]	*ee* [%]
1	**homogeneous**	1.0	49	93
		1.5	52	>99
2	**2d**	1.0	48	83
		2.0	55	>99
3	**2a**	1.0	50	90
		1.5	54	>99
4	**2b**	1.0	54	>99
5	**2c**	1.0	54	>99

Table 2 Recycle of the poly(styrene) supported Co(III) salen complex **2b** in the HKR of *rac*-epichlorohydrin.

Cycle	Time [h]	Conv. [%]	*ee* [%]
1	1.0	54	>99
2	1.5	55	>99
3	2.0	55	>99
4	2.0	53	98

Compared to existing salen systems, our new polymer-immobilized complexes perform better than the homogeneous complex due to the increased probability for bimolecular interactions among the salen complexes (Table 1). However, the cyclic oligomeric salen complexes previously described still display superior reactivity in the epoxide ring-opening reactions (36). A disadvantage of the cyclic oligomer systems is that they are more difficult to recycle via the method described here, as the low molecular weight of these catalysts makes precipitation problematic. Normally, these low molecular weight catalysts have been recycled on the bench scale by removing the volatile reactants followed by the addition of more starting material to the reaction vessel and without the isolation of the catalytic species (47). In contrast, our polymeric Co(salen) complexes, besides the desirable catalytic performance in the HKR, hold the advantage of more facile product separation and catalyst recycling (Table 2). Hence, these supported catalysts are particularly suitable for the kinetic resolution of epoxides (e.g., epichlorohydrin) that are prone to racemization in the presence of the catalysts. For larger scale production, it is anticipated that both the polymeric systems could be utilized in a membrane reactor to achieve high turnover numbers.

Experimental Section

Immobilized Pd(II) pincer complexes were prepared as described previously (20-22). Heck couplings of iodoarenes and acrylates were carried out in DMF using tertiary amines as base and reaction kinetics were monitored as previously reported (20-22).

Poly(norbornene)-supported Co(III)-Salen complexes (44) and poly(styrene)-supported Co(III)-Salen complexes (45,46) were synthesized via newly developed procedures.   In particular, a new, high-yielding, one-pot synthesis of non-symmetrical salens was developed (46). Hydrolytic kinetic resolutions were carried out at room temperature and the products were characterized by chiral GC.

Conclusions

In this work we presented two case studies of immobilized metal complex catalysts in organic synthesis. It was demonstrated that immobilization of homogeneous complexes is neither a general solution to the problem of achieving high turnovers when using expensive or toxic metals, nor does immobilization always result in compromised catalytic properties. In the study of Pd(II) pincer complexes in Heck couplings, we show that the precatalyst decomposes to liberate soluble palladium species that are the true catalysts. In contrast, we show that immobilization of Co(III) salen complexes can lead to improved catalytic rates and excellent enantioselectivities in the HKR of epoxides. Thus, in one case, immobilization leads to the same or worse catalytic properties compared to the homogeneous case, whereas in the other, proper immobilization gives a better overall catalyst.

Acknowledgements

The U.S. DOE Office of Basic Energy Sciences is acknowledged for financial support through Catalysis Science Contract No. DE-FG02-03ER15459. The authors thank Profs Peter J. Ludovice and C. David Sherrill of Georgia Tech and Prof. Robert J. Davis at the University of Virginia for helpful discussions.

References

1. C. Reyes, S. Tanielyan and R. Augustine, Chemical Industries (CRC Press), **104**, (*Catal. Org. React.*), 59-63 (2005).
2. M. Ohff, A. Ohff, M. vanderBoom and D. Milstein, *J. Am. Chem. Soc.* **119**, 11687 (1997).
3. F. Miyazaki, K. Yamaguchi and M. Shibasaki, *Tetrahedron Lett.* **40**, 7379 (1999).
4. D. Morales-Morales, R. Redon, C. Yung and C. M. Jensen, *Chem. Commun.* 1619 (2000).

5. A. S. Gruber, D. Zim, G. Ebeling, A. L. Monteiro and J. Dupont, *Org. Lett.* **2**, 1287 (2000).
6. D. E. Bergbreiter, P. L. Osburn, A. Wilson and E. M. Sink, *J. Am. Chem. Soc.* **122**, 9058 (2000).
7. R. Chanthateyanonth and H. Alper, *J. Mol. Catal. A.* **201**, 23 (2003).
8. B. L. Shaw, *New J. Chem.* **22**, 77 (1998).
9. E. N. Jacobsen, *Acc. Chem. Res.* **33**, 421 (2000).
10. J. Dupont, C. S. Consorti and J. Spencer, *Chem. Rev.* **105**, 2527 (2005).
11. A. Zapf and M. Beller, *Chem. Commun.* 431 (2005).
12. W. A. Herrmann, V. P. W. Bohm and C. P. Reisinger, *J. Organomet. Chem.* **576**, 23 (1999).
13. M. Nowotny, U. Hanefeld, H. van Koningsveld and T. Maschmeyer, *Chem. Commun.* 1877 (2000).
14. I. P. Beletskaya, A. N. Kashin, N. B. Karlstedt, A. V. Mitin, A. V. Cheprakov and G. M. Kazankov, *J. Organomet. Chem.* **622**, 89 (2001).
15. R. B. Bedford, C. S. J. Cazin, M. B. Hursthouse, M. E. Light, K. J. Pike and S. Wimperis, *J. Organomet. Chem.* **633**, 173 (2001).
16. C. Rocaboy and J. A. Gladysz, *Org. Lett.* **4**, 1993 (2002).
17. C. S. Consorti, F. R. Flores and J. Dupont, *J. Am. Chem. Soc.* **127**, 12054 (2005).
18. R. B. Bedford, C. S. J. Cazin and D. Holder, *Coord. Chem. Rev.* **248**, 2283 (2004).
19. I. P. Beletskaya and A. V. Cheprakov, *J. Organomet. Chem.* **689**, 4055 (2004).
20. K. Yu, W. J. Sommer, M. Weck and C. W. Jones, *J. Catal.* **226**, 101 (2004).
21. K. Yu, W. J. Sommer, J. M. Richardson, M. Weck and C. W. Jones, *Adv. Synth. Catal.* **347**, 161 (2005).
22. W. J. Sommer, K. Yu, J. S. Sears, Y. Y. Ji, X. L. Zheng, R. J. Davis, C. D. Sherrill, C. W. Jones and M. Weck, *Organometallics* **24**, 4351 (2005).
23. D. E. Bergbreiter, P. L. Osburn and J. D. Frels, *Adv. Synth. Catal.* **347**, 172 (2005).
24. M. R. Eberhard, *Org. Lett.* **6**, 2125 (2004).
25. K. Takenaka and Y. Uozumi, *Adv. Syn. Catal.* **346**, 1693 (2004).
26. D. Olsson, P. Nilsson, M. El Masnaouy and O. F. Wendt, *Dalton Trans.* 1924 (2005).
27. G. Rodriguez, M. Lutz, A. L. Spek and G. van Koten, *Chem. Eur. J.* **8**, 46, (2002).
28. J. Dupont, M. Pfeffer, and J. Spencer, *Eur. J. Inorg. Chem.* 1917 (2001).
29. E. Diez-Barra, J. Guerra, V. Hornillos, S. Merino and J. Tejeda, *Organometallics* **22**, 4610 (2003).
30. M. T. Reetz and J. G. de Vries, *Chem. Commun.* 1559 (2004).
31. C. R. LeBlond, A. T. Andrews, Y. K. Sun and J. R. Sowa, *Org. Lett.* **3**, 1555 (2001).
32. S. S. Prockl, W. Kleist, M. A. Gruber and K. Kohler, *Angew. Chem. Int. Ed.* **43**, 1881 (2004).
33. A. F. Littke and G. C. Fu, *Angew. Chem. Int. Ed.* **41**, 4176 (2002).

34. R. G. Konsler, J. Karl and E. N. Jacobsen, *J. Am. Chem. Soc.* **120**, 10780 (1998).
35. R. Breinbauer and E. N. Jacobsen, *Angew. Chem. Int. Ed.* **39**, 3604 (2000).
36. J. M. Ready and E. N. Jacobsen, *J. Am. Chem. Soc.* **123**, 2687 (2001).
37. D. A. Annis and E. N. Jacobsen, *J. Am. Chem. Soc.* **121**, 4147 (1999).
38. H. Sellner, J. K. Karjalainen and D. Seebach, *Chem. Eur. J.* **7**, 2873 (2001).
39. L. Canali, E. Cowan, H. Deleuze, C. L. Gibson and D. C. Sherrington, *J. Chem. Soc., Perkin Trans. 1* 2055 (2000).
40. M. D. Angelino and P. E. Laibinis, *Macromolecules* **31**, 7581 (1998).
41. T. S. Reger, and K. D. Janda, *J. Am. Chem. Soc.* **122**, 6929 (2000).
42. L. L. Welbes, R. C. Scarrow and A. S. Borovik, *Chem. Commun.* 2544 (2004).
43. M. A. Kwon and G. J. Kim, *Catal. Today* **87**, 145 (2003).
44. M. Holbach and M. Weck, *J. Org. Chem.* **71**, in press.
45. X. L. Zheng, C. W. Jones and M. Weck, *Chem. Eur. J.* **12**, 576 (2006).
46. M. Holbach, X. L. Zheng, C. Burd, C. W. Jones and M. Weck, *J. Org. Chem.* **71**, in press (2006).
47. M. Tokunaga, J. F. Larrow, F. Kakiuchi, and E. N. Jacobsen, *Science*, 277, 936 (1997).

2. Supported Re Catalysts for Metathesis of Functionalized Olefins

Anthony W. Moses,[a] Heather D. Leifeste,[a] Naseem A. Ramsahye,[a] Juergen Eckert[b] and Susannah L. Scott[a]

[a] Department of Chemical Engineering and [b] Department of Materials, University of California, Santa Barbara CA 93106-5080

sscott@engineering.ucsb.edu

Abstract

The molecular role of organotin promoters, which confer functional group tolerance on supported Re catalysts for olefin metathesis, was explored through spectroscopic and computational analysis, as well as kinetic studies. On dehydrated silica and silica-alumina, the addition of $SnMe_4$ results in two surface reactions: (i) *in situ* generation of $MeReO_3$; and (ii) capping of Brønsted acid sites. The former is responsible for catalytic activity towards polar α-olefins; thus, an independently-prepared sample of $MeReO_3$/silica-alumina catalyzed the homometathesis of methyl-3-butenoate. The latter stabilizes the catalyst: in sequential batch reactor tests involving propylene homometathesis, $MeReO_3$ deposited on silica-alumina capped with hexamethyldisilazane (to eliminate Brønsted acidity) showed activity identical to that of the perrhenate/silica-alumina catalyst promoted with $SnMe_4$. Thus, a completely Sn-free catalyst performs metathesis as efficiently as the organotin-containing perrhenate catalyst.

Introduction

Catalysts for olefin metathesis are used in relatively few large-scale industrial processes (e.g., SHOP, OCT). A few more applications are found in specialty chemicals (e.g., neohexene) and engineering plastics (e.g., PDCD). The economics of practicing metathesis on a commercial scale are impacted by the low activation efficiency and rapid deactivation of known heterogeneous catalysts, typically Mo, W or Re dispersed as oxides on supports such as silica and alumina. Furthermore, these catalysts are intolerant of polar functional groups, making it impossible to extend metathesis processing to biorenewable feedstocks such as seed oils. One notable exception is Re-based catalysts promoted by alkyltin or alkyllead reagents, which show modest activity for metathesis of functionalized olefins (1). However, once these catalysts deactivate, they are not regenerable by calcination. Thus there is considerable need for longer-lived, highly active heterogeneous catalysts that tolerate polar groups.

The mechanism of the catalytic metathesis reaction proceeds via reaction of the olefin substrate with a metal carbene intermediate, which may be generated *in situ* (as is the case for heterogeneous catalysts based on supported metal oxides and early homogeneous catalysts based on mixtures of metal halide and a main group alkylating agent), or prior to addition of the substrate (as is the case for 'well-defined' homogeneous catalysts such as those of Grubbs' and Schrock). A supported organometallic catalyst, $MeReO_3$ on silica-alumina, has also been reported to show activity in olefin metathesis. In solution, $MeReO_3$ does not react with α-olefins, nor does the silica-alumina support catalyze olefin metathesis. However, $MeReO_3$ supported on silica-alumina is effective for the metathesis of both simple and functionalized olefins at room temperature, without further thermal or chemical activation (2-4).

Deposition of white, air-stable $MeReO_3$ either by sublimation or from solution onto calcined, dehydrated silica-alumina generates a brown, air-sensitive solid. Evidence from both EXAFS and DFT calculations suggest that Lewis acidic aluminum centers on silica-alumina represent the most favorable chemisorption sites (5). One Re=O bond is substantially elongated due to its interaction with a distorted four-coordinate Al site. Coordination of Re to an adjacent bridging oxygen also occurs, creating the rigidly-bound surface organometallic fragment shown in Scheme 1. Interaction with a Lewis acid is known to promote tautomerization of $MeReO_3$ (6), leading (at least transiently) to a carbene. However, the participation of this carbene tautomer in initiating metathesis has not been established.

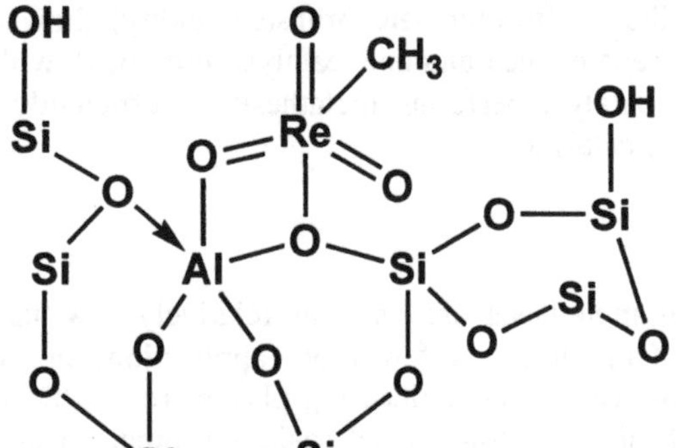

Scheme 1 Structure of $MeReO_3$ dispersed on the surface of dehydrated silica-alumina, as established by EXAFS and DFT (5).

The role of SnR_4 promoters in increasing activity and conferring functional group tolerance on supported perrhenate catalysts was originally suggested to involve *in situ* formation of organorhenium species such as $RReO_3$ (2); however, direct evidence for their participation has not been previously sought. When a perrhenate/silica-alumina catalyst is treated with $SnMe_4$, methane is evolved. This has been interpreted as evidence for carbene formation via double methylation of perrhenate, followed by α-H elimination. The formation of tin perrhenates has also been discussed (7). The question of the presence (or not) of tin in the active site is crucial, in view of the detrimental effect of tin on the ability to regenerate the deactivated catalyst.

Experimental Section

Materials. Methyltrioxorhenium, NH_4ReO_4 and Re_2O_7 were purchased from Aldrich and used as received. The silica-alumina was Davicat 3113 (7.6 wt.% Al, BET surface area 573 m^2/g, pore volume 0.76 cm^3/g), provided by Grace-Davison (Columbia, MD). For reactions involving $MeReO_3$, silica-alumina was pretreated by calcination for 12 h under 350 Torr O_2 at 450°C to remove adsorbed water, hydrocarbons, and carbonates, then allowed to cool to room temperature under dynamic vacuum. The silica was Aerosil 200 (BET surface area 180 m^2/g, with no significant microporosity) from Degussa (Piscataway, NY).

Hexamethyldisilazane (>99.5%, Aldrich) and tetramethyltin (>99%, Aldrich) were both subjected to several freeze-pump-thaw cycles to remove dissolved gases and stored over P_2O_5 in evacuated glass reactors. A solution of 0.25 mL methyl-3-butenoate (Aldrich) in 2 mL dry hexanes (Fisher) was prepared under N_2 and dried by stirring over P_2O_5 before use. Propylene (99.5%, polymer grade, Matheson Tri-Gas) was purified by passing through a trap containing BTS catalyst (Fluka) and activated molecular sieves (4 Å, Aldrich), and was stored in a Pyrex bulb over activated sieves.

Catalyst preparation. All operations were performed with strict exclusion of air and moisture, either on a high vacuum line (base pressure 10^{-4} Torr) or in a N_2-filled glove box equipped with O_2 and moisture sensors. Silica-alumina-supported perrhenate catalysts were prepared by stirring silica-alumina with aqueous ammonium perrhenate, air-drying at 80°C for >1 hr then drying under dynamic vacuum at 450°C for 4 hrs, followed by calcination in 250 Torr O_2 at 450°C for 16 h, heating under dynamic vacuum at 450°C for 1 hr, and cooling to room temperature under vacuum. $MeReO_3$ was deposited on calcined silica-alumina either by vapor deposition under reduced pressure, or from a solution of dry hexanes (1 mg/mL) prepared under N_2. In both methods, the white silica-alumina acquired a light brown color and became highly air-sensitive.

Re loadings were determined by quantitative extraction, followed by UV spectrophotometric analysis. Samples containing ca. 30 mg silica-alumina were first weighed in an inert atmosphere. (The mass of calcined silica-alumina increases up to 15% upon exposure to air, due to adsorption of atmospheric moisture.) Re was extracted as perrhenate by stirring overnight in air with 5 mL 3 M NaOH. Samples were diluted to 25 mL with 3 M H_2SO_4 and filtered. The Re concentration was determined at 224 nm, using a calibration curve prepared with NH_4ReO_4.

Capping. Capped silica-alumina was prepared by vapor phase transfer of hexamethyldisilazane ($\geq$ 99.5%, Aldrich) onto calcined silica-alumina until there was no further uptake, as indicated by stabilization of the pressure. The reactor was evacuated and the material heated to 350°C under dynamic vacuum for 4 h to remove ammonia produced during the capping reaction.

Kinetics. The catalysts were loaded into a glass batch reactor (volume ca. 120 mL) in a glovebox. The reactor was removed from the glovebox and evacuated. The section of the reactor containing the catalyst was immersed in an ice bath at 0°C in order to control the rate of the reaction on a readily-monitored timescale, as well as to maintain isothermal reaction conditions. Propylene was introduced at the desired pressure via a high vacuum manifold. Aliquots of 1.9 mL were expanded at timed intervals into an evacuated septum port that was separated from the reactor by a stopcock. 50 µL samples of the aliquot were removed with a gas-tight syringe via a septum. Gases were analyzed by FID on a Shimadzu GC 2010 equipped with a 30 m Supelco® Alumina Sulfate PLOT capillary column (0.32 mm i.d.). Quantitation was achieved using the peak area of the small propane contaminant present in the propene as an internal standard.

Computational analysis. Calculations were performed on an Intel Xeon computer running Linux, as well as the VRANA-5 and VRANA-8 clusters at the Center for Molecular Modeling of the National Institute of Chemistry (Ljubljana, Slovenia), using the DFT implementation in the Gaussian03 code, Revision C.02 (8). The orbitals were described by a mixed basis set. A fully uncontracted basis set from LANL2DZ was used for the valence electrons of Re (9), augmented by two f functions (ζ = 1.14 and 0.40) in the full optimization. Re core electrons were treated by the Hay-Wadt relativistic effective core potential (ECP) given by the standard LANL2 parameter set (electron-electron and nucleus-electron). The 6-31G** basis set was used to describe the rest of the system. The B3PW91 density functional was used in all calculations.

Results and Discussion

Reaction of silica-supported perrhenate with SnMe$_4$. Computational analysis of the reaction of oxide-supported perrhenates with SnMe$_4$ was accomplished using cube models to represent the oxide surface. Cage-like structures, such as the partially and fully condensed silsesquioxanes (10), are good computational models for silicon-based oxide surfaces because of their constrained Si-O-Si angles (11,12), and because of their oxygen-rich nature. Perrhenate was attached to a silsesquioxane monosilanol cube, Scheme 2, to represent the grafted site $\equiv$SiOReO$_3$. The optimized geometry displays a single SiO-Re attachment. Transmetalation of the perrhenate cube by SnMe$_4$, resulting in displacement of MeReO$_3$ and attachment of a trimethyltin fragment to the silsesquioxane framework, is slightly exothermic (by 4 kJ/mol).

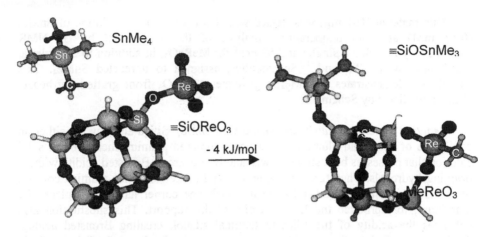

Scheme 2 Reaction of perrhenate attached to silsesquioxane cube (a computational model for the silica surface) with SnMe₄ is predicted to liberate MeReO₃.

Although perrhenate/silica is not itself active as an olefin metathesis catalyst, the model reaction shown in Scheme 2 is of interest because the expected product, $MeReO_3$, does not chemisorb onto silica, and can therefore be recovered. To investigate this prediction, perrhenate/silica was prepared according to a literature method (13). A sample of silica was first calcined at 1100 °C for 23 h to generate strained siloxane-2 rings (0.12/nm²), eq 1. This material was treated with Re_2O_7 vapor at 250°C under 250 Torr O_2, to generate cleanly the silica-supported perrhenate in the absence of water, eq 2.

$$(1)$$

$$(2)$$

This material (167 mg) was treated with 1 mL of a CDCl$_3$ solution of SnMe$_4$ (0.17 mM) at room temperature. Analysis of the supernatant by ^{1}H NMR spectroscopy revealed a singlet at 2.63 ppm for MeReO$_3$, in addition to a singlet at 0.081 ppm with ^{117}Sn and ^{119}Sn satellites, assigned to unreacted SnMe$_4$. This experiment demonstrates that Me$_4$Sn generates MeReO$_3$ from grafted perrhenate sites, as predicted by Scheme 2.

Reaction of perrhenate/silica-alumina with SnMe$_4$. Silica-alumina is neither an admixture of silica and alumina, nor a poorly ordered aluminosilicate, but a solid solution that possesses both strong Lewis acidity and strong Brønsted acidity without bridging hydroxyls. Therefore, our computational model for silica-alumina consists of a silsesquioxane monosilanol cube in which one corner has been replaced by aluminum, to reproduce the Lewis acidity of the support. This substitution also enhances the acidity of the adjacent terminal silanol, creating Brønsted acidity. Perrhenate grafted to the aluminosilsesquioxane cube is shown in Scheme 3. In contrast to the simple C$_{3v}$ symmetry of the perrhenate/silsesquioxane model (Scheme 2), the optimized structure obtained for silica-alumina adopts a lower symmetry in order to accommodate a Lewis acid-base interaction between an oxo ligand of the grafted perrhenate fragment and the Al center.

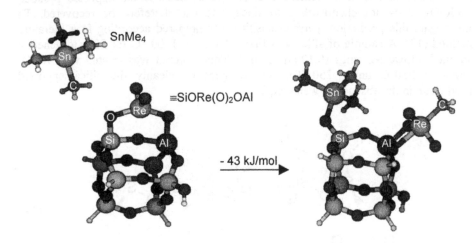

Scheme 3 Reaction of perrhenate on aluminosilsesquioxane cube (model for silica-alumina) with SnMe$_4$ is predicted to form grafted MeReO$_3$.

Transmetalation of the perrhenate/aluminosilsesquioxane cube model with SnMe$_4$ is considerably more exothermic than for the perrhenate/silsesquioxane cube model. A similar grafted trimethyltin fragment is formed, as is MeReO$_3$; however, the latter is not liberated. It remains bound to the aluminosilsesquioxane cube via the Lewis acid-base interaction with the Al center. The optimized structure also contains a Lewis acid-base interaction between Re and an adjacent framework oxygen

(AlOSi) of the cube. This predicted structure is practically the same as that determined experimentally by direct deposition of MeReO$_3$ on amorphous silica-alumina, as demonstrated by EXAFS (5).

Since MeReO$_3$ chemisorbs irreversibly on silica-alumina, we did not expect to be able to retrieve it as a soluble species upon SnMe$_4$ treatment of perrhenate/silica-alumina. However, exposure of this material to SnMe$_4$ resulted in an immediate color change from white to brown, similar to that observed upon treatment of unmodified silica-alumina with MeReO$_3$. Evidence for the possible involvement of the *in situ*-generated MeReO$_3$ in olefin metathesis was then sought via reactivity studies.

Metathesis of functionalized olefins. SnR$_4$-promoted perrhenate catalysts are known to promote the metathesis of olefins bearing polar functional groups (1). The activity of silica-alumina-supported MeReO$_3$ towards functionalized olefins was also tested, in the homometathesis of methyl-3-butenoate, eq 3. The reaction was conducted at 15°C in pentane. The progress of the reaction was followed by monitoring the evolution of ethylene in the head space, Figure 1, yielding the pseudo-first-order rate constant $k_{obs} = (1.3 \pm 0.3) \times 10^{-3}$ s^{-1}. The presence of the non-volatile dimethyl ester of hex-3-enedoic acid in the liquid phase was confirmed at the end of the reaction by GC/MS.

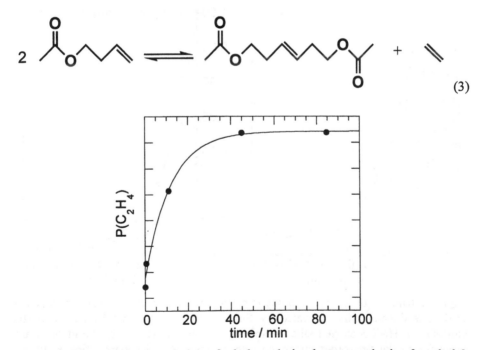

(3)

Figure 1 Time-resolved evolution of ethylene during homometathesis of methyl-3-butenoate, catalyzed by 10 mg MeReO$_3$/SiO$_2$-Al$_2$O$_3$ (8.8 wt% Re) in pentane at 15°C. The solid line is the curve-fit to the first-order integrated rate equation.

Metathesis activity. A quantitative comparison of metathesis activities was made in the gas phase homometathesis of propylene. The reaction kinetics are readily monitored since all olefins (propylene, ethylene, *cis-* and *trans-*2-butylenes) are present in a single phase. Metathesis of 30 Torr propylene was monitored in a batch reactor thermostatted at 0 °C, in the presence of 10 mg catalyst. The disappearance of propylene over perrhenate/silica-alumina (0.83 wt% Re) activated with SnMe$_4$ is shown in Figure 2a. The propylene-time profile is pseudo-first-order, with k_{obs} = $(1.11 \pm 0.04) \times 10^{-3}$ s^{-1}. Subsequent additions of propylene to the catalyst gave a slightly lower rate constant, $(0.67 \pm 0.02) \times 10^{-3}$ s^{-1}. The pseudo-first-order rate constants are linearly dependent on Re loading, Figure 3. The slope yields the second-order rate constant $k = (13.2 \pm 0.2)$ s^{-1} (g Re)$^{-1}$ at 0°C.

A similar experiment was performed with MeReO$_3$ supported on silica-alumina. However, prior to deposition of the organometallic catalyst, the support was first treated with hexamethyldisilazane (HMDS), eliminating Brønsted acid sites on the silica-alumina surface by converting them to unreactive silyl ethers (14). This serves to reproduce one of the effects of treating perrhenate/silica-alumina with SnMe$_4$, since the latter is protonated by the hydroxyl groups to generate stannyl ethers and methane.

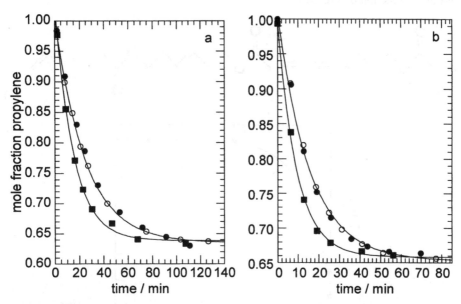

Figure 2 Kinetics of gas-phase propylene homometathesis at 0°C, catalyzed by (a) perrhenate/silica-alumina activated by SnMe$_4$ (10 mg, 0.83 wt % Re); and (b) MeReO$_3$ on HMDS-capped silica-alumina (10 mg, 1.4 wt % Re). Solid lines are curve-fits to the first-order integrated rate equation. Solid squares: first addition; solid circles: second addition; open circles: third addition of propylene (30 Torr) to the catalyst.

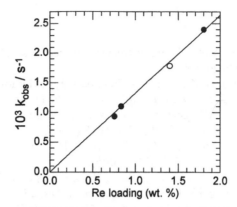

Figure 3 Dependence of pseudo-first-order rate constants measured at 0°C for propylene homometathesis, on the Re loading in 10 mg samples of two kinds of supported Re catalysts: $SnMe_4$-promoted perrhenate/silica-alumina (solid circles) and $MeReO_3$ on HMDS-capped silica-alumina (open circle).

The rate of reaction of propylene over the $MeReO_3$/HMDS/silica-alumina catalyst (1.4 wt% Re) is shown in Figure 2b. The profile is similar to that of the Sn-promoted perrhenate catalyst, with $k_{obs} = (1.78 \pm 0.09) \times 10^{-3}$ s^{-1}, and the activity responds similarly to subsequent additions of propylene. In fact, the pseudo-first-order rate constant for the organometallic catalyst lies on the same line as the rate constants for the Sn-promoted perrhenate catalyst, Figure 3. Therefore we infer that the same active site is generated in both organometallic and promoted inorganic catalyst systems.

Conclusions

The molecular role of the $SnMe_4$ promoter, which activates supported perrhenate metathesis catalysts and confers functional group tolerance, appears to be to generate $MeReO_3$ *in situ*. The promoted inorganic catalyst is kinetically indistinguishable from an organometallic catalyst made directly from $MeReO_3$. The organotin reagent simultaneously caps the surface hydroxyls, by a mechanism analogous to the reaction of HMDS with Brønsted sites. We conclude that a bimetallic (Sn/Re) active site is *not* required for the metathesis of polar olefins; consequently design of regenerable catalysts without Sn is feasible. Understanding the detailed mechanism of olefin metathesis by $MeReO_3$ will be key to creating highly active and robust solid catalysts for the metathesis of functionalized olefins.

Acknowledgements

This work was funded by the U.S. Department of Energy, Basic Energy Sciences, Catalysis Science Grant No. DE-FG02-03ER15467.

References

1. K. J. Ivin and J. C. Mol, *Olefin Metathesis and Metathesis Polymerization*, Academic, San Diego, 1997.
2. W. A. Herrmann, J. G. Kuchler, J. K. Felixberger, E. Herdtweck and W. Wagner, *Angew. Chem., Int. Ed. Engl.*, **27**, 394-396 (1988).
3. W. A. Herrmann, W. Wagner, U. N. Flessner, U. Volkhardt and H. Komber, *Angew. Chem., Int. Ed. Engl.*, **30**, 1636-1638 (1991).
4. A. M. J. Rost, H. Schneider, J. P. Zoller, W. A. Herrmann and F. E. Kühn, *J. Organomet. Chem.*, **690**, 4712-4718 (2005).
5. A. W. Moses, N. A. Ramsahye, C. Raab, H. D. Leifeste, S. Chattopadhyay, B. F. Chmelka, J. Eckert and S. L. Scott, *Organometallics*, **25** (2006).
6. C. Zhang, I. A. Guzei and J. H. Espenson, *Organometallics*, **19**, 5257-5259 (2000).
7. J. C. Mol, *Catal. Today*, **51**, 289-299 (1999).
8. M. J. Frisch et al., *Gaussian 03, Revision C.02*, Gaussian, Inc.: Wallingford, CT, 2004.
9. M. A. Pietsch, T. V. Russo, R. B. Murphy, R. L. Martin and A. K. Rappe, *Organometallics*, **17**, 2716-2719 (1998).
10. F. J. Feher and T. A. Budzichowski, *Polyhedron*, **14**, 3239-3253 (1995).
11. B. Civalleri, E. Garrone and P. Ugliengo, *Chem. Phys. Lett.*, **299**, 443-450 (1999).
12. J. Sauer and J.-R. Hill *Chem. Phys. Lett.*, **218**, 333-337 (1994).
13. S. L. Scott and J. M. Basset, *J. Am. Chem. Soc.*, **116**, 12069-12070 (1994).
14. D. J. Rosenthal, M. G. White and G. D. Parks, *AIChE J.*, **33**, 336-340 (1987).

3. Catalytic Hydrogenation of a Schiff's Base over Pd/Carbon Catalysts: Kinetic Prediction of Impurity Fate and Byproduct Formation

Steve S.Y.Wang, William F. Merkl, Hyei-Jha Chung, Wendel Doubleday and San Kiang

Process Research and Development, Bristol-Myers Squibb Company, One Squibb Drive, New Brunswick, NJ 08903-0191
steve.y.wang@bms.com

Abstract

In situ reduction of Schiff's bases is a common reaction used in the preparation of pharmaceutical intermediates. Muraglitazar (PPAR α/γ dual agonist) is being evaluated for the treatment of type II diabetes. We have recently carried out a large scale hydrogenation of an imine, prepared through the condensation of an aromatic aldehyde with an amine, to produce the final intermediate in the muraglitazar synthesis. Studies were carried out for better process understanding. We present a kinetic analysis using the Langmuir-Hinshelwood approach to model the complex reaction network. This knowledge has led to process conditions that minimize the formation of potential impurities, resulting in a more robust process.

Introduction

Intermediate C is prepared by a two-step telescoped reductive amination as depicted in the reaction scheme below. A solution of glycine methyl ester free base in methanol is generated from the corresponding hydrochloride and triethylamine. Schiff's base B is formed by condensation of the free base with the aldehyde A. Catalytic hydrogenation of Schiff's base is subsequently carried out at 40°C using 5%Pd/C and a hydrogen pressure of 45psig. With these process conditions, numerous impurities were identified. The impurity profile strongly depends on hydrogenation performance parameters related to the type of reactor, operating conditions, catalyst loading and reaction time. It is essential to limit impurity formation in order to maximize the yield and to minimize isolation and purification complexity.

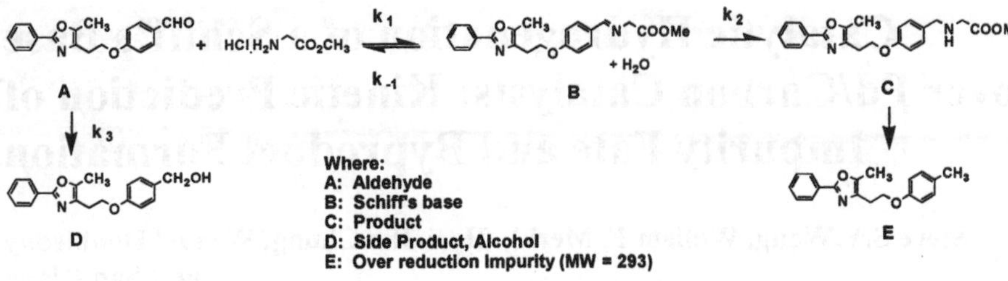

Where:
A: Aldehyde
B: Schiff's base
C: Product
D: Side Product, Alcohol
E: Over reduction Impurity (MW = 293)

The objective of the present study is to provide a kinetic evaluation for the rate of formation of byproduct D, impurity E and possibly other impurities related to the intermediate Schiff's base B.

Kinetic Analysis

Liquid phase hydrogenation catalyzed by Pd/C is a heterogeneous reaction occurring at the interface between the solid catalyst and the liquid. In our one-pot process, the hydrogenation was initiated after aldehyde A and the Schiff's base reached equilibrium conditions (A⇔B). There are three catalytic reactions A ⇒ D, B ⇒ C, and C ⇒ E, that occur simultaneously on the catalyst surface. Selectivity and catalytic activity are influenced by the ability to transfer reactants to the active sites and the optimum hydrogen-to-reactant surface coverage. The Langmuir-Hinshelwood kinetic approach is coupled with the quasi-equilibrium and the two-step cycle concepts to model the reaction scheme (1,2,3). Both A and B are adsorbed initially on the surface of the catalyst. Expressions for the elementary surface reactions may be written as follows:

$A \Leftrightarrow A^*$
$A^* + H_2 \Rightarrow D + *$
$B \Leftrightarrow B^*$
$B^* + H_2 \Rightarrow C + *$

The equilibrium constants for the adsorption of the aldehyde and Schiff's base are $K_A = k_{a1}/k_{a-1}$ and $K_B = k_{b1}/k_{b-1}$ respectively. Product C may adsorb but is less competitive than the surface adsorption of the Schiff's base B and aldehyde A. The total surface coverages are expressed as the sum of the adsorbed species and empty sites $\Theta_A + \Theta_B + \Theta^\circ = 1$.

The expressions for Θ_A and Θ_B are:

$$\Theta_B = K_B [B] / (1 + K_A [A] + K_B [B]) \tag{1}$$
$$\Theta_A = K_A [A] / (1 + K_A [A] + K_B [B]) \tag{2}$$

The rate expressions for formation of product C and alcohol D are:

Rate (product C formation) = k_2 K_B [B] [H2] Θtotal / (1 + K_A [A] + K_B [B])   (3)

Rate (alcohol D formation) = k_3 K_A [A] [H2] Θtotal / (1 + K_A [A] + K_B [B])   (4)

Where

k_2 is the rate constant for product C formation.

K_B is equal to k_{b1}/k_{b-1}, the adsorption equilibrium constant for Schiff's base B.

k_3 is the rate constant for product D formation.

K_A is equal to k_{a1}/k_{a-1}, the adsorption equilibrium constant for aldehyde.

Θtotal is the total surface fractional coverage (equals one).

By dividing equation (4) by equation (3), the ratio of alcohol to product formation is:

Ratio=$k_3 K_A$[A]/k_2 K_B [B]   (5)

Separate hydrogenation kinetic studies were performed using the same Engelhard Escat 142 catalyst (edge coated, reduced) (4) to compare the adsorption strength and hydrogenation rate between the Schiff's base and aldehyde. The kinetics are modeled using a conventional Langmuir-Hinshelwood expression to correlate the initial rate with the substrate concentration (5).

Experimental Section

(a) Catalytic hydrogenation

In our development studies, Endeavor (5 mL) and Buchi (1L) reactor systems were used to screen catalysts and to evaluate the impurity profile under various process conditions. Hydrogenation kinetic studies were carried out using a 100 mL EZ-seal autoclave with an automatic data acquisition system to monitor the hydrogen uptake and to collect samples for HPLC analysis. Standard conditions of 5 g of aldehyde in 25 mL ethyl acetate and 25 mL methanol with 0.5 g of 5%Pd/C Engelhard Escat 142 were used in this investigation. For the Schiff's base formation and subsequent hydrogenation, inline FT-IR was used to follow the kinetics of the Schiff's base formation under different conditions. Tables 1 and 2 show the changes in the substrate concentration under different conditions. Both experiments were carried without any limitations of gas-liquid mass transfer.

Table 1. Hydrogenation of aldehyde A to form the alcohol D
             40°C, 30psig, 2500 rpm in autoclave reactor

Time (min)	H2 uptake (mL)	A (mmole)	dA/dt (mmole/min)

0	0	16.28	--
15	63.5	14.14	0.142
30	120.0	11.8	0.147
45	174.6	9.48	0.155
60	217.8	8.75	0.049
75	256.0	8.62	0.027
90	287.6	8.23	0.017
190	316.0	5.42	0.028

Table 2. Hydrogenation of Schiff's base B, 38 °C, 30 psig, 700 rpm in a Parr
reactor

Time (min)	Schiff's Base Conc. (mmole)	dB/dt (mmole/min)
0	16.0	-
10	5.67	1.034
15	2.34	0.664
20	0.53	0.362
25	0.029	0.101
30	0.019	0.002

Based on the Langmuir-Hinshelwood expression derived for a unimolecular reaction system (6); Rate $=k'$ Ks (substrate) $/[1 +$ Ks (substrate)] , Table 3 shows boththe apparent kinetic rate and the substrate concentration were used to fit against the model. Results show that the initial rate is zero-order in substrate and first order in hydrogen concentration. In the case of the Schiff's base hydrogenation, limited aldehyde adsorption on the surface was assumed in this analysis. Table 3 shows a comparison of the adsorption equilibrium and the rate constant used for evaluating the catalytic surface.

Table 3. Surface catalytic behavior comparison

Reaction 40°C, 30 psig	Adsorption Equilibrium Constant K (mmole^{-1})	Rate constant k' (min $^{-1}$)
A → D	$K_A = 0.24$	$k_3 = 0.28$
B → C	$K_B = 0.55$	$k_2 = 1.32$

Calculated adsorption equilibrium constants indicate the Schiff's base is adsorbed more favorably on the catalyst surface than the aldehyde. This observation is consistent with "situation kinetics " occurring during the initial stage of the hydrogenation. The apparent rate constant shows that the product C formation is much faster than the alcohol formation.

(b) Schiff's base formation

Schiff's base formation occurs by condensation of the free amine base with aldehyde A in EtOAc/MeOH. The free amine base solution of glycine methyl ester in methanol is generated from the corresponding hydrochloride and triethylamine. Table 4 shows the reaction concentration profiles at 20-25°C. The Schiff's base formation is second order with respect to both the aldehyde and glycine ester. The equilibrium constant (ratio k(forward)/ k(reverse)) is calculated to be 67.

Table 4. Reaction kinetics of Schiff base formation

Time (min)	Aldehyde A (molar)	Glycine methyl. ester·HCL(molar)	Schiff's base B (molar)
0.65	0.365	0.294	0.031
2.42	0.303	0.232	0.093
5.29	0.242	0.171	0.154
7.52	0.211	0.14	0.185
10.8	0.18	0.109	0.216
16.18	0.149	0.078	0.247
27.21	0.118	0.047	0.278
44.28	0.10	0.029	0.296
72.2	0.09	0.019	0.306

k(forward) = 0.006695 L/mol.sec, k(reverse) = 0.0001 L/mol.sec

Impurity Fate and Byproduct Formation

The Schiff's base hydrogenation is the second step of a telescoped reductive amination and is carried out in the presence of the aldehyde. When the Schiff's base is initially prepared, the magnitude of the equilibrium concentration of aldehyde A is two orders lower than the Schiff's base B. In the reaction network, catalytic hydrogenation of A and B occur simultaneously. Based on the adsorption strength and catalytic activity comparison between A and B shown in Table 4, k_2 x K_B is ten times higher than k_3 x K_A. Therefore, the ratio of alcohol to product formation, Eq (5), is about 10^{-3}. This result indicates that the alcohol formation is not significant in the reaction network. Since the Schiff's base is the dominant adsorbed species on the catalyst during hydrogenation, the product C molecules do not compete strongly with the Schiff's base for the palladium surface adsorption. No E from the reduction of product C is expected until the Schiff's base transformation is almost complete.

The Schiff's base catalytic hydrogenation rate can be expressed as

dC/dt = k $[H_2]_L$ [Schiff base*]

$= k_2 K_B [B] [H_2]_L / (1 + K_B [B])$

where

[B] is the Schiff 's base concentration in the reaction solution

$[H_2]_L$ is the hydrogen concentration in the reaction mixture

The formation rate of product C is dependent on the concentration of adsorbed Schiff's base and the molecular hydrogen dissolved in the liquid phase. The rate constant was determined in the temperature range of 10 to 45 °C. From the results an apparent activation energy (Eap) of 40.2 kJ/mole and a pre-exponential factor A of 2.64 x 10^5 mol $min^{-1}g\cdot cat^{-1}$ were calculated. The proposed kinetic model suggests that a lower hydrogen concentration in the solution may slow down the desired transformation. The Schiff's base is the most strongly bound component on the palladium and has a high turnover frequency. In the case of gas-liquid mass transfer limitation (7), the reaction takes longer or may not even proceed to completion. The reduced Schiff's base C has the potential to undergo a second reductive alkylation with starting material A to form a dimer and subsequently produce other hydrogenated dimer impurities.

Maintaining the hydrogenation under kinetic control provides limited alcohol formation and avoids over reduction of product C. The performance of a hydrogenator depends on the gas-liquid mass transfer characteristics Kla (8). Possible operating scenarios with their observed impurity profiles are summarized in Table 5.

Table 5.   Impurity formation related to process conditions

Mass Transfer Characteristics	High catalyst loading	Low catalyst loading
high Kla	a. form over-reduction impurity E b. decrease process yield	a. less impurities E & D b. less palladium residue
low Kla	a. form side product D b. over reduction impurity E increased with time and catalyst loading	a. reaction will not go to completion b. C reacts with A to form dimers (not shown in the reaction scheme)

All experiments were carried out in a 1 L Buchi hydrogenator.
The mass transfer coefficient Kla range was 0.01 – 0.9 L/sec.

Conclusion

Using the quasi-equilibrium and two-step reaction concepts in the catalytic cycle, the hydrogenation kinetics of Schiff's base B were investigated. The analysis showed that strong Schiff's base adsorption provided rapid reduction and led to limited byproduct and impurity formation. The proposed mechanism suggested that lower catalyst loading or hydrogen diffusion limitations would slow down the desired transformation and lead to enhanced impurity formation. This knowledge led to the design of a more robust process and a successful scale up.

Acknowledgements

We gratefully acknowledge the support provided by the project team for sharing the analytical methods, in-process monitoring and process scale-up experience, and Engelhard Corporation for providing catalysts and characterization supports. We thank also Dr. W.L. Parker for fruitful discussions.

References

1. M. Boudart and G. Djega-Mariadassou, *Catal. Lett.*, 29, 7 (1994).
2. M. Boudart and K. Tamaru., *Catal.Lett.*, 9, 15 (1991).
3. M. Boudart, *J. AIChE*, 18, 3 (1972).
4. S.Y.Wang, J.Li, K. TenHuisen, J. Muslehiddinoglu, S. Tummala, S. Kiang and J.P.Chen, *Catalysis of Organic Reactions*, Marcel Dekker, Inc., New York p.499 (2004)
5. M. Boudart and G. Djega-Mariadassou, *Kinetics of Heterogeneous Catalytic Reactions*, Princeton University Press, (1984)
6. A. Stanislaus and B. H. Cooper, Catal. *REV-SCI. ENG.*, 36, 1 (1994)
7. A. Deimling, B.M. Karandikav, Y.T. Shah, N.L.Carr, *Chem Eng. J.* 29, 140 (1984)
8. F. Baier, The Advanced Buss Loop Reactor Diss. ETH No. 14351.

4. Halophosphite Ligands for the Rhodium Catalyzed Low-Pressure Hydroformylation Reaction

Thomas A. Puckette

Eastman Chemical Company, Texas Eastman Division,
P.O. Box 7444, Longview TX 75607-7444

Tapucket@eastman.com

Abstract

The discovery and use of fluorophosphites and chlorophosphites as trivalent phosphorus ligands in the rhodium catalyzed, low-pressure hydroformylation reaction are described. The hydroformylation reaction with halophosphite ligands has been demonstrated with terminal and internal olefins. For the hydroformylation of propylene, the linear to branched ratio of the butyraldehyde product shows a strong dependency on the ligand to rhodium molar ratios, the reaction temperature, and the carbon monoxide partial pressure.

Introduction

The hydroformylation reaction, also known as the oxo reaction, is used extensively in commercial processes for the preparation of aldehydes by the reaction of one mole of an olefin with one mole each of hydrogen and carbon monoxide. The most extensive use of the reaction is in the preparation of normal- and iso-butyraldehyde from propylene. The ratio of the amount of the normal aldehyde product to the amount of the iso aldehyde product typically is referred to as the normal to iso (N:I) or the normal to branched (N:B) ratio. In the case of propylene, the normal- and iso-butyraldehydes obtained from propylene are in turn converted into many commercially-valuable chemical products such as n-butanol, 2-ethyl-hexanol, trimethylol propane, polyvinylbutyral, n-butyric acid, iso-butanol, neo-pentyl glycol, 2,2,4-trimethyl-1,3-pentanediol, the mono-isobutyrate and di-isobutyrate esters of 2,2,4-trimethyl-1,3-pentanediol.

Slaugh and Mullineaux (1) disclosed a low pressure hydroformylation process using trialkylphosphines in combination with rhodium catalysts for the preparation of aldehydes as early as 1966. Trialkylphosphines have seen much use in industrial hydroformylation processes but they typically produce a limited range of products and frequently are very oxygen sensitive.

In 1970, Pruett and Smith (2) described a low pressure hydroformylation process which utilizes triarylphosphine or triarylphosphite ligands in combination with rhodium catalysts. The ligands, although used in many commercial applications, have limitations due to oxidative and hydrolytic stability problems. Since these early disclosures, numerous improvements have been made to increase the catalyst stability, catalyst activity and the product ratio with a heavy emphasis on yielding linear aldehyde product. As a result of many years of research work in academic and industrial labs, a wide variety of monodentate phosphite and phosphine ligands, bidentate ligands such as bisphosphites and bisphosphines as well as tridentate and polydentate ligands have been prepared and disclosed in the literature (3,4). The early patents are still very significant today as all large scale commercial applications of the low pressure hydroformylation reaction are based on the triorganophosphine or triorganophosphite technology that was initially disclosed over thirty years ago.

Results and Discussion

The evaluation of novel trivalent phosphorus compounds as ligands for the low pressure hydroformylation reaction is an integral part of an ongoing program to develop and test new hydroformylation catalysts. Thus, when Klender et al. (5) of Albemarle Corporation published data demonstrating that the fluorophosphite, Ethanox 398™, is surprisingly stable to refluxing aqueous isopropanol, we were intrigued as to whether or not this material would have sufficient stability to serve as a trivalent phosphorus ligand in a hydroformylation catalyst. A search of the chemical literature revealed numerous antioxidant applications but did not reveal any use of fluorophosphite compounds as ligands with transition metals. Our personal experiences with compounds containing halogen phosphorus bonds

Ethanox 398™

1

were that these materials are highly reactive, easily hydrolyzed, and subject to secondary reactions such as disproportionation.

The hydrolytic decomposition of a potential fluorophosphite ligand would generate free fluoride ions which would be expected to be detrimental to the activity of a hydroformylation catalyst. The patent literature contains abundant references to the detrimental effects of halogens (6) on hydroformylation catalysts, and based on the patent information, one could not reasonably expect a halophosphite to be a successful hydroformylation ligand. However, a second publication by Klender (7) shows that exposure of **1** and other fluorophosphites to moisture at temperatures of 250°C to 350°C does not generate fluoride, even at part per million levels.

Despite our personal skepticism, a sample of **1** was obtained (8) and tested in a bench test unit that simulates continuous vapor stripped reactor operation (9). The initial results were both surprising and successful. A catalyst mixture composed of 2.11 grams of the fluorophosphite *1* and rhodium dicarbonyl acetylacetonate dimer (15 milligrams of Rh) in 190 milliliters of Texanol® (2,2,4-trimethyl-1,3-pentanediol monoisobutyrate) solvent successfully converted a mixture of propylene, hydrogen and carbon monoxide into a mixture of butyraldehyde isomers with a normal to branched ratio of 3.1 and at a catalyst activity rate of 5.9 kilograms of butyraldehyde per gram of rhodium-hour. Analysis of the recovered catalyst showed that the fluorophosphite was not decomposed under reactor conditions.

The focus of our research immediately turned to the exploration of compounds with the fluorophosphite functional group as hydroformylation ligands. The synthesis of these halophosphite compounds is well documented (10) and we set about to design and prepare a group of ligands which would explore the limits of this class of ligands in the hydroformylation reaction.

Fluorophosphites

Fluorophosphites are prepared by a two step sequence. The initial step is the reaction of phenolic materials with phosphorus trichloride to prepare a chlorophosphite intermediate. The chlorophosphite is then treated with a fluoride source to convert the chloro- intermediate into the desired fluorophosphite product. Many different fluoride sources have been described in the literature including anhydrous hydrogen fluoride, anhydrous potassium fluoride, and antimony trifluoride. While any of these fluoride sources can be used successfully, we have found that antimony trifluoride (10) works well for small scale, lab preparations.

Compounds *1* - *5* are representative of the types of fluorophosphite compounds that can be used successfully as hydroformylation ligands.

The ligands, *1* - *5* include acyclic and cyclic examples with alkyl and electron-donating groups on the aromatic rings. Notably absent from this group are any ligands with electron-withdrawing groups as substituents.

Chlorophosphites

Efforts to prepare fluorophosphites with electron withdrawing groups on the phenolic aromatic rings were unsuccessful. The reactions to convert the chlorophosphite intermediates with electron withdrawing groups failed and yielded only the unreacted chlorophosphite starting materials. Decomposition of the chlorophosphites did not occur to any significant extent. More forcing conditions were attempted but the chlorophosphite intermediates were recovered unscathed. The chemical stability of these intermediates with electron withdrawing groups was unexpected and prompted the question, "Just how stable are these compounds – stable enough to serve as ligands?"

Bench unit and autoclave testing of the electron withdrawing substituted chlorophosphite ligands demonstrated that these intermediate compounds can serve as viable hydroformylation ligands (11). Compounds *6 – 10* are representative of the types of chlorophosphite compounds that can be used successfully as hydroformylation ligands. Compound *10* is particularly stable and operates very well in heavy ester solvents such as bis-2-ethylhexyl phthalate.

Effect of Reaction Parameters on Catalyst Performance

The molar ratio of the phosphorus ligand to rhodium has pronounced effects on catalyst activity and selectivity. It is well established that increasing the molar ratio of the ligand to the rhodium leads to a higher linear to branched isomer ratio at the expense of catalyst activity (12). The change in the linear to branched ratio in the products can be understood as the effect of shifting the equilibriums that exist in the catalyst solution from a predominance of the monoligated, catalytically active rhodium species to bis ligated rhodium species as shown in Eq. 1.

$$L \quad + \quad RhH(CO)_3L \quad \underset{+ \, CO, \, - \, L}{\overset{-CO, \, +L}{\rightleftarrows}} \quad RhH(CO)_2L_2 \quad + \quad CO \qquad (1)$$

Mono Ligated	Bis Ligated
Higher activity	Lower Activity
Lower N/I	Higher N/I

The halophosphite ligands show the same relationship between activity and the preference for the more linear aldehyde isomer as a function of ligand concentration. A series of bench unit studies utilizing halophosphite catalysts were conducted in which propylene was allowed to react to form butyraldehyde. Table 1 presents bench unit data on the effects of the ligand to rhodium molar ratios.

Table 1 Effects of Ligand to Rhodium Molar Ratio on Activity and Selectivity

Run[a]	Ligand	Ligand to Rhodium Molar Ratio	N/I Ratio	Catalyst Activity[b]
1	*1*	14	1.4	15.7
2	*1*	30	3.1	6.6
3	*1*	50	3.8	3.3

[a] All runs were conducted at 260 psig, 115°C, 190 mL of bis-2-ethylhexylphthalate solvent (DOP) with 15 mg Rh, 1:1 H_2/CO and 54 psia C_3H_6.
[b] Catalyst activity is expressed as kilograms of butyraldehyde per gram-Rh-hour.

The effects of temperature on catalysts derived from traditional triorganophosphorus ligands has been studied and reported previously (13). In general, as the temperature of the reaction increases, the catalyst activity increases while the selectivity to the linear isomer decreases. Temperature effects on halophosphite catalysts follow the expected trend. Table 2 presents supporting bench unit data.

Table 2 Temperature Effects on Butyraldehyde Production

Run[a]	Ligand	Temperature, °C	N/I Ratio	Catalyst Activity[b]
1	*2*	95	5.0	3.7
2	*2*	105	3.9	6.0
3	*2*	115	3.8	6.2

[a] All runs conducted at 260 psig and specified temperature utilizing 7.7 mg of Rh, 190 mL of DOP, 1.2 grams of ligand with 1:1 H_2/CO and 54 psia C_3H_6
[b] Catalyst activity is expressed as kilograms of butyraldehyde per gram-Rh-hour.

The effect of the hydrogen to carbon monoxide molar ratios in the feed gas to the reaction is significant. Changes in the feed gas ratio will strongly affect the equilibrium of Equation 1 and thus impact the performance and selectivity of the catalyst. A series of bench unit runs were performed and the data is summarized in Table 3.

Table 3 Reactant Partial Pressure Effects

Run[a]	Ligand	H_2/CO Ratio	N/I	Catalyst Activity[b]
1	*10*	0.5:1	4.5	1.43
2	*10*	1:1	5.9	2.13
3	*10*	2:1	7.2	1.63

[a] All runs at 115°C, 260 psig utilizing 15 mg Rh, 2.06 grams of ligand in 190 mL of DOP with a C_3H_6 partial pressure of 54 psia.
[b] Catalyst activity is expressed as kilograms of butyraldehyde per gram-Rh-hour.

The hydroformylation capabilities of halophosphite catalysts are not limited to propylene or alpha olefins. A variety of other olefins have been examined and representative examples are presented in Table 4.

Table 4 Hydroformylation of Various Olefins

Run[a]	Ligand	Substrate	Products	Ratio
1	*1*	Methyl Methacrylate	Methyl isobutyrate, Methyl(2-methyl-3-formyl)propionate, & methyl (2-methyl-2-formyl)propionate	L/B = 2.63

2	*1*	Mixed 2-octenes	71% Nonyl Aldehydes	L/B = 0.91
3	*7*	1-Octene	98% Nonyl Aldehydes	L/B = 1.96
4	Bis(2-methylphenyl)chlorophosphite	Trans-2-Octene	2-Methyl-1-octanal, 2-Ethyl-1-heptanal, & 2-Propyl-1-hexanal.	L/B = 0.15
5	Bis(2-methylphenyl)chlorophosphite	1,7-Octadiene	Four Dialdehydes (99.3% total; 40.4% 1,10-decanedialdehyde). No monoaldehydes.	

[a] All runs performed in autoclaves as described in references 9 and 11.

In summary, chlorophosphites and fluorophosphites represent a new and viable class of ligands for the hydroformylation reaction and behave much like traditional triorganophosphorus ligands. The halophosphites are easy to obtain and surprisingly stable under process conditions.

Experimental Section

Ligands – 2,2'-Ethylidenebis (4,6-di-tert-butylphenyl) fluorophosphite (*1*) was purchased from Aldrich Chemical Company. The remaining chlorophosphite and fluorophosphite ligands were prepared by literature procedures or by minor modifications of the published procedures (10). All ligands were characterized by ^1H NMR, ^{31}P NMR and mass spectroscopy.

Bench Unit Testing – The physical description of the bench unit and operation of the unit has been described in reference 9.

Acknowledgements

Although many people have contributed to the success of this project, a few deserve special mention: Ginette Struck Tolleson as a co-worker, Tom Devon as a co-worker and mentor, Jimmy Adams and Sue Gray for bench unit operations and Eastman Chemical Company for permission to publish this work.

References

1. L. H. Slaugh and R. D. Mullineaux, U.S. Pat. No. 3,239,566, to Shell Oil Company (1966).
2. R. L. Pruett and J. A. Smith, US Pat. 3,527,809 to Union Carbide Corporation (1970).

3.   P. W. N. M. van Leeuwen, C. P. Casey, and G. T. Whiteker in *Rhodium Catalyzed Hydroformylation*, P. W. N. M. van Leeuwen and C. Claver, Ed., Kluwer Academic Publishers, Boston, 2000, p. 63 – 96.

4.   P. C. J. Kamer, J. N. H. Reek, and P. W. N. M. van Leeuwen in *Rhodium Catalyzed Hydroformylation*, P. W. N. M. van Leeuwen and C. Claver, Ed., Kluwer Academic Publishers, Boston, 2000, p. 35 – 59.

5.   G. J. Klender, V. J. Gatto, K. R. Jones, and C. W. Calhoun, *Polym. Preprints*, **24**, 156 (1993).

6.   A. B. Abatjoglou and D. R. Bryant, US Pat. 5,059,710, to Union Carbide (1988); A. A. Oswald, T. G. Jermasen, A. A. Westner and I. D. Huang, US Pat. 4,595,753, to Exxon Research and Engineering (1986); and K. D. Tau, US Pat. 4,605,781, to Celanese Corp. (1986).

7.   G. J. Klender, "Polymer Durability" in *Advances in Chemistry*, Vol. 249, R. L. Clough, Ed., American Chemical Society, Washington D. C., 1996, p. 396- 423.

8.   A sample was obtained from Aldrich Chemical Company, catalog number 370487-100G.

9.   T. A. Puckette and G. E. Struck, U.S. Patent 5,840,647, to Eastman Chemical Co., (1998) describes the bench unit and the operation of the unit in detail.

10. L. P. J. Burton, US Pat. 4,912,155, to Ethyl Corporation (1990) and E. A. Burt, L. P. J. Burton, M. S. Ao, and B. C. Stahly, US Pat. 5,049,691 to Ethyl Corporation (1991).

11. G. S. Tolleson and T. A. Puckette, US Pat. 6,130,358 to Eastman Chemical Company (2000).

12. K. L. Olivier and F. B. Booth, *Hydrocarbon Processing*, **49** (4), 112 (1970) and P. W. N. M. van Leeuwen and C. F. Roobeek, *J. Orgmet. Chem.*, **258**, 343 (1983).

13. I. Wender and P. Pino, *Organic Syntheses via Metal Carbonyls*, Vol. 2, John Wiley & Sons, New York, 1977, p. 176 -179.

5. Development of a Monolithic Bioreactor: Tailor-Made Functionalized Carriers

Karen M. de Lathouder, Freek Kapteijn and Jacob A. Moulijn

Delft University of Technology, Faculty of Applied Sciences, DelftChemTech, Section R&CE, Julianalaan 136, 2628 BL Delft, The Netherlands

k.m.delathouder@tnw.tudelft.nl

Abstract

The use of a monolithic stirred reactor for carrying out enzyme-catalyzed reactions is presented. Enzyme-loaded monoliths were employed as stirrer blades. The ceramic monoliths were functionalized with conventional carrier materials; carbon, chitosan, and polyethylenimine (PEI). The different nature of the carriers with respect to porosity and surface chemistry allows tuning of the support for different enzymes and for use under specific conditions. The model reactions performed in this study demonstrate the benefits of tuning the carrier material to both enzyme and reaction conditions. This is a must to successfully intensify biocatalytic processes. The results show that the monolithic stirrer reactor can be effectively employed in both mass transfer limited and kinetically limited regimes.

Introduction

Ceramic honeycomb monoliths are porous macro-structured supports consisting of parallel channels. On the walls a thin layer of active material can be applied (Figure 1). Honeycomb catalyst supports were originally developed for use in automotive emission control systems where the combination of low pressure drop and high surface area are important (1). For liquid systems, the advantages of structured reactors compared to fixed-bed or slurry reactors include a high available surface area, a low pressure drop over the reactor, ease of product separation, absence of maldistribution problems, and easy scale-up (2,3).

Figure 1. Monoliths

Immobilized enzymes have a wide range of practical applications. Although activity usually decreases slightly upon immobilization, they possess important advantages over dissolved enzymes, of which the possibility to recover and reuse the enzyme is the most important. Most conventional enzyme carriers are inorganic particles or porous beads of synthetic polymers or gel-like materials such as chitosan, agarose or alginate. If one uses large beads, intraparticle limitations are bound to occur (4). Note that in enzymatic systems, not only substrate diffusion can be limiting: intraparticle pH gradients or ionic strength gradients can be equally problematic. An alternative to large beads in a fixed-bed reactor is a stirred slurry of beads that can be as small as 100 µm (5). However, the soft support-material lacks the mechanical strength for high intensity contacting. Also, the density of the support material is often close to that of the solvent and, as a consequence, an (often cumbersome) separate filtering step is required. The use of structured support materials could provide an interesting alternative for conventional enzyme support materials.

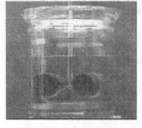

The monolithic stirrer reactor (MSR, Figure 2), in which monoliths are used as stirrer blades, is a new reactor type for heterogeneously catalyzed liquid and gas-liquid reactions (6). This reactor is thought to be especially useful in the production of fine chemicals and in biochemistry and biotechnology. In this work, we use cordierite monoliths as stirrer blades for enzyme-catalyzed reactions. Conventional enzyme carriers, including chitosan, polyethylenimine and different carbonaceous materials are used to functionalize the monoliths. Lipase was employed in the acylation of vinyl acetate with butanol in toluene and immobilized trypsin was used to hydrolyze N-benzoyl-L-arginine ethyl ester (BAEE).

Figure 2. MSR

Results and Discussion

Results of enzyme adsorption and tests in the MSR are given in Table 1.

Table 1. Results of enzyme immobilization and catalyst performance

Carrier	Enzyme	Yield [mg]	Initial activity [mol/s*m$^3_{monolith}$]	Activity [mmol/s*g$_E$]
-	Lipase			3.6
CNF	Lipase	350	9.2	0.94
Sucrose-carbon	Lipase	65	2.0	1.1
PFA-carbon	Lipase	70	1.9	0.81
PEI	Lipase	35	1.0	1.22
-	Trypsin			0.2
Chitosan	Trypsin	100	0.034	0.019

For lipase, initial activity corresponds to the amount of protein that was adsorbed. Specific activity is constant at 1 mmol/s*g$_E$ for this carrier-enzyme system, which compares to 27% of the free enzyme activity. The trypsin system shows a lower specific activity that is only 10% of the free enzyme. The reason for the lower recovered activity of this system is not known. To rule out possible internal diffusion limitations, the Wheeler-Weisz modulus was estimated, assuming a carrier layer thickness of 0.1 mm for all carriers. Using the data of the experiments performed at 150 rpm, one finds:

$$\Phi = \left(\frac{n+1}{2}\right) \cdot \frac{r_{v,obs} \cdot L^2}{D_{eff} \cdot C_b} \approx 10^{-2} \text{ for the lipase system and } 10^{-4} \text{ for the trypsin system,}$$

which is below the threshold value of 0.15, indicating the absence of diffusion limitations. To investigate any external mass transfer limitations present in the system, the stirrer rate was varied between 50 and 400 rpm. The results for the immobilized lipase are plotted in Figure 3.

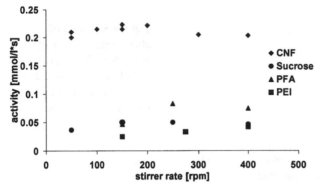

Figure 3. Initial activity of immobilized lipase in the acylation of vinyl acetate in butanol at varying stirrer rate. Comparison of different carbon carriers.

For these biocatalysts, no profound influence of stirrer rate could be detected. Apparently no external mass transfer limitations are present in the system. This was confirmed by calculating the Carberry number (Ca = $\dfrac{r_{v,obs}}{a^{'} \cdot k_{s} \cdot C_{b}}$), the ratio of the observed rate and the maximal mass transfer rate. For all experiments, Ca << 0.05, so no external mass transfer limitations are present. Temperature stability of this system was studied at 150 rpm. In Figure 4 results for free lipase and lipase on CNF are presented.

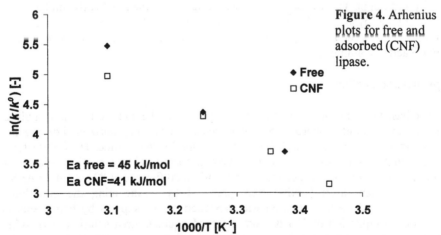

Figure 4. Arhenius plots for free and adsorbed (CNF) lipase.

The catalysts were found to be stable up to 323 K without any deactivation. An activation energy of 45 kJ/mol was found for the free enzyme, 41 kJ/mol for the immobilized lipase, indeed confirming the absence of mass transfer limitations.

When immobilized trypsin was used in the mass transfer limited regime, a clear effect of stirrer rate was observed (Figure 5). By fitting a first order Michaelis

Menten equation, the observed reaction rate was calculated assuming a mass transfer limited regime in the substrate concentration range 0.1-0.2 g/l.

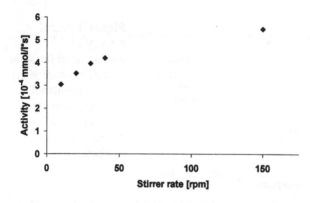

Figure 5. Effect of stirrer rate on trypsin activity in the hydrolysis of BAEE at under the mass transfer limited conditions at 308 K.

From these results, the effective mass transfer coefficient k_s [s^{-1}] can be calculated and was found to vary from $8*10^{-6}$ m/s (10 rpm) to $1.2*10^{-5}$ m/s (40 rpm). Around 100 rpm, the system enters the kinetically limited regime (Ca=0.05). These results show that immobilized trypsin is very active and useful for determining mass transfer rates for liquid solid systems.

In summary, it can be concluded that the monolithic stirrer reactor is a convenient reactor type both for the laboratory and the production plant. It is user-friendly and can be used to compare different catalysts in the kinetically limited regime or hydrodynamic behavior in the mass transfer controlled regime. Stirrers or monolith samples can be easily exchanged and reloaded to suit the desired enzyme and/or reaction conditions.

Experimental Section

Sucrose-based carbon carriers were prepared following the method of (7). Monoliths were coated with a 65% sucrose solution in water, followed by horizontal drying and carbonization for two hours at 823 K in a horizontal furnace under H_2. Polyfurfuryl alcohol (PFA) based carbon coatings were prepared (8) by coating with PFA solution. Carbonization was performed in a horizontal furnace at 823 K under Ar for 2 h. Carbon nanofibers (CNFs) were prepared by washcoating the monolithic supports with a silica layer (9). Ni was deposited on the support by homogeneous deposition precipitation at 353 K from a 0.5 M aqueous urea solution (10). After reduction under H_2 for 1 h at 823 K, carbon nanofibers were grown at 823 K under methane and H_2 in N_2. Polyethylenimine-functionalized supports were prepared (11) via a direct coupling through (3-glycidoxypropyl)trimethoxysilane (GPTMS). The monoliths were washcoated with a colloidal silica layer, followed by reaction at room temperature for 24 h in a 5wt% solution of silane in toluene, containing 0.1% v/v triethylamine. The polymer was attached from a 10 wt% PEI solution in water (pH 10) under ambient conditions for 24 h. Chitosan coatings were applied by mixing 100 ml of a 10 g/l chitosan solution with 4.1 ml of 25% glutaraldehyde in water. Monoliths were dipped for 1 min and cleaned with pressurized nitrogen. After drying for 90 min, the gel was rehydrated and washed with water. Enzymes were adsorbed from a 3 g/l solution in 0.01 M phosphate buffer pH7 at 293 K. Enzyme concentration in solution was followed by UV-VIS.

Catalytic tests were performed in a glass vessel equipped with a stirrer motor. Two monoliths (diameter 4.3 cm, length 4 cm) were mounted in plane on the stirrer axis. The total reaction volume was 2.5 l. Lipase was assayed in the acylation of vinyl acetate with butanol in toluene. Initial reaction rate was followed by GC analysis. Immobilized trypsin was used in the hydrolysis of N-benzoyl-l-arginine ethyl ester (BAEE) in a 0.01 M phosphate buffer pH 8 at 308 K. The reaction was followed by UV-VIS at 253 nm, and reaction rate was calculated in the mass transfer limited situation.

Acknowledgment is made to the Donors of **The American Chemical Society Petroleum Research Fund,** for the partial support of this work through the award of a travel grant.

References

1. A. Cybulski, *Structured Catalysts and Reactors*. Second Edition. J.A. Moulijn, Ed. CRC Taylor & Francis, Boca Raton. (2006)
2. F. Kapteijn, J.J. Heiszwolf, T.A. Nijhuis and J.A. Moulijn, *CATTECH*, **3**(1), 24-41 (1999).
3. M.T. Kreutzer, F. Kapteijn, J.A. Moulijn and J.J. Heiszwolf, *Chem Eng Sci*, **60**, 5859-5916 (2005).
4. M.R. Benoit and J.T. Kohler, *Biotechnol Bioeng*, **17**, 1616-1626 (1975).
5. R.J. Barros, E. Wehtje and P. Adlercreutz, *Biotechnol Bioeng*, **59**, 364-373 (1997).
6. I. Hoek, T.A. Nijhuis, A.I. Stankiewicz and J.A. Moulijn, *Chem Eng Sci* **59**, 4975-4981 (2004).
7. T. Valdés-Solís, G. Marbán and A.B. Fuertes, *Microporous Mesoporous Mater*, **43**, 113-126 (2001).
8. Th. Vergunst, M.J.G. Linders, F. Kapteijn and J.A. Moulijn, *Catal Rev- Sci Eng* **43**, 291-314 (2001).
9. T.A. Nijhuis, A. Beers, T. Vergunst, I. Hoek, F. Kapteijn and J.A. Moulijn, *Catal Rev*, **43** 345-380 (2001).
10. M.L. Toebes, J.H. Bitter, A.J. van Dillen and K.P. de Jong, *Catal Today*, **76**, 33-42 (2002).
11. C. Mateo, O. Abian, R. Fernandez-Lafuente and J.M.Guisan, *Biotechnol Bioeng*, **68**, 98-105 (2000).

6. Highly Selective Preparation of *trans*-4-Aminocyclohexane Carboxylic Acid from *cis*-Isomer over Raney® Nickel Catalyst

Sándor Gőbölös, Zoltán Banka, Zoltán Tóth, János Szammer and
József L. Margitfalvi

*Chemical Research Center, Institute of Surface Chemistry and Catalysis,
H-1025 Budapest, Pusztaszeri út 59-67, Hungary*

gobolos@chemres.hu

Abstract

4-Amino-benzoicacid was hydrogenated to 4-aminocyclohexane carboxylic acid over different alumina supported 5 wt.% Ru and Rh catalysts Complete ring saturation was achieved in 2 wt. % NaOH-H_2O at 80-100 °C, 10 MPa H_2, and 5 h however, the ratio of *trans/cis* stereo-isomers of the product was only between 1/3-1/1. Raw reaction mixture was further processed in the presence of a commercial Raney® nickel catalyst at 130°C, 100 bar H_2 for 5 h. In this alkali mediated isomerization the *trans/cis* isomer ratio was 7/3. The *cis* isomer was isolated by fractional crystallization, and then reacted on Raney® nickel catalysts in 2%NaOH-H_2O at 120-140°C, 1 MPa H_2 for 5 h to obtain the *trans* isomer with a yield of ca. 70 %. The two-step synthesis resulted in *trans*-4-aminocyclohexane carboxylic acid with a yield above 90%. Catalytic tests were performed in a high-throughput reactor system equipped with 16 mini autoclaves (SPR16 AMTEC GmbH, Germany).

Introduction

Recent years, stereochemically pure drugs have increasingly dominated the global pharmaceutical industry. The need for single isomers has fueled the development of stereochemical reactions with the aim to synthesize compounds of desired configuration, rather than mixtures of stereoisomers that must be separated later. Stereochemically pure amino acid derivatives containing N-terminal *trans*-4-alkylcyclohexanoyl fragment are reported as precursors of physiologically active agents in the treatment of various diseases (cancer, osteoporosis). The hydrogenation of disubstituted aromatics, particularly amino benzoic acids to the corresponding cyclohexane carboxylic acids, has therefore attracted the interest of the industry (1). These compounds are used as key intermediates for the manufacture of pharmaceuticals. It is noteworthy that trans-isomer of 4-aminocyclohexane carboxylic acid esters are reported as the physiologically active agents. *Trans*-4-substituted cyclohexane carboxylic acids can be obtained by the hydrogenation of the corresponding benzoic acid compounds followed by the isomerization of the resulting mixture of *trans*- and *cis*-isomers.

Due to the absence of electron withdrawing groups the aromatic amines are hydrogenated with difficulty. Supported noble metal catalysts (Pd, Pt, Ru, Rh, Ir) are usually active at elevated temperature and pressure. The perhydrogenation of aromatic amines is mainly accompanied by two side reactions, i.e. the hydrogenolysis and reductive coupling, the latter increasing in the sequence Ru<Rh<<Pd<<Pt (1). In the hydrogenation of various substituted anilines *cis* isomer is formed in acidic medium, and mainly *trans* compound is obtained in neutral or alkaline medium (2). It is noteworthy that the *cis* product predominates on the most frequently employed Ru and Rh catalysts (3).

In the hydrogenation of benzoic acid derivatives the desired main reaction is the conversion of aromatic skeleton to a cyclohexane ring, the functional groups remaining unchanged. As regards the mechanism of perhydrogenation, the benzoic acid derivatives, similarly to other disubstitued benzenes, are hydrogenated via olefinic intermediates, being the main source of the formation of *trans* product (4). In the hydrogenation of dialkylbenzenes the main product is the *cis*-dialkylcyclohexane. The formation of *cis*-cycloalkanes can be interpreted through stereospecific superficial hydrogenation, while *trans*-cycloalkanes are probably formed via olefinic intermediates (5). The proportion of trans isomers increases as temperature rises (1)

The increase of the quantity of catalyst enhances the rate, but it does not influence the stereochemistry in the hydrogenation of phenol derivatives (6). The *cis* product formation is favored in acidic medium, and the *trans* product formation in neutral or alkaline medium (7). On Ru and Rh, about twice as much *cis* isomer is formed as *trans* isomer, whereas on Pt and Pd, the isomers are obtained in approximately equivalent amounts. Isomerization during the hydrogenation can be excluded (8).

The role of the metal catalyst in the hydrogenation reaction and its stereoselectivity has been widely studied (9). Group VIII metals are generally used, but different behaviours are observed. In the hydrogenation of disubstituted phenols palladium mainly gives cyclohexanone derivatives, rhodium and platinum are very selective for the *cis* isomer, whereas nickel is more selecive for the *trans* isomer (9).
It was reported that ruthenium-containing catalysts are more active in the hydrogenation of aromatic ring than platinum- or palladium-based catalytic systems (10). Recently, a carbon supported 5%Ru-Ni(Ru:Ni=9:1) has been developed for the hydrogenation of 4-alkylbenzoic acids in aqueous NaOH at 150°C, 3-4MPa and 1h. A mixture of *trans* and *cis*-4-alkylcyclohexanecarboxylic acid was obtained in a ratio ca. 2:3. Alkali mediated isomerization of the mixture was done in the absence of catalyst at 260-280°C, 3-4MPa for 2h. After cooling to RT the mixture was acidified with HCl to pH=2. The precipitate formed was filtered off and dried to afford a mixture, which contained ca. 75% of the *trans*-isomer (11).
Cis-4-aminocarboxylic acid was transformed to *trans*-isomer with 70% yield by treating a mixture of isomers (ca. 6wt.%) in 2% aqueous NaOH over Raney® nickel (3wt.%) at 130°C, 15MPa H_2 for 6 h (12).

Results and Discussion

As seen from the data given in Table 1, 2%NaOH-H_2O was the only appropriate solvent for the hydrogenation of 4-aminobenzoicacid over alumina supported Ru and Rh catalysts. Despite the higher solubility of H_2 in organics than in H_2O (1), the pure activity of catalysts in organic solvents can probably be explained by the absence of salt formation and lower solubility of substrate. As expected, both Ru and Rh are highly active in the hydrogenation of aromatic ring, and the conversion of substrate was practically complete both at 80 and 100°C. The ratio of *cis/trans* isomer estimated from TLC results was ca. 1:1.

Table 1 Hydrogenation of 4-amino-benzoicacid on Ru/Al_2O_3 and Rh/Al_2O_3 catalysts. Effect of catalyst, solvent and temperature

No.	Catalyst	Solvent	T, °C	Yield %	*trans*-isomer %
1	5%Ru	NaOH	80	100	~50
2	4%Rh	NaOH	80	100	~50
3	5%Ru(E44)	NaOH	80	100	~50
4	5%Rh(E34)	NaOH	80	100	~50
5[a]	5%Ru	NaOH	100	100	~50
6[a]	4%Rh	NaOH	100	100	~50
7[a]	5%Ru(E44)	NaOH	100	100	~50
8[a]	5%Rh(E34)	NaOH	100	>90	~50
9	5%Ru	EtOAc	100	0	-
10	4%Rh	EtOAc	100	0	-
11	5%Ru(E44)	EtOAc	100	<10	-
12	5%Rh(E34)	EtOAc	100	<10	-
13	5%Ru	1,4-Dioxane	100	<10	-
14	4%Rh	1,4-Dioxane	100	~10	-
15	5%Ru(E44)	1,4-Dioxane	100	~10	-
16	5%Rh(E34)	1,4-Dioxane	100	~10	-

Reaction condition: $m_{catalyst}$=0.1g; Solvent: V=8ml; Substrate: 0.1g; P=10MPa; t=5h.
Abbreviations: NaOH=2%NaOH-H_2O, EtOAc=ethyl acetate,
[a]0.4g substrate was used; only traces of by-products were detected.

It is interesting to note that larger amount of 4-amino-cyclohexane carboxylic acid (ca. 10-15 g) was prepared using also 2%NaOH-H_2O as a solvent. The reaction was carried out in a Parr reactor (V=300ml) at 100°C, 10MPa H_2, for 5h. After filtering off the noble metal catalyst the liquid product was further reacted in the presence of Raney® nickel catalyst at 100°C, 10MPa H_2, for 5h. After this reaction according to both ^{1}H-NMR and TLC measurement the *trans/cis* isomer ratio in the reaction mixture was 3/7. As known the basic media is advantageous for the formation of *trans* isomer, therefore the B113W Raney® Ni was modified with potassium ions (2wt. %) using K_2CO_3 as precursor compound (13). However, the modification did not increase further the *trans/cis* ratio.

Conversion data listed in Table 2 indicate that in the hydrogenation of 4-amino-ethyl benzoate using cyclohexane as a solvent the Al_2O_3 supported Ru and Rh catalysts are more active than the carbon supported ones. In addition, the conversion of aromatic ester is smaller on Pd/C catalyst than on the carbon supported Ru and Rh. In ethyl acetate the hydrogenation proceeded slower than in cyclohexane, and Rh being more active than Ru. The *trans/cis* isomer ratio estimated from TLC results varied between 1/3 and 1/1. The UV active by-products were formed in coupling reactions.

Table 2 Hydrogenation of 4-amino-ethyl benzoate on alumina and carbon supported Ru, Rh and catalysts. Effect of catalyst, solvent and temperature

No.	Catalyst	Solvent	Yield %	By-products
1	5%Ru/Al	CH	<100	few
2	4%Rh/Al	CH	<100	few
3	5%Ru/Al(E44)	CH	<100	few
4	5%Rh(E34)	CH	<100	few
5	5%Pd/C(E10)	CH	30	very much
6	5%Ru/C(E40)	CH	50	significant
7	5%Rh/C(E30)	CH	50	significant
8	5%Ru/Al	EtOAc	~30	much
9	4%Rh/Al	EtOAc	~60	much
10	5%Ru/Al(E44)	EtOAc	~20	much
11	5%Rh/Al(E34)	EtOAc	~60	much

Reaction condition: $m_{catalyst}$=0.1g; Solvent: V=8ml; Substrate: 0.1g; T=100°C; P=10MPa; t=5h. Abbreviations: CH=cyclohexane, EtOAc=ethylacetate.

Table 3 Isomerization of *cis*-4-aminocyclohexane carboxylic acid on Raney® nickel catalyst (Metalyst). Effect of the amount of catalyst, temperature and H_2 pressure

No.	$m_{catalyst}$, mg	T, °C	P MPa	*trans*- isomer,%
1	50	100	3	30
2	50	110	3	30
3	50	120	3	45
4	50	130	3	50
5	100	100	7	40
6	100	110	7	40
7	100	120	7	50
8	100	130	7	45
9	150	100	7	45
10	150	120	7	50
11	150	130	7	55
12	0	130	7	0
8	100	130	7	45
11	150	130	7	55
13	0	100	1	0
14	150	100	1	50
9	150	100	7	45
15	50	120	1	40
3	50	120	3	45
16	100	120	1	65
7	100	120	7	50

Reaction condition: Solvent: 2 %NaOH-H_2O, V=6ml; Substrate: 0.15g; t=5h.
In addition *cis*-isomer of 4-aminocyclohexane carboxylic acid isolated by fractional crystallization from the reaction mixture obtained in the Parr reactor was also reacted over Raney® nickel catalyst to prepare the desired *trans*-isomer with high yield. It is

noteworthy that, in contrast to ref. (11), no isomerization was observed in the absence of catalyst (see experiments No. 12 and No. 13 in Table 3). In the catalytic isomerization of pure *cis* isomer the yield of the *trans* isomer estimated from TLC results are listed in see Table 3. It is noteworthy that upon increasing the reaction temperature and the amount of Raney® nickel catalyst (Metalyst, 35 % nickel content) the yield of trans isomer increases (compare experiments No.1 - No.11 in Table 3). For clarity, some experiments were listed and grouped twice in Table 3.The yield of trans isomer increases, indeed, with the amount of catalyst (see experiments at the bottom of Table 3). Careful analysis of data given in Table 3 suggests that 120-130°C is required to reach 45-55% *trans* isomer yield on the Metalyst Raney® nickel catalyst. The H_2 pressure had also an influence on the trans isomer yield. However, the highest yield was obtained at 1 MPa and 120°C (experiment No. 16 in Table 3). It is worth to remind that the *trans*-isomer was probably formed via olefinic intermediates on the surface of Raney® nickel catalyst (5). This suggests that the hydrogen pressure should have an optimum value.

Table 4 Isomerization of *cis*-4-aminocyclohexane carboxylic acid on Raney® nickel catalyst (B 113 W). Effect of the amount of catalysts, temperature and H_2 pressure

No.	$m_{catalyst}$, mg	T, °C	P MPa	*trans*- isomer,%
1	150	100	1	45
2	150	120	1	70
3	150	140	1	70
4	150	100	2	45
5[a]	150	120	2	50
6	150	140	2	60
7	150	100	4	40
8	150	120	4	50
9	150	140	4	60
10	100	100	1	40
11	100	140	1	70
12	100	100	4	30
13	100	140	4	50
14	200	100	2	45
15	200	120	2	55
16	200	140	2	60

Reaction condition: Solvent: 2 %NaOH-H_2O, V=6ml; Substrate: 0.15g; t=5h.

In the catalytic isomerization of pure *cis* isomer the yield of the *trans* isomer estimated from ^{1}H-NMR results are listed in Table 4. Results obtained on Raney® nickel catalyst (B 113 W, nickel content 90%) indicate that the yield of the *trans* isomer significantly increases with the temperature and also affected by the hydrogen pressure (compare experiment No. 1 – No. 9 in Table 4). However the highest yield of the *trans* isomer was reached at 1 MPa H_2 pressure (see experiments No. 10 – No. 13 in Table 4). Upon increasing the amount of catalyst from 100 to 150 mg the yield of the *trans* isomer slightly increased.

The ^{1}H-NMR spectrum of pure *cis*-4-aminocyclohexane carboxylic acid ethyl ester is seen in Figure 1. Signals with chemical shift at δ=1.2-1.3 ppm can be assigned to the protons of the methyl group, whereas the signals of the -CH_2- protons in the ester group appeared at δ=4.1 ppm. Signals of the -CH$_2$- protons in the cyclohexane ring originating from axial hydrogen atoms are at δ=1.4 ppm, whereas those of equatorial atoms are at δ=1.6 ppm. The proton next to ester group gives a signal with a chemical shift of δ=2.43 ppm. The nitrogen atom of the amino group shifted the signal of nearby proton in the cyclohexane ring to δ=2.84 ppm.

Figure 2 shows the ^{1}H-NMR spectrum of the mixture of *cis-* and *trans-4-* aminocyclohexane carboxylic acid ethyl ester. The assignation of the signals of protons in the ester group and those of the -CH$_2$- groups in the cyclohexane ring are as given above. However, the signals belonging to the protons next to the ester and amino groups in the *trans* isomer appeared at lower chemical shifts, δ=2.19 ppm and δ=2.64 ppm, respectively, than those of the *cis* isomer.

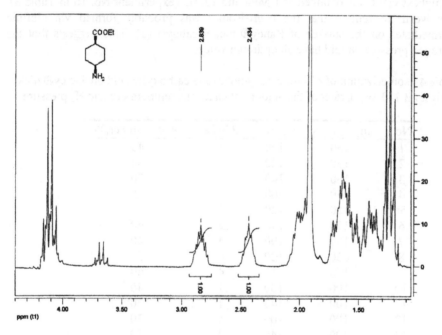

Figure 1 NMR spectrum of *cis-*4-aminocyclohexane carboxylic acid ethyl ester.

The ^{1}H-NMR spectrum of *trans-*4-aminocyclohexane carboxylic acid ethyl ester is enriched in a mixture containing ca. 10 % of the *cis* isomer is seen in Figure 3. The signals belonging to the protons next to the ester and amino groups in the *trans* isomer appeared at chemical shifts of δ=2.20 ppm and δ=2.65 ppm, respectively. The intensity of signals in the NMR spectra allows estimating the ratio of stereoisomers.

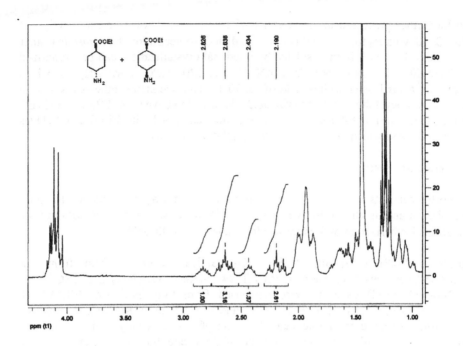

Figure 2 NMR spectrum of the mixture of stereoisomers of esters.

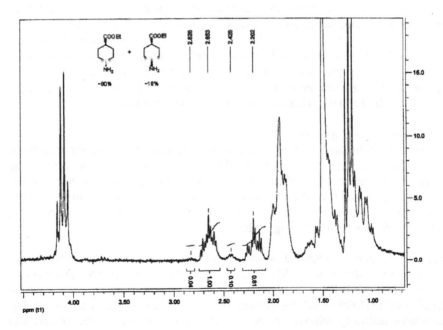

Figure 3 NMR spectrum of a mixture containing ~90% of *trans*-isomer of ester.

In conclusion, 4-Amino-benzoicacid can be hydrogenated to 4-aminocyclohexane carboxylic acid on 5wt.% Ru- and Rh/Al$_2$O$_3$ catalysts with 100% yield in 2 wt. % NaOH-H$_2$O at 80-100 °C, 10 MPa H$_2$ for 5h. However, the ratio of *trans/cis* stereoisomers of the product does not exceed a value of 1/1. Raw reaction mixture can be further processed in the presence of a commercial Raney nickel catalyst at 130 °C, 10 MPa H_2 for 5 h. In this alkali mediated isomerization the *trans/cis* isomer ratio was 7/3. The *cis* isomer isolated by fractional crystallization, can be isomerized on Raney® nickel catalysts in 2%NaOH-H_2O at 120-140°C, 1 MPa H_2 for 5 h to obtain the *trans* isomer with a yield of ca. 70 %. The two-step synthesis resulted in *trans*-4-aminocyclohexane carboxylic acid with a yield above 90%. SPR16 (AMTEC GmbH, Germany) high-throughput reactor system equipped with 16 mini autoclaves is useful for fast screening of liquid-phase catalytic reactions.

Experimental Section

Supported catalysts were used for the hydrogenation of the aromatic ring of 4-amino-benzoicacid and its ethyl ester, whereas Raney® nickel was applied to catalyze the cis→trans isomerization of 4-amino-cyclohexane carboxylic acid.

Catalysts: 5%Ru/Al_2O_3 and 4%Rh/Al_2O_3 catalysts were received from Air Poducts&Chemicals Co., whereas ESCAT 44 5%Ru/Al_2O_3 (E44) and ESCAT 34 5%Rh/Al_2O_3 (E34) catalysts were purchased from Engelhard. Co. ESCAT 10 5%Pd/C (E10), ESCAT 40 5%Ru/C (E40) and ESCAT 30 5%Rh/C (E30) catalysts were also purchased from Engelhard Co. Raney® nickel catalyst Metalyst-alpha-1301, and B113 W with nickel content of 35% and 90 %, respectively were purchased from Degussa GmbH.

Activity test: Catalyst screening and optimization of process parameters for both steps, i.e. aromatic ring saturation and stereo-isomerization, were performed using a high throughput (HT) fully automated batch reactor system equipped with 16 mini-autoclaves (SPR16, AMTEC GmbH Germany). In the hydrogenation of the aromatic ring the effect of catalyst, solvent and reaction temperature was investigated. In the isomerization step and the effect of reaction temperature (100-140 °C), hydrogen pressure (1-7 MPa) and the substrate/catalyst ratio was screened.

Separation of cis/trans isomers
The separation of the stereoisomers was done by fractional crystallization modifying the method described in ref. (14). After filtering the catalyst and evaporating the majority of water ethanol was added and pure *cis* isomer was precipitated. Then the precipitate was filtered and washed with the ethanol. After cooling the mother liquor the *cis* isomer precipitate was removed again, and the solvent evaporation, precipitation, filtration and washing cycle was repeated twice.

Analysis: Fast quality control of both the product of hydrogenation and stereo-isomerization and was done by thin layer chromatography (TLC) (15). The analysis of the hydrogenation product was done by using a mixture $CHCl_3$:CH_3OH:NH_4OH=5:5:1 for elution. o-Chloro-toluidine was used to develop the product 4-amino-cyclohexane carboxylic acid. The ratio of *trans/cis* isomers was also estimated by using TLC: Sil/G UV_{254}, (detection at 254 nm) water for elution, and the same compound as developing agent. The ratio of *trans/cis* isomers of selected samples was determined by ^{1}H-NMR analysis of the ester derivative of 4-amino-cyclohexane carboxylic acid. NMR spectra were taken in $CDCl_3$ at 30°C using a Varian Unity Inova 400 spectrometer. Chemical shifts are reported relative to internal TMS in ppm, and 512 transients were accumulated to achieve the appropriate signal/noise ratio.

References

1. P.N. Rylander, *Catalytic Hydrogenation in Organic Syntheses*, Academic Press, New York, (1979).
2. A. Skita, *Chem. Berichte*, **56** (1923) 1014.
3. Kh. Friedlin, E.F. Litvin, V.V. Yakubenok, I.L. Vaisman, *Izv. Akad. Nauk SSSr, Ser. Khim.*, **1976**, (1976) 976.
4. S. Siegel, G.S. McCaleb, *J. Am. Chem. Soc.*, **81** (1959) 3655.
5. O.V. Bragin, A.L. Liberman, *Prevrashcheniya Uglevodorodov, na Metallsoderzhaahchikh Katalizatorakh*, Khimiya, Moscow, (1981).
6. H.A. Smith, R.G. Thompson, *Adv. Catal.*, **9** (1957) 727.
7. R.J. Wicker, *J. Chem. Soc.*, **1956**, (1957) 3299.
8. M. Huang, C.Y. Chen, *J. Sin. Chem. Soc.*, 27 (1980) 151.
9. M. Bartók, *Stereochemistry of Heterogeneous Metal Catalysis*, Wiley, Chichester, (1985).
10. R.N. Gurskii, R.V. Istratova, A.V. Kirova, S.A. Kotlyar, O.V. Ivanov, N.G. Luk'yanenko, *J. Org. Chem. USSR (Engl. Transl.)* **24** (1988) 543.
11. A.A. Bazurin, S.V. Krasnikov, T.A. Obuchova, A.S. Danilova, K.V. Balakin, *Tetrahedron Lett.*, **45** (2004) 6669.
12. K.I. Karpavitsys, A.I. Palayma., I.L. Knunyanits, *Izv. Akad. Nauk SSSr*, **20** (1977) 2374.
13. S. Gőbölös, A. Fási, J.L.Margitfalvi, L. Milliánné, European Patent to Clariant GmbH, EP0301407 (2003).
14. Johnston, *J. Med. Chem.* **20** (1977) 279
15. J. Podaky, *J. Chromatogr.* **12** (1963) 541.

7. Gas-Phase Acetone Condensation over Hydrotalcite-like Catalysts

Francisco Tzompantzi[1], Jaime S. Valente[2], Manuel S. Cantú[2] and Ricardo Gómez[1]

[1]*Universidad Autónoma Metropolitana-Iztapalapa, San Rafael Atlixco No 152, México 09340, D.F.*
[2]*Instituto Mexicano del Petróleo, Eje Central #152, México, D.F., 07730.*

gomr@xanum.uam.mx

Abstract

Self-condensation of acetone is considered a reaction catalyzed typically by bases. In this study a flow microreactor has been used to evaluate the performance of several calcined MgZnAl hydrotalcite-like materials (HTs). The main products were mesityl oxide and aldol compounds. The nature of the catalyst's composition, i.e., the effect of Zn loading on the reaction intermediates and by-products was reported. In order to give an overall description of the base catalyzed oligomerization of acetone in gas phase, a suitable mechanism was proposed. When increasing the Zn content in the hydrotalcite structure, the activity for the acetone's oligomerization increased and the selectivity to mesityl oxide reached values about 88% mol. The most active catalyst was the one with the higher Zn content. An interpretation in function of the adsorption-desorption equilibrium of products was proposed, which discloses that the product's desorption is slow in strong basic sites. Therefore, the activity and selectivity on calcined HTs could be easily regulated by controlling the nature and amount of cations.

Introduction

The increasing demand for the acetone aldol (dimer) and mesitylene (trimer), employed as important intermediate reagents for a large variety of fine chemicals, renewed the research efforts in these reactions by using heterogeneous catalytic processes (1,2). Indeed, gas-phase acetone condensation reaction has been carried out over basic oxides like MgO (3). Good activity for aldolization reactions has been reported also over doped MgO, and MgAl hydrotalcite (4,5). In fact, basic mixed oxides obtained from hydrotalcite-like materials (HTs) have showed good activities in many organic reactions (6-9). The structure of HTs resembles that of brucite, Mg $(OH)_2$, in which magnesium is octahedrally surrounded by hydroxyls. A HT structure is created replacing some of the M^{2+} divalent cations by M^{3+} trivalent cations turning the layered array positively charged. These layers are electrically compensated for by anions which are located in the interlaminar region. A large variety of synthetic HTs can be prepared and are represented by the general formula: $[M^{2+}_{1-x} M^{3+}_x (OH)_2]^{x+}$

$[A^{n-}]_{x/n} \cdot m$ H_2O, where $M^{2+} = Mg^{2+}$, Ni^{2+}, Zn^{2+}, etc; $M^{3+} = Al^{3+}$, Fe^{3+}, Ga^{3+}, etc; $A^{n-} = (CO_3)^{2-}$, Cl^-, $(NO_3)^-$, etc.

The physicochemical properties of HTs and the solid solutions produced after their calcination can be easily tuned by changing the nature and amount of metal cations and anions (10). Therefore, to elucidate the role of the chemical composition on the gas phase acetone self-condensation reaction, a series of bimetallic and trimetallic HTs were prepared, characterized and tested.

Results and Discussion

Table 1 summarizes the different chemical composition of the fresh HTs, where the molar ratios M^{2+}/M^{3+} varied from 2 to 3. Even if from synthesis it was established a nominal molar ratio $M^{2+}/M^{3+}=3$, the real values varied from 3.0 for ZnAl to 2.0 for MgZnAl-12, these differences can be attributed to an incomplete incorporation of the cations inside the layers. The increasing of Zn loading can be stated from the formulae, it ranges from 7.56 wt.% for MgZnAl-7 up to 43.62 wt.% for ZnАІ3.

Table 1. Chemical composition of the fresh hydrotalcite-like compounds.

Sample	Chemical Formulae	M^{2+}/M^{3+}
MgAl3	$[Mg_{0.744}Al_{0.256}(OH)_2]$ $(CO_3)_{0.128} \cdot 0.850$ H_2O	2.9
MgZnAl-7	$[Mg_{0.614}Zn_{0.093}Al_{0.293}(OH)_2]$ $(CO_3)_{0.146} \cdot 0.731 H_2O$	2.4
MgZnAl-12	$[Mg_{0.493}Zn_{0.165}Al_{0.342}(OH)_2]$ $(CO_3)_{0.171} \cdot 0.978$ H_2O	2.0
ZnAl3	$[Zn_{0.75}Al_{0.25}(OH)_2]$ $(CO_3)_{0.125} \cdot 0.840$ H_2O	3.0

Figure 1 exhibits the XRD powder patterns of fresh HT′s; they present the characteristic reflections corresponding to a hydrotalcite-like structure.

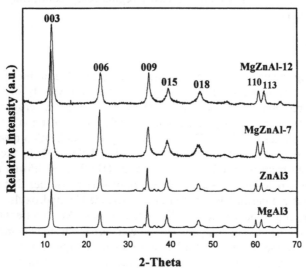

Figure 1. XRD powder patterns of fresh HT′s.

The unit cell parameters were calculated assuming a 3R stacking sequence, therefore $a = 2d_{110}$ and $c = 3d_{003}$ where c is the interlayer distance regulated by the size and charge of anion placed between the layers and the cell parameter a is the average metal-metal distance inside the layers. A way to verify the isomorphic substitution of Mg^{2+} by Zn^{2+} is analyzing the change of cell parameter a, the a value varied from 3.054 Å for MgAl3 to 3.072 Å for ZnAl3, this variance can be explained by the differences of cation sizes, since the ionic radius of Zn^{2+} is bigger than Mg^{2+}, 0.88 and 0.86Å respectively. The a values for MgZnAl samples rely on their different chemical composition and are between the range > 3.054 and < 3.072 Å. These results confirm the isomorphic substitution of Zn in the HTs layers.

All samples showed type IV isotherms, which correspond to mesoporous solids. The specific surface area (SSA) of the calcined solids were 62 and 254 m^2/g for ZnAl3 and MgAl3, both with pore volumes of 0.478 and 1.148 cc/g, and an avg. pore diameter of 225 and 181 Å, respectively. The SSA of the MgZnAl samples decreased as the Zn content rose, but they displayed remarkable bigger SSA than the ZnAl3 sample, for instance, the MgZnAl-12 sample presented a SSA of 197 m^2/g, a pore volume of 0.664 cc/g, and an avg. pore diameter of 135 Å.

Figure 2 shows illustrated mechanism for acetone self-condensation over calcined hydrotalcites, where the enolate ion is formed in a first step followed by two possible kinetic pathways: 1) In the first case the subtracted proton is attracted by the basic sites and transferred to the oxygen of the enolate ion to form an enol in equilibrium. 2) In the second case the enolate ion reacts with an acetone molecule in the carbonyl group, to produce the aldol (diacetone alcohol). Finally, the β carbon is deprotonated to form a ternary carbon and then loses an OH⁻ group to obtain the final products.

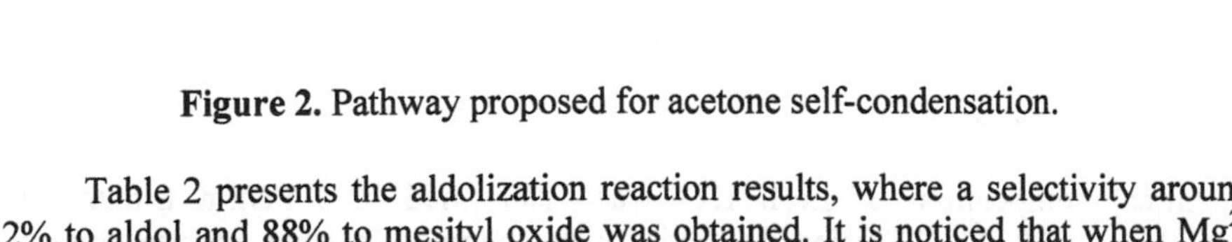

Acetone   Aldol   Mesityl Oxide

Figure 2. Pathway proposed for acetone self-condensation.

Table 2 presents the aldolization reaction results, where a selectivity around 12% to aldol and 88% to mesityl oxide was obtained. It is noticed that when Mg^{2+} was partially or totally substituted by Zn^{2+}, the reaction rate was improved. For instance, the reaction rates in MgZnAl-12 and ZnAl3 were 1.3 and 2.4 times higher than MgAl3, respectively. It is stated that the addition of Zn^{2+} inside the layers of calcined hydrotalcite improves the aldolization process, increasing the oligomerization conversion of acetone to mesityl oxide, due to charge defects generated by the Zn^{2+} insertion, thus changing the basic sites strength. It has been reported (10) that when Mg^{2+} is replaced by Zn^{2+} a substantial decrease in basic strength is observed, so a lower basic strength is expected in the ZnAl3 calcined sample in regard to MgAl3, however the activity of the former one is higher than in the last. This implies that the activity of the solids on the acetone self-condensation reaction is probably driven by the adsorption-desorption equilibrium between the intermediate species and the basic sites. Thus, it is suggested that a high interaction occurs between strong basic sites and the reaction intermediates, leading to a slow intermediate desorption to form the final products. On the other hand, the interaction of the reaction intermediates with weaker basic sites leads to an easier desorption of the products which is reflected in higher reaction rates.

Table 2. Results of self-condensation of acetone over various calcined hydrotalcite-like compounds.

Catalyst	%Se Aldol	%Se Mesityl Oxide	Reaction rate $(x10^7 \text{ mol/g s})$	% Conversion
MgAl3	11.9	88.1	8.1	2.3
MgZnAl-7	12.9	87.1	6.8	1.9
MgZnAl-12	12.3	87.7	10.4	2.9
ZnAl3	11.5	88.5	19.5	5.6

Experimental Section

Catalyst Preparation

The HTs were prepared by coprecipitation at 60 °C and constant pH according to procedures described earlier (10). For the MgAl3 HT for instance, an aqueous solutions containing 0.75 mol/L of Mg $(NO_3)_2.6H_2O$ and 0.25 mol/L of Al $(NO_3)_3.9H_2O$, and the second one 1.5 mol/L of KOH and 0.5 mol/L of K_2CO_3, were introduced by two electric pumps to a 4 L flask and mixed under vigorous stirring. The mixture was aged at 60 °C for 18 h under stirring. The precipitate was washed several times until the solution was free of excess ions, and then dried at 100 °C fresh HTs. For the synthesis of MgZnAl HTs, $Zn(NO_3)_2.6H_2O$ was used as Zn source. During the synthesis the percent of zinc was varied keeping the ratio $M^{2+}/M^{3+}=3$. Hereinafter, the solids thus obtained will be referred to as MgZnAl-wt.%Zn.

Characterization of Solids

Chemical composition of fresh HTs was determined in a Perkin Elmer Mod. OPTIMA 3200 Dual Vision by inductively coupled plasma atomic emission spectrometry (ICP-AES). The crystalline structure of the solids was studied by X-ray diffraction (XRD) using a Siemens D-500 diffractometer equipped with a CuKα radiation source. The average crystal sizes were calculated from the (003) and (110) reflections employing the Debye-Scherrer equation. Textural properties of calcined HTs (at 500°C/4h) were analyzed by N_2 adsorption-desorption isotherms on an AUTOSORB-I, prior to analysis the samples were outgassed in vacuum (10^{-5} Torr) at 300°C for 5 h. The specific surface areas were calculated by using the Brunauer-

Emmett-Teller (BET) equation and the pore size distribution and total pore volume were determined by the BJH method.

Catalytic Test

The activity of calcined HTs was determined in self-condensation reaction of acetone (J.T. Baker) by using a fixed bed catalytic reactor with an on-line GC. Prior to the catalytic test, catalysts were pretreated *in-situ* under nitrogen atmosphere at 450°C for 5h. Acetone was supplied to the reactor by bubbling nitrogen gas through the acetone container at 0 °C. The reaction temperature was established at 200°C. The products were analyzed by means of GC (Varian CP-3800) using a WCOT Fused silica column, equipped with a FID detector.

Acknowledgements

Thanks to Conacyt and IMP (project D.00285) for their financial support.

References

1. C.D. Chang, W.H. Lang and W.K. Bell, Chemical Industries (Marcel Dekker), **5**, (*Catal. Org. React.*), 73-94 (1981).
2. Y.-C. Son, S.L. Suib and R.E. Malz, *Catalysis of Organic Reactions*, D.G. Morrel, Ed. Marcel Dekker Inc., New York, p. 559-564. 2003
3. J.I. DiCosimo and C.R. Apesteguia, *Appl. Catal. A: General*, **137**, 149 (1996).
4. M. Zamora, T. Lopez, R. Gomez, M. Asomoza and R. Melendez, *Catal. Today*, **107-108**, 289, (2005).
5. C. Noda-Perez, C.A. Perez, C.A. Henriques and J.L.F. Monteiro, *Appl. Catal. A: General*, **272**, 229 (2004).
6. M. Bolognini, F. Cavani, C. Felloni and D. Scagliarini, *Catalysis of Organic Reactions*, D.G. Morrel, Ed. Marcel Dekker Inc., New York, p. 115-127. 2003
7. K. K. Rao, M. Gravelle, J. S. Valente and F. Figueras, *J. Catal.*, **173**, 115 (1998).
8. P.S. Kumbhar, J. S. Valente, J. Lopez and F. Figueras, *Chem. Commun.*, **7 (5)**, 535 (1998).
9. P.S. Kumbhar, J. S. Valente and F. Figueras, *Chem. Commun.*, **10**, 1091 (1998).
10. J. S. Valente, F. Figueras, M. Gravelle, P.S. Kumbhar, J. Lopez and J.P.Besse, *J. Catal.* **189**, 370 (2000).

8. Gas Phase Trimerization of Isobutene Using Green Catalysts

Angeles Mantilla[1], Francisco Tzompantzi[2], Miguel Torres[1] and Ricardo Gómez[2]

[1]*Universidad Autónoma Metropolitana Azcapotzalco. Av. San Pablo 180, Reynosa Tamaulipas, Azcapotzalco, México 02200, D.F.*
[2]*Universidad Autónoma Metropolitana-Iztapalapa, San Rafael Atlixco No 152, México 09340, D.F.*

E-mail: gomr@xanum.uam.mx

Abstract

In this work, the performance of two green acid catalysts, a TiO_2 synthesized by the sol-gel method and sulfated in situ was compared with a traditional NiY zeolite in the trimerization of isobutene. The reaction was carried out at mild conditions: atmospheric pressure and 40°C of temperature. The results obtained in the catalytic evaluation showed higher conversion and stability as well as a better selectivity to tri-isobutylene for the sulfated titania catalyst with respect to the NiY zeolite.

Introduction

The oligomerization of isobutene is carried out firstly in order to get diisobutylene as a product of the dimerization of isobutene, because of the high octane number of this compound, which could be an interesting alternative to the MTBE plants substitution. However, this process, like other petrochemical processes, use strong mineral acids HF, H_2SO_4 and H_3PO_4 as catalysts, in liquid form or impregnated over a solid support, which are environmentally corrosive and pollutant. It is desirable the substitution of these materials with green, environmental friendly catalyst. Several works focused in this subject has been published, including thermodynamic studies [1], kinetic studies [2-4], new acid catalyst evaluation and characterization [5-13], process design [14-16] and uses of the product [17-19]. However, in the present paper our attention will be devoted to the study of the trimerization of isobutene, due to the economical interest in the industrial application of triisobutylene, as a source of neoacid compounds [20], for the production of dodecylbencene used in the fabrication of biodegradable detergents or for gas-oil additives [21].

Several authors suggest the use of the NiY zeolite and Ni/Si-Al as catalysts for the oligomerization of isobutene [7-9]. The aim of this work is to compare the performance of two different green acid catalysts: in situ sulfated TiO_2 synthesized by the sol-gel method [22] with NiY zeolite, in the gas phase trimerization of isobutene at mild pressure and temperature conditions: atmospheric pressure and 40°C, in terms of activity, selectivity and stability.

Results and Discussion

The oligomerization of isobutene in function of the time is shown in Figures 1 and 2 for the sultated TiO_2 and NiY zeolite catalysts respectively. After a stabilization time of 2 hours on stream, the sulfated TiO_2 catalyst showed initial conversion values after 3 h in stream of 100, 70, 65 and 42% at GHSV of 8, 16, 32 and 64 h^{-1}, respectively (Fig. 1). For the NiY zeolite catalyst the activities were 56, 43, 33 and 17% for the previous GHSV values respectively (Fig. 2).

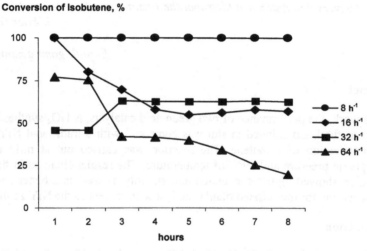

Figure 1 Conversion of isobutene over "in situ" sulfated TiO_2 catalyst at different GHSV values.

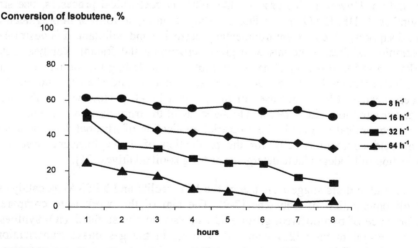

Figure 2 Conversion of isobutene over NiY catalyst at different GHSV values.

With respect to deactivation, sulfated TiO_2 catalyst showed a high stability, mostly at GHSV of $8h^{-1}$, where no deactivation occurs after 6 hours on stream (Figure 1). On the other hand, at high GHSV $64\ h^{-1}$ the sulfated TiO_2 deactivation in function of time and the conversion goes down from 42 to 23%. Stable activity was observed for GHSV of 16 y 32 h^{-1}. In the NiY catalyst, the lost of activity is notably pronounced for GHSV values of 32 and 64 h^{-1}, and after 7 h on stream the catalyst was totally deactivated (Fig. 2). The isobutane was fed to the reactor as a diluent and it did not show conversion in any experiment. The selectivity to triisobutylene was higher for the "in situ" sulfated titania than for the NiY catalyst in all the cases (Fig. 3).

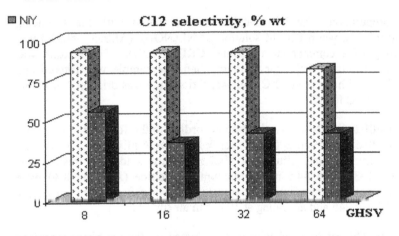

Figure 3 Selectivity to $C_{12}^=$ fraction over sulfated TiO_2 and NiY catalysts

The isobutene oligomerization is a highly exothermic reaction, carried out via the carbenium ion mechanism, which is thermodynamically favoured at low temperature. The kind of products obtained as well as the conversion and stability at constant temperature and pressure will depend on the reaction GHSV, which determine the intermediate carbenio ion formed during the first steps.

Another important factor in the behaviour of the catalysts is the nature and force of their acid sites. Both Lewis and Brönsted acid sites are necessary to carry out the reaction, but an important fact is the relative abundance of them as it has been recently reported for sulfated TiO_2 [12].

Figure 4 shows the reaction pathway for the isobutene oligomerization. After the dimer formation, the addition of another one butene molecule will depend mainly of Brönsted acidity to stabilize the formation of the carbocation. The oligomerization to heavier olefins will be favored on catalyst showing a low L/B acids sites ratio [12].

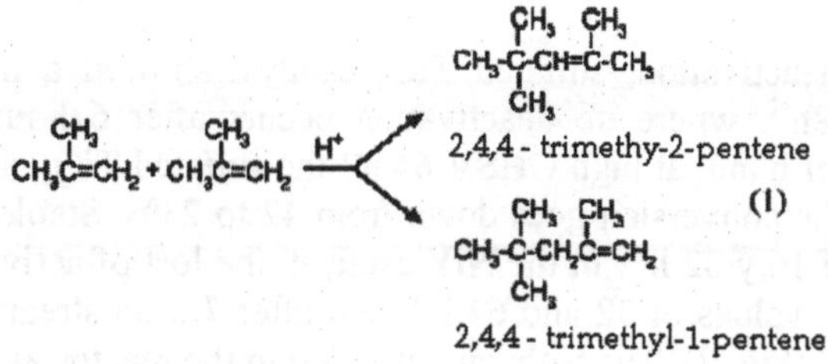

Figure 4 Reaction pathways for the isobutene oligomerization

Experimental Section

Catalysts preparation. The NiY catalyst was prepared with the following procedure: 300 ml of a warm 0.33 M solution of Ni $(NO_3)_2$, (Aldrich, 99.9%) was mixed with 10 g of a commercial Y zeolite (CRITERION). The mixture was maintained under reflux at 70°C for 5 hours; then, the sample was filtered and washed three times with water at 50°C. The washed sample was dried at 110°C and then annealed at 400°C for 12 h..

The in situ sulfated TiO_2 was synthesized according to the following procedure: 200 ml of bidistilled water and 200 ml of ter-butanol were mixed in a glass flask under reflux and stirring, then sulfuric acid (Baker 99%) was added to this solution until adjust at pH=3. Then 84.5 ml of titanium n-butoxide (Aldrich 98%) were slowly added, maintaining the solution under reflux for 24h. After gelling the sample was dried at 70°C for 24h and annealing at 400°C in air for 12 h.

Catalytic test. The catalytic behavior was evaluated for the gas phase isobutene trimerization reaction using a fixed bed reactor, with dimensions of 2 cm of diameter and 55 cm of length respectively. The operation conditions and evaluation procedure were as follows: the catalyst was activated at 400°C in flowing air (1 ml/s) during 8 hours. After the activation treatment, temperature was lowered to 40°C and a mixture of isobutane/isobutene 72:28 w/w was feed. The GHSV value was varied to 8, 16, 32 and 64 h^{-1} respectively. The average time of reaction was 11 h. The time of reactor stabilization after the beginning of the catalytic evaluation was 2 h.

The analysis of products was made in all the cases by FID-gas chromatography (Varian Mod. CX3400) equipped with a PONA column of 50 m and coupled to a workstation.

Acknowledgements

Thanks Gabriel Pineda Velázquez for the technical support in the chromatographic analysis.

References

1. R. J. Quann, L.A. Green, S.A. Tabak and F.J. Krambeck. *Ind. Eng. Chem. Res.*, **27**, 565-570 (1988).
2. J.F. Izquierdo, M.Vila, J.Tejero, F.Cunill and M.Iborra. *Appl.Catal. A: General*, **106**, 155-165 (1993).
3. A.P.Vogel, A.T. O´Connor and M.Kojima. *Clay Minerals*, **25**, 355-362 (1990).
4. A. Reffinger and U. Hoffman, *Chem. Eng. Technol;* **13**, 150 (1990).
5. G. Busca, G. Ramis and V. Lorenzelli; *J.Chem. Soc. Trans.*, **85(1)**, 137-146 (1998).
6. A. Saus and E. Schmidt; *J. Catal.*, **94**, 187-194 (1998).
7. B. Nkosi, F.T.T. Ng and G.L. Rempel; *Appl.Catal.A: General*, **158**, 225-241 (1997).
8. B. Nkosi, F.T.T. Ng and G.L. Rempel; *ApplCatal. A: General*, **161**, 153-166 (1997).
9. J. Heveling, C.P. Nicolaides and M.S. Scurrell; *ApplCatal.A: General*, **248**, 239-248 (2003).
10. W. Zhang, Ch. Li, R. He, X. Han and X. Bao. *Catal.Letters*, **64**, 147-150 (2000).
11. A.S. Chellapa, R.C. Millar and W.J.Thompson; *Appl.Catal. A: General*, **209**, 359-374 (2001).
12. A. Mantilla, G. Ferrat, A. López-Ortega, E. Romero, F. Tzompantzi, M. Torres, E. Ortíz-Islas and R. Gómez; *J. Mol. Cat. A: Chemical*, **228**, 333-338 (2005).
13. A. Mantilla, G. Ferrat, F. Tzompantzia , A. López-Ortega, E. Romero, E. Ortiz-Islas, R.Gómez, and M. Torres; *Chemm. Com., 13,1498-1499 (2004)..*
14. M. Golombok and J. de Brujin; *Ind. Eng. Chem. Res.*, **39**, 267-271(2000).
15. L.O. Stine and S.C. Gimre, US Pat 5,856,604, to UOP LLC (1999).
16. M. Torres, L. Lopez, J. Domínguez, A. Mantilla, G. Ferrat, M. Gutiérrez and M. Maubert. *Chem. Eng. J.*, **92**, 1-6 (2003).
17. M. Golombok and J. de Brujin. *Ind. Eng. Chem. Res.*, **39**, 267-271(2000).
18. M. Marchionna, M. Di Girolamo and R. Patrini, R., *Catal. Today*, **65**, 397-403 (2001).
19. M. Golombok and J. De Bruijn, *J.App. Catal. A: General*, **208,** 47-53 (2001).
20. E. López, A. Mantilla, G. Ferrat and A. López y Mario Vera. *Petrol. Sci. and Technol.*, 177-187 (2004.).
21. R. Alcántara, E. Alcántara, L. Canoira, M.J. Franco. M. Herrera and A. Navarro. *Reactive & Functional Polymers*, **45**, 19-27 (2000).
22. X. Bokhimi, A. Morales, E. Ortíz, T. López, R. Gómez and J. Navarrete, *J. Sol-Gel Sci. Technol.*, **29**, 31 (2004).

9. Synthesis of Methyl Isobutyl Ketone over a Multifunctional Heterogeneous Catalyst: Effect of Metal and Base Components on Selectivity and Activity

Arran S. Canning, Jonathan J. Gamman, S. David Jackson and Strath Urquart

WestCHEM, Dept. of Chemistry, The University, Glasgow G12 8QQ, Scotland.

sdj@chem.gla.ac.uk

Abstract

Four catalyst systems, Pd-KOH/silica, Pd-CsOH/silica, Ni-KOH/silica, and Ni-CsOH/silica, have been investigated for the conversion of acetone to MIBK. Nickel catalysts were generally less selective, with MIBK being further hydrogenated to MIBC, and isophorone becoming a major product at high temperatures. KOH-based catalysts were found to be less active than their CsOH-based counterparts. The activation energy for MIBK production suggested that the rate determining step was within the base catalyzed part of the reaction sequence and this was supported by the product distributions.

Introduction

Condensation reactions are an important route to the industrial synthesis of a wide range of compounds including solvents such as methyl isobutyl ketone (MIBK) and Guerbet alcohols for use as lubricants and surfactants. The synthesis of MIBK from acetone proceeds via a base extraction of the α-proton catalyzed by aqueous NaOH and intermolecular attack of the carbonyl group to produce diacetone alcohol. Subsequent acid catalyzed dehydration of diacetone alcohol (DAA) to mesityl oxide (MO) by H_2SO_4 at 373 K is followed by hydrogenation of MO to MIBK using Cu or Ni catalysts at 288 – 473 K and 3-10 bar (1). Recently however new one-step technology has been developed and commercialized using palladium on acidic ion-exchange resins (2). The change from the three-step process to a single-step significantly reduces the amount of waste generated at the base catalysis stage and removes the need for concentrated sulphuric acid and its residues. Research in this area is nevertheless continuing with a view to improving the process further (3, 4).

In this study we have investigated utilizing a bi-functional catalyst with a solid base function for the aldol condensation reaction and a metal function for the hydrogenation. This work is a continuation of the study that examined the supported base catalyzed aldol condensation of acetone (5, 6). In those studies we examined the aldol condensation step to MO. In this study we have added palladium or nickel to facilitate the hydrogenation of MO to MIBK. Other groups have also used copper (7) as well as palladium (3, 8) and nickel (9).

The possible reactions that can take place are outlined in Figure 1 (6). Typical by-products of the aldol reaction are phorone and isophorone, however as can be seen in the figure other by-products can be formed, especially in the presence of a hydrogenating functionality.

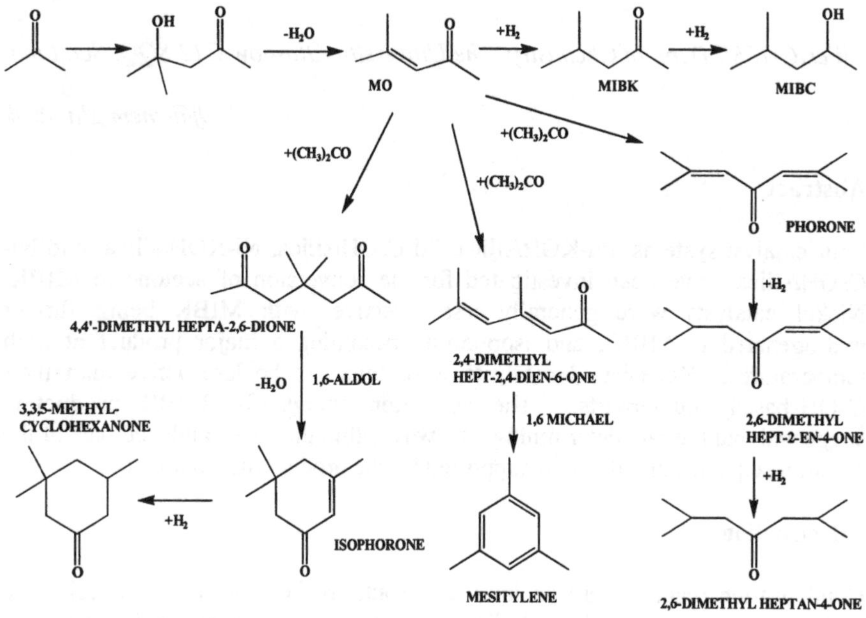

Figure 1. Mechanistic scheme for main reaction and by-products.

Results and Discussion

The Pd-KOH/silica, Pd-CsOH/silica, Ni-KOH/silica and Ni-CsOH/silica catalysts were tested at various temperatures. The selectivity to MIBK, MIBC, and isophorone as a function of temperature after 24 h on-line is shown in Figure 2. The conversion at the three temperatures is shown in Figure 3.

All the catalysts showed high selectivity until 673 K where there was a dramatic breakdown in selectivity over the nickel catalysts. Note that this loss in selectivity was not associated with a dramatic increase in conversion. Indeed

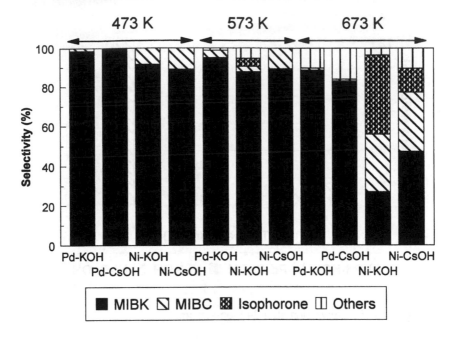

Figure 2. Selectivity after 24 h on-line.

the Pd-CsOH/silica catalyst had the highest activity and also maintained its high selectivity. After 24 h on-stream at 473 K only two products (MIBK and methyl isobutyl carbinol (MIBC, 4-methyl pentan-2-ol)) were detected for all the catalysts except Pd-CsOH/silica, which produced only MIBK. There were no secondary aldol by-products, the hydrogenation of MIBK to MIBC was the sole route to an undesired product. As expected palladium was poorer at hydrogenating the carbonyl group compared to nickel and hence gave the higher selectivity. The product yields for the four catalysts at 673 K are shown in Figure 4.

The range of species produced reflects the overall mechanistic scheme shown in Figure 1. The detail however gives insights into the different processes and the effectiveness of the catalyst components in catalysing a specific transformation.

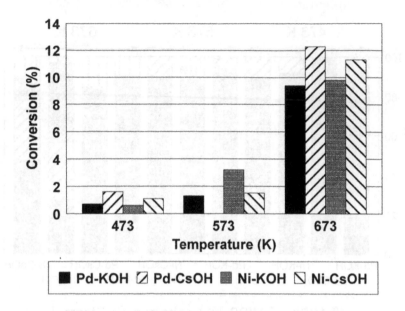

Figure 3. % Conversion after 24 h on-stream.

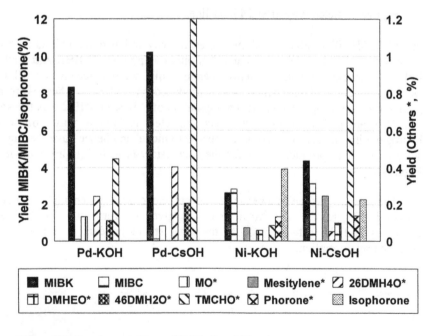

Figure 4. Product yields at 673 K after 24 h on-stream.
2,6DMH4O - 2,6-dimethyl-4-heptanone; DMHEO - 2,6-dimethyl hept-2-en-4-one; TMCHO - 3,3,5-methyl-cyclohexanone; 46DMH2O - 4,6-dimethyl heptan-2-one.

At 473 K the system was at its most selective for all the catalysts. The reaction from acetone to MO was sufficiently slow, relative to the hydrogenation step, that all MO produced was hydrogenated selectively to MIBK and only small quantities of MIBK were further hydrogenated to MIBC (Figure 5). Similar behaviour was observed with a Ni/alumina catalyst (9), which gave high selectivity to MIBK at 373 K. Interestingly this catalyst was made by co-precipitation using sodium hydroxide, however no sodium analysis was performed (9). Hence residual sodium acting as a base catalyst cannot be discounted in this study and indeed would fit very well with the results obtained.

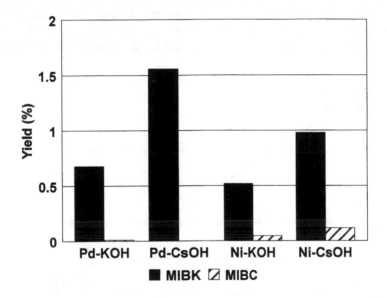

Figure 5. Yield of MIBK and MIBC at 473 K.

The data in Figure 5 shows the effect of the base component on the system for the 473 K condition. It is clear that the stronger base was more effective for the aldol condensation reaction and hence more MIBK was produced. The MIBK:MIBC ratio for the nickel catalysts was 10:1 reflecting the ease of hydrogenation of the C=C bond compared to the C=O bond at this temperature. For the Pd catalysts, the ratio was much higher, at $\geq 65:1$

At 673 K the product distribution over the palladium catalysts (Figure 4) was still highly selective to MIBK (> 80%) however further aldol condensation of acetone with MO to form the three intermediates, phorone, 4,4'-dimethyl hepta-2,6-dione and 2,4-dimethyl hept-2,4-dien-6-one was observed. These species were formed by aldol condensation of acetone with MO at different points in the molecule (10). All can continue to react either by subsequent aldol condensation, Michael addition, or hydrogenation as per Figure 1. There was no detectable isophorone produced, the product of internal 1,6-aldol reaction of

4,4'-dimethyl hepta-2,6-dione (6) but the production of 3,3,5-methyl-cyclohexanone revealed that 4,4'-dimethyl hepta-2,6-dione had indeed undergone an internal aldol condensation to produce isophorone which had then been hydrogenated. Similarly no phorone was observed but the hydrogenated product, 2,6-dimethyl-4-heptanone was detected. The half-hydrogenated form of 2,4-dimethyl hept-2,4-dien-6-one, 2,6-dimethyl hept-2-en-4-one was also seen although at low levels.

The high MIBK selectivity over the palladium catalysts suggested that the hydrogenation of MO was facile and that the overall rate-determining step may lie in the aldol part of the sequence rather than with the hydrogenation. If this was the case the case then we may have expected to have similar activation energies for the formation of MO and MIBK. In a previous study (11) an activation energy of 23 ± 4 kJ.mol^{-1} was calculated for the formation of MO from

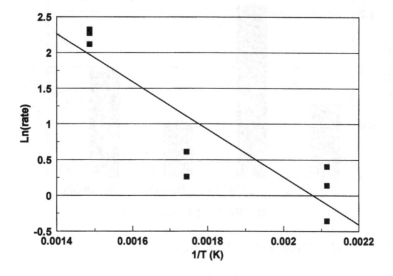

Figure 6. Activation energy plot for production of MIBK.

acetone over a CsOH/silica catalyst. In the present study the activation energy calculated for MIBK was 27 ± 5 kJ.mol^{-1} (Figure 6) in keeping with the suggestion that the rate determining step is within the base catalyzed part of the reaction sequence.

In comparison with Pd/SAPO catalysts (8) our systems are much more selective, with selectivities of ~100 % at 473 K and ~95 % at 573 K compared with 65 % at 473 K and ~10 % at 573 K for Pd/SAPO. However this comparison was not at equal conversion. Indeed it is difficult to compare activity directly as the catalysts were only run for a few hours (8) whereas ours were on-stream for over 24 h.

At 673 K the nickel catalysts had low selectivity to MIBK (< 50 %) and a significant contribution to this loss in selectivity was due to hydrogenation of MIBK to MIBC. Indeed with the Ni-KOH/silica catalyst the yield of MIBC was greater than the yield of MIBK. This behaviour was to be expected as nickel is used industrially to hydrogenate 2-ethyl hexenal, the intermediate in the OXO-alcohol process to make 2-ethyl hexanol (1), where both alkene and carbonyl are required to be hydrogenated. The ratio of MIBK:MIBC at 673 K was 1.4:1 for the Ni-CsOH catalyst and 0.9:1 for the Ni-KOH catalyst compared to ~10:1 for both catalysts at 473 K. This change in ratio with temperature is likely to be due to different activation energies for C=C and C=O hydrogenation.

It is noticeable that over the nickel catalysts, isophorone was a major product. The production of isophorone and the small quantities of other by-products once again revealed formation of the three intermediate aldol products of the reaction between MO and acetone. However it is clear that hydrogenation was not as facile over the nickel catalysts as it was over the palladium catalysts. Hence there was more secondary and tertiary aldol condensation. To further investigate this a reaction was performed at 573 K using the Ni-KOH/silica catalyst with nitrogen as the carrier gas rather than hydrogen. The results are shown in Table 1.

Table 1. Comparison of selected reaction product yields (%) in N_2 carrier compared to H_2 carrier over Ni-KOH/silica at 573 K.

Ni-KOH/SiO$_2$	MO	MIBK	MIBC	Isophorone
Nitrogen	2.1	0	0	1.2
Hydrogen	0.1	2.8	0.1	0.2

Clearly there is a switch from MO to MIBK when hydrogen was made the carrier gas. The yield of isophorone dropped as at this temperature the hydrogenation intercepted the MO before it could undergo another aldol condensation. At 673 K the aldol condensation reaction became kinetically more competitive with the hydrogenation reaction.

Over the temperature range studied the activity was controlled by the base component of the catalyst. The conversion observed with the Ni-KOH catalyst was approximately the same as that observed with the Pd-KOH catalyst; while the CsOH catalysts were more active at both temperatures.

In conclusion, four catalyst systems, Pd-KOH/silica, Pd-CsOH/silica, Ni-KOH/silica, and Ni-CsOH/silica, have been investigated for the conversion of acetone to MIBK. Systems highly selective to MIBK have been obtained (Pd-CsOH/silica, 100 % at 473 K). Both the metal and the base affected the product distributions. The nickel catalysts were generally less selective, with MIBK being further hydrogenated to MIBC and isophorone becoming a major product at high temperatures. Over Ni-CsOH/silica, a selectivity of around 30% was obtained for MIBC compared with <1% for Pd-CsOH/silica under identical conditions. For isophorone a selectivity of 40% was observed from Ni-KOH/silica whereas Pd-KOH/silica produced no isophorone. The KOH-based catalysts were found to be less active than their CsOH-based counterparts. The activation energy for MIBK production suggested that the rate determining step was within the base catalyzed part of the reaction sequence and this was supported by the product distributions.

Experimental Section

The catalysts were prepared by impregnation using metal nitrates. Potassium nitrate (Hopkin and Williams), cesium nitrate (Aldrich), nickel nitrate (Avocado) and palladium nitrate (Alfa Aesar) were all used as received. The silica used was Fuji-Silysia CARiACT Q-10, with a surface area of 296 m^2g^{-1}. The metal weight loading of the catalysts was 1% Pd and 4% alkali metal or 15 % Ni and 4% alkali metal. The combined group VIII/alkali metal catalysts were prepared by dissolving both nitrates in the least amount of distilled water and spraying onto the silica support. The resulting material was dried at 323 K for 24 h and calcined at 723 K for 4 h. A particle size of 250µm was used throughout the study.

The aldol condensation/hydrogenation reaction was carried out in a continuous flow microreactor. The catalysts (0.5 g) were reduced *in situ* in a flow of H_2 at atmospheric pressure at 723 K for 1 h for the palladium systems and 2 h for the nickel systems. The liquid reactant, acetone (Fisher Scientific HPLC grade >99.99%), was pumped via a Gilson HPLC 307 pump at 5 mL hr^{-1} into the carrier gas stream of H_2 (50 cm^3 min^{-1}) (BOC high purity) where it entered a heated chamber and was volatilised. The carrier gas and reactant then entered the reactor containing the catalyst. The reactor was run at 6 bar pressure and at reaction temperatures between 373 and 673 K. Samples were collected in a cooled drop out tank and analyzed by a Thermoquest GC-MS fitted with a CP-Sil 5CB column

References

1. G. J. Kelly, F. King and M. Kett, *Green Chem.* **4**, 392 (2002).
2. K. Schmitt, J. Distedorf, W. Flakus, U.S. Patent 3,953,517 to Veba Chemie AG (1976); J. J. McKetta, and W. A. Cunningham, *Encyclopedia of Chemical Processing and Design*, Vol. 30, p. 61. Dekker, New York, 1988, p. 61.
3. W. K. O'Keefe, M. Jiang, F. T. T. Ng and G. L. Rempel, Chemical Industries (CRC), **104**, (*Catal. Org. React.*), 261-266 (2005).
4. A. A. Nikolopoulos, G. B. Howe, B. W.-L. Jang, R. Subramanian, J. J. Spivey, D. J. Olsen, T. J. Devon and R. D. Culp, Chemical Industries (Dekker), **82** (*Catal. Org. React.*), 533 (2001).

5. A. S. Canning, S. D. Jackson, E. McLeod and E. M. Vass, Chemical Industries (CRC), **104**, (*Catal. Org. React.*), 363-372 (2005).
6. A. S. Canning, S. D. Jackson, E. McLeod and E. M. Vass, *Appl. Catal., A*, **289**, 59 (2005).
7. V. Chikan, A. Molnar and K. Balazsik, *J. Catal.*, **184,** 134 (1999)
8. S.-M. Yang and Y. M. Wu, *Appl. Catal. A*, **192**, 211 (2000).
9. S. Narayan and R. Unnikrishnan, *Appl. Catal. A,* **145**, 231 (1996).
10. W.T. Reichle. *J. Catal.*, **63**, 295 (1980).
11. A. S. Canning, S. D. Jackson, E. McLeod and G. M. Parker, manuscript in preparation.

10. The Control of Regio- and Chemo-Selectivity in the Gas-phase, Acid-Catalyzed Methylation of 1,2-Diphenol (Catechol)

Mattia Ardizzi, Nicola Ballarini, Fabrizio Cavani, Luca Dal Pozzo, Luca Maselli, Tiziana Monti and Sara Rovinetti

University of Bologna, Dipartimento di Chimica Industriale e dei Materiali, Viale Risorgimento 4, 40136 Bologna, Italy. INSTM, Research Unit of Bologna. A member of Concorde CA and of Idecat NoE (6FP of the EU).

fabrizio.cavani@unibo.it

Abstract

The catalytic properties of materials having either (i) Brønsted-type acidity (H-mordenite and Al phosphate), or (ii) Lewis-type acidity (Al trifluoride), were investigated in the reaction of gas-phase catechol methylation with methanol. H-mordenite and AlF_3 gave high selectivity to guaiacol (the product of *O*-methylation) only under mild reaction conditions, that is for very low catechol conversion. In fact, the increase of temperature led to the transformation of guaiacol to phenol and cresols, and to considerable catalyst deactivation. The use of a mildly acidic catalyst, $AlPO_4$, instead made it possible to obtain a stable catalytic performance, with high selectivity to guaiacol at 42% catechol conversion.

Introduction

Phenol and diphenols (catechol, resorcinol and hydroquinone, the ortho, meta and para isomers, respectively), are versatile starting material for the synthesis of many compounds. However, most of the corresponding industrial transformations still have drawbacks that make the processes environmentally unfriendly, such as:

(1) The use of alkylhalides, or dimethylsulphate or acylhalides as co-reactants, with co-production of waste effluents containing inorganic salts, which have to be disposed.

(2) The high number of synthetic steps that sometimes are necessary, with consequent low overall yield, due to the by-products formed in each step.

(3) The use of stoichiometric conditions instead than catalytic, or of such large amounts of catalyst, often not recoverable, that the process can be regarded as a stoichiometric one. Moreover, the use of homogeneous catalysts often implies the development of corrosive reaction media, and problems in the purification of the product.

(4) The use of solvents that may have themselves unfriendly characteristics.

Therefore, the development of new processes having lower environmental impact represents one important target (1,2). In the present work we report about the use of heterogeneous acid catalysts in the gas-phase methylation of catechol, for the synthesis of guaiacol.

Results and Discussion

Figure 1 plots the results obtained in the gas-phase methylation of catechol, with the H-mordenite catalyst. The main reaction product at low temperature and low catechol conversion was guaiacol, the selectivity of which however decreased when the reaction temperature was increased, with a corresponding higher formation of phenol and, at a minor extent, of 3-methylcatechol, o-cresol and p-cresol. This trend is substantially different from the one observed in the methylation of phenol catalyzed by zeolites (3,4). In the latter case, in fact, anisole is the prevailing product at low temperature, but the decrease of its selectivity for increasing temperatures is mainly due to the intramolecular rearrangement to o-cresol (5). The latter becomes the prevailing product, together with polymethylated phenols, at high temperature. Also the direct C-alkylation contributes to the formation of cresols and polymethylated phenols. On the contrary, in our case the product of intramolecular rearrangement of guaiacol (3-methylcatechol), was produced with low selectivity. Phenol formed by loss of one formaldehyde molecule from guaiacol. The selectivity to veratrol was nil, and that to 4-methylcatechol was less than 0.5%.

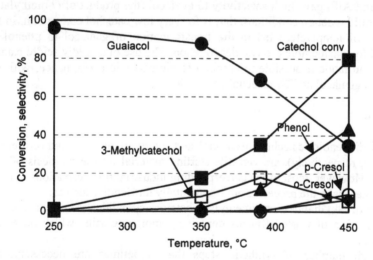

Figure 1. Catechol conversion and selectivity to guaiacol, o-cresol, phenol, p-cresol and 3-methylcatechol as functions of temperature. Catalyst H-mordenite. Feed: catechol and methanol (1/10 molar feed ratio).

Analogous results were obtained with AlF_3; the prevailing product at low temperature was guaiacol, the selectivity of which decreased in favour of phenol, and of other minor products. It is worth noting that the similar catalytic performance obtained for the H-mordenite and AlF_3 suggests similarity also in the nature of active sites. This is in favour of the generation, under reaction conditions, of a Brønsted-type acidity at the surface of AlF_3, either by interaction of water or of catechol with the Lewis sites. With this catalyst, however, the deactivation was more relevant than with the zeolite. One reason for the deactivation effect may be related to the strong interaction of catechol and guaiacol with the active sites. This also may be the reason for the low stability of guaiacol; the high effective permanence time at the adsorbed state, or its slow diffusion, may favor its further transformation to phenol. This hypothesis was confirmed by directly feeding guaiacol, in the absence of methanol.

Figure 2 compares the distribution of products obtained when guaiacol was fed in the absence of methanol, at two levels of temperature and with the AlF_3 catalyst. Results were taken at the very beginning of the reaction time, that is, before deactivation caused a rapid decline of conversion. At 300°C, the main products of guaiacol transformation were catechol, veratrol and methylguaiacols. The amount of de-oxygenated compounds (phenol and cresols) was negligible. Therefore, under these conditions the main reaction was the methylation of guaiacol onto another molecule of guaiacol, to generate catechol and either veratrol or methylguaiacols. Moreover, either the intramolecular alkylation, or the intermolecular alkylation of guaiacol on catechol, generated methylcatechols (mainly 3-methylcatechol). It is worth noting that the sum of selectivity to veratrol and methylguaiacols was equal to the selectivity to catechol; this confirms that bimolecular alkylation reactions occurred, and that catechol was the co-product in the formation of veratrol and methylguaiacols. The situation was different at 390°C; besides catechol, main products were those of C-alkylation (methylcatechols and methylguaiacols), rather than veratrole. The selectivity to de-oxygenated compounds (mainly phenol, with minor amounts of cresols) was relevant; therefore under such conditions the transformation of guaiacol to phenol + formaldehyde gave an important contribution to guaiacol conversion.

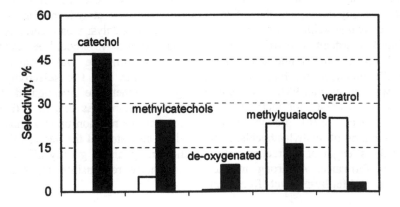

Figure 2. Initial selectivity to the products at 300°C (white bars) and at 390°C (black bars), in the transformation of guaiacol. Catalyst AlF_3.

The main reactions that occur in catechol methylation over strongly acid catalysts (H-mordenite and AlF$_3$), as inferred from catalytic tests, are summarized in Scheme 1. The main indication is that it is possible to achieve a good selectivity to guaiacol only under mild reaction conditions, that is, at low catechol conversion. If the temperature is increased to reach a higher conversion, the selectivity to guaiacol drops, with an increase in the formation of phenol and cresols, rather than of methylated catechol. Therefore, the reaction network is different from that obtained in the case of phenol methylation over zeolites. This difference is due to the low stability of guaiacol, which at temperatures higher than 300°C is transformed to catechol and to phenols.

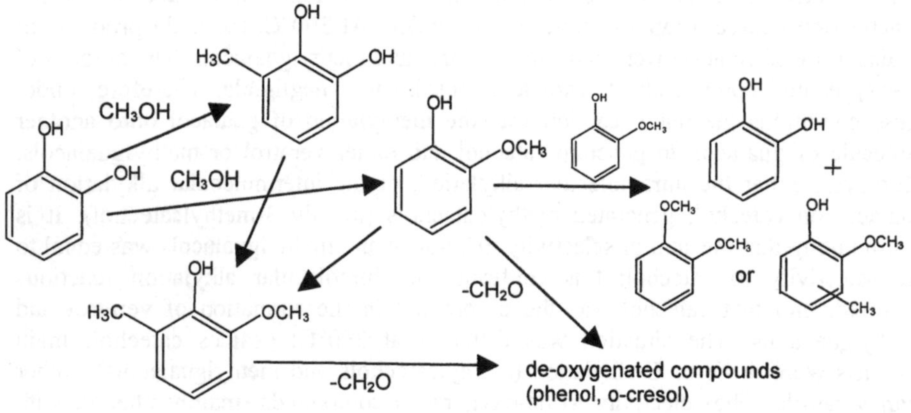

Scheme 1. Reaction scheme in catechol methylation with strongly acid catalysts.

Al/P/O catalyst, containing crystalline tridimyte AlPO$_4$, possesses mildly acidic properties (6). Figure 3 plots the effect of temperature on catalytic performance. Despite the weaker acidity as compared to zeolites, the catalyst was clearly more active than H-mordenite and AlF$_3$. A guaiacol conversion of 42% could be reached at 275°C; even more remarkably, the catalytic performance was maintained substantially unaltered for hundreds hours reaction time (7). The selectivity to guaiacol was higher than 95% in the entire range of temperature investigated. The slight fall of selectivity for increasing reaction temperature was due to both the formation of veratrole (the product formed by consecutive reaction upon guaiacol), and the formation of methylcathecols. Data obtained indicate that the nature of acid sites and the strength of interaction with catechol and guaiacol greatly affects the catalytic performance. A strong interaction is finally responsible for catalyst deactivation, and for by-products formation.

Experimental Section

The following catalysts were used for reactivity tests: (a) H-mordenite zeolite, H-MOR40, shaped in extrudates, supplied by Süd-Chemie AG, and having SiO$_2$/Al$_2$O$_3$ ratio equal to 40. (b) AlF$_3$, supplied by Aldrich. (c) Al/P/O, prepared following the procedure described in (7), and having an Al/P atomic ratio equal to 0.74. Catalytic tests in the gas phase have been performed in a tubular flow glass reactor. The following reaction conditions were used: (a) with the H-mordenite and AlF_3 catalysts, catechol/methanol 1/10 molar ratio, residence time 0.6 s for the total feed flow, including N_2 as the carrier gas (0.1 l/min). (b) with the $AlPO_4$ catalyst, catechol/methanol 1/3.4 molar ratio, residence time 7.5 s.

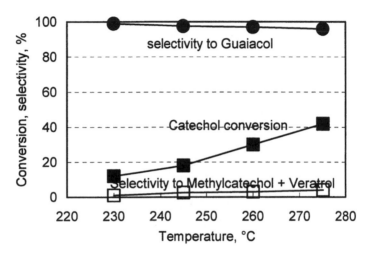

Figure 3. Catechol conversion, selectivity to guaiacol and to methylcatechol + veratrol as functions of temperature. Catalyst Al/P/O. Feed: catechol and methanol (1/3.4 molar feed ratio).

References

1. P. Marion, R. Jacquot, S. Ratton and M. Guisnet, *Zeolites for Cleaner Technologies*, M. Guisnet and J.-P. Gilson, Eds., ICP, London, p. 281. 2002
2. P. Métivier, *Fine Chemicals through Heterogeneous Catalysis*, R.A. Sheldon and H. van Bekkum, Eds., Wiley-VCH, p. 161 and 173. 2001
3. M. Marczewski, J.-P. Bodibo, G. Perot and M. Guisnet, *J. Mol. Catal.*, **50**, 211 (1989).
4. S. Balsamà, P. Beltrame, P.L. Beltrame, P. Carniti, L. Forni and G. Zuretti, *Appl. Catal.*, **13**, 161 (1984).
5. R.F. Parton, J.M. Jacobs, D.R. Huybrechts and P.A. Jacobs, *Stud. Surf. Sci. Catal.*, **46**, 163 (1989).
6. F.M. Bautista, J.M. Campelo, A. Garcia, D. Luna, J.M. Marinas and A.A. Romero, *Appl. Catal.*, **96**, 175 (1993).
7. F. Cavani, T. Monti, P. Panseri, B. Castelli and V. Messori, WO Pat. 74,485, to Borregaard Italia (2001).

11.   The Environmentally Benign, Liquid-phase Benzoylation of Phenol Catalyzed by H-β Zeolite. An Analysis of the Reaction Scheme.

Mattia Ardizzi, Nicola Ballarini, Fabrizio Cavani, Massimo Cimini, Luca Dal Pozzo, Patrizia Mangifesta and Diana Scagliarini

University of Bologna, Dipartimento di Chimica Industriale e dei Materiali, Viale Risorgimento 4, 40136 Bologna, Italy. INSTM, Research Unit of Bologna. A member of Concorde CA and Idecat NoE (6FP of the EU)

fabrizio.cavani@unibo.it

Abstract

The reaction of phenol benzoylation catalyzed by a H-β zeolite (Si/Al ratio 75) has been studied, with the aim of determining the reaction scheme. The only primary product was phenylbenzoate, which then reacted consecutively to yield o- and p-hydroxybenzophenone and benzoylphenylbenzoate isomers, via both intermolecular and intramolecular mechanisms.

Introduction

Friedel-Crafts acylation of aromatic substrates is widely employed for the synthesis of aryl ketones, intermediates of relevant industrial importance for the production of drugs, fragrances and pesticides. The acylating compounds are acyl halides, and homogeneous Lewis-type acids (typically, metal halides) are the catalysts employed (1). The latter are, however, used in large amounts, often larger than the stoichiometric requirement; also, recovery of the metal halides at the end of the reaction is not possible, and big amounts of waste streams are co-generated. Therefore, the environmental impact of these processes makes highly desirable the development of new technologies that employ heterogeneous, recyclable catalysts, and of reactants that generate more environmentally friendly co-products (2).

With this purpose, several different types of solid acid catalysts have been investigated for the acylation of aromatics, but the best performances have been obtained with medium-pore and large-pore zeolites (3-9). In general, however, the use of acylating agents other then halides, e.g., anhydrides or acids, is limited to the transformation of aromatic substrates highly activated towards electrophilic substitution. In a previous work (10), we investigated the benzoylation of resorcinol (1,3-dihydroxybenzene), catalyzed by acid clays. It was found that the reaction mechanism consists of the direct O-benzoylation with formation of resorcinol monobenzoate, while no primary formation of the product of C-benzoylation (2,4-dihydroxybenzophenone) occurred. The latter product formed exclusively by consecutive Fries rearrangement upon the benzoate. In the present paper, we report about the benzoylation of phenol catalyzed by a H-β zeolite; the study was carried out in order to check the feasibility of the reaction when catalyzed by zeolites, and determine the reaction scheme.

Results and Discussion

Figure 1 plots the effect of the reaction time on phenol conversion, at 200°C; the reaction was carried out in the absence of solvent.

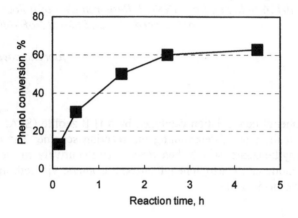

Figure 1. Phenol conversion as a function of the reaction time. Catalyst H-β 150.

A maximum phenol conversion of 65% was reached, due to the fact that the consumption of benzoic acid was higher than that of phenol. Indeed, despite the 1/1 load ratio, the selectivity to those products the formation of which required two moles of benzoic acid per mole of phenol, made the conversion of benzoic acid approach the total one more quickly than phenol. A non-negligible effect of catalyst deactivation was present; in fact, when the catalyst was separated from the reaction mixture by filtration, and was then re-loaded without any regeneration treatment, together with fresh reactants, a conversion of 52% was obtained after 2.5 h reaction time, lower than that one obtained with the fresh catalyst, i.e., 59% (Figure 1). The extraction, by means of CH_2Cl_2, of those compounds that remained trapped inside the zeolite pores, evidenced that the latter were mainly constituted of phenol, benzoic acid and of reaction products, with very low amount of heavier compounds, possible precursors of coke formation.

Figure 2 plots the effect of phenol conversion on the distribution of products. The following compounds were obtained: (i) phenylbenzoate, (ii) o- and p-hydroxybenzophenone, and (iii) o- and p-benzoylphenylbenzoate. The first product was obtained with 100% selectivity only at moderate conversion, for low reaction times. In fact, when the conversion of phenol increased due to longer reaction times, the selectivity to phenylbenzoate rapidly declined, in favor of the formation of the ortho and para isomers of hydroxybenzophenone and of benzoylphenylbenzoate.

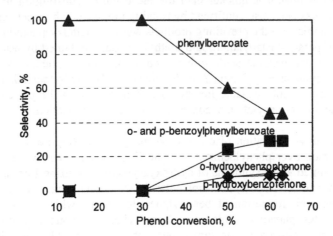

Figure 2. Selectivity to products as functions of phenol conversion. Catalyst H-β 150. T 200°C.

The data clearly indicate that the only primary product was phenylbenzoate, while all the other compounds formed by consecutive reactions upon the ester. Therefore, the scheme of reaction is that one summarized in Figure 3. The formation of the ester as the only primary product represents one important difference with respect to the Friedel-Crafts benzoylation of phenol with benzoylchloride or benzyltrichloride, catalyzed by AlCl₃. In the latter case, in fact, the product of para-*C*-acylation (p-hydroxybenzophenone) is the main product of reaction; this is due to the fact that AlCl₃ coordinates with the O atom of the hydroxy group in phenol, and makes it less available for the ester formation, due to both electronic and steric reasons.

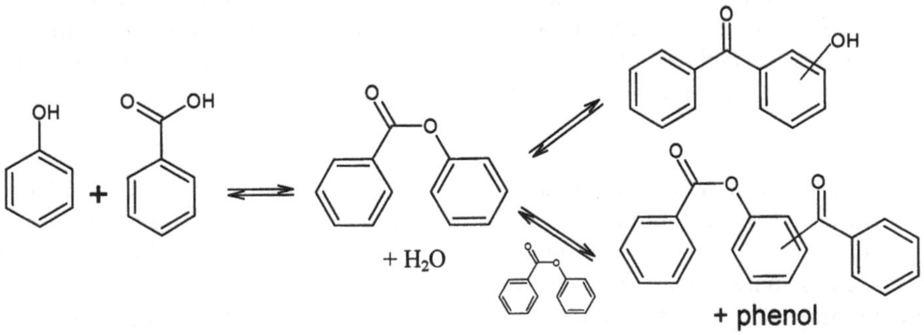

Figure 3. Reaction scheme for the reaction between phenol and benzoic acid catalyzed by the H-β zeolite.

This does not occur in the case of catalyst and reactants here described. With Brønsted-type catalysis, the reaction between the benzoyl cation, $Ph-C^+=O$, and the hydroxy group in phenol is quicker than the electrophilic substitution in the ring. This hypothesis has been also confirmed by running the reaction between anisole and benzoic acid; in this case the prevailing products were (4-methoxy)phenylmethanone (the product of para-C-benzoylation) and methylbenzoate (obtained by esterification between anisole and benzoic acid, with the co-production of phenol), with minor amounts of phenylbenzoate, phenol, 2-methylphenol and 4-methylphenol. Therefore, when the O atom is not available for the esterification due to the presence of the substituent, the direct C-acylation becomes the more favored reaction.

The consecutive formation of o-hydroxybenzophenone (Figure 3) occurred by Fries transposition over phenylbenzoate. In the Fries reaction catalyzed by Lewis-type systems, aimed at the synthesis of hydroxyarylketones starting from aryl esters, the mechanism can be either (i) intermolecular, in which the benzoyl cation acylates phenylbenzoate with formation of benzoylphenylbenzoate, while the $Ph-O-Al^-Cl_3$ complex generates phenol (in this case, hydroxybenzophenone is a consecutive product of phenylbenzoate transformation), or (ii) intramolecular, in which phenylbenzoate directly transforms into hydroxybenzophenone, or (iii) again intermolecular, in which however the benzoyl cation acylates the $Ph-O-Al^-Cl_3$ complex, with formation of another complex which then decomposes to yield hydroxybenzophenone (mechanism of monomolecular deacylation-acylation). Mechanisms (i) and (iii) lead preferentially to the formation of p-hydroxybenzophenone (especially at low temperature), while mechanism (ii) to the ortho isomer. In the case of the Brønsted-type catalysis with zeolites, shape-selectivity effects may favor the formation of the para isomer with respect to the ortho one (11,12).

In our case, all the compounds obtained by transformation of the intermediate, phenylbenzoate, were primary products. This indicates that the following parallel reactions occurred on phenylbenzoate: (i) the intramolecular Fries transposition generated o-hydroxybenzophenone, (ii) phenylbenzoate acted as a benzoylating agent on phenol, to yield p-hydroxybenzophenone (with also possible formation of the ortho isomer) and phenol; and (iii) phenylbenzoate acylated a second molecule of phenylbenzoate to generate benzoylphenylbenzoates, with the co-production of phenol.

Reactivity tests were carried out by directly feeding phenylbenzoate, in order to confirm the hypothesis formulated about the reaction network between phenol and benzoic acid. Surprisingly, phenylbenzoate yielded benzoylphenylbenzoate and phenol as the only primary products of reaction, indicating the exclusive presence of the intermolecular mechanism of acylation. Therefore, when a high concentration of phenylbenzoate is adsorbed on the catalyst, the bimolecular mechanism is quicker that the intramolecular one.

Experimental Section

The catalytic tests were carried out in a batch-wise three-neck flask, with mechanical stirring and internal thermocouple. The reaction mixture contained 0.05 mol phenol, 0.05 mol benzoic acid and 1.5 g of catalyst, in the absence of solvent. The reaction was carried out by first mixing phenol, benzoic acid and the catalyst in the powder form (this corresponds to the starting time for the reaction) and then by heating the solid mixture up to 200°C. The result taken at 0.5 h reaction time corresponds to the beginning of the isothermal period, after the heating step. The reaction mixture was analyzed by gas chromatography (Carlo Erba GC 6000 Vega Series 2 instrument), equipped with a HP-5column. The zeolite used for catalytic tests is a commercial H-β sample (Si/Al ratio 75), provided by Süd-Chemie.

Acknowledgements

We thank Süd-Chemie for providing us the zeolite sample.

References

1. H.W. Kouwenhoven and H. van Bekkum, *Handbook of Heterogeneous Catalysis,* G. Ertl, H. Knözinger, J. Weitkamp, Eds., VCH, Weinheim, Vol. 5, p. 2358, 1997.
2. A. Corma, *Chem. Rev.*, **95**, 559 (1995).
3. B. Chiche, A. Finiels, C. Gauthier, P. Geneste, J. Graille and D. Pioch, *J. Org. Chem.*, **51**, 2128 (1986).
4. A. Corma, M.J. Climent, H. Garcia and J. Primo, *Appl. Catal.*, **49**, 109 (1989).
5. A.J. Hoefnagel and H. van Bekkum, *Appl.Catal. A*, **97**, 87 (1993).
6. I. Neves, F. Jayat, P. Magnoux, G. Perot, F.R. Ribeiro, M. Gubelmann and M. Guisnet, *J. Molec. Catal.*, **93**, 164 (1994).
7. C. Castro, A. Corma and J. Primo, *J. Molec. Catal. A*, **177**, 273 (2002).
8. P. Marion, R. Jacquot, S. Ratton and M. Guisnet, *Zeolites for Cleaner Technologies*, M. Guisnet, J.-P. Gilson, Eds., ICP, 2003, p. 281.
9. F. Bigi, S. Carloni, C. Flego, R. Maggi, A. Mazzacani, M. Rastelli and G. Sartori, *J. Molec. Catal. A*, **178**, 139 (2002).
10. M. Bolognini, F. Cavani, M. Cimini, L. Dal Pozzo, L. Maselli, D. Venerito, F. Pizzoli and G. Veronesi, *Comptes Rendus Chimie*, 7, 143 (2004).
11. A. Vogt, H.W. Kouwenhoven and R. Prins, *Appl. Catal A*, **123**, 37 (1995).
12. Y. Pouilloux, J.P. Bodibo, I. Neves, M. Gubelmann, G. Perot and M. Guisnet, *Heterogeneous Catalysis and Fine Chemicals II*, Elsevier, Amsterdam, 1991, p. 513.

II. Symposium on Catalytic Oxidation

II. Symposium on Catalytic Oxidation

12.

Biphasic Catalytic Oxidation of Hydrocarbons Using Immobilized Homogeneous Catalyst in a Microchannel Reactor

Jianli Hu, Guanguang Xia, James F. White, Thomas H. Peterson and Yong Wang

Pacific Northwest National Laboratory
902 Battelle Boulevard, Richland, WA 99352

Jianli.hu@pnl.gov

Abstract

This research relates to developing of immobilized homogeneous Co(II) catalyst and a demonstration of its performance in hydrocarbon oxidations. Although the catalyst material is normally a solid, it performs as if it were a homogeneous catalyst when immobilized by dissolution in an ionic liquid and deposited on a catalyst support. Evidence shows that after immobilization, the typical homogeneous catalytic properties of Co(II) catalyst are still observed. Surprisingly, the immobilized catalyst is also thermally stable up to 360°C and essentially no metal leaching has been observed under oxidation reaction conditions up to 250°C, and 220 psig. Because the obvious mass transport limitation between organic phase and catalyst phase, integration of such a catalyst system into microchannel reactor becomes extremely attractive. Microchannel reaction technology is an advanced chemical process technology which has not been previously explored for biphasic catalytic oxidation, yet it appears to have the essential features required for step changes in this emerging chemical process area. One significant characteristic of such an immobilized homogeneous catalyst in a microchannel reactor is the high mass transport efficiencies, enhancing reaction rates and space time yields while minimizing side reactions that can be enhanced by mass transfer limitations. The typical high ratio of geometric surface area of the reactor walls to total reactor volume in microchannel devices should be a strong benefit to the concept. In the cyclohexane oxidation reaction, preliminary experimental results have shown 100% selectivity towards desired products. Compared to conventional homogeneous catalyst, turnover frequency of the immobilized Co(II) catalyst is almost 7 times higher than reported for conventional hydrocarbon oxidation reaction schemes. The uniqueness of such a catalyst-reactor integration allows homogeneous oxidation reactions to be operated in heterogeneous mode therefore enhancing productivity and improving overall process economy while at the same time reducing the environmental impact of operating the process in the conventional fashion with a water based reaction media.

Introduction

Although it has been long recognized that homogeneous reactions are often not commercially viable due to catalyst recovery difficulties, today some chemical processes are still operated in homogeneous mode. Hydroformylation catalysis, for example, is one of the largest volume process in the chemical industry (1). Significant efforts have been made on immobilizing organometallic species responsible for catalysis (2). One problem encountered in using conventional immobilized catalyst has been the poor product selectivity. Over the last 15 years, biphasic catalysis based on using ionic liquids has attracted significant attention in the scientific community as an alternative reaction medium for homogeneous catalysis (3). On the basis of their highly charged nature, ionic liquids are well suited for biphasic reactions with organic substrates. Chauvin, et. al., utilized water soluble phosphine ligands to retain active rhodium complex in ionic liquid phase and used them successfully in biphasic hydrformylation reactions (4). Although liquid-liquid biphasic catalysis has been successfully demonstrated, heterogeneous catalysts are still preferred by industry because of the ease of product separation and catalyst recovery. This research was directed towards immobilization of ionic liquid phases containing catalytic specie onto high surface solid supports (5,6). The active species dissolved in ionic liquids performs as a homogeneous catalyst, therefore, making it is possible to operate a homogeneous reaction in a heterogeneous mode.

This research is related to exploring application of biphasic catalysis in hydrocarbon oxidation reactions which are commercially operated using homogeneous catalysts. Particularly, the oxidation reactions we are interested include cyclohxane oxidation to cyclohexanol and cyclohexanone, intermediates for adipic acid production (7). Also interested in our research is oxidation of aromatics. Essentially, hydrocarbon oxidation using biphasic catalysis is a new approach in this field. Biphasic catalysis is known to occur at interfaces between ionic liquid and organic reactant phases. For hydrocarbon oxidation reactions, oxygen penetration and dissolution in ionic liquids may be limiting steps. Solving such a mass transport limitation is critical for commercial applications. In our research, we use a thin film of ionic liquid to immobilize homogeneous Co(II) catalyst onto high surface area supports, and integrate such a catalyst system into in a microchannel reactor. The advantage of such an integration is high mass transport efficiencies, enhancing reaction rates and space time yields while minimizing side reactions associated with mass transfer limitations. The typical high ratio of geometric surface area of the reactor walls to total reactor volume in microchannel devices should be a strong benefit to the concept. Additional features related to utilizing a microchannel reactor is the possibility of operating oxidation reaction near or inside a flammable regime (8).

Results and Discussion

Metal Leaching Test

The leaching behavior of immobilized homogeneous catalyst was measured before commencing activity testing. It is highly desirable that the immobilized Co(II) catalyst, designated as Co(II)-IL/SiO$_2$, exhibits minimum or zero metal leaching behavior. Leaching tests were carried out under the following three conditions in the presence of toluene and cyclohexane, respectively:

1. Static Leaching Test at room temperature, no agitation
2. Reflux Leaching Test at boiling points of the solvents under agitation
3. High Temperature Leaching Test at 150°C and 85 psig in a closed bomb reactor with shaking

In each leach test, IL containing dissolved cobalt source was exposed to organic phase under the time and temperature conditions shown.

All of organic phase leaching solutions were analyzed using a UV-Vis spectrometer for Co(II) concentration. In our investigation, the level of Co (II) in leaching solution remained below detection limit of 0.1 ppm. Figure 1 illustrates comparison between leaching solution and standards. It provides a visual judgment on whether metal is leached significantly into organic. In general, the successful immobilization of the Co(II) catalyst is accomplished by means of strong adhesion of ionic liquid on SiO$_2$ support and the insolubility of the Co(II)-IL complex in the organic media.

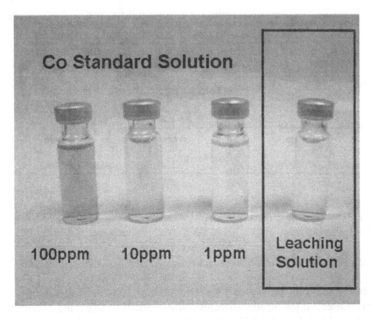

Figure 1. Comparison of leaching solution with standards using UV-Vis spectrometer. (Leaching test was carried out at 150°C, and 85 psig pressure).

Thermal Stability

Ionic liquids are characterized by low vapor pressure, permitting operation in a wide temperature range without causing detrimental effect to the catalyst or the ionic liquid. All of Co(II)-IL/Support catalysts were characterized by TGA. Figure 2 illustrates that the novel catalyst system was stable up to almost 360°C. This has allowed us to utilize this catalyst system for applications other than just hydrocarbon oxidation reactions.

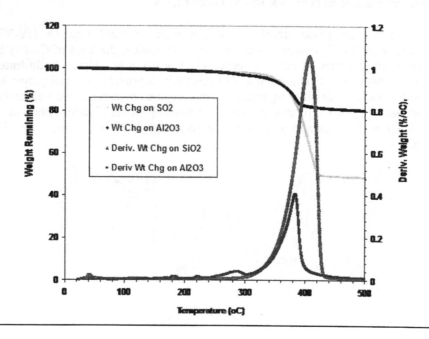

Figure 2. TGA analysis of supported Co(II)-ionic liquid catalysts (thermal stability of ionic liquid phase).

Retention of Homogeneous Property in Ionic Liquid

Immobilization of homogeneous catalysts has been practiced by many chemical researchers. Often, after immobilization, the normally expected homogeneous behavior diminishes or largely disappears and heterogeneous characteristics predominate. One problem encountered in using heterogeneous catalysts to replace homogeneous is selectivity. At the best, the performance of heterogeneous catalysts can only approximate those of their homogeneous counterparts. In contrast, when an ionic liquid is used in immobilizing homogeneous catalyst on SiO_2 support, homogeneous behavior predominates. XRD analysis was used to measure if any crystalline Co(II) was formed. Background analysis on silica gel alone didn't show crystalline, which is commonly known in the open publications (9). Figure 3 shows that after immobilization on SiO_2, no crystalline Co(II) was formed. This is different from conventional supported Co catalysts where crystalline Co phases can be easily detected. Based on the XRD analysis, we believe that with Co(II)-IL/SiO_2 catalyst, the normal expected homogeneous property of Co(II) is fully retained in ionic liquid.

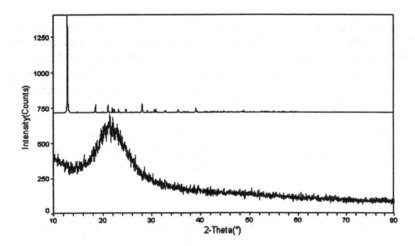

Figure 3. XRD analysis of Co(II)-IL/SiO_2 catalyst (top spectra: reference spectra of crystalline of Co(OAc)$_2$·4H$_2$O salt. Bottom spectra: Co(II) homogeneously dissolved in ionic liquid on SiO_2.

Oxidation Activity

The supported ionic liquid catalysis combines the advantages of ionic liquid media with solid support materials which enables the application of fixed-bed technology and the usage of significantly reduced amounts of ionic liquid in comparing to liquid-liquid biphasic catalysis. We are able to either pack a fixed bed reactor with solid catalyst consisting of Co(II)-IL deposited on SiO_2 beads or alternatively immobilize Co(II)-IL on microchannel reactor wall coated with SiO_2. The latter configuration allows us to manipulate the diffusion at interfaces by varying the thickness of ionic liquids, the gap of the channel and the organic feed flow rate. Oxidation of cyclohexane was chosen as proof-of-concept demonstration reaction. Under the conditions of O_2 concentration = 5% T=150°C, P=250 psig, LHSV=23 h^{-1}, we found we could obtained 2% cyclohexane conversion with 100% selectivity to desired products consisting of cyclohexanol and cyclohexanone. In this investigation, we haven't observed any CO_2 or CO by-products associated with over oxidation. Commercially, cyclohexane oxidation is catalyzed by homogeneous Co catalyst where low single pass conversion of 4-8% is maintained to minimize CO_2

formation (10). Keep in mind that the catalyst formulation of Co(II)-IL/SiO$_2$ developed in this research has not been optimized as to composition, co-catalysts or concentration. It is anticipated that with proper improvement, the Co(II)-IL/SiO$_2$ can achieve the same level of conversion comparable with commercial homogeneous catalyst with at least same level of selectivity. One impediment to higher conversion in these studies is likely the low O$_2$ concentration. For operational safety we had to operate below the O$_2$ flammability limit of 5% O$_2$. The observed low level of cyclohexane conversion is likely due to these two combined effects.

The turnover frequency (TOF) of Co(II)-IL/SiO$_2$ catalyst was calculated and compared with available conventional homogeneous Co (II) catalyst. Typically, the TOF of homogeneous Co(II) catalyst varies from 0.0024 to 0.011 s^{-1} (11). It appears that the present Co(II)-IL/SiO$_2$ catalyst exhibits TOF= 0.018s^{-1}. Although slightly higher than the published results (by at least 60%), this is within the range of expected activity for homogeneous Co oxidation catalysts.

Furthermore, the Co(II)-IL/SiO$_2$ catalyst was operated in oxidation mode for a total of 41 hours, during which startup and shut down were performed several times. No change in performance was observed over that time.

Experimental Section

Three ionic liquids were purchased from Aldrich: 1-butyl-3-methylimidazolium chloride, 1-butyl-3-methylimidazolium hexafluorophosphate and 1-butyl-3-methylimidazolium tetrafluoroborate. Homogeneous Co (II) catalyst precursors used in our experiments include Co(BF$_4$)$_2$, Co(OAc)$_2$, and Co(ClO$_4$)$_2$ each of which have high solubilities in above ionic liquids. High surface area catalyst supports SiO$_2$ and Al$_2$O$_3$ were obtained from Davison and Engelhard, respectively.

During catalyst preparation, the Co(II) homogeneous catalyst component was dissolved in ionic liquids to form a clear pink colored solution. Then, the solution was impregnated on catalyst support using incipient wetness methods. The catalysts were vacuum dried at 110°C overnight before activity testing. The experiments were carried out in a microchannel reactor (316 stainless steel), with the dimensions of 5.08 cm x 0.94 cm x 0.15 cm. Experiments were conducted at temperatures from 150-220°C and pressure from 0.5-1 MPa with oxygen concentration of 5-8 vol% . All the experiments were carried out under isothermal conditions as indicated by the uniform temperature distribution along catalyst bed. A schematic flow diagram of experimental apparatus is shown in Figure 4.

The reaction products were analyzed by on-line gas chromatography (HP 5890 GC) equipped with both TCD and FID detectors. GC column used is GS-Q 30 m manufactured by JW Scientific. Temperature program of 5°C/min to 300°C was chosen for the analysis. Liquid products were collected in a cold trap at −3°C and were also analyzed by GC-mass spectrometry.

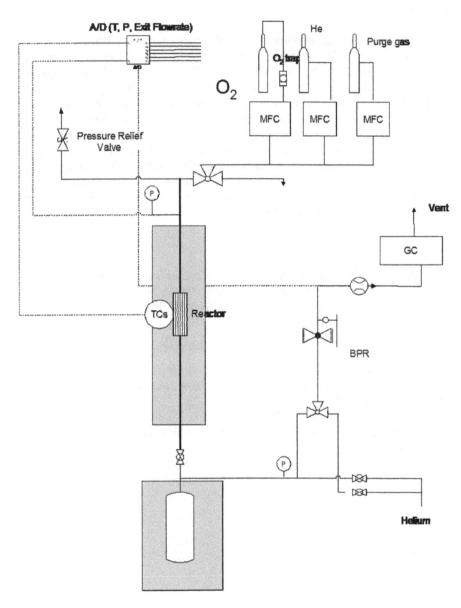

Figure 4. Schematic flow diagram of microchannel reactor testing apparatus

Conclusions

From a commercial application point of view, the catalyst immobilization technique based on using ionic liquids desirably allows homogeneous reactions to be operated in the heterogeneous mode. In addition to the retention of homogeneous characteristic, the advantages of this novel system include the high surface area provided by the solid, and thus the increased interfacial area between the organic and ionic liquid phases, and the possibility of selectivity variations from bulk equilibrium product distribution through the effect of the interface. When such a catalyst system is integrated into the design of microchannel reactor, it is possible to conduct normally homogeneous hydrocarbon oxidation reactions in a homogeneous mode. We also speculate that the beneficial features of mictochannel reactor systems may also allow oxidations with a stoichiometric ratio of oxygen to substrate to achieve high product yields and selectivity.

References

1.   S. N. Bizzari, S. Fenelon, M. Ishikawa-Yamaki, *Chemical Economics Handbook*, SRI International, Menlo Park, 682.7000A (1999)
2.   D.C. Bailey, J.H. Langer, *Chem. Rev.*, **81**, 109-148 (1981)
3.   T. Welton, *Chem. Rev.*, **99**, 2071 (1999)
4.   Y. Chauvin , L. Mussmann, H. Olivier, *Angew. Chem., Ind. Ed. Engl.*, **34**, 2698(1995)
5.   C.P. Mehnert, E.J. Mozeleski, R. A. Cook, *Chem. Commun.*, 3010-3011 (2002)
6.   J. P. Arhancet, M. E. Davis, J.S. Merola, B. E. Hanson, *Nature*,**339**, 454(1989)
7.   U.S. Patent 6, 888, 034
8.   M.T. Janicke, *J. Cat., 191, 282-293 (2000)*
9.   P. Li, C. Ohtsuki, T. Kokubo, *J. Material Science, Materials in Medicine* 4, 127-131, (1993)
10. S. Liu, Z. Liu, S. Kawi, *Korean J. Chem. Eng.*, **15**(5), 510-515 (1998)
11. *Short Communication, Pak J Sci Ind Res 42(6) 336-338*(1999)

13. Reaction Pathways for Propylene Partial Oxidation on Au/TiO₂

Rahul Singh and Steven S. C. Chuang

Department of Chemical and Biomolecular Engineering, the University of Akron,
Akron, OH, 44325

schuang@uakron.edu

Abstract

The surface OH (i.e., hydroxyl) group plays an important role in the formation of C_3 oxygenates. Propionaldehyde and acetone are proposed to be formed via a common intermediate because they have the same maximum formation rates at 160 °C on Au/TiO₂-SiO₂ and 140 °C on Au/TiOₓ-SiO₂. The propylene oxide formation occurs at higher temperature than that of propionaldehyde and acetone, suggesting its unique formation pathway which may involve different form of adsorbed oxygen and the active site.

Introduction

The activity and selectivity of propylene partial oxidation is very sensitive to the catalyst composition and reaction conditions (1, 2). Extensive studies have shown that Au supported on TiO₂ is one of the most active and selective catalysts for CO oxidation in the presence of H₂ and propylene epoxidation producing propylene oxide (i.e., PO) from propylene and oxygen in presence of hydrogen (3-8).

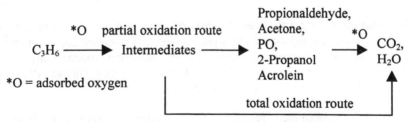

Figure 1. General reaction scheme for propylene partial oxidation.

Fig. 1 shows the general reaction scheme for propylene partial oxidation. Depending on the surface and structure of catalysts, adsorbed propylene species can react with various forms of adsorbed oxygen, producing a wide range of C_3 oxygenated products including propanal (propionaldehyde), acetone, 2-propanol, PO, and acrolein. A general catalyst requirement for catalytic partial oxidation includes (i) the capability of the catalyst to activate adsorbed olefin and adsorbed oxygen and (ii) the low activity for oxidation of desired reaction product.

The most valuable product for propylene partial oxidation is PO. Increasing temperature usually enhances the PO yield and propylene conversion, further increasing temperature causes PO yield and selectivity to decrease. High PO selectivity was only observed at low propylene conversion. The highest reported PO yield was reported at the level of 2 -3 % (3). A better understanding of the mechanism of the reaction is needed for the design of a selective catalyst to give high PO yields. We have employed *in situ* infrared (IR) spectroscopy to study the nature and structure of adsorbed species and mass spectrometry (MS) for product formation profile during partial oxidation. Our objective is to determine the reaction pathway for the formation of specific oxygenated products including propionaldehyde, acetone and PO from propylene partial oxidation over Au-based catalysts.

Experimental Section

Two types of Au (nominal 2 wt%) catalyst were prepared using two different gold precursors as well as support (9). The specific procedure for preparing Au/TiO_x-SiO_2 involved (i) dispersion of Silica gel (Alfa Products) in methanol/titanyl acetylacetonate (Sigma-Aldrich), (ii) calcination of resulting powder obtained in first step at 800°C for 4 h, (iii) deposition-precipitation of Au on the supports, (iv) drying at 100°C for 2 h and calcination at 300 °C for 4 h (10). The catalyst prepared is denoted as Au/TiO_x-SiO_2 because the stoichiometry of oxygen associated with Ti on the SiO_2 is not known.

Preparation of Au/TiO_2-SiO_2 involved (i) intermixing of Silica gel (Alfa Products) and TiO_2 (Degussa) in appropriate amount keeping Ti/Si atomic ratio at 2/100, (ii) introduction of powdered mixture obtained in first step into a solution of $AuCl_3$ at 70 °C, adjusting the pH to 7, (iii) drying at 80 °C for 12 h and calcination at 400 °C for 5 h. The catalysts prepared were characterized by DR (diffuse reflectance) UV-VIS (Hitachi).

The experimental reactor system consists of three sections: (i) a gas metering section with interconnected 4-port and 6-port valves, (ii) a reactor section including an *in situ* diffused reflectance infrared Fourier transform spectroscopy reactor (DRIFTS, Thermo-Electron) connected to tubular quartz reactor, (iii) an effluent gas analysis section including a mass spectrometer (11).

Temperature-programmed reaction (TPR) studies of partial oxidation of propylene was carried out by flowing C_3H_6 (Praxair), O_2 (Praxair), H_2 (Praxair) and Ar (Praxair) through DRIFTS and stainless steel tubular reactor (3/8" OD) loaded with catalyst. Feed gas at 40 ml/min and 1 atm consists of C_3H_6 (10%), O_2 (10%), H_2 (10%) and Ar (70%) for temperature program reaction studies. Prior to each experiment, the catalyst was pretreated in H_2 (10 vol%) and O_2 (10 vol%) simultaneously at 250°C. Temperature was monitored with a K type thermocouple connected to an omega temperature controller.

During TPR with $Ar/C_3H_6/H_2/O_2$ (28/4/4/4 ml/min), the composition of the reactor effluent was monitored by Omnistar MS (Pfeiffer Vacuum). The m/e signals monitored by the mass spectrometer were 2 for H_2, 18 for H_2O, 41 for propylene, 32 for O_2, 40 for Ar, and 44 for CO_2, 29 for propanal, 43 for acetone, 45 for iso-propanol and 58 for propylene oxide. The assignment of a specific m/e value to a C_3 oxygenate is based on the extensive calibration results listed in Table 1 as well as comparison with the m/e profile of the TPR. Table 1 lists the intensity factor which was obtained by (i) injecting a steady stream of a specific C_3 oxygenate, which emulates TPR condition and (ii) determining the intensity of its fragmentation pattern. The highest intensity (i.e., 100) is used for acetone (m/e= 43), propanal (m/e=29), and isopropanol (m/e=45). Although both propanal and propane have the same highest intensity at m/e=29, the significant difference in the maximum consumption of hydrogen and the maximum peak of m/e=29 in Fig. 5 suggests that the majority of m/e=29 is due to propanal instead of propane. This is because the maximum point for the formation of propane (i.e., a hydrogenation product) should correspond to the lowest point in the H_2 profile.

Table 1. Intensity factor for mass fragments of C_3 oxygenates calculated from the mass spectrometer signals (Intensity factor = the intensity ratio of a m/e peak to the largest peak; the intensity of the largest peak being 100).

m/e	Fragments	PO	$(CH_3)_2CO$	CH_3CH_2CHO	$(CH_3)_2CHOH$	C_3H_8
2	H_2^+	0.1	0.7	0.7	3.0	-
		-	-	-	-	-
18	H_2O^+	4.8[a]	3.6	4.3	18.0	-
		0.0[b]	0.4	1.1	0.3	0.0
27	$C_2H_3^+$	58.4	11.4	52.3	30.9	-
		49.7	5.5	60.0	11.6	41.9
28	$C_2H_4^+/CO^+$	100	18.0	91.0	7.2	-
		100	1.1	90.9	1.8	59.0
29	$C_2H_5^+/CHO^+$	79.8	6.7	100	20.6	-
		32.6	3.4	100	6.2	100
43	$C_2H_3^+=O$	22.0	100	4.7	43.0	-
		36.5	100	0.0	10.9	24.0
45	$(C_2H_4=OH^+)$	1.3	4.8	3.7	100	-
		0.2	0.4	0.0	100	1.1
58	$(C_2H_3=OCH_2^+)$	38.7	25.2	30.8	9.3	-
		82.0	64.0	85.0	0.0	0.0

a. Data from this study; b. Data from NIST (www.nist.gov) listed in the second row.

Results and Discussion

DRUV-VIS absorption spectroscopy

Fig. 2 shows the DRUV-VIS spectra of the catalysts, pure SiO_2 and TiO_2 in the wavelength range of 260-800 nm. TiO_2 exhibits an absorption band in the wavelength region 200-400 nm while SiO_2 does not show any distinct absorption band. The spectrum of Au/TiO_2-SiO_2 appears to be combination of TiO_2 and SiO_2 spectra. No obvious plasmon absorption band of Au was observed on Au/TiO_2-

SiO_2, suggesting Au particles size is less than 7 nm. In contrast, a marked surface plasmon absorption band of Au was observed on Au/TiO_x-SiO_2 at 522 nm which corresponds to Au particle sizes between 7.5 and 33 nm (12). It is interesting to note that the support material (i.e., TiO_2-SiO_2 and TiO_x-SiO_2) and catalyst precursors have a significant effect on Au particle size. The used Au/TiO_x-SiO_2 catalysts show a broadening of peak width in comparison to the fresh Au/TiO_x-SiO_2 due to isolation of Au particles and an increase in its particle size (13). Fig. 2 also shows Au/TiO_x-SiO_2 spectrum lacks a sharp TiO_2 absorption band, suggesting that TiO_x is not in a crystalline form.

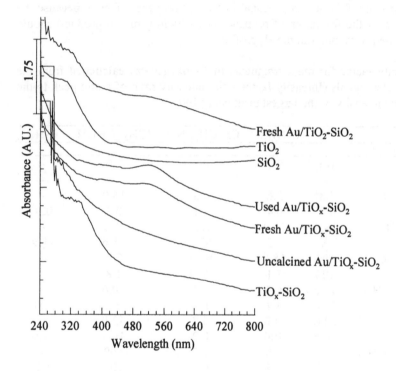

Figure 2. DRUV-VIS spectroscopy of catalysts.

IR characterization of catalysts

Fig. 3 shows both Au/TiO_x-SiO_2 and Au/TiO_2-SiO_2 exhibit an OH band at 3737 cm^{-1}. Au/TiO_x-SiO_2 adsorbs significantly less amount of H_2O than Au/TiO_2-SiO_2 as indicated by the broad H-bonding band at 2500 – 3800 cm^{-1}. High temperature calcination appears to cause TiO_x-SiO_2 to lose the OH moieties and hydrophilicity.

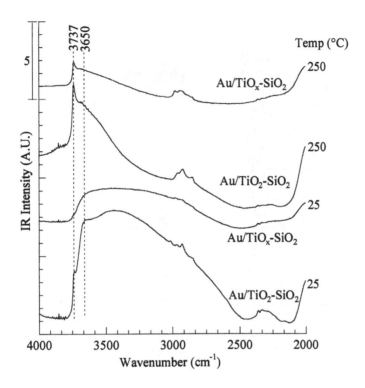

Figure 3. IR analysis of Au/TiO$_x$-SiO$_2$ and Au/TiO$_2$-SiO$_2$.

TPR results over Au/TiO$_2$-SiO$_2$

Fig. 4 shows IR spectra as a function of temperature during TPR study of propylene partial oxidation over Au/TiO$_2$-SiO$_2$. Exposure of the catalyst to Ar/C$_3$H$_6$/H$_2$/O$_2$ (28/4/4/4 ml/min) at 25 °C led to the decrease in the intensity of the 3741 cm^{-1} (OH stretching) band and the formation of bands associated with gaseous and adsorbed propylene at 3104 cm^{-1} (υ_{as}=CH$_2$ of gas phase propylene), 3079 cm^{-1} (υ_{as}=CH$_2$ of π-complexed propylene), 2983 cm^{-1} (υ_s=CH$_2$ of π-complexed propylene), 2954 cm^{-1} (υ_sCH). 1633 cm^{-1} (C=C of π-complexed propylene on M^{n+}), 1475 and 1441 cm^{-1} (δCH of adsorbed propylene) (14). Adsorbed propylene species appears to interact with adsorbed oxygen on the Au surface and the OH on oxide surface, producing the 1708 cm^{-1} (C=O of adsorbed propionaldehyde) band and the 1666 cm^{-1} (C=O of adsorbed acetone) band. The decrease in the OH intensity at 3741 cm^{-1} provides the direct evidence of interaction between OH and propylene and C$_3$ oxygenated products (i.e., propionaldehyde and acetone). (15).

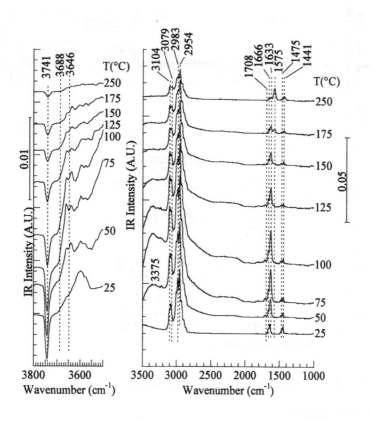

Figure 4. In situ IR spectroscopy of TPR of partial oxidation of propylene over Au/TiO$_2$-SiO$_2$.

Fig. 5 shows the calibrated MS profiles depicting flow rates of reactants and products during TPR over Au/TiO$_2$-SiO$_2$. The mass spectrometry analysis shows the consumption of H$_2$ (m/e=2), O$_2$ (m/e=32) and C$_3$H$_6$ (m/e=41) indicated by the negative profiles as well as the formation of products indicated by the positive profiles for propionaldehyde (m/e=29), acetone (m/e=43), iso-propanol (m/e=45), PO (m/e=58), and CO$_2$ (m/e=44). The formation of propionaldehyde, acetone, iso-propanol, and CO$_2$ reached a maximum at 160 °C. The occurrence of the maximum formation at the same temperature suggests the formation of these various C$_3$ oxygenates may share a common rate-limiting step. It is important to note that the m/e=58 profile behaved significantly differently from other C$_3$ oxygenates. The formation of PO began to occur at temperatures above 160 °C where a 1575 cm^{-1} emerged, as shown in Fig. 4. This 1575 cm^{-1} band is in good agreement with adsorbed acetate/formate which has been suggested to be formed from bidentate propoxy species on Au/TiO$_{22}$ (19). Propylene oxide can further oxidize to give such species. These observations further justified the assignment of m/e=58 to PO even though other C$_3$ oxygenates's fragments contribute to m/e=58.

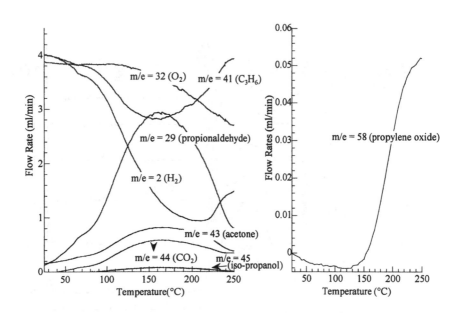

Figure 5. Calibrated MS profiles of reactants and products during TPR of propylene partial oxidation over Au/TiO$_2$-SiO$_2$ (Catalyst weight = 0.14 g).

TPR results over Au/TiO$_x$-SiO$_2$

Fig. 6 shows the IR-TPR of C$_3$H$_6$ partial oxidation over Au/TiO$_x$-SiO$_2$. The interaction of Ar/C$_3$H$_6$/H$_2$/O$_2$ with the catalyst surface at 50 °C led to decrease in the band intensity of OH vibration at 3741 cm^{-1} and the formation of a 1662 cm^{-1} (C=O of adsorbed acetone) band. Mass spec. results in Fig. 7 shows the reaction produced propionaldehyde, acetone, and CO$_2$. The rate of product formation peaked at 140 °C.

Table 2 shows the activity of Au/TiO$_x$-SiO$_2$ and Au/TiO$_2$-SiO$_2$ for propylene partial oxidation at 200 °C under transient condition.

Table 2 Activity and selectivity of the catalysts during TPR studies at 200 °C

Catalyst	Weight (gm)	Conversion (%) C$_3$H$_6$/H$_2$/O$_2$	Selectivity (%) CO$_2$/C$_2$H$_5$CHO/(CH$_3$)$_2$CO/(CH$_3$)$_2$COH/PO
Au/TiO$_x$ -SiO$_2$	0.085	11.7/46/4.8	10.2/74.8/13.6/0/0
Au/TiO$_2$ -SiO$_2$	0.140	22.7/76.2/19	4.8/71.4/21/1.8/0.91

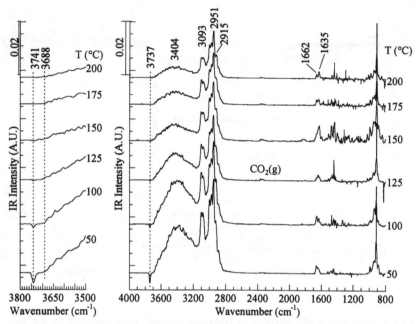

Figure 6. *In situ* IR spectroscopy of TPR of propylene partial oxidation over Au/TiO$_x$-SiO$_2$.

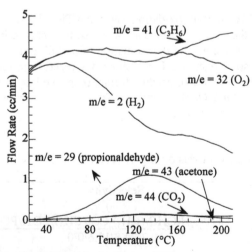

Figure 7. MS analysis during TPR of propylene partial oxidation over Au/TiO$_x$-SiO$_2$ (catalyst weight = 0.085 g).

In summary, hydroxyl plays an important role in the formation of C_3 oxygenates. The low surface concentration of OH on Au/TiO_x-SiO_2 appears to relate to its low activity for catalyzing the C_3 oxygenate formation, as shown in Table 1. Fig. 8 illustrates the proposed reaction pathway for the formation of the major C_3 oxygenates. Propionaldehyde and acetone are proposed to be formed via a common intermediate because their formation rates reach a maximum at the same temperature. Propionaldehyde and acetone adsorbed on the surface in the $\eta^1(C,O)$ form on the catalyst surface. The PO formation occurs at higher temperature than propionaldehyde and acetone, suggesting its unique formation pathway which may involve different form of adsorbed oxygen and the active site. Atomic oxygen is a likely candidate for electrophilic addition while oxygen with negative charge would behave as the nucelophile. Adsorbed oxygen could release from H_2O_2 on Au site, interacting with adsorbed propylene, leading to the formation of PO. PO formed could further adsorb on the catalyst surface as a propoxy-bidentate complex.

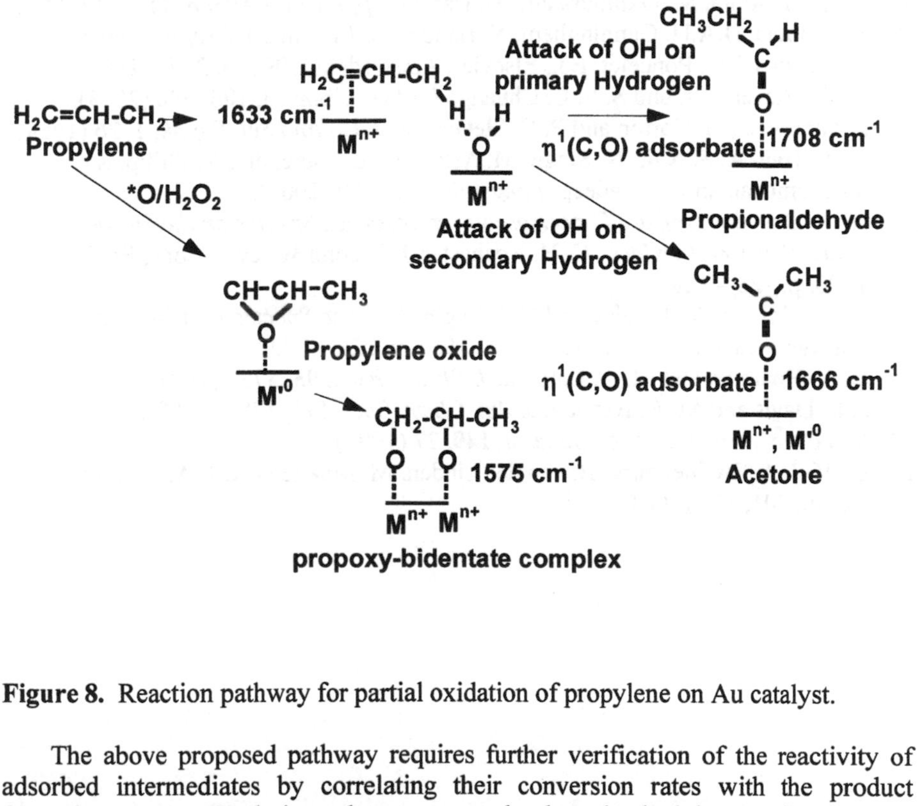

Figure 8. Reaction pathway for partial oxidation of propylene on Au catalyst.

The above proposed pathway requires further verification of the reactivity of adsorbed intermediates by correlating their conversion rates with the product formation rates. Work is underway to study the adsorbed intermediates under transient condition to determine the reactivity of the IR-observable specie on the Au catalyst surface and to identify the nature of active sites (i.e., M^{n+}, $M^{,0}$).

Acknowledgements

We thank the Ohio Board of Regents for their partial financial support.

References

1. G. J. Hutchings and M. S. Scurrell, *CATTECH*, **7**, 90 (2003).
2. R. A. Khatri, R. Singh, S. S. C. Chuang and R. W. Stevens, Chemical Industries (CRC Press), **104** (*Catal. Org. React.*), 403-411 (2005).
3. T. Hayashi, K. Tanaka and M. Haruta, *J. Catal.*, **178**, 566 (1998).
4. B. Uphade, Y. Yamada, T. Akita, T. Nakamura and M. Haruta, *Appl. Catal. A: General*, **215**, 137 (2001).
5. F. Moreau, C. G. Bond, and O. Adrian, *J. Catal.* **231**, 105 (2005).
6. C. Rossignol, S. Arrii, F. Morfin, L. Piccolo, V. Caps and J. Rousset, *J. Catal.*, **230**, 476 (2005).
7. T. A. Nijhuis, B. J. Huizinga, M. Makkee and J. A. Moulijn, *Ind. Eng. Chem. Res.*, **38**, 884 (1999).
8. A. K. Sinha, S. Seelan, S. Tsubota and M. Haruta, *Topics Catal.*, **29**, 95 (2004).
9. C. Qi, T. Akita, M. Okumura and M. Haruta, *Appl. Catal A:Gen*, **218**, 81 (2001).
10. S. Tsubota, D.A.H. Cunningham, Y. Bando and M. Haruta, *Preparation of Catalysts VI*, G. Poncelet, Ed., Elsevier, Amsterdam, 1995, p. 227 – 235.
11. R. W. Stevens, Jr. and S. S. C. Chuang, *J. Phys. Chem. B*, **108**, 696 (2003).
12. J. Turkevich, G. Garton and P. C. Stevenson, *J. Colloid Sci. Suppl.*, **1**, 26 (1954).
13. R. T. Tom, A. S. Nair, N. Singh, M. Aslam, C. L. Nagendra, R. Philip, K. Vijayamohan and T. Pradeep, *Langmuir*, **19**, 3439 (2003).
14. A. A. Davydov, *Infrared Spectroscopy of Adsorbed Species on the Surface of Transition Metal Oxides.*, C. H. Rochester, Ed., John Wiley & Sons, England, 1990, p. 128 – 168.
15. L. H. Little, A. V. Kiselev and V. I. Lygin, *Infrared Spectra of Adsorbed Species*, Academic Press, New York, 1966, p. 228 – 272.
16. C. A. Houtman and M. A. Barteau, *J. Phys. Chem.*, **95**, 3755 (1991).
17. J. L. Davis and M. A. Barteau, *J. Am. Chem. Soc.*, **111**, 1782 (1989).
18. V. Ponec, *Appl. Catal. A: General*, **149**, 27 (1997).
19. G. Mul, A. Zwijnenburg, B. van der Linden, M. Makkee and J. A. Moulijn, *J. Catal.*, **201**, 128 (2001).

14. The Transformation of Light Alkanes to Chemicals: Mechanistic Aspects of the Gas-phase Oxidation of n-Pentane to Maleic Anhydride and Phthalic Anhydride

Mirko Bacchini[1], Nicola Ballarini[1], Fabrizio Cavani[1], Carlotta Cortelli[1], Stefano Cortesi[1], Carlo Fumagalli[2], Gianluca Mazzoni[2], Tiziana Monti[2], Francesca Pierelli[1], Ferruccio Trifirò[1]

[1]Dipartimento di Chimica Industriale e dei Materiali, Viale Risorgimento 4, 40136 Bologna, Italy. INSTM, Research Unit of Bologna. A member of the Concorde CA and Idecat NoE (6FP of the EU) [2]Lonza SpA, Via E. Fermi 51, 24020 Scanzorosciate (BG), Italy

fabrizio.cavani@unibo.it

Abstract

The effect of the main characteristics of vanadyl pyrophosphate, $(VO)_2P_2O_7$, on catalytic performance in the gas-phase oxidation of n-pentane to maleic and phthalic anhydrides, has been investigated. A relationship was found between the distribution of products, and the oxidation state of V in catalysts. Higher concentrations of more oxidant V^{5+} sites in vanadyl pyrophosphate led to the preferred formation of maleic anhydride, while samples having average V valence state close to 4.0+ were the most selective to phthalic anhydride. The latter samples were also those characterized by the higher Lewis-type acid strength. In the reaction mechanism, including the competitive pathways leading either to maleic anhydride through O-insertion onto activated n-pentane, or to phthalic anhydride through the acid-assisted dehydrocyclodimerization to an intermediate C_{10} alkylaromatic compound, precursor of phthalic anhydride, the balance of oxidizing sites and Lewis-type acid sites controlled the selectivity to the main products.

Introduction

One of the most important challenges in the modern chemical industry is represented by the development of new processes aimed at the exploitation of alternative raw materials, in replacement of technologies that make use of building blocks derived from oil (olefins and aromatics). This has led to a scientific activity devoted to the valorization of natural gas components, through catalytic, environmentally benign processes of transformation (1). Examples include the direct exoenthalpic transformation of methane to methanol, DME or formaldehyde, the oxidation of ethane to acetic acid or its oxychlorination to vinyl chloride, the oxidation of propane to acrylic acid or its ammoxidation to acrylonitrile, the oxidation of isobutane to methacrylic acid, and others as well. In some cases, this activity has led to the successful development of new industrial processes.

One reaction that has been investigated in recent years is the direct oxidation of n-pentane to maleic anhydride (MA) and phthalic anhydride (PA) (2-4). The wide availability of this alkane as a raw material is a consequence of the removal of lighter hydrocarbons from gasoline. A new process for the transformation of this hydrocarbon to valuable chemicals would be of industrial interest and might compete with current industrial technologies for the production of organic anhydrides. The reaction starts from a C_5 alkane and yields oxygenated C_4 (MA) and C_8 (PA) compounds, in a single catalytic passage. The catalyst able to promote this complex reaction is a V/P mixed oxide, having composition $(VO)_2P_2O_7$ (vanadyl pyrophosphate, VPP). The latter is also the industrial catalyst for the oxidation of n-butane to MA; it represents an example of how heterogeneous systems for the selective oxidation of alkanes must combine different types of active sites, which cooperate in order to obtain the final desired product. The aim of the present work is to study how the surface properties of VPP affect the distribution of products in n-pentane selective oxidation.

Results and Discussion

Vanadium is present as V^{4+} in stoichiometric VPP; however, the latter can host V^{3+} or V^{5+} species as defects, without undergoing substantial structural changes (5,6). Therefore, the role of the different V species in the catalytic behavior of VPP in n-butane oxidation has been the object of debate for many years (7-9). Moreover, the catalyst may contain crystalline and amorphous vanadium phosphates other than $(VO)_2P_2O_7$ (10); for instance, outer surface layers of V^{5+} phosphates may develop in the reaction environment, and play active roles in the catalytic cycle. This is particularly true in the case of the fresh catalyst, while the "equilibrated" system (that one which has been kept under reaction conditions for 100 hours at least) contains only minor amounts of compounds with V species other than V^{4+}.

Much less studied has been the role of V species in n-pentane oxidation to MA and PA. Papers published in this field are aimed mainly at the determination of the reaction mechanism for the formation of PA (2-4,11). Moreover, it has been established that one key factor to obtain high selectivity to PA is the degree of crystallinity of the VPP; amorphous catalysts are not selective to PA, and the progressive increase of crystallinity during catalyst equilibration increases the formation of this compound at the expense of MA and carbon oxides (12). Also, the acid properties of the VPP, controlled by the addition of suitable dopants, were found to play an important role in the formation of PA (2).

In order to understand the role of V^{5+} species in the formation of MA and PA, we carried out oxidizing treatments on a fully equilibrated VPP. These V species undergo quick transformation under reaction conditions, but their transient catalytic performance may furnish indications on their role in the formation of products. Table

1 summarizes the main chemical features, surface area and distribution of V species as determined by chemical analysis, for catalysts employed in the present work. Sample eq was a fully equilibrated catalyst, while samples ox$n$ were obtained by controlled oxidizing treatments of sample eq. The procedure adopted for this treatment is also given in the Table; more severe conditions led to the generation of a higher amount of V^{5+}, while sample eq did not contain V^{5+} at all.

Table 1. Main chemical-physical features of catalysts before catalytic tests.

Sample	T of treatment in air (°C), time (h)	S. area, m^2/g	n in V^{n+}
eq	none	21	3.99
ox1	450, 4	17	4.035
ox2	500, 4	nd	4.052
ox3	550, 4	14	4.47

Figure 1 reports the corresponding UV-Vis Diffuse Reflectance spectra. The band at 400 nm is relative to the CT between O^{2-} and V^{5+}; its intensity was proportional to the amount of V^{5+}. On the other hand, the intensity of the band at 900 nm decreased, relative to a d-d transition for V^{4+} in the vanadyl moiety. The XRD pattern of all samples, but ox3, corresponded to that of well-crystallized VPP; in the case of sample ox3, additional reflections were due to the presence of δ-VOPO$_4$ (JCPDS 37-0809). Therefore, the treatment of the VPP led to the generation of surface V^{5+} in samples ox1 and ox2, and to the growth of crystalline VOPO$_4$ in sample ox3.

Table 2 compares the chemical analysis of the spent samples, downloaded after reaction (samples eqsp and ox$n$sp); Figure 2 reports the corresponding UV-Vis DR spectra. It is shown that catalysts after reaction were more reduced than the corresponding samples before reaction. Sample ox1sp was fully reduced, and also contained a low amount of V^{3+}; samples ox2sp and ox3sp, instead, yet retained some fraction of V^{5+}. In electronic spectra, the intensity of the band at 400 nm was lower than in corresponding fresh samples (Figure 1); in the case of catalysts eqsp and ox1sp, this band was absent. Therefore, the V^{5+} species generated by the oxidizing treatment of the VPP was not stable in the reaction environment, and was progressively reduced to V^{4+}. In addition, a weak band at $\approx$ 500 nm, attributed to the V^{3+} species (13), was present in samples eqsp and ox1sp.

Figure 3 reports the amount of residual pyridine adsorbed on acid sites as a function of temperature, for samples eqsp, ox1sp and ox3sp. Tests were carried out by adsorbing pyridine at room temperature, and by recording the FT-IR spectrum after desorption at increasing temperatures. The amount of pyridine retained was proportional to the total number of acid sites (Lewis + Brønsted). The Figure also shows the corresponding fraction of Lewis-type sites; the latter were the predominant acid sites in all samples. However, the residual amount of adsorbed pyridine after evacuation at 400°C was very low; this indicates that the acid sites were of medium strength.

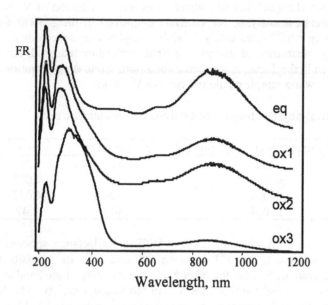

Figure 1. UV-Vis DR spectra of catalysts before reaction (fresh samples).

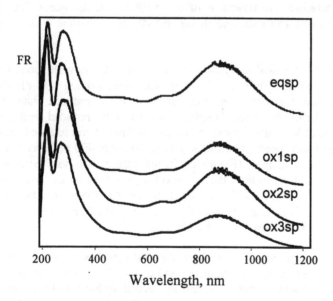

Figure 2. UV-Vis DR spectra of catalysts after reaction (spent samples).

The overall number of acid sites was the highest in sample ox1sp, and the lowest in the most oxidized sample, ox3sp. On the other hand, samples eqsp and ox3sp retained the greatest amount of pyridine after desorption at 400°C, and therefore these samples had the largest fraction of sites with the higher strength. Sample eqsp did not possess Brønsted acidity at all, while sample ox3sp had Brønsted sites which still retained pyridine after evacuation at high T, and that therefore can be classified as strong ones. In the case of sample ox1sp, no pyridine remained adsorbed on Brønsted sites after evacuation at 300°C.

Table 2. Main chemical-physical features of catalysts after reactivity tests.

Sample	S. area, m²/g	n in V^{n+}
eqsp	21	3.99
ox1sp	21	3.98-3.99
ox2sp	nd	4.020-4.025
ox3sp	16	4.28

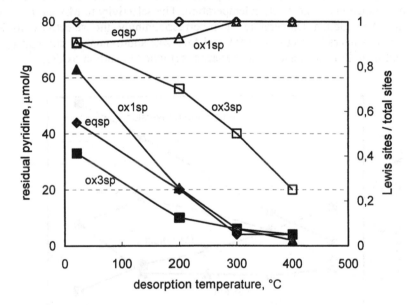

Figure 3. Total amount of residual adsorbed pyridine (full symbols) and fraction of Lewis sites with respect to the total number of acid sites (open symbols) as a function of the desorption temperature (after adsorption of pyridine at room temperature).

The stronger Lewis-type acidity in the equilibrated catalyst agrees with the many literature evidences about the existence of these sites in well-crystallized VPP (14). Also, a discrete amount of V^{3+} ions may contribute to an enhancement of Lewis-type acidity, due to the presence of defects associated to the anionic vacancies (5). The Brønsted acidity in the oxidized catalyst (sample ox3sp) can be associated to the presence of P-OH groups arising from the transformation of pyrophosphate into orthophosphate groups.

Catalytic tests were carried out following a protocol that allowed establish relationships between the characteristics of fresh and spent catalysts and the catalytic performance. Specifically, within the first 5-10 hours reaction time, the catalytic performance was not stable, due to variations in samples characteristics. Therefore, two different sets of results were taken, the first after less than 1 hour (representative of fresh samples), and the second one after 3-4 hours (representative of spent samples). Only in the case of the equilibrated sample, no difference was found between the characteristics of samples eq and eqsp (Tables 1 and 2); for this catalyst, also the catalytic performance did not vary at all during catalytic tests.

Figure 4 shows the catalytic performance of sample ox1sp; the conversion of n-pentane, and the selectivity to the main products, MA, PA and carbon oxides, are reported as functions of the reaction temperature. The selectivity to MA increased on increasing the reaction temperature, and correspondingly the selectivity to PA decreased. The overall selectivity to PA and MA was approximately constant up to 400°C, but then decreased, due to the preferred formation of carbon oxides.

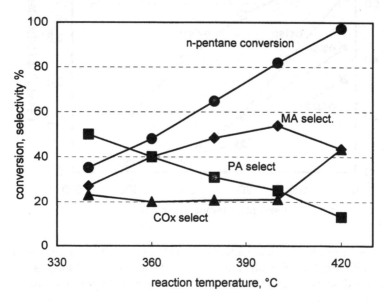

Figure 4. Effect of temperature on catalytic performance. Catalyst ox1sp.

Figure 5 summarizes the results obtained for fresh and spent samples; the selectivity to MA and that to PA at approximately 43-50% n-pentane conversion are plotted as a function of the average oxidation state of V. The most oxidized the catalyst was, the most preferred was the formation of MA with respect to that of PA, with a MA/PA selectivity ratio equal to 7 at 50% n-pentane conversion for sample ox3. The opposite was true for most reduced samples, with a MA/PA selectivity ratio around 1. The decrease of selectivity to PA, for increasing values of V oxidation state, was much more relevant than the corresponding increase of MA; therefore, the overall selectivity to the products of partial oxidation (MA+PA) decreased for increasing extents of V oxidation, from 80% for sample ox1sp to 59% for sample ox3sp. The exception was sample ox3 (the most oxidized one), which gave an overall selectivity to MA+PA equal to 68%.

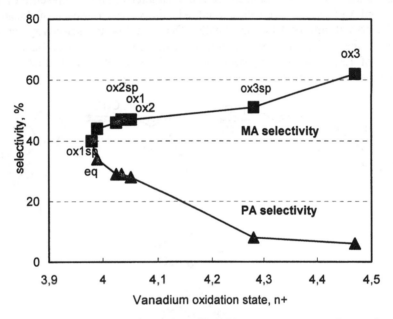

Figure 5. Selectivity to MA and to PA at 43-50% n-pentane conversion as functions of the average oxidation state of V in fresh and spent samples.

The mechanism for the formation of PA in n-pentane oxidation has been widely discussed in the literature. The main controversial point is whether MA and PA are formed through parallel reactions starting from a common intermediate, or if, on the contrary, PA is consecutive to the formation of MA. In regard to the former hypothesis, it has been proposed that n-pentane is first oxydehydrogenated to pentene and pentadiene; the latter can either be oxidized to MA (with the loss of one C atom), or be transformed to cyclopentadiene, which then dimerizes to a cyclic template and is finally oxidized to PA (15,16). An alternative route to PA is through an acid-assisted dimerization and dehydrocyclization of intermediate pentadiene to yield a cyclic 6C-membered ring aliphatic C_{10} hydrocarbon, precursor of an alkylaromatic compound, that is finally oxidized to PA and 2 CO_x (3,17). This latter mechanism is illustrated in Figure 6. Other authors suggest that n-pentane is either transformed to isopentane (precursor of citraconic anhydride), or to butene and butadiene; the latter is oxidized to MA, and finally a Diels-Alder reaction between MA and adsorbed butadiene leads to the formation of PA (2). In this case, therefore, PA is consecutive to the formation of MA.

In the mechanism illustrated in Figure 6, the combination of the redox and acid properties of the catalyst determines the relative contribution for the formation of MA and PA. It is generally accepted that the higher the crystallinity of the VPP, the more selective to PA is the catalyst (3,4,10-12,17,18). Poorly crystalline VPP, like that one formed after the thermal treatment of the precursor (especially when it is carried out under oxidizing conditions), is selective to MA, but non-selective to PA. On the contrary, a fully equilibrated catalyst, characterized by the presence of a well-crystallized VPP, yields PA with a good selectivity. The presence of dopants that alter the crystallinity of VPP may finally affect the MA/PA selectivity ratio (19). Moreover, the surface acidity also influences the distribution of products (17); an increase of Lewis acidity improves the selectivity to PA, while that to MA is positively affected by Brønsted acidity (2).

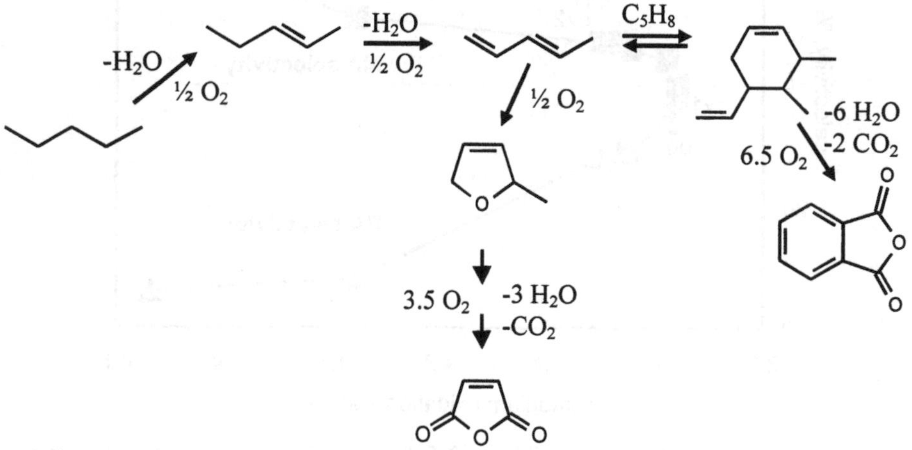

Figure 6. Mechanism proposed for the one-pot oxidation of n-pentane to MA and to PA (3,17).

Data reported in the present work demonstrate that the degree of crystallinity and the acid properties are related the amount of V^{5+} present at the surface of VPP. When the VPP is not fully equilibrated, and hence may contain discrete amounts of V^{5+}, it is more selective to MA and less to PA. The reason is that in oxidized catalysts, the olefinic intermediate is preferentially oxidized to MA, rather then being subjected to the acid-catalyzed condensation with a second unsaturated molecule, to yield the precursor of PA. When instead the catalyst is more crystalline, and hence it does contain less oxidized V sites, its surface acid properties predominate over O-insertion properties, and the catalyst becomes more effective in PA formation. In this case, the selectivity to PA at $\approx 50\%$ n-pentane conversion becomes comparable to that one of MA.

Experimental Section

V/P/O samples were prepared according to the following procedure: the catalyst precursor, $VOHPO_4 \cdot 0.5H_2O$, was precipitated starting from a suspension of 8.2 g V_2O_5 (Aldrich, purity 99.6+%) and 10.1 g H_3PO_4 (Aldrich 98+%) in 75 ml isobutanol (Aldrich, 99%+); the suspension was heated under reflux conditions for 6 hours. The precipitate obtained was filtered, washed with a large excess of isobutanol and dried at 125°C overnight. After a spray-drying conglomeration of particles into fluidizable material (20), the compound was thermally treated to develop the VPP structure, following the hydrothermal-like procedure described in detail elsewhere (21). The thermal treatment was aimed at the development of a well-crystallized VPP, which took shorter time to reach the equilibrated state under reaction conditions, with respect to conventional thermal treatments. Then, the catalyst was run in fluidized-bed pilot reactor for three months, under the following conditions: temperature 400°C, feed 4% n-butane in air, pressure 3 atm. After this period the catalyst was down-loaded by cooling the reactor under reaction feed. This sample is denoted as sample eq. Sample eq was then oxidized for 4 h in flow of air, in a dedicated furnace, at different temperatures (see Table 1), in order to develop discrete amounts of V^{5+}. Samples obtained were denoted ox1, ox2 and ox3; they were then used for catalytic tests in a lab reactor. Corresponding samples after catalytic tests are denoted with the suffix sp.

Catalysts were characterized by means of X-ray diffraction (Phillips diffractometer PW3710, with CuKα as radiation source), UV-Vis-DR spectroscopy (Perkin-Elmer Lambda 19) and chemical analysis. Measurements of surface acidity were carried out by recording transmission FT-IR spectra of samples pressed into self-supported disks, after adsorption of pyridine at room temperature, followed by stepwise desorption under dynamic vacuum at increasing temperature (Perkin-Elmer mod 1700 instrument). The procedure for chemical analysis is described in detail in ref. (13).

Catalytic tests of n-pentane oxidation were carried out in a laboratory glass flow-reactor, operating at atmospheric pressure, and loading 3 g of catalyst diluted with inert material. Feed composition was: 1 mol% n-pentane in air; residence time was 2 g s/ml. The temperature of reaction was varied from 340 to 420°C. The products were collected and analyzed by means of gas chromatography. A HP-1 column (FID) was used for the separation of C_5 hydrocarbons, MA and PA. A Carbosieve SII column (TCD) was used for the separation of oxygen, carbon monoxide and carbon dioxide.

Acknowledgements

Acknowledgement is made to the Donors of the American Chemical Society Petroleum Research Fund, for the partial support of this activity.
Dr Cristina Flego from EniTecnologie SpA is gratefully acknowledged for performing FT-IR measurements.

References

1. P. Arpentinier, F. Cavani and F. Trifirò, *The Technology of Catalytic Oxidations*, Editions Technip, Paris (2001).
2. V.A. Zazhigalov, J. Haber, J. Stoch and E.V. Cheburakova, *Catal. Comm.*, **2**, 375 (2001).
3. F. Cavani, A. Colombo, F. Giuntoli, E. Gobbi, F. Trifirò and P. Vazquez, *Catal. Today*, **32**, 125 (1996).
4. Z. Sobalik, S. Gonzalez Carrazan, P. Ruiz and B. Delmon, *J. Catal.*, **185**, 272 (1999).
5. P.L. Gai and K. Kourtakis, *Science*, **267**, 661 (1995).
6. P.T. Nguyen, A.W. Sleight, N. Roberts and W.W. Warren, *J. Solid State Chem.*, **122**, 259 (1996).
7. K. Aït-Lachgar, M. Abon and J.C. Volta, *J. Catal.*, **171**, 383 (1997).
8. U. Rodemerck, B. Kubias, H.-W. Zanthoff, G.-U. Wolf and M. Baerns, *Appl. Catal. A*, **153**, 217 (1997).
9. G. Koyano, F. Yamaguchi, T. Okuhara and M. Misono, *Catal. Lett.*, **41**, 149 (1996).
10. S. Albonetti, F. Cavani, F. Trifirò, P. Venturoli, G. Calestani, M. Lopez Granados and J.L.G. Fierro, *J. Catal.*, **160**, 52 (1996).
11. Z. Sobalik, S. Gonzalez and P. Ruiz, *Stud. Surf. Sci. Catal.*, **91**, 727 (1995).
12. C. Cabello, F. Cavani, S. Ligi and F. Trifirò, *Stud. Surf. Sci. Catal.*, **119**, 925 (1998).
13. F. Cavani, S. Ligi, T. Monti, F. Pierelli, F. Trifirò, S. Albonetti and G. Mazzoni, *Catal. Today*, **61**, 203 (2000).
14. G. Centi, F. Trifirò, J.R. Ebner and V. Franchetti, *Chem. Rev.*, **88**, 55 (1988).
15. G. Centi, F. Trifirò, *Chem. Eng. Sci.*, **45**, 2589 (1990).
16. G. Centi, J. Lopez Nieto, F. Ungarelli and F. Trifirò, *Catal. Lett.*, **4**, 309 (1990).
17. F. Cavani and F. Trifirò, *Appl. Catal. A*, **157**, 195 (1997).
18. M. Lopez Granados, J.M. Coronado, J.L.G. Fierro, F. Cavani, F. Giuntoli and F. Trifirò, *Surf. Interf. Anal.*, **25**, 667 (1997).
19. F. Cavani, A. Colombo, F. Trifirò, M.T. Sananes-Schulz, J.C. Volta and G.J. Hutchings, *Catal. Lett.*, **43**, 241 (1997).
20. D. Suciu, G. Stefani and C. Fumagalli, U.S. Pat 4,654,425, to Lummus and Alusuisse Italia (1987).
21. G. Mazzoni, G. Stefani and F. Cavani, Eur Pat. 804,963, to Lonza SpA (1997).

15.
Transition Metal Free Catalytic Aerobic Oxidation of Alcohols Under Mild Conditions Using Stable Nitroxyl Free Radicals

Setrak K. Tanielyan[1], Robert L. Augustine[1], Clementina Reyes[1], Nagendranath Mahata[1], Michael Korell[2], Oliver Meyer[3]

[1]Center for Applied Catalysis, Seton Hall University, South Orange, NJ 07079
[2]Degussa Corp., BU Building Blocks, Parsippany, NJ 07054
[3]Degussa AG, BU Building Blocks, Marl 45764, Germany

tanielse@shu.edu

Abstract

The oxidation of alcohols to the corresponding aldehydes, ketones or acids represents one of the most important functional group transformations in organic synthesis. While this is an established reaction, it still presents an important challenge (1). Although there are numerous oxidation methods reported in the literature, only a few describe the selective oxidation of primary or secondary alcohols to the corresponding aldehydes or ketones utilizing TEMPO based catalyst systems in combination with clean oxidants such as O_2 and H_2O_2 (2). The most efficient, common, TEMPO based systems require the use of substantial amounts of expensive and/or toxic transition metal complexes, which makes them unsuitable for industrial scale production. Here, we report a highly effective aerobic oxidation of primary and secondary alcohols by a catalyst system, based on 4-acetamido-2,2,6,6-tetramethylpiperidin-1-oxyl (AA TEMPO), $Mg(NO_3)_2$ and N-bromosuccinimide (NBS) in an acetic acid solvent.

Introduction

The oxidation of alcohols to the corresponding aldehydes, ketones or acids certainly represents one of the more important functional group transformations in organic synthesis and there are numerous methods reported in the literature (1-3). However, relatively few methods describe the selective oxidation of primary or secondary alcohols to the corresponding aldehydes and ketones and most of them traditionally use a stoichiometric terminal oxidant such as chromium oxide (4), dichromate (5), manganese oxide (6), and osmium or ruthenium oxides as primary oxidants (7).

A convenient procedure for the oxidation of primary and secondary alcohols was reported by Anelli and co-workers (8,9). The oxidation was carried out in CH_2Cl_2 with an aqueous buffer at pH 8.5-9.5 utilizing 2,2,6,6-tetramethylpiperidine-1-oxyl (TEMPO, **1**) as the catalyst and KBr as a co-catalyst. The terminal oxidant in this system was NaOCl. The major disadvantage of using sodium hypochlorite or any other hypohalite as a stoichiometric oxidant is that for each mole of alcohol oxidized during the reaction one mole of halogenated salt is formed. Furthermore, the use of hypohalites very frequently leads to the formation of undesirable halogenated by-products thus necessitating further purification of the oxidation product. Extensive review of TEMPO based oxidation methods can be found in the literature (10,11,12) and the industrial application of a highly efficient solvent- and bromide-free system has been reported (13,14).

In recent years, much effort has been spent on developing both selective and environmentally friendly oxidation methods using either air or oxygen as the ultimate, oxidant. One of the most selective and efficient catalyst systems reported to date is based on the use of stable nitroxyl radicals as catalysts and transition metal salts as co-catalysts (15). The most commonly used co-catalysts are $(NH_4)_2Ce(NO_3)_6$ (16), CuBr$_2$-2,2'-bipiridine complex (17), $RuCl_2(PPh_3)_3$ (18,19), Mn(NO$_3$)$_2$-Co(NO$_3$)$_2$ and Mn(NO$_3$)$_2$-Cu(NO$_3$)$_2$ (20). However, from an economic and environmental point of view, these oxidation methods suffer from one common drawback. They depend on substantial amounts of expensive and/or toxic transition metal complexes and some of them require the use of halogenated solvents like dichloromethane, which makes them unsuitable for industrial scale production.

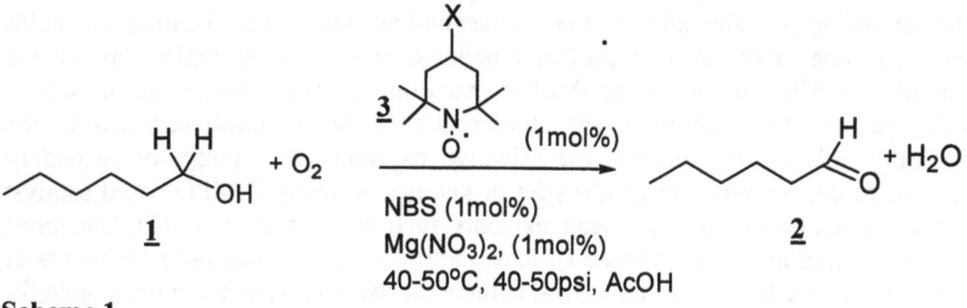

Scheme 1.

X = H, TEMPO (3); X = OCH$_3$, MeO-TEMPO (4); X = NHCOCH$_3$, AATEMPO (5)

Very recently, Hu et al. claimed to have discovered a convenient procedure for the aerobic oxidation of primary and secondary alcohols utilizing a TEMPO based catalyst system free of any transition metal co-catalyst (21). These authors employed a mixture of TEMPO (1 mol%), sodium nitrite (4-8 mol%) and bromine (4 mol%) as an active catalyst system. The oxidation took place at temperatures between 80-100 °C and at air pressure of 4 bars. However, this process was only successful with activated alcohols. With benzyl alcohol, quantitative conversion to benzaldehyde was achieved after a 1-2 hour reaction. With non-activated aliphatic alcohols (such as 1-octanol) or cyclic alcohols (cyclohexanol), the air pressure needed to be raised to 9 bar and a 4-5 hour of reaction was necessary to reach complete conversion. Unfortunately, this new oxidation procedure also depends on the use of dichloromethane as a solvent. In addition, the elemental bromine used as a co-catalyst is rather difficult to handle on a technical scale because of its high vapor pressure, toxicity and severe corrosion problems. Other disadvantages of this system are the rather low substrate concentration in the solvent and the observed formation of bromination by-products.

We wish to report here on a new and highly efficient catalyst composition for the aerobic oxidation of alcohols to carbonyl derivatives (Scheme 1). The catalyst system is based on 2,2,6,6-tetramethylpiperidine N-oxyl (TEMPO), $Mg(NO_3)_2$ (MNT) and N-Bromosuccinimide (NBS), utilizes ecologically friendly solvents and does not require any transition metal co-catalyst. It has been shown, that the described process represents a highly effective catalytic oxidation protocol that can easily and safely be scaled up and transferred to technical scale.

Results and Discussion

Our initial work on the TEMPO / $Mg(NO_3)_2$ / NBS system was inspired by the work reported by Yamaguchi and Mizuno (20) on the aerobic oxidation of the alcohols over aluminum supported ruthenium catalyst and by our own work on a highly efficient TEMPO-[$Fe(NO_3)_2$ / bipyridine] / KBr system, reported earlier (22). On the basis of these two systems, we reasoned that a supported ruthenium catalyst combined with either TEMPO alone or promoted by some less elaborate nitrate and bromide source would produce a more powerful and partially recyclable catalyst composition. The initial screening was done using hexan-1-ol as a model substrate with MeO-TEMPO as a catalyst (4.1mol %) and 5%Ru/C as a co-catalyst (0.3 mol% Ru) in acetic acid solvent. As shown in Table 1, the binary composition under the standard test conditions did not show any activity (entry 1). When either N-bromosuccinimide (NBS) or $Mg(NO_3)_2$ (MNT) was added, a moderate increase in the rate of oxidation was seen especially with the addition of MNT (entries 2 and 3).

Table 1. Oxidation of **1** to **2** using a TEMPO based catalyst composition.[a]

#	Catalyst Composition, mol%				Efficiency, %			Comments
	5%Ru/C	4	MNT	NBS	Rate[b]	Cnv[b]	Sel[b]	
1	0.3	4.1	-	-	0.00	0	-	Long induction (4h+)
2	0.3	4.1	-	2.6	0.01	5	93	No effect
3	0.3	4.1	1.9	-	0.05	39	94	No effect
4	0.3	4.1	1.9	2.6	0.21	98	86	All 4 components
5	0.3	4.1	1.9	2.6	0.22	100	93	Re-used Ru
6	-	4.1	1.9	2.6	0.25	99	95	No need of Ru

[a] [1-Hexanol]= 2mL (16mmol), CH_3COOH =10mL , [MeO-TEMPO, 3] = 0.656mmol, [NBS]=0.416 mmol and $Mg(NO_3)_2$ = 0.30 mmol, T= 46°C, P (O_2) 15psi. Reaction time 60 min
[b] Oxidation rate, mmol O_2/min, conversion of **1** (Cnv), selectivity to **2** (Sel).

On the other hand, when all four components were used at the concentration levels listed in the table, a rapid oxidation took place at a high rate to give hexanal in high selectivity (entry 4). An even more surprising result was achieved when we tried to re-use the heterogeneous Ru/C component with a fresh portion of the homogeneous MeO-TEMPO / $Mg(NO_3)_2$ / NBS composition (entry 5). Full conversion of hexan-1-ol to hexanal, at much higher selectivity of 93% was observed within a 60 min reaction time. The data in these two last examples indicated that, most likely, the Ru/C component in the catalyst was not taking part in the overall oxidation cycle.

The real breakthrough was achieved when the test oxidation was initiated with just the homogeneous MeO-TEMPO/Mg(NO$_3$)$_2$/NBS catalyst system, used in the

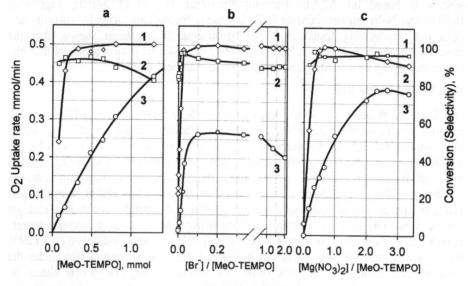

Fig 1. Effect of the [MeO-TEMPO] (a), the [NBS] (b) and the [Mg(NO$_3$)$_2$] (c) on the rate of oxidation (○), on the conversion of Hexanol-1 (◇) and on the selectivity to Hexanal (□) at T= 46°C, P (O$_2$) 15psi, solvent CH$_3$COOH =10mL, [Hexanol-1] = 2mL (16mmol). [NBS]=0.96 mmol and Mg(NO$_3$)$_2$ = 0.48 mmol (for a), [M-TEMPO]=0.48mmol and Mg(NO$_3$)$_2$ = 0.48 mmol (for b) , [M-TEMPO]=0.48mmol and [NBS] = 0.048 (for c).

same concentration levels (entry 6). Under the test conditions the reaction was completed at near full conversion of the starting alcohol (99%), at very high aldehyde selectivity (95%) and at turnover frequencies of 45 h^{-1} at 46°C. This result, without any optimization, was much higher than the TOF's reported in the literature for the TEMPO / Mn(NO$_3$)$_2$ / Cu(NO$_3$)$_2$ system of 1.7 h^{-1} at 40°C (20) and for the TEMPO / NaNO$_2$ / Br system of 27.8 h^{-1} at 80°C (21).

Effect of the MeO-TEMPO, NBS and the MNT on the Reaction Efficiency.

The initial study on the MeO-TEMPO / Mg(NO$_3$)$_2$ / NBS triple catalyst system in the oxidation of **1** indicated the necessity of all three components; the TEMPO based catalyst, the nitrate source (MNT) and the bromine source (NBS). A large number of metal nitrates and nitrites were screened initially and the highest activity and aldehyde selectivity under comparable reaction conditions were recorded using Mg(NO$_3$)$_2$ as the nitrate component. A number of organic and inorganic bromides soluble in HOAc were also screened and high reaction rates were found when NBS was used as the bromide source. The effect of the concentration of the individual components of the new triple catalyst system on the reaction rate, on the conversion of **1** and on the selectivity to **2** over 60 min reaction time is shown in Figure 1.

Fig 1a shows the near first order relationship between the MeO-TEMPO concentration and the reaction rate (curve 3) and at the same time it shows a gradual deterioration of the reaction selectivity to the aldehyde **2** at higher MeO-TEMPO concentrations in the region 0.5-1.0 mmol (3 - 6 mol %), (curve 2). The only product detected along with the desired aldehyde **2** is the product of the over-oxidation, hexanoic acid.

In Fig 1b, the [Br⁻]/[MeO-TEMPO] ratio was varied over a relatively wide range at equimolar concentrations for MeO-TEMPO and $Mg(NO_3)_2$ of 0.48mmol (3 mol%). Three distinct regions are clearly seen: i) a region of low Br concentration where increasing the NBS concentration leads to dramatic rate increase and where the conversion of **1**, measured at 60 min reaction time reached 100% at selectivity to **2** of 97-98%, ii) a region, where increasing the concentration of the NBS did not produce a significant change in the reaction effectiveness and iii) a region, where further increase in the [Br]/[MeO-TEMPO] ratio led to a rate reduction (curve 3).

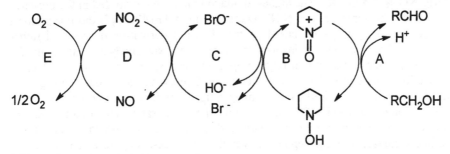

Scheme 2.

In the next set of experiments, the MeO-TEMPO concentration was fixed at 0.48 mmol (3 mol%) and the NBS concentration at 0.048 mmol (0.3 mol%) while the $Mg(NO_3)_2$ / [MeO-TEMPO] ratio was gradually increased in the range shown in Figure 1c. The first region of relatively low concentration of $[Mg(NO_3)_2]$ is characterized by a first order relationship between the rate of oxidation and the nitrate concentration and in the same region, at $Mg(NO_3)_2$ / [MeO-TEMPO] $\cong$ 1, a complete conversion of **1** was observed with high aldehyde selectivity. Further increase in the nitrate concentration, however, leads to severe inhibition, resulting in gradual deactivation of the catalyst system and rapid decline in the reaction rates (curve 2 and 3).

At the current initial concentrations of **1** in the acetic acid solvent, (1.33M, 16.6 v/v %), the optimum ratio between the three components of the catalyst system seams to be MeO-TEMPO:$Mg(NO_3)_2$:NBS = 1:1:0.1. As we will show in the next section, the higher the initial concentration of the alcohol substrate, the more NBS is needed for achieving high reaction rates in the selective oxidation of **1** to **2**.

A plausible reaction sequence including a multi step cascade of four redox cycles is shown in Scheme 2. Since we have no experimental data to support either the nature of the intermediate species involved or their oxidation state, nor have we any direct observation regarding the location of each redox cycle in the overall cascade, the schematic is given for illustration purposes only. What we only know is that removing any of the links in this sequence leads to complete shutdown of the oxidation reaction and as a result, one can speculate that a complete catalytic cycle should involve transferring of the α-hydrogen from the alcohol molecule in cycle A to the oxygen in cycle E. In an alternative scenario, the hydrogen and an activated form of oxygen could interact in either the bromine (C) or the nitrate cycle (D) to produce HO⁻. Studies are currently underway to determine the nature of the intermediate species and possibly, the position of the individual cycles in the cascade. Based on the data in Figure 1, one can also say that the bromine cycle proceeds faster than the rate of the TEMPO (B) or the nitrate (D) cycles.

Screening for an efficient TEMPO catalyst.

A number of commercially available TEMPO derivatives were screened at 16 mmol substrate scale at fixed concentrations for X-TEMPO (3.0 mol%), NBS (0.3 mol%) and $Mg(NO_3)_2$ (7.5 mol%). The highest oxidation rates were recorded in the presence of **5**, **3** and **4** (Scheme 3). The N,N - Dimethylamino TEMPO (**7**) did not perform well and the catalyst system based on this TEMPO derivative deactivated rather rapidly. The three best performing systems were tested extensively and the results for the changes in the initial rate of oxygen uptake (curve 3), the conversion of **1** at 60 min reaction time (curve 1) and the selectivity to **2** (curves 2) are plotted in Figure 2. In all three instances the initial rate of oxygen uptake is first order dependant on the concentration of the X-TEMPO catalyst in the 0-5 mmol range and shows a trend to saturation at high concentrations, clearly indicating that the TEMPO cycle (D) is most likely the rate limiting step in the overall reaction cascade, presented in Scheme 2. At comparable catalyst concentrations, the AA-TEMPO based catalyst composition (Fig 2c) is nearly twice as active than the one based on MeO-TEMPO (2a) or TEMPO (2b). For example, to attain the maximum rate in the oxygen uptake for a full conversion of **1**, the MeO-TEMPO and the TEMPO based systems require 1.25mmol catalyst (7.8 mol%) while the AA-TEMPO system reached its maximum performance at 0.6mmol level (3.7 mol%).

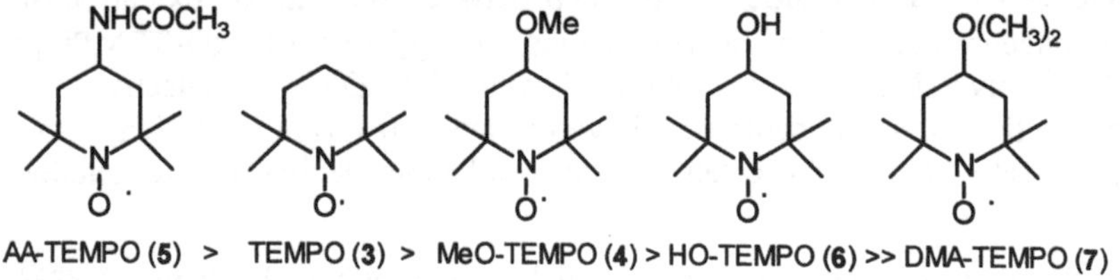

AA-TEMPO (**5**) > TEMPO (**3**) > MeO-TEMPO (**4**) > HO-TEMPO (**6**) >> DMA-TEMPO (**7**)

Scheme 3.

Another interesting observation from the data in Figure 2 was the effect of the catalyst concentration on the aldehyde selectivity (curves 2 in 2a-c). As mentioned earlier, at this moderate reaction temperature, the only by-product present in measurable quantities was hexanoic acid, formed as a product of the over-oxidation of **2**. Contrary to what was reported in the literature for other TEMPO based oxidations of alcohols (20,21), the current catalyst system, particularly at higher

TEMPO concentrations is able to promote the further oxidation to an acid once the alcohol is completely converted to **2**. The same is true for the best performing system in Figure 2c. At the optimum concentration for the AA-TEMPO catalyst of 0.6 mmol (3.7 mol%), the alcohol substrate is completely consumed at an aldehyde selectivity of 95+ %. Further increase in the AA-TEMPO concentration doesn't lead to further rate gains and the only visible result is the deterioration in the aldehyde selectivity.

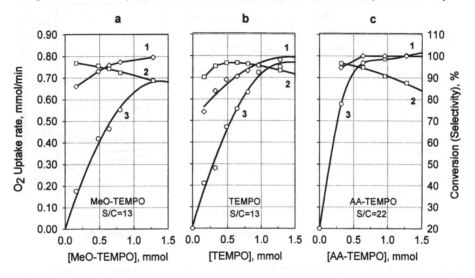

Fig 2. Effect of the [MeO-TEMPO] (a), the [TEMPO] (b) and the [AA-TEMPO] (c) concentrations on the rate of oxidation (○), on the conversion of the Hexanol-1 (◇) and on the selectivity to Hexanal (□) at T= 46°C, P (O$_2$) 15psi, solvent CH$_3$COOH =10mL, [Hexanol-1] = 2mL (16mmol), [NBS]=0.048 mmol and Mg(NO$_3$)$_2$ = 1.2 mmol.

Pressure-Temperature Matrix

Under these standard reaction conditions, the effect of the temperature and the oxygen pressure was also investigated. The reaction temperature was varied in a relatively narrow range of 34-43°C by an incremental increase of 3°C at each step and the oxygen pressure was sequentially increased in the range 5-30 psi (Fig 3). Increasing the oxygen pressure led to a linear increase in the initial reaction rate for all four temperatures (Figure 3a) clearly showing that the rate limiting step (at the concentration of the AA-TEMPO currently employed) has shifted to one of the cycles of the cascade in which the concentration of the dissolved oxygen affects the rate. For two of the pressures in this matrix, (10 psi and 30 psi) activation energies of 32.2 kJ/mol and 31.4 kJ/mol, respectively, were calculated. These values are remarkably low for an overall oxidation process and are an indication for a multi step cascade type interaction, such as that shown in Scheme 2.

The changes in the conversion of **1** at different temperatures as a function of the oxygen pressure are shown in Figure 3b. As expected, the oxygen pressure (or the concentration of the dissolved oxygen in the liquid phase), becomes more important at higher temperatures. At 43°C, the reaction is fast enough, the starting alcohol is completely converted to **2** over the 60 min reaction time and the aldehydes selectivity remains high in the 92-94% range (3c).

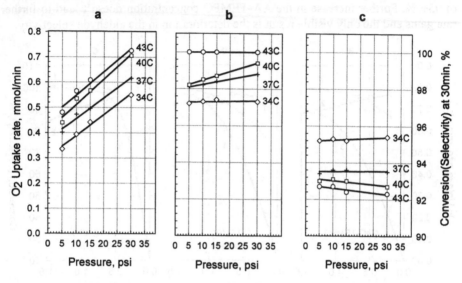

Fig 3. Effect of the oxygen pressure and the temperature on the rate of oxidation (a), on the conversion of Hexanol-1 (b) and on the selectivity to Hexanal (c) at 34°C (◊), 37°C (+), 40°C (□) and 43 °C (○). Conditions: [Hexanol-1] = 2mL (16mmol), CH_3COOH =10mL, [AA TEMPO] = 0.64mmol, [NBS]=0.048 mmol and $Mg(NO_3)_2$ = 1.2 mmol.

Reduction of the Volume of HOAc at High Substrate to Catalyst (S/C) Ratios
The next attempt to further improve the reaction efficiency was to reduce the volume of the acetic acid solvent and to proportionally increase the initial concentration of the alcohol substrate while keeping the total reaction volume at constant level. The main purpose of these studies was to determine the minimum amount acetic acid needed to maintain a homogeneous system until complete conversion of hexan-1-ol to **2**. Since the oxidation reaction produces stoichiometric amounts water, it was felt, that the formation of a second aqueous phase along with the hydrophobic aldehyde phase would lead to the creation of a two-phase reaction system with the inevitable partition of the catalyst system between the two phases. In addition it was also important to determine the highest possible S/C ratio while maintaining a reasonable reaction rate.

The starting point here was the 16 mmol scale oxidation reaction using the standard conditions shown in Figure 4a (curve 1). Full conversion was achieved over 22 min at a TOF of 68h[-1]. In the next experiment (curve 2), the hexan-1-ol concentration was doubled (32mmol scale, 2.7M, 33.3% v/v) while the concentration of the AA-TEMPO-MNT-NBS composition was kept at the same level. Again, complete conversion was achieved over 105 min reaction time at significantly lower TOF of 28.6 h^{-1}. In the third run, at three-fold increase in the substrate concentration (48mmol scale, 4.0M, 50.0 % v/v) the oxidation surprisingly entered into severe inhibition at 44.2% conversion level (curve 3). When the initial concentration of hexan-1-ol was further increased to 64 mmol scale (5.3 M, 66.7% v/v), the catalyst system was completely deactivated at 7.4% conversion over 300 min reaction time (curve 4).

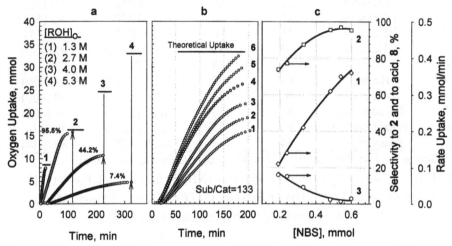

Fig 4. Effect of the initial concentration of the Hexanol-1 on the conversion (a), the concentration of the NBS on the oxygen uptake (b) and the NBS concentration on the rate of oxygen uptake (c-curve1), selectivity to **2** (c-curve 2) and the amount of the acid **8** by-product (c-curve 3) at T= 48°C, P (O$_2$)=15psi, [AA TEMPO] = 0.64mmol and Mg(NO$_3$)$_2$ = 1.2 mmol.
Conditions for (a): [NBS] = 0.048mmol, [Hexanol-1] = 2mL (16mmol) (1), 4mL (32mmol) (2), 6mL (48mmol) (3) and 8mL (64mmol) (4), CH$_3$COOH = balance to 12mL total liquid phase.
Conditions for (b,c): [Hexanol-1] = 8mL (64mmol), CH$_3$COOH = 4mL, [AA TEMPO] = 0.48mmol (0.75 mol%), [Mg(NO$_3$)$_2$] = 0.48mmol, [NBS] = 0.191mmol (1), 0.231mmol (2), 0.287mmol (3), 0.48mmol (4), 0.55mmol (5) and 0.61mmol (6).

The data in Figure 4a indicated that at least one of the components of the ternary catalyst had been gradually consumed or slowly degraded at high substrate concentrations. To determine the true reason of this deactivation phenomenon, the same 64 mmol scale reaction was repeated using either 10% excess AA-TEMPO or 10% excess MNT. Both reactions (not shown in the plot) produced the same negative result. When the concentration of the NBS was incrementally increased from 0.29 mol% to 0.95 mol%, a rapid and complete conversion of **1** was achieved at TOF = 33 h^{-1} and at S/C ratio of 133. The data in Figure 4c show that the amount of the NBS used at these high S/C ratios affects not only the overall rate of oxidation (curve 1) but, to a large extent, controls the level of the acid by-product (curve 3). Using the same optimized substrate to catalyst ratios of hexan-1-ol : AA TEMPO :

MNT : NBS of 133 : 1 : 1 :1.19, a twelve fold scaled up reaction was carried out at

Table 2. Oxidation of alcohols using the X-TEMPO-NBS-Mg(NO$_3$)$_2$ system[a].

Run	Substrate	Scale	Substrate/ Catalyst	X-TEMPO	Yield (time)
		Mmol	Mol/mol		% (min)
1	1-Hexanol	16	33	MeO- TEMPO	91 (47)
2	1-Hexanol	16	33	TEMPO	92 (60)
3	1-Hexanol	16	33	HO TEMPO	92 (36)
4	1-Hexanol	16	33	AA-TEMPO	95 (25)
5	1-Hexanol	64	133	AA-TEMPO	92 (280)
6	2-Hexanol	64	133	AA-TEMPO	33 (600)
7	1-Heptanol	64	133	AA-TEMPO	99 (400)
8	1-Octanol	64	133	AA-TEMPO	99 (400)
9	1-Dodecanol	64	133	AA-TEMPO	89 (400)
10	Benzyl alcohol	64	133	AA-TEMPO	93 (30)
11	1 Phenylethanol	64	133	AA-TEMPO	89 (240)

Conditions for 1-4: [1-Hexanol] = 16mmol (2 cc), [CH$_3$COOH] = 10 cc, [X-TEMPO] = 3 %mol, Mg(NO$_3$)$_2$ = 0.48 mmol, [NBS] = 0.048 mmol. For 5-11: [1-Hexanol]= 64 mmol (8cc),[CH$_3$COOH]=4cc, [X-TEMPO] = 0.75 %mol, Mg(NO$_3$)$_2$ = 0.48mmol, [NBS] = 0.55 mmol. T= 46°C, Pressure O$_2$ =15 psi

48°C and an air pressure of 200 psi. The 800 mmol reaction was completed in 3 hours for full conversion of **1** at 93.1% selectivity to **2**.

Scope of Application

The data in Table 2 show the potential of the triple X-TEMPO / Mg(NO$_3$)$_2$ / NBS based catalyst system tested over large number of representative alcohols at low (entry 1-4) and at high C/S ratio (5-11).The primary alcohols were oxidized to the corresponding aldehydes at complete conversion of the alcohol and at 90-93% selectivity with the only by-product detected were the corresponding acid and some minor amounts of the symmetrical ester. The best performing TEMPO derivative under the optimized conditions (T 46°C and [MgNO$_3$]$_2$ = 3mol%), was the AA-TEMPO catalyst (entry 4) in the presence of which complete conversion of **1** was achieved over a 25 min reaction time with a turnover frequency of 80h[-1], which is the highest reported to date for a TEMPO based aerobic oxidation. At high S/C ratios, the triple catalyst system shows the characteristic for other TEMPO based oxidation systems; preference in oxidation of primary over secondary alcohols (19). For example the oxidation of hexan-1-ol was nearly an order of magnitude faster as the oxidation of hexan-2-ol under comparable conditions (entry 5, 6). Benzyl alcohol and 1-Phenylethanol were both quantitatively converted to the respective benzaldehyde and acetophenone (entry 10,11).

Conclusions

It has been shown that the aerobic oxidation of alcohols takes place only when all three components of the catalyst are present; AA TEMPO, $Mg(NO_3)_2$, NBS (23). The reaction can be run under mild conditions: temperature 310-330K; oxygen pressure of 0.1-0.5Mpa; alcohol concentration of 5.4M in HOAc; ratio of substrate : AA TEMPO:$Mg(NO_3)_2$:NBS = 133:1:1:1.19. Non-activated primary alcohols are quantitatively converted to the corresponding aldehydes at high selectivity. The only by-product observed in some instances is the corresponding carboxylic acid. Some secondary alcohols were also oxidized with high selectivity to the corresponding ketones but at lower conversions. Both benzyl alcohol and the sec-phenylethanol were smoothly converted to their respective aldehyde or ketone oxidation products.

Experimental Section

A. Experimental Setup. The oxidation reactions were performed using an in-house made Multi Autoclave Glass Volumetric system (24). All reactions were carried out in Ace Glass reaction flasks with Teflon heads, equipped with Swagelock based injection and thermocouple ports. Digital stirrers and Fisher cross stir bars were used for providing an efficient and controlled stirring. The Multi Autoclave reactor system allows conducting five simultaneous oxidations with five independent variables. The only common parameter for all five reactions is the reaction temperature. The reactor system permits the recording of the oxygen uptake with the time, thus allowing precise monitoring the progress of the oxidation. The system was described in detail in our earlier work (24).

B. Oxidation procedure for the small-scale experiments. The required amounts of hexan-1-ol, AA-TEMPO, $Mg(NO3)2$, NBS and HOAc were loaded in the reactor and the flask was connected to the volumetric manifold. The flask was flushed three times with oxygen and immersed in the thermostated water bath held at the reaction temperature for two minutes to equilibrate the system. The oxygen was admitted to the reaction pressure and the continuous monitoring of the oxygen uptake initiated and recorded against the time. After the reaction was completed the product composition was analyzed by GC using dioxane as an internal standard.

C. Large-scale oxidation protocol. The large-scale oxidations reactions were carried out in a 300mL Parr autoclave equipped with an injection port, a thermocouple port, a septa sealed addition port and port connected to the volumetric measurement and gas supply module. The module consists of a forward pressure regulator and a calibrated ballast reservoir. The pressure in the reactor and in the ballast reservoir is monitored constantly and the pressure drop in the ballast reservoir is constantly converted into moles of oxygen uptake recorded vs. the time.

In a typical oxidation reaction, the autoclave was charged with hexan-1-ol (100mL, 800 mmol), HOAc (50 mL, 0.87mol), AA-TEMPO (1.28g, 6.0 mmol, 0.75 mol%), $Mg(NO_3)_2$.$6H_2O$ (1.54g, 6mmol) and NBS (1.21g, 6.8mmol). The system was purged three times with nitrogen, the nitrogen pressure increased to 150 psi and the autoclave heated to the desired temperature. When the target temperature was reached, oxygen was admitted at 200 psi and the computer monitoring of the oxygen uptake started. After the consumption of 390 mmol of oxygen, which under these conditions was completed in 180 min, the autoclave was cooled to ambient temperature, the pressure released and a sample taken for GC analysis.

Acknowledgements

This work was supported by Degussa Building Blocks.

References

1. S. Kirk-Othmer, *Encyclopedia of Chemical Technology, 4th ed.*, Wiley – Interscience, New York, Vol. 2, p. 481. 1992
2. M. Hudlicky, *Oxidations in Organic Chemistry*, **ACS Monograph No. 186**, American Chemical Society, Washington DC ,1990.
3. R.A. Sheldon, J.K. Kochi, *Metal-Catalysed Oxidations of Organic Compounds*; Academic Press, New York, 1981.
4. J.R. Holum, *J.Org.Chem.*, **26**, 4814, (1961).
5. D.G. Lee, U.A. Spitzer, *J. Org. Chem.*, **35**, 3589, (1970)
6. R.J. Highet,; W.C. Wildman, *J. Am. Chem. Soc.*, **78**, 6682, (1958)
7. S.-I. Murahashi, T. J. Naota, *Synth. Org. Chem. Jpn.*, **46**, 930, (1988).
8. P. Anelli, C. Biffi, F. Montanari, and S. Quici, *J. Org. Chem.*, **52**, 2559, (1987)
9. P. Anelli, S. Banfi, F. Montanari and S. Quici, *J. Org. Chem.*, **54**, 2970, (1989)
10. A. E. J. de Nooy, A. C. Besemer, H. van Bekkum, *Synthesis*, 1153, (1996)
11. P. L. Bragd, H. van Bekkum and A. C. Besemer, *Topics in Catalysis*, **27**, 49, (2004)
12. R. A. Sheldon, I. W. C. E. Arends, G. Brink and A. Dijksman, *Acc. Chem. Res.*, **35**, 774, (2002)
13. S.K. Tanielyan, R.L. Augustine, I. Prakash, K.E. Furlong, R.C.Scherm, H.E. Jackson, PCT Int. Appl. WO 2004067484 (2004); I. Prakash, S.K. Tanielyan, R.L. Augustine, K.E. Furlong, R.C.Scherm, H.E. Jackson, US Pat. # 6,825,385, (2004).
14. S.K. Tanielyan, R.L. Augustine, I. Prakash, K. Furlong and H.E. Jackson, This volume.
15. R.A. Sheldon, *In Dioxygen Activation and Homogeneous Catalytic Oxidation*, L.L. Simandi, Ed., Elsevier, Amsterdam, 1991, p. 573.
16. S.S. Kim, H.C. Jung, *Synthesis*, **14**, 2135, (2003)
17. P. Gamez, I.W.C.E. Arends, J. Reedijk,; R.A. Sheldon, *Chem. Commun.*, **19**, 2414-2415, (2003)
18. T. Inokuchi, K. Nakagawa, S. Torii, *Tetrahedron Letters*, **36**, 3223.
19. A. Dijksman, A. Marino-Gonzalez, A.M. Payeras, I.W.C.E. Arends, R.A. Sheldon, *J. Am. Chem.Soc.*, **123**, 6826, (2001)
20. A. Cecchetto, F. Fontana, F. Minisci, F. Recupero, *Tetrahedron Letters*, **42**, 6651, (2001).
21. R. Liu, X. Liang, C. Dong, X. Hu, *J. Am. Chem. Soc.*, **126**, 4112, (2004).
22. S.K. Tanielyan, R.L. Augustine, I. Prakash, K.E. Furlong, H.E. Jackson, PCT Int. Appl. WO 2005082825 (2005).
23. S.K. Tanielyan, R.L. Augustine, O. Meyers, M. Korell, U.S. Pat. Appl. (2004).
24. R. Augustine, S Tanielyan, Chemical Industries (Dekker), **89**, (Catal. Org. React.), 73 (2003).

16. The Conversion of Aminoalcohols to Aminocarboxylic Acid Salts over Chromia-Promoted Skeletal Copper Catalysts

Dongsheng S. Liu, Noel W. Cant and Andrew J. Smith

*School of Chemical Engineering and Industrial Chemistry,
The University of New South Wales, Sydney NSW 2052. Australia*

andrew.smith@unsw.edu.au

Abstract

The rates of oxidative dehydrogenation of a range of monoalcohols and diols, with and without amino groups, has been studied over skeletal copper catalysts in an autoclave that allows the reactants to be preheated to reaction temperature prior to mixing. Chromia, introduced by deposition during leaching, improves activity and stability. Some aspects of the reactivity order are consistent with the inductive effect of NH_2 and OH groups in weakening CH bonds but steric effects on adsorption via the nitrogen atom appear necessary to explain the low rates observed for alkyl substitution at this position. NMR spectra of samples taken at different conversions show that the two OH groups in diethanolamine react in sequence forming 2-hydroxyethylglycine as an intermediate. Much of the progress of the reaction can be modeled assuming first order kinetics but the final stages proceed faster than predicted.

Introduction

The trail-blazing patent of Goto *et al.* (1) for the oxidative dehydrogenation of aminoalcohols to the corresponding aminocarboxylic acid salts over Raney® copper catalysts in strongly alkaline solutions was cast in terms of the general reaction

$$R_1R_2NCH_2CH_2OH + OH^- \rightarrow R_1R_2NCH_2COO^- + 2 H_2 \qquad [1]$$

Published work since then has largely concentrated on two specific reactions, the conversion of ethanolamine to glycinate e.g. (2, 3)

$$H_2NCH_2CH_2OH + OH^- \rightarrow H_2NCH_2COO^- + 2 H_2 \qquad [2]$$

and that of diethanolamine to iminodiacetate for use in the production of the herbicide, glyphosate, e.g. (4, 5)

$$HN(CH_2CH_2OH)_2 + 2 OH^- \rightarrow HN(CH_2COO^-)_2 + 4 H_2 \qquad [3]$$

Patents describing improved catalysts for this class of reactions often claim suitability for a wide range of substrates (6-9). However there is no data comparing rates of reaction for alcohols with different structure. In part this is because the emphasis is usually on yield but also because the reaction is normally carried out batchwise in stirred autoclaves under severe conditions ($\geq 160°C$, ≥ 9 bar, $\geq 6M$ NaOH). Kinetics are then difficult to measure since considerable reaction takes place during warm-up and the catalyst commonly deactivates during the course of reaction.

In recent work we have developed a modified autoclave which solves these difficulties by allowing the starting components, aminoalcohol and aqueous NaOH containing the catalyst, to be preheated separately to reaction temperature before mixing (10). Here we have made use of this modified reactor to determine rate constants for a range of alcohols with different structures.

In the context of mechanism, substrates with two or more OH groups are of particular interest since the question arises as to whether the groups will react in sequence or in parallel. If the reaction is wholly or partially sequential then an intermediate containing both OH and COO^- groups will be present at partial conversion. In order to determine if this happens, a method for analysis of the intermediate(s) at different stages of reaction is needed. In the present work we have developed ^1H NMR spectroscopy for this purpose and made use of it to follow the reaction of diethanolamine, reaction [3].

Results and Discussion

More than 20 catalyst samples were prepared, pretreated and tested during the present work. Their characteristics, averaged over all samples at various stages are shown in Table 1. Pretreatment reduces the Al content of the unpromoted catalyst to near zero accompanied by a large fall in surface area. The Cr-promoted catalyst retains significant Al after pretreatment and the loss in surface area is proportionately much less. The Cr_2O_3 content is not changed. With both catalysts, subsequent use has little effect on properties beyond that of pretreatment.

Table 1. Composition and BET surface area of unpromoted and promoted catalysts

Catalyst	Stage	wt % Al	wt% Cr_2O_3	BET area (m^2/g)
unpromoted copper	as prepared	1.01	-	18
	after pretreatment	0.015	-	2.3
	after one use	0.011	-	1.1
chromia-promoted copper	as prepared	2.2	0.9	40
	after pretreatment	0.64	0.9	13
	after one use	0.63	0.9	13

For catalysed liquid phase reactions that are first order overall, as found for ethanolamine (2, 11), the integrated equation for a batch reactor is

$$-\ln (1-X) = k (W/V) t \qquad\qquad [4]$$

where X is the conversion, W the mass of catalyst, V the liquid volume and k the first order rate constant. If the stoichiometry follows [1], then the conversion can be calculated from the ratio of the volume of H_2 evolved at time t, v_t, to that calculated for complete conversion, v_∞, i.e. $X_{H2} = v_t/v_\infty$.

Figure 1 shows plots of $-\ln(1-X_{H2})$ versus t for some of the monoalcohols tested, in comparison with ethanolamine, using the chromia-promoted catalyst. The plots are reasonably linear, but with some upward curvature indicating a deviation from first order behaviour at high conversion. In all cases the slopes (i.e. the rate) are less than that for the control ethanolamine. The structures and the first order rate constants, k_{H2}, calculated from the slopes in Figure 1 are listed in Table 2.

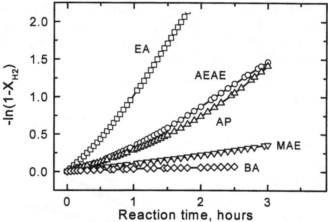

Figure 1. First order plots based on hydrogen evolution for the oxidative dehydrogenation of ethanolamine (EA), 2-(2-aminoethylamino)ethanol (AEAE), 3-amino-1-propanol (AP), 2-(methylamino)ethanol (MAE) and benzyl alcohol (BA) over chromia-promoted copper.

Table 2. Reactants and first order rate constants for reaction over promoted catalysts

reactant	structure	k_1, L g^{-1} h^{-1}
ethanolamine	$NH_2CH_2CH_2OH$	0.0290
3-amino-1-propanol	$NH_2CH_2CH_2CH_2OH$	0.0084
2-(2-aminoethylamino)ethanol	$NH_2CH_2CH_2NHCH_2CH_2OH$	0.0120
2-(methylamino)ethanol	$CH_3NHCH_2CH_2OH$	0.0028
benzyl alcohol	$C_6H_5CH_2OH$	0.0007

Interpretation of the activity pattern is uncertain since the mechanism of the reaction is not known. Deuterium isotope studies of the dehydrogenation of methanol over skeletal copper show that the alcohol is reversibly adsorbed with loss of the hydroxyl hydrogen to form a surface methoxy species (12). The rate-determining step is the subsequent detachment of an adjacent CH hydrogen to form an aldehyde. If a similar route was followed here, then the inductive effect of the NH_2 group might be expected to aid dissociation which would explain why the rate is higher for ethanolamine than for 3-amino-1-propanol where the effect of the NH_2 is attenuated by an additional CH_2 group. However the much lower rates with benzyl alcohol and 2-(methylamino)ethanol is clearly inconsistent with this interpretation so other effects are present as well. One possibility is that the reactant is bound to the catalyst surface partly via the nitrogen lone pair so that if this is absent, or the H of NH_2 is replaced by a more bulky alkyl group, then the strength of attachment is reduced. This might also explain why 2-(2-aminoethylamino)ethanol, which has a second nitrogen to aid attachment, is relatively reactive.

If a diol is oxidatively dehydrogenated to form a diacid via an intermediate with one OH group, then a first order plot based on hydrogen evolution can exhibit some curvature. This is because the slope at any time will reflect the instantaneous concentrations of the diol and intermediate as well as their intrinsic reactivities. First order plots for the reaction of ethylene glycol, 1,4-butanediol and diethanolamine are shown in Figure 2. All plots are reasonably linear, consistent with reaction via an intermediate with a rate constant rather similar to that of the starting diol (or a direct reaction with no intermediate whatsoever).

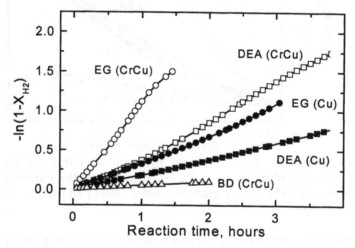

Figure 2. First order plots based on hydrogen evolution for the reaction of ethylene glycol (EG), diethanolamine (DEA) and 1,4-butanediol (BD) over unpromoted copper (Cu) and chromia-promoted copper (CrCu).

Apparent rate constants, k_{H2}', calculated from the slopes of these plots are shown in Table 3. Comparison against Table 2 shows that the reactivity of ethylene glycol over the promoted catalyst is about 60% that of ethanolamine, while that of 1,4-butanediol is more than an order of magnitude lower. The implication is that the reaction is facilitated through activation by neighbouring groups, with an OH substituent somewhat less activating than an NH_2 one, and that this activation is greatly attenuated by intervening CH_2 groups. As found previously for the reaction of ethanolamine (10, 11), the chromia-promoted catalyst is two or three times as active as the unpromoted one. However with both catalysts, the CH_2OH groups in diethanolamine react at less than one-half the rate of those in ethanolamine. The implication is that either a CH_2CH_2OH side chain is less activating than is H alone, or that a second bulky group hinders reaction or adsorption.

Table 3. Apparent rate constants (k_{H2}') for diols based on H_2 production

reactant	structure	k_{H2}', L g^{-1} h^{-1}	
		Cu catalyst	Cu-Cr catalyst
ethylene glycol	$HOCH_2CH_2OH$	0.0065	0.020
1,4-butanediol	$HOCH_2CH_2CH_2CH_2OH$	-	0.001
diethanolamine	$HN(CH_2CH_2OH)_2$	0.0050	0.011

Figure 3 shows the NMR spectra of samples taken at several stages during the reaction of diethanolamine (DEA) over unpromoted copper. DEA exhibits triplets at ~2.55 ppm due to the protons designated 'a' below and at ~3.5 ppm due to the 'd' ones. The early stages of reaction are accompanied by the appearance of a singlet at 3.02 ppm. This first grows in intensity and then declines as a second singlet appears at 2.99 ppm and becomes dominant at high conversions. The 2.99 ppm line clearly arises from the CH protons in the final product, iminodiacetic acid (IDA), denoted 'b' below. The 3.02 ppm line can be attributed to the 'c' protons in the intermediate, 2-hydroxyethylglycine (HEG). Its a′ and d′ protons are responsible for the superposition of additional lines on both of the DEA triplets as reaction proceeds.

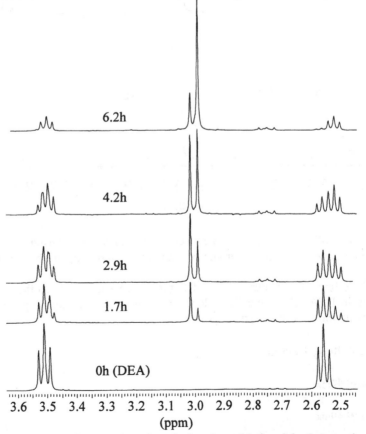

Figure 3. NMR spectra for samples taken during the oxidative dehydrogenation of diethanolamine over unpromoted copper.

$$HN(CH_{2a}CH_{2d}OH)_2 \quad HN(CH_{2a'}CH_{2d'}OH)(CH_{2c}COO^-) \quad HN(CH_{2b}COO-)_2$$
$$\text{(DEA)} \qquad\qquad \text{(HEG)} \qquad\qquad\qquad \text{(IDA)}$$

Reaction over the chromia-promoted catalyst proceeded in a similar way. The spectrum at an overall conversion of 71% based on H_2 evolution is shown in Figure 4. The second spectrum here is that for the same sample but with glycine (NH_2CH_2COOH) added. The CH_2 lines in glycine give rise to the additional singlet at ~3.00 ppm. Clearly glycine is negligible as a byproduct of reaction of diethanolamine. Indeed, no side products could be seen by NMR (the weak lines evident near 2.75 ppm in all spectra arise from the DSS internal standard in the D_2O solvent). Some oxalate may be formed but that is invisible by ^1H NMR.

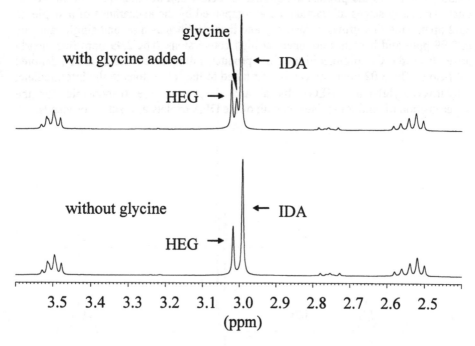

Figure 4. NMR spectra for a sample taken at 71% conversion during the reaction of diethanolamine over chromia-promoted copper with and without glycine added.

If the lines in the four regions of Figure 3 are attributable to the a, b, c and d protons in DEA, HEG and IDA alone, then the areas (A) are proportional to the number of moles of each type of proton present:

$$A_a \propto 4n_{DEA} + 2n_{HEG} \qquad\qquad\qquad [5]$$

$$A_b \propto 4n_{IDA} \qquad\qquad\qquad\qquad [6]$$

$$A_c \propto 2n_{HEG} \qquad\qquad\qquad\qquad [7]$$

noting that the proportionality constant is identical for these equations when using the same spectrum. Substituting Equations [5-7] into the usual equations for relative mole fractions (Y), for example $Y_{IDA} = n_{IDA} / (n_{DEA} + n_{HEG} + n_{IDA})$, gives:

$$Y_{DEA} = (A_a - A_c) / (A_a + A_b + A_c) \qquad [8]$$

$$Y_{HEG} = 2A_c / (A_a + A_b + A_c) \qquad [9]$$

$$Y_{IDA} = A_b / (A_a + A_b + A_c) \qquad [10]$$

Given that the number of a and d protons are the same, an identical set of equations with Ad in place of Aa can also be used. Here we have used an average of Aa and Ad with the areas assessed after Lorentzian deconvolution in order to provide better separation of the 2.99 and 3.02 ppm lines. Mole fractions derived using this NMR method for reaction over the unpromoted copper catalyst are shown as a function of time in Figure 5. Clearly IDA is formed largely via HEG as intermediate since the concentration of the latter passes through a well-defined maximum.

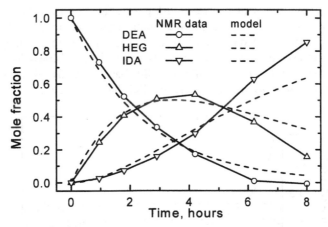

Figure 5. Mole fractions versus time for the reaction of diethanolamine over unpromoted copper.

The reaction was modeled in an attempt to obtain rate constants for the reactions of DEA and HEG separately. This was done on the assumption that the system was entirely sequential, with both reacting according to first order kinetics, i.e.

$$\text{DEA} \xrightarrow{k_1} \text{HEG} \xrightarrow{k_2} \text{IDA}$$

In this case the equations for the concentrations of the three components as a function of time are as follows when DEA alone is present initially (13)

$$[\text{DEA}] = [\text{DEA}]_0 \exp(-k_1 t) \quad \text{or} \quad \ln[\text{DEA}] = \ln[\text{DEA}]_0 - k_1 t \qquad [11]$$

$$[\text{HEG}] = \{k_1/(k_2 - k_1)\} [\text{DEA}]_0 \{\exp(-k_1 t) - \exp(-k_2 t)\} \qquad [12]$$

$$[IDA] = [DEA]_0 [1 + \{k_1 \exp(-k_2t) - k_2 \exp(-k_1t)\}/(k_2 - k_1)] \tag{13}$$

The value of k_1 can be calculated from a plot of ln[DEA] versus time. The calculation needs to be confined to low conversion since examination of equation [8] shows that Y_{DEA} depends on the difference between A_a and A_c and this becomes imprecise at high conversion. The second rate constant, k_2, cannot be calculated directly but its ratio relative to k_1 can be estimated from the maximum mole fraction attained by HEG in the following way. The time at which this occurs, obtained by differentiation of equation [12], is given by (13)

$$t(max) = \ln(k_1/k_2) / (k_1 - k_2)$$

Substitution of this in place of t in [12] and simplification leads to

$$Y_{HEG}(max) = R^{R/(1-R)} \tag{14}$$

where R is the ratio, k_2/k_1.

Figure 6 shows a plot of this relationship as a function of k_2/k_1 together with horizontal lines corresponding to the maximum values attained by Y_{HEG} during reaction over the unpromoted catalyst (~0.53) and for the promoted catalyst (~0.43). These values correspond to R = 0.41 and 0.71 respectively. Table 4 lists the values for k_1 and k_2 obtained for the two catalysts on this basis. The CH_2OH groups in DEA appear to be about twice as active as those in HEG (but see below).

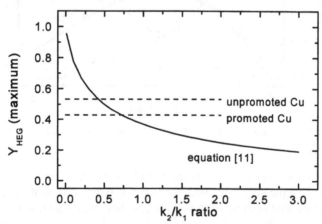

Figure 6. Theoretical dependence of Y_{HEG} on k_2/k_1 in comparison with values observed for the reaction of DEA over unpromoted and chromia-promoted copper catalysts

The dashed lines in Figure 5 show mole fractions for the three components calculated on the basis that the ratio of k_2 to k_1 is 0.5. The fit is reasonably good until well past the maximum in Y_{HEG}. However the growth in Y_{IDA} and fall-off in Y_{HEG} at high conversions is clearly underestimated. A ratio of 0.41 as estimated above gave an exact fit at the maximum but at the expense of worse fit at the extremes. Only a small fraction of the apparent increase in rate at high conversion can be explained in terms of an increase in the ratio of catalyst weight to liquid volume as liquid samples are removed for analysis. Most of the discrepancy must arise from another cause. One possibility is that the difference in rate constant between DEA and HEG arises, at least in part, because they compete for catalyst surface with HEG being disfavoured when its mole fraction is low, perhaps because it lacks the second OH as an attachment point to the surface. In this case the reduced concentration of DEA at high conversion would permit greater access by HEG leading to a faster end stage reaction.

Table 4. Estimated values for k_1 and k_2 for the reaction of diethanolamine

	Cu catalyst	Cu-Cr catalyst
k_1, L g^{-1} h^{-1}	0.009	0.013
k_2, L g^{-1} h^{-1}	0.004	0.009

Conclusions

The oxidative dehydrogenation of alcohols over skeletal copper catalysts exhibits a wide range of rates. Inductive effects can account for some of the differences but other factors, possibly related to binding between nitrogen atoms and the surface, are also present. Diethanolamine reacts sequentially with 2-hydroxyethylglycine as an intermediate. Changes in mole fraction, measured by ^{1}H NMR, during much of the reaction can be modeled according to sequential first order kinetics but the final stages of reaction are faster than predicted.

Experimental Section

Unpromoted and chromia-promoted skeletal copper catalysts were prepared as described in detail previously (10, 11, 14, 15) by leaching a CuAl$_2$ alloy, sieved to 106-211µm, in a large excess (500 mL) of 6.1 M NaOH, either alone or containing Na$_2$CrO$_4$ (0.004 M), for 24 hours at 5°C.

Each batch of catalyst was characterised as to BET surface area, Al content, Cr$_2$O$_3$ content, and tested in the modified autoclave described earlier (10). This comprised a 300 mL stainless steel autoclave (Parr Inst. Co., series 4560) modified with a side valve connected to an independently heated stainless steel cylinder (150 mL). Each sample of catalyst (~9 g) was first pretreated in the autoclave in ~9.7 M NaOH at 200°C for 6 hours. After cooling it was washed with distilled water and a second quantity of NaOH solution (identical to the above) added. The autoclave was flushed with N$_2$ and pressurised to ~2 bar at room temperature. The temperature was brought to near the desired reaction temperature (160°C, ~6 bar) on a linear ramp over a period of 45 minutes with the stirrer operating at 80 rpm. 60-70mL of alcohol was preheated in the reservoir to a matching temperature and then introduced using N$_2$ at >10 bar as a driving gas. The alcohols tested were obtained from commercial sources with minimum stated purities of at least 95% (97% or more in most cases).

The reaction was followed continuously by measurement of H_2 vented, through an electronic pressure controller (Rosemount Inst, model 5866) set at 9 bar, using a wet gas meter. Liquid samples, of 0.5 mL each, preceded by 0.5 mL to flush the line, were removed for NMR analysis at times corresponding to a spread of conversions from 10 to 90% as estimated from the cumulative hydrogen evolution. Spectra were recorded on 0.1 mL samples accurately diluted with 0.5 mL of D_2O containing 0.05% DSS (2,2-dimethyl-2-silapentane-5-sulfonate, sodium salt) as an internal standard using a Bruker DPX-300 spectrometer operating at 300 MHz for protons.

Acknowledgements

This work was supported by grants from the Australian Research Council. The authors are grateful to Drs. B. Messerle and J. Hook for advice concerning the NMR studies and to Dr. P. Stiles for assistance with some aspects of the modelling.

References

1. T. Goto, H. Yokoyama and H. Nishibayashi, U.S. Pat. 4,782,183, to Nippon Shokubai Kagaku Kogyo Company Limited (1988).
2. Y. Yang, Z. Duan, W. Liu, G. Li and X. Xiong, *Huaxue Fanying Gongcheng Yu Gongyi*, **17**, 210 (2001).
3. Z.-Y. Cai, Z.-G. Lu and C.-W. Bai, *Jingxi Huagong*, **20**, 187 (2003).
4. X. Zeng, G. Yang and W. Liao, *Nongyao*, **41**, 19 (2002).
5. H.-P. Ye, Z.-K. Duan, Y.-Q. Yang, J. Wu and J. Yu, *Gongye Cuihia*, **12**, 48 (2004).
6. T. S. Franczyk, US Pat. 5,367,112 , to Monsanto Company (1994).
7. J. R. Ebner and T. S. Franczyk, U.S. Pat. 5,627,125, to Monsanto Company (1997).
8. T. S. Franczyk, Y. Kadono, N. Miyagawa, S. Takasaki and H. Wakayama, US Pat. 5,739,390, to Monsanto Company (1998).
9. D. A. Morgenstern, H. C. Berk, W. L. Moench and J. C. Peterson, US Pat. 6,376,708, to Monsanto Technology LLC (2002).
10. D. Liu, N. W. Cant, A. J. Smith and M. S. Wainwright, *Appl. Catal. A,* in press.
11. D. Liu, N. W. Cant, L. Ma and M. S. Wainwright, Chemical Industries (CRC Press), **104,** (*Catal. Org. React.*) 27-36 (2005).
12. N. W. Cant, S. P. Tonner, D. L. Trimm and M. S. Wainwright, *J. Catal.*, **91,** 197 (1985).
13. P. W. Atkins, *Physical Chemistry, 4th edition*, Oxford University Press, Oxford, 1990, p. 798-799.
14. L. Ma and M. S. Wainwright, *Appl. Catal. A: General*, **187,** 89 (1999).
15. L. Ma and M. S. Wainwright, Chemical Industries (Dekker), **89,** (*Catal. Org. React.*) 225-245 (2003).

17. Bromine-Free TEMPO-Based Catalyst System for the Oxidation of Primary and Secondary Alcohols Using NaOCl as the Oxidant

Setrak K. Tanielyan[1], Robert L. Augustine[1], Indra Prakash[2],
Kenneth E. Furlong[2] and Handley E. Jackson[2]

[1]Center for Applied Catalysis, Seton Hall University, South Orange, NJ 07079
[2]The NutraSweet Corporation, Chicago, IL 60654

tanielse@shu.edu

Abstract

One of the most efficient methods for oxidation of primary alcohols to either aldehydes or carboxylic acids is the one, commonly known as the Anelli oxidation. This reaction is carried out in a two-phase (CH_2Cl_2/aq.buffer) system utilizing TEMPO/NaBr as a catalyst and NaOCl as the terminal oxidant The new system described here is an extension of the Anelli oxidation, but surprisingly, does not require the use of any organic solvents and replaces the KBr co-catalyst with the more benign, $Na_2B_4O_7$ (Borax). The use of the new cocatalyst reduces the volume of the buffer solution and eliminates completely the need of a reaction solvent. The new system was successfully applied in the industrial synthesis of the 3,3-Dimethylbutanal, which is a key intermediate in the preparation of the new artificial sweetener Neotame.

Introduction

The selective oxidation of alcohols to the corresponding carbonyl compounds is one of the more important transformations in synthetic organic chemistry. A large number of oxidants have been reported in the literature but most of them are based on the use of transition metal oxides such as those of chromium and manganese (1,2). Since most of the oxidants and the products of their transformation are toxic species, their use creates serious problems concerning their handling and disposal. For these reasons, the search for an efficient, easily accessible catalysts and "clean" oxidants such as hydrogen peroxide or molecular oxygen for industrial applications is still a challenge (3-6). A particularly convenient procedure for the oxidation of primary and secondary alcohols is reported by Anelli (4). The oxidation has been carried out in a two-phase system (CH_2Cl_2-water) utilizing 2,2,6,6-tetramethylpiperidinyl-1-oxyl (TEMPO) as a catalyst, NaBr as a cocatalyst and cheap and readily accessible NaOCl as an oxidant. The aqueous phase is buffered at pH 8.5-9.5 using $NaHCO_3$ (Scheme 1). In another development of the same procedure, the ethyl acetate-toluene mixture has been used to replace the CH_2Cl_2 (7)

Despite the extensive work reported in the area of the selective oxidation of primary alcohols, there is still a continuous need for developing efficient and economical oxidation methods which do not require bromine based catalysts, can be carried out with environmentally friendly oxidants and avoid the use of organic solvents.

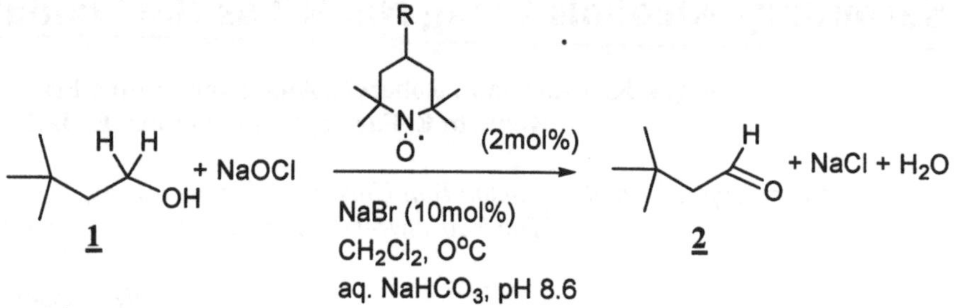

Scheme 1

R=H, TEMPO (3); R=OCH$_3$, MeO TEMPO (4); R= NHCOCH$_3$, AATEMPO (5)

Here we report on a new TEMPO based catalyst system for the oxidation of primary and secondary alcohols selectively to aldehydes and ketones using NaOCl as a terminal oxidant.

Results and Discussion

In the standard Anelli protocol for the oxidation of a primary alcohol such as 3,3-dimethylbutanol (1) to form an aldehyde, i.e. 3,3-dimethylbutanal (2) the oxidation takes place in a two-phase system (4). The alcohol substrate and the MeO-TEMPO catalyst, 4, are dissolved in the dichloromethane phase while the KBr co-catalyst is in the aqueous phase buffered at pH=8.6 with sodium bicarbonate. The most critical parameters for an efficient oxidation to take place are the continuous control of the pH, maintaining the temperature in a relatively narrow range between 0 and –2°C, efficient stirring and the metered addition of the NaOCl solution. The aldehyde yields, plotted against the time at various Br⁻: 4 ratios are shown in Figure 1. At high concentrations of the KBr co-catalyst, the alcohol is completely oxidized within the first 5 minutes, but the selectivity to aldehyde was surprisingly low, due to the over-oxidation to the corresponding carboxylic acid. On lowering the Br⁻: 4 ratios, the aldehyde selectivity passes through a maximum, which is slowly shifted to longer reaction times. At a Br⁻: 4 ratio of 1:1, full conversion of the alcohol at 92% aldehyde selectivity was attained over a 20 min reaction time.

A profile of a hypothetical reactor containing all the reactants is shown in Figure 1b. The top section of the bar is the volume of the bleach to be introduced, the hatched dark section is the volume of the CH$_2$Cl$_2$ solvent, the gray area is the volume of the aqueous buffer and at the bottom of the bar graph is the volume of the alcohol substrate, which, under these standard conditions is just 2.3% of the total volume of the liquid phase. The deficiencies in the standard reaction protocol are obvious and they make the application of the oxidation procedure not feasible economically. Particularly, it could be pointed out that the low substrate to catalyst ratio (less than 100), the large volume of dichloromethane solvent, the large amount of buffer and the need of KBr as a co-catalyst are all detrimental to commercial applications. A significantly improved method would have to be far more efficient in the use of the

TEMPO catalyst, eliminate the bromine-based co-catalyst, significantly reduce the volume of the buffer solution and avoid the use of organic solvents. With just a ten-fold buffer reduction, the volume which the substrate occupies is now close to 40% of the initial charge (bar **B** in Figure 1b).

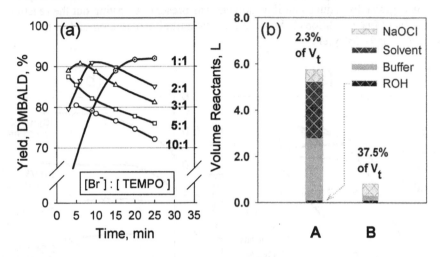

Fig 1. Effect of the KBr:[MeO TEMPO] ratio on the yield of **3** (a) and graphical presentation of the volumes of each reactant used (b). Reaction conditions: [MeO TEMPO] = 0.08mM, [3,3-Dimethylbutanol] = 8mM, [NaOCl] = 10mM, CH_2Cl_2 20mL, T = 0°C

Screening for an efficient TEMPO catalyst

The various commercially available TEMPO's were screened for activity in this reaction with the following order the result.

MeO-TMP ≅ AA-TMP > TMP > HO-TMP > DMA-TMP

Solvent effect

Attempts were made to replace the CH_2Cl_2 solvent without sacrificing the reaction rates and the selectivities to the target aldehyde, **2**. The screening was done at [MeO-TEMPO] :Br⁻ =1. High reaction rates and complete conversion of **1** was recorded in acetone and toluene. When we carried the reaction in acetone, even after complete addition of the hypochlorite solution, the reaction composition remained homogeneous and no phase separation occurred. The reaction rates were high and full conversion to aldehyde was recorded within the first 5 minutes of the reaction. The last option was to test the system by eliminating the solvent completely. The graph 2b shows the progress of the reaction when the toluene fraction in the organic phase was gradually reduced until no solvent was present. Carrying out the reaction under solvent free conditions led to an almost quantitative oxidation of **1** to the aldehyde, **2** in 91.6% yield.

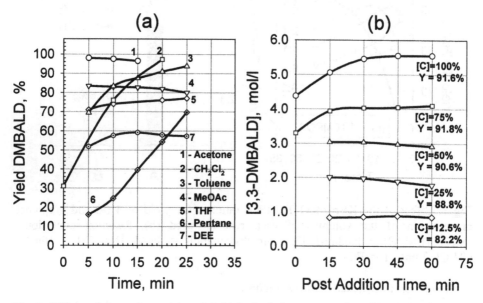

Fig 2. Effect of the solvent (a) and initial alcohol concentration (b) on the reaction efficiency. Reaction conditions, same as for Fig 1 but the reaction solvent is toluene (20mL).

Finding a viable alternative to the KBr co-catalyst

A number of soluble oxymetal salts were screened, particularly those of Mo, W, V, Ti and Zr, which are known to act via peroxometal and/or oxometal mechanisms (8). Unfortunately, no alternative to the existing KBr based protocol was found, so it was decided to, at least, improve the aqueous buffer system to significantly reduce the volume of the aqueous phase.

When the optimized reference oxidation was run without solvent, using MeO-TEMPO, KBr and $NaHCO_3$ buffer a 91% aldehyde yield was obtained over a 90 min reaction time (Table 1). In the second run, the $NaHCO_3$ buffer was replaced with $Na_2B_4O_7$, at ten times lower concentration, using the catalyst composition of MeO-TEMPO/KBr. The aldehyde yield was 95% over the same reaction time. This favorable result was achieved mostly through gains in the product selectivity.

Next, the oxidation reaction was run under identical conditions just with a $Na_2B_4O_7$ additive and a $NaHCO_3$ buffer, but, this time, in the absence of the KBr co-catalyst. Surprisingly, a high yield of **2** was recorded, indicating that, quite likely, the sodium borate is acting as a co-catalyst to the MeO-TEMPO, replacing the KBr. When a reaction was run with only $Na_2B_4O_7$ and in the absence of the $NaHCO_3$ buffer, the yield dropped to 91%, showing the need to buffer the system (see entry 4). With the new co-catalyst and the $NaHCO_3$ but with twice reduced MeO-TEMPO concentration, the alcohol was almost quantitatively converted to the corresponding **2** in 94-99% Yield. When we run the reaction only with the catalyst and the buffer, the result was completely deteriorated (entry 6).

Table 1. Oxidation of alcohol, **1** selectively to aldehyde **2** using KBr and $Na_2B_4O_7$ based systems[a]

	4	KBr	$Na_2B_4O_7$	$NaHCO_3$	Yield[b]	Comments
	mol%	Mol%	mol%	mol %	%	
1	0.35	0.10	-	6.8	91	Reference
2	0.35	0.10	0.85	-	95	Remove buffer
3	0.35	-	0.85	6.8	96	Absence of KBr
4	0.35	-	0.85	-	91	Need of buffer
5	0.17	-	0.85	6.8	99	Less TEMPO
6	0.17	-	-	7.6	67	New co-catalyst!

[a] 3,3-Dimethylbutanol (117mmol), [3] (0.4mmol), [NaOCl] (126mmol) added at 1.4mmol/min, T=0°C, Bleach addition time 90 min, post addition time 120 min.
[b] Determined by GC.

The data in Table 1 clearly show that the $Na_2B_4O_7$ most likely serves as a co-catalyst effectively replacing the KBr, shows superior performance compared to the routinely used KBr and it allows for more efficient use of the MeO-TEMPO catalyst.

Table 2. Selected results on the oxidation of some model alcohols using the $Na_2B_4O_7$ based catalyst system[a].

Run[a]	Substrate	R-TEMPO	Product	Yield
1	Heptan-1-ol	MeO-TEMPO	Heptanal	93
2	Octan-1-ol	MeO-TEMPO	Octanal	85
3	Hexan-1-ol	MeO-TEMPO	Hexanal	90
4	Benzyl alcohol	MeO-TEMPO	Benzaldehyde	100
5	4-Me-cyclohexanol	MeO-TEMPO	4-Me Cyclohexanone	94
6	4-Me-2-pentanol	MeO-TEMPO	4-Me-2-pentanone	92
7	3,3-DM Butanol[1]	MeO-TEMPO	3,3-DM Butanal	92
8	3,3-DM Butanol	TEMPO	3,3-DM Butanal	91
9	3,3-DM Butanol	AA-TEMPO	3,3-DM Butanal	88

[a][ROH] (73mmol), toluene 10mL, [3] (0.25mmol), [$Na_2B_4O_7$] (0.34mmol), $NaHCO_3$ (0.43g dissolved in 15mL water), [NaOCl] (84mmol) added at 1.4mmol/min, T 0°C, Bleach addition time 90 min, post addition time 120 min

Scope of application of the new $Na_2B_4O_7$ based catalyst system

The data in Table 2 show the potential of the $Na_2B_4O_7$ based catalyst system tested over large number of representative alcohols. The primary alcohols were oxidized to the corresponding aldehydes at complete conversion of the alcohol and at 90-93% selectivity. The only by-products observed were the corresponding acid and minor amounts of the symmetrical ester (Entry 2, 3). Benzyl alcohol was quantitatively converted to benzaldehyde. The secondary alcohols, 4-methyl cyclohexanol and 4-methylpentanol were converted to the corresponding ketones at room temperature.

Experimental Section

Reactions were carried out in a 100mL jacketed reaction flask (ACE Glass Inc.) equipped with a Teflon coated magnetic stir bar and Teflon screw cap, accommodating a pH probe, a thermocouple and an addition line for the metered injection of the bleach solution. The stirring rate was controlled by a magnetic digital stirrer (VWR 565) with electronic speed control. The bleach solution was metered into the reactor using a Master flex PTFE tubing pump. All substrates and reagents were commercial grade and were used without further purification. The concentrated bleach solution (12.3%) was purchased from Kuehne Chemical Company and the TEMPO based catalysts were supplied by Degussa Building Blocks Parsippany, NJ.

In a typical procedure, the reaction flask is charged with a solution of the alcohol, such as **1**, either neat or in an appropriate solvent (10 mL, 0.8M), a solution of MeO-TEMPO in the alcohol or reaction solvent (10 mL, 0.008M), the $NaHCO_3$ buffer solution (10 mL, pH = 8.6) , aqueous $Na_2B_4O_7$ (0.68mL, 0.1 M). The resulting mixture is cooled under stirring at 1400 RPM. When the temperature reached 0°C, the bleach solution (28.6 mL, 0.35M) was added over 30 min. Aliquots were taken from the organic phase and analyzed by GC using *tert*- Butyl benzene as an internal standard.

Acknowledgements

This work was supported by the Nutrasweet Company.

References

1. S. Kirk-Othmer, *Encyclopedia of Chemical Technology*, 4th ed., Wiley – Interscience, New York, Vol.2, p 481, 1992
2. M. Hudlicky, *Oxidations in Organic Chemistry*, ACS Monograph No.186, American Chemical Society, Washington, D.C. 1990
3. A. Dijksman, I.W.C.E. Arends, and R. Sheldon, *Chem. Commun.*, 1591, (1999)
4. P. Anelli, C. Biffi, F. Montanari, and S. Quici, *J. Org. Chem.*, **52**, 2559, (1987)
5. P. Anelli, S. Banfi, F. Montanari and S. Quici, *J. Org. Chem.*, **54**, 2970, (1989)
6. P. Anelli, F. Montanari and S. Quici, *Org. Synth.*, **69**, 212, (1990)
7. M. R. Leanna, T. Sowin and H. Morton, *Tetrahedron Letters*, **33**, 5029, (1992)
8. R. Sheldon, I.W.C.E. Arends, D. Dijksman, *Catalysis Today*, **57**, 157, 2000

18.   *In situ* Infrared Study of Catalytic Oxidation over Au/TiO$_2$ Catalysts

Duane D. Miller and Steven S.C. Chuang

Dept. of Chemical and Biomolecular Engineering, University of Akron,
Akron, OH, 44325-3906

duane1@uakron.edu, schuang@uakron.edu

Abstract

In situ infrared study of catalytic CO oxidation over Au/TiO$_2$ shows that the catalyst prepared from AuCl$_3$ exhibits higher activity than those prepared from HAuCl$_4$. The high activity of Au appears to be related to the presence of reduced and oxidized Au sites as well as carbonate/carboxylate intermediates during CO oxidation. Addition of H$_2$O$_2$ further promotes the oxidation reaction on Au/TiO$_2$ catalysts.

Introduction

Supported Au catalysts have been extensively studied because of their unique activities for the low temperature oxidation of CO and epoxidation of propylene (1-5). The activity and selectivity of Au catalysts have been found to be very sensitive to the methods of catalyst preparation (i.e., choice of precursors and support materials, impregnation versus precipitation, calcination temperature, and reduction conditions) as well as reaction conditions (temperature, reactant concentration, pressure). (6-8) High CO oxidation activity was observed on Au crystallites with 2-4 nm in diameter supported on oxides prepared from precipitation-deposition. (9) A number of studies have revealed that Au^0 and Au^{+3} play an important role in the low temperature CO oxidation. (3, 10) While Au^0 is essential for the catalyst activity, the Au^0 alone is not active for the reaction. The mechanism of CO oxidation on supported Au continues to be a subject of extensive interest to the catalysis community.

The reaction pathway, reactivity of the active sites, and the nature of adsorbed intermediates constitute the catalytic reaction mechanism. Our study has been focused on the investigation of the nature of adsorbed intermediates under reaction conditions. We report the results of *in situ* infrared study of CO and ethanol oxidation on Au/TiO$_2$ catalysts. This study revealed the high activity of Au/TiO$_2$ is related to the presence of reduced Au and oxidized Au sites which may promote the formation of carbonate/carboxylate intermediates during CO oxidation.

Experimental Section

Two 1% Au/TiO$_2$ catalysts, designated as HAuCl$_4$ and AuCl$_3$ were prepared by deposition-precipitation of HAuCl$_4$ (Aldrich) and AuCl$_3$ (Alfa Asar) onto Degussa-

P25 TiO_2, respectively (16, 17). The specific procedure involves (i) adding NaOH solution in an appropriated amount of aqueous solution (150ml) of $AuCl_3$ or $HAuCl_4$-$4H_2O$ with 2 g TiO_2 at 343 K to adjust the mixture to pH = 7, (ii) washing the resulting solid five times with warm distilled water, (iii) centrifuging to remove Na^+ and Cl^- ions, (iv) drying the sample at 353 K for 12 h, and then (v) calcining the sample at 673 K for 5 h. The experimental apparatus is explained elsewhere (*18*) but briefly described here. The experimental apparatus consists of (i) a gas flow system with a four port and six port valve, (ii) a DRIFTS (Diffuse Reflectance Infrared Spectroscopy) reactor, (iii) an analysis section with Mass Spectrometer (MS). The CO oxidation was performed from 298 K to 523 K. The reactor temperature was varied at a rate of 10 K/min. The gas species consists of He/CO (90/10 Vol%), He/CO/O_2 (72/14/14 Vol%), He/CO/H_2O_2/O_2 (72/13.3/13.3/1.4 Vol%), He/O_2 (86/14 Vol%), He/H_2O_2/O_2 (84/2/14 Vol%), He/CH_3CH_2OH (83/17 Vol%), and He/CH_3CH_2OH/O_2 (72/14/14 Vol%), He/CH_3CH_2OH/H_2O_2/O_2 (75/8/2/15 Vol%) at a total flow rate of 35 cm^3/min; it takes 13 s for the gases to reach the DRIFTS reactor and 27 s to reach the MS from the four port valve. CH_3CH_2OH and H_2O_2 species added by flowing He through a saturator. Transient IR Spectra were collected by a Digilab FTS4000 FT-IR. The effluent gases of the DRIFTS reactor were monitored by a Pfeiffer Omnistar^{TM} Mass Spectrometer.

Results and Discussion

Fig. 1(a) shows both Au catalysts give very similar XRD patterns; Fig. 1(b) shows both catalysts exhibit a UV peak in the 500-600 nm region; the Au particle size was determined to be 88 nm for $HAuCl_4$ and 86 nm for $AuCl_3$ by XRD.

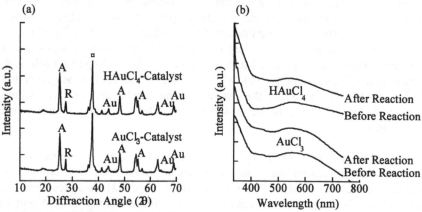

Figure 1. (a) XRD pattern of $AuCl_3$ and $HAuCl_4$/TiO_2; A denotes anatase; R indicates rutile. □ denotes the Instrument peak. (b) UV-Vis spectra of $HAuCl_4$ and $AuCl_3$ catalysts before and after CO oxidation.

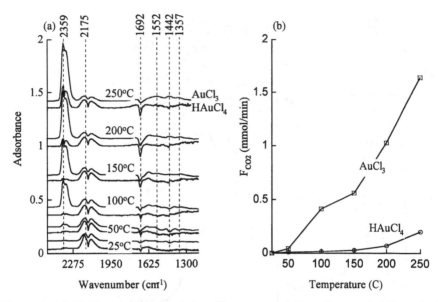

Figure 2. (a) IR Spectra Temperature Programmed Reaction on AuCl₃ and HAuCl₄-Catalyst, (b) Rate of formation of CO₂ on AuCl₃ and HAuCl₄-Catalyst.

In situ infrared spectra of CO oxidation on both catalysts as a function of temperature, shown in Fig. 2(a) reveal that $AuCl_3$ catalysts (Au/TiO_2 prepared from $AuCl_3$ precursor) is more active than $HAuCl_4$ (Au/TiO_2 prepared from $HAuCl_4$) as indicated by CO_2 intensity. Furthermore, $AuCl_3$ catalyst gave higher carbonate and carboxylate intensities in the 1350 - 1560 cm^{-1} region than $HAuCl_4$. The negative band at 1692 cm^{-1} is a result of the incomplete cancellation of the IR background before and during CO oxidation. Comparison of CO_2 formation rate in Fig. 2(b) shows $AuCl_3$ catalyst exhibit low temperature CO oxidation activity. Rate of CO_2 formation over $AuCl_3$ is more sensitive to temperature than over $HAuCl_4$.

Fig. 3 (a) shows CO adsorbed on $HAuCl_4$ catalyst as linear CO on Au^0 site at 2110 cm^{-1}; Fig 3(b) shows weakly adsorbed linear CO on $AuCl_3$ at 2180 cm^{-1} on Ti^{4+} (1), CO adsorbed as linear CO on Au^+ at 2122 cm^{-1}. 2046 and 2004 cm^{-1} bands resemble those reported in the literature on gold electrodes during electro-oxidation of CO at negative potentials. (11, 12) The intensity of these adsorbed CO decreased upon switching of the CO/He flow to the He flow. Removal of gaseous CO by switching of the flow from CO/He to the pure He allowed the direct observation of the contour of the adsorbed CO band and their binding strength to the catalyst surface. Although the 2046 and 2004 cm^{-1} bands were strongly bonded on the Au^0 surface, they were not observed during CO oxidation in Fig. 2 (a). The difference in IR spectra between CO adsorption (Fig 3) and CO oxidation (Fig, 2(a)) suggests that the Au/TiO_2 catalyst surface state varies with its gaseous environment.

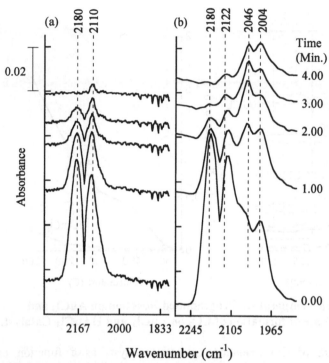

Figure 3. (a) IR spectra of adsorbed CO on HAuCl₄ Catalyst at 298 K, (b) IR spectra of adsorbed CO on AuCl₃ Catalyst at 298 K during switch of the flow (35cm³/min) from CO/He (90/10 Vol%) to He (100 Vol%).

The presence of various types of Au sites and carbonate/carboxylate species as well as variation in the OH intensity of TiO_2 (not shown here) during the reaction suggests that CO oxidation over $AuCl_3$ catalyst could follow the carboxylate mechanism which involves the reaction of adsorbed CO with OH to produce a carbonate/carboxylate species on Au cations and the decomposition of carboxylate to CO_2.(3, 10) Transient infrared study needs to be employed to further verify the role of carbonate/carboxylate species in the reaction pathway. (13)

CO and CH₃CH₂OH Pulses with and without H₂O₂

Hydrogen peroxide (H_2O_2) is used as an oxidizing agent to produce adsorbed oxygen species to gain a better understanding of the mechanisms of CO_2 formation. Addition of H_2O_2 promoted CO oxidation. H_2O_2 addition also enhanced ethanol oxidation as indicated by the increase in IR intensity of CO_2 at 2351 cm⁻¹, as shown in Fig. 4(b). Fig. 4 also shows pulsing ethanol into $He/H_2O_2/O_2$ causes (i) emergence of the ethanol's C-H stretching bands at 3973, 2923, and 2893 cm⁻¹, (ii) the formation of acetate and carboxylate in 1290 - 1570 cm⁻¹ (14, 15) (iii) and displacement of H_2O as indicated by the decrease in H_2O intensity at 1633 cm⁻¹ and the increase H-bonding intensity of ethanol in the 3380 cm⁻¹ region. The variation of acetate and carboxylate intensity with the change in CO_2 intensity suggests that acetate and carboxylate species may be a reaction intermediate.

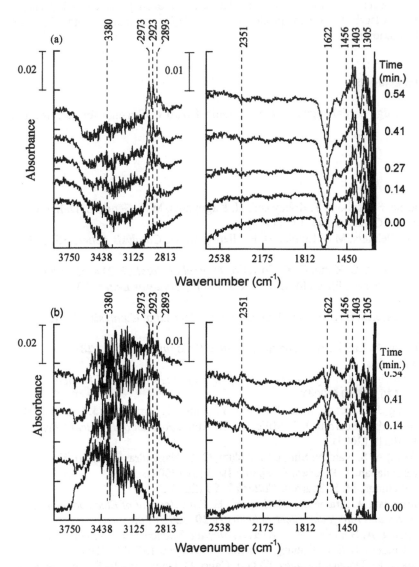

Figure 4. (a) Pulsing 1 cm³ CH₃CH₂OH/He (83/17 Vol%) into flowing He/O₂ (86/14 Vol%) at 35 cm³/min and 323 K on HAuCl₄. (b) Pulsing 1 cm³ CH₃CH₂OH/He (83/17 Vol%) into flowing He/H₂O₂/O₂ (84/2/14 Vol%) at 35 cm³/min and 323 K on HAuCl₄.

Summary

Au/TiO₂ catalyst prepared from AuCl₃ possesses reduced Au and oxidized Au sites which exhibit high CO oxidation activity. The catalyst produce carbonate and carboxylate species which are potential intermediates during CO and CH_3CH_2OH oxidation. Adsorbed oxygen produced from H_2O_2 promoted both CO and CH_3CH_2OH oxidation.

Acknowledgements

We acknowledge partial support from the Ohio Board of Regents for carrying out this research.

References

1. M. Haruta, S. Tsubota, T. Kobayashi, H. Kageyama, M. J. Genet, B. Delmon, *Journal of Catalysis* **144**, 175 (1993).
2. M. A. P. Dekkers, M. J. Lippits, B. E. Nieuwenhuys, *Catalysis Letters* **56**, 195 (1999).
3. H. H. Kung, M. C. Kung, C. K. Costello, *Journal of Catalysis* **216**, 425 (2003).
4. B. Schumacher, V. Plzak, M. Kinne, R. J. Behm, *Catalysis Letters* **89**, 109 (2003).
5. S. S. Pansare, A. Sirijaruphan, J. G. Goodwin, *Journal of Catalysis* **234**, 151 (2005).
6. M. Valden, X. Lai, D. W. Goodman, *Science (Washington, D. C.)* **281**, 1647 (1998).
7. S. Derrouiche, P. Gravejat, D. Bianchi, *Journal of the American Chemical Society* **126**, 13010 (2004).
8. J. H. Yang, J. D. Henao, M. C. Raphulu, Y. Wang, T. Caputo, A. J. Groszek, M. C. Kung, M. S. Scurrell, J. T. Miller, H. H. Kung, *Journal of Physical Chemistry B* **109**, 10319 (2005).
9. R. T. Tom, A. S. Nair, N. Singh, M. Aslam, C. L. Nagendra, R. Philip, K. Vijayamohanan, T. Pradeep, *Langmuir* **19**, 3439 (2003).
10. G. J. Hutchings, M. S. Scurrell, *Cattech* **7**, 90 (2003).
11. K. Kunimatsu, A. Aramata, H. Nakajima, H. Kita, *Journal of Electroanalytical Chemistry and Interfacial Electrochemistry* **207**, 293 (1986).
12. S. C. Chang, A. Hamelin, M. J. Weaver, *Surface Science* **239**, L543 (1990).
13. M. V. Konduru, S. S. C. Chuang, *Journal of Catalysis* **187**, 436 (1999).
14. J. Arana, J. M. Dona-Rodriguez, C. G. i. Cabo, O. Gonzalez-Diaz, J. A. Herrera-Melian, J. Perez-Pena, *Applied Catalysis, B: Environmental* **53**, 221 (2004).
15. G. Jacobs, B. H. Davis, *Applied Catalysis, A: General* **285**, 43 (2005).
16. S. Tsubota, M. Haruta, T. Kobayashi, A. Ueda, Y. Nakahara, *Studies in Surface Science and Catalysis* **63**, 695 (1991).
17. K. Ruth, M. Hayes, R. Burch, S. Tsubota, M. Haruta, *Applied Catalysis, B: Environmental* **24**, L133 (2000).
18. K. Almusaiteer, S. S. C. Chuang, *Journal of Catalysis* **180**, 161 (1998).

III. Symposium on Catalytic Hydrogenation

19.   **2006 Murray Raney Award Lecture:**
                 Synthesis and Features of New Raney®
                 Catalysts from Metastable Precursors

Isamu Yamauchi

Osaka University, Yamadaoka 2-1 Suita, Osaka, 565-0871 Japan
yamauchi@mat.eng.osaka-u.ac.jp

Abstract

Principles of skeletal structure formation of Raney® catalysts are discussed, first from the perspective of phase transformation by chemical leaching. Some ideas are then proposed for making new Raney catalysts. Rapid solidification and mechanical alloying (MA) are described as potential processes for preparing particulate precursors. A rotating-water-atomization (RWA) process developed by the author and co-workers is shown as an example of rapid solidification.

Specific examples of RWA are further described. Raney copper precursor with a small amount of Pd was prepared by this process. Rapid solidification was effective in keeping most of the added Pd dissolved in the precursor. The specific surface area of the leached specimen increased by about 3 times in comparison with that of ordinary Raney copper catalyst. The conversion from acrylonitrile to acrylamide by the hydration reaction was about 60%, or more than 20% higher than that from ordinary Raney copper catalyst. In case of Ti or V addition, the conversion increased to 70-80%. Rapid solidification was quite effective in decreasing the defect rate of Raney catalysts from some precursors. The potential for design of new catalysts may be widely extended by rapid solidification.

Some examples of ternary alloy precursors are also shown. Their general properties were examined and their microstructures were directly observed by transmission electron microscopy. Thus, catalysts synthesized from multi-system alloys had high solubility of additional elements into a major element, and might be expected to work as new catalysts.

The relation between the size of fine particles in skeletal structure and X-ray broadening was also discussed. These results may bring important information to designing new Raney catalysts.

Large particles (diskettes) were formed by applying a spark plasma sintering process to leached catalyst particles. Almost all of the catalyst specific surface area was retained after sintering. This may open a way to manufacture pellet type catalysts or electrodes with high specific surface area.

Introduction

Raney® catalyst is a very interesting material for metallurgists (1-5). The external appearance of a precursor particle is not significantly changed by leaching. An example is shown in Fig.1 for Raney Cu. However, the crystallographic structure was changed from tetragonal Al_2Cu to face centered cubic (fcc) Cu due to

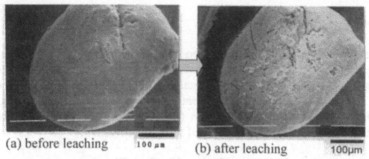

(a) before leaching $100 \mu m$ (b) after leaching $100\mu m$

Figure 1. Shape retention during leaching (Al_2Cu precursor)

composition change. It means that the structure change was not caused by the deposition of a new phase from leaching solution. Cu atoms may be retained on original lattice sites at intermediate stages of leaching. Many vacancy sites may be introduced by removal of Al atoms during leaching as shown in Fig.2(a)-(b). The rearrangement of remained Cu atoms can be easily due to the vacancies Fig.2 (c). This process is viewed as a sort of phase transformation.

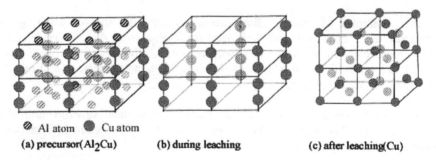

⊘ Al atom   ● Cu atom

(a) precursor(Al_2Cu)   (b) during leaching   (c) after leaching(Cu)

Figure 2. Crystallographic structure changes during leaching

Figure 3 shows an equilibrium phase diagram of Al-Cu system (6). This is the basis for the phase transformation. The skeletal fcc Cu was finally formed from the Al_2Cu (θ) phase by leaching at a constant temperature. The process of the transformation is schematically illustrated by an arrow at a constant temperature as imposed in Fig.3. A conventional phase transformation may be achieved by change in temperature at constant composition, but in this leaching process, it is achieved by change in the composition at constant temperature.

The majority of metallurgists may not accept the above evidence. They usually suppose that the mobility of atoms at leaching temperature is too low to rearrange for the phase transformation. However, transformation has clearly occurred except in a few cases. One of the motivations for our research studies on Raney catalysts is to clarify what happens during the leaching process.

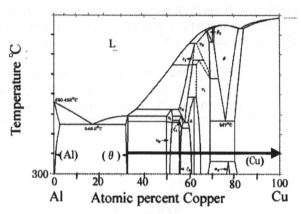

Figure 3. Principle of phase transformation

From a general phase transformation theory, the crystallographic structure and the specific surface area may depend on kinetics and thermodynamics. Therefore, if we can control these factors, new Raney catalysts can be developed.

We have developed a process to make metastable materials by rapid solidification (7) and have applied it to various alloy systems (8). There were few reports on such a metastable processing to make precursors for Raney catalyst. The metastable processes maybe extend the probability of precursor design by decreasing the limit of additive amount to precursor.

In this paper, a few examples of Raney catalysts produced by metastable processes and their catalytic properties are discussed. Then, some examples of multi alloy systems, their microstructures and general properties will be shown. Finally, we will discuss the possibility of forming large particle materials with high specific surface area.

Results and Discussion

Rapid Solidification and Mechanical Alloying for Preparing Super-saturated Precursors

As mentioned above, the formation of skeletal structure is a sort of phase transformation by rearrangement of retained atoms. Therefore, it was important for the skeletal structure formation to control the kinetics of rearrangement. In the past few decades, some researchers have intended to produce new Raney catalysts by adding a third element (9). In many cases, a part of the additional element was dissolved in a precursor. However, if the amount of the additive exceeded a limit, most of the additive was not dissolved in the precursor and the excess additive often crystallized or precipitated as a secondary phase. The secondary phase that is enriched with the third element often remains after leaching. In this case, the third element in the secondary phase may not contribute to the skeletal structure formation. The research studies on development of new catalyst by additives have been limited.

If the solubility of the third element into the precursor can be expanded, the above limitation may be decreased. By applying rapid solidification to prepare the precursor, the solubility may often be increased significantly. The other well-known process for increasing the solubility is mechanical alloying (MA).

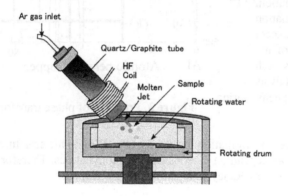

Figure 4. Schematic illustration of Rotating-Water-Atomization Process Unit

One ordinary rapid solidification process, the single roller method, forms thin ribbon of about $20\,\mu$ m in thickness (10). On the other hand, we have developed a new method called the "Rotating-Water-Atomization (RWA) Process" for particulate specimens (7). A schematic illustration of the process is shown in Fig.4. In this process, the molten jet was injected to the rotating water layer through a nozzle. The molten jet impinges on the rotating water surface layer and is broken into fine droplets. The fine droplets are immediately cooled by water. Thus fine and rapidly solidified particles can be easily obtained. The cooling rate was estimated about 10^4 K/s -10^5 K/s and the mean particle size was about $200\,\mu$ m depending on the hole size of the nozzle and the rotating water layer velocity (7).

In most Al-containing alloys, the shape of the particles was tear-drop like due to the tight surface oxide film. The typical shape was shown in Fig.1. The effect of rapid solidification on microstructures is shown in Fig.5 for Al$_2$Cu (precursor for Raney Cu) with a small amount of Pd (11). In the case of slowly solidified (conventional) precursor, most of the added Pd was solidified as a secondary Pd rich phase shown by white dendritic structure in Fig.5 (a). On the other hand, no such secondary phase was observed in a rapidly solidified precursor as shown in Fig.5 (b).

The X-ray diffraction patterns also showed no evidence of a secondary phase. In both cases, the matrix (major phase) was the θ (Al₂Cu) phase. The θ phase of slowly

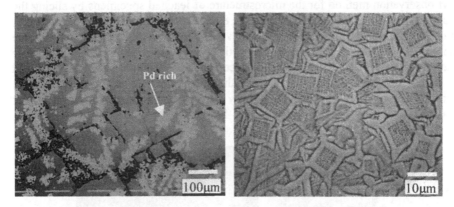

(a) slowly solidified   (b) rapidly solidified

Figure 5. SEM views of slowly and rapidly solidified precursors of Al₂Cu+ 1.5%Pd

solidified specimen has low solubility of Pd. On the other hand, the θ phase of rapidly solidified specimen dissolved most of the Pd in it. The solubility of Pd in the θ phase caused the differences in the specific surface area and the catalytic properties.

The effect of rapid solidification on the specific surface area is shown in Fig. 6. The specific surface area of RWA specimens increased with increase of Pd content. It was up to 3 times larger than that of the slowly solidified specimens. The addition of Pd and rapid solidification for Pd added Al-Cu alloy was quite effective in increasing the specific surface area. Figure 7 shows the conversion ratio from to acrylonitrile to acrylamide. Similarly, it was about 1.5 times higher than that of

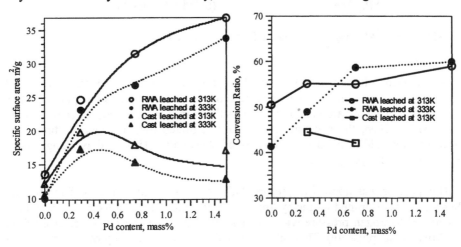

Figure 6. Effect of Pd and cooling rate on specific surface area

Figure 7. Conversion from acrylonitrile to acrylamide

slowly solidified specimen. In this case, the direct effect of Pd dissolved in the Cu skeleton on the conversion ratio is hard to explain clearly, because high specific surface area contributes to the high conversion ratio. We have developed a direct TEM observation method for the microstructure of leached specimens by slicing the specimen using an ultra-microtome. Figure 8 shows an example of TEM images for Cu of Pd free and 1.5mass % Pd. They have a skeletal structure composed of fine granular Cu particles. The size of Cu particles was about 30-40nm for a Pd free material and 10nm for a 1.5 mass %Pd added material. It suggests that the dissolved Pd in the precursor suppressed the growth of Cu particles as obstacles for the rearrangement of Cu. The size obtained by direct observation by TEM was almost the same as that evaluated by the X-ray line- broadening method.

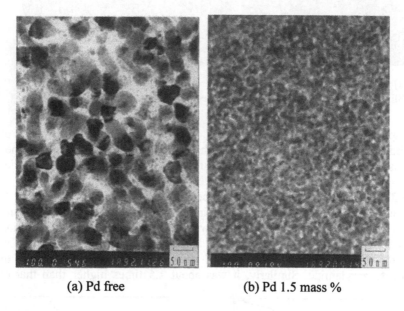

(a) Pd free   (b) Pd 1.5 mass %

Figure 8. TEM image of as-leached state (Raney Cu).

Mechanical alloying (MA) is another process solution in addition to rapid solidification for making a supersaturated solid.   In this process, the pure powders of different elements are mixed and encapsulated in a metallic container with some steel balls. Then, the container is rotated or vibrated for an extended time. The powders are cold worked by collisions of steel balls. This cold working makes mechanically alloyed powder.

This process was applied to prepare Al_3Ni precursor with 2at% Ti. The catalytic behavior was examined for the hydrogenation of acetone in an autoclave. The rate of change of hydrogen pressure with time was evaluated, i.e. the rate of decrease from 0.3MPa (initial pressure). This value is an indicator of the activity of the catalyst. It was -0.0009, -0.0032 and -0.0030 (MPa/s) for conventional Raney Ni, RWA and MA, respectively.

Although the activity obtained by the MA route was not the highest in this case, the effectiveness of MA was still much higher than that of conventional catalyst. In some cases, MA is more effective than rapid solidification for making super-saturated precursors. An example is the Al-Co-Cu ternary system (11). The ranking of effectiveness of these methods may in general depend on the alloy system.

Extension of Rapid Solidification to Ternary Alloy

Ni and Co are immiscible elements with Ag or Cu. Especially Ag is not miscible with Ni or Co even in liquid state. Therefore, there were no previous attempts to make Raney Ni or Co with dissolved Ag and no studies on their catalytic properties. Here, we will show an approach to make Raney Ni or Co with Cu or Ag. In the case of Al-Co-Cu system precursor, MA was an effective process to prepare a super-saturated single solid solution as shown in Fig.9 (11). The crystallographic structure after leaching of this specimen showed a broad fcc phase and no other phases were observed. To examine the as-leached state, the variation of magnetization was measured with the temperature (12). Figure 10 shows the result. The magnetization

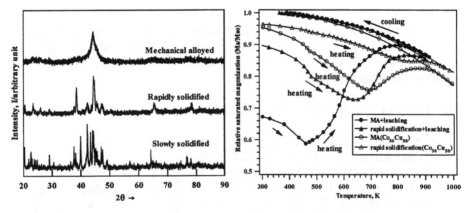

Figure 9. XRD patterns of precursors of Al$_{75}$Co$_{12.5}$Cu$_{12.5}$

Figure 10. Variation of saturated magnetization with temperature

change by heating was a maximum for MA + leaching specimen. It suggests that some amount of Co dissolved in the Cu phase as-leached state. The precipitation of the ferro-magnetic Co phase by heating increased the magnetization. The as-leached state was a metastable state which can not be attained by conventional methods. However, another possible explanation for Fig.10 is that Co particles already exist in the as-leached state, but they are very small and may show super-paramagnetic behavior that is often observed in ferro-magnetic materials with a few nano-size particles. The Co particle size became coarser by heating. This is the subject of a future study. These results strongly suggest that the chemical leaching route is an excellent candidate for development of new materials.

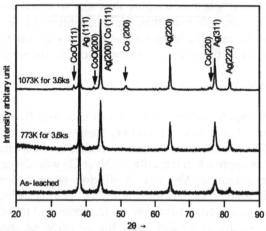

Figure 11. XRD patterns of annealed specimens of leached $Al_{75}Co_{10}Ag_{15}$

In the case of Al-Co-Ag, rapid solidification was effective to make a super-saturated solid solution (13). In this system, the magnetization also increased with increase of temperature and the diffraction lines of fcc Co appeared upon annealing at 1073K for 3.8ks as shown in Fig. 11 (14).

Improvement of defect rate by rapid solidification

In producing some Raney catalysts, their defect rates strongly depend on crystallographic structure of precursors. For an example, if Ag_2Al phase was used as a precursor for the formation of skeletal Ag, the Ag's resulting defect rate may be very high, because the Ag_2Al phase is stable against leaching. For the skeletal Ag formation, Ag should be dissolved in the α-Al solid solution. The equilibrium solubility of Ag in the α-Al solid solution is negligibly small at room temperature. Most of the Ag exists as Ag_2Al phase at room temperature. Therefore, it is difficult to form skeletal Ag by the ordinary route. However, more than 25 at % of Ag can be dissolved in the α phase particles produced by RWA. The formation ratio of skeletal Ag to the total Ag in the precursor exceeded 80% as

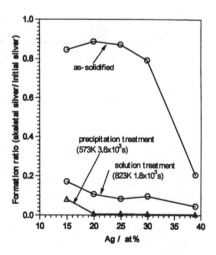

Figure 12. Effect of rapid solidification on formation ratio

shown in Fig. 12. It was less than 20% for the conventional precursors such as precipitated or solution treated ones, where the Ag_2Al phase was stable and the solubility of Ag in the α-Al phase is negligibly small.

Similar result was obtained for skeletal Si[15]. The solubility of Si in the α-Al solid solution is negligibly small so that the two phases α-Al and relatively coarse Si existed in the conventionally prepared Al-Si alloy. By hydrochloric acid leaching, only the coarse stable Si remains after leaching, so that it is difficult to form fine Si particles by leaching. In this case, the rapid solidification increases the solubility to about 12 at%. The specific surface area of skeletal Si formed from RWA was in the range of 65-75m^2/g and the mean particles size was about 30-50nm by direct observation of the skeletal Si.

Bulk material formation from leached particles by SPS process

One of the excellent characteristics of Raney catalyst is a porous skeletal structure composed of ultra fine particles. From the macroscopic view, the size of the precursor particles is not changed after leaching and they can be treated like conventional particles in powder metallurgy. To extend the application field as catalysts in fixed bed or electrodes, bodies with large regular geometric size and shape are often desired. In the conventional powder metallurgical consolidation process, compaction has occurred during consolidation and it leads to decrease of the specific surface area. Recently, the spark plasma sintering process (SPS) as shown in Fig.13 has been developed for the consolidation of ordinary non-skeletal particles. This process can be

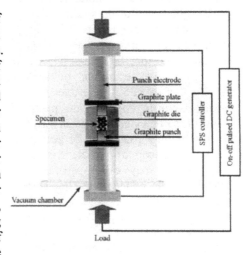

Figure 13. Schematic illustration of SPS equipment

operated at relatively low temperature in

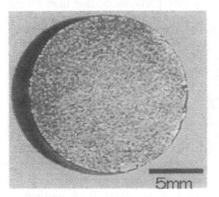

Figure 14. Example of disk consolidated by SPS (Ni-Ag, 15mm x 2mm, specific SA = 28m^2/g)

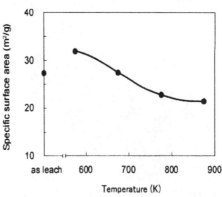

Figure 15. Variation of specific surface area with SPS temperature for Ni-Ag

vacuum. An example of consolidated disks of skeletal Ni-Ag by SPS at 673K under 32MPa is shown in Fig.14. The apparent density of the disk was about 3.7g/cm³. This is less than 50% lower than the density of a non-skeletal Ni-Ag mixture. This means many pores remain in the disk. The specific surface area of disks consolidated at various temperatures is shown in Fig. 15. The specific surface area is first increased and then decreased. The specific surface area of the disk consolidated at 673K was almost the same as that of the as-leached specimen. Thus it was possible to form bulk material by SPS without decreasing the specific surface area.

The reason why the specific surface area increased or was kept at high value after consolidation is considered as follows. The sintering may occur at some primary particle boundaries and not at the boundary of nano-size skeletal particles boundaries. So the skeletal structure was kept after consolidation. The mechanical strength of the disk was supported by sintering each primary boundary. The contribution of the primary boundary to the specific surface area was negligibly small in comparison with the skeletal boundaries. Therefore, the decrease of the specific surface area with the SPS temperatures was minimal. The increase of the specific area at 573K may come from the separation of skeletal particles during SPS. The skeletal metal contains some hydrogen in the particles. When the particles are heated in vacuum, then the hydrogen gas may be generated. The large volume of rapidly expanding hydrogen gas may cause separation at skeletal particle boundaries. New surface area may be formed by the separation at the boundaries. One point of evidence for this was shown in Fig.16. In this case, a primary particle was mounted on a carbon mesh and directly observed by

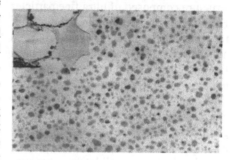

Figure 16. Dispersion of ultra-fine particles by heating a primary leached particle in electron beam in vacuum

TEM. When the particle is irradiated by the electron beam in vacuum, it is heated suddenly. The generated hydrogen gas forms fine particles as shown in Fig.16. In this figure, the top left shows generated skeletal size particles on a carbon mesh at low magnification. The enlargement shows very fine particles of less than 10 nm.

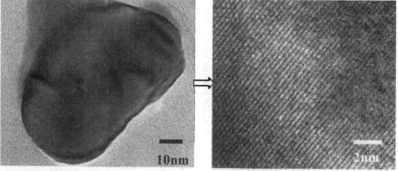

Figure 17. High resolution TEM image of skeletal structure. Each particle is composed of single crystal (left: low magnification; right: high magnification)

These fine particles were not newly formed but already existed as skeletal particles. Similar dispersion of ultra fine particles may occur in SPS by heating in vacuum. Each particle was composed of a single crystal as shown in high resolution transmission microscopy (Fig.17).

Concluding remarks

Some examples of new Raney® skeletal materials have been introduced by applying metastable processing to form their precursors. Applying these new materials to various organic reactions can lead to better catalytic results. The potential for routes to improved Raney® catalysts is thereby demonstrated.

Acknowledgements

This work was supported by the Japan Ministry of Education and Science, the Hosokawa Powder Technology Foundation, Samsung Co. Ltd. and Japan Science and Technology Agency.

References

1. D.J.Young, M.S. Wainwright, R.B. Anderson, *J. Catal.* **64**, 116 (1980)
2. M.L.Bakker, D.J.Young, M.S. Wainwright, *J. Mat. Sci.* **23**, 3921 (1988)
3. M.S.Wainwright, *Chem. Ind.* **68**, 213 (1996)
4. E.Ivanov, S.A.Makholu, K.Sumiyama, K.Suzuki, G.Golubukuval, *J. Alloys Comp.* **185**, 25 (1192)
5. F.Matsuda, *Chemitech* **7**, 306 (1977)
6. T.B.Massalsky, *Binary Phase Diagram,* ASM, 103 (1986)
7. I.Ohnaka, I.Yamauchi, S.Kawamoto, *J. Mater. Sci.* **20**, 2148 (1985)
8. I.Ohnaka, I.Yamauchi, *Mater. Sci. Eng.* **A 182**, 1190 (1994)
9. A.B.Fasman, N.J.Molyukova, T.Kabiev, D.V. Sokol'skii, K.T.Chernousova, *Zh. Fiz.Khim,* **40**,1758(1966)
10. *"Rapidly Solidified Alloys"* ed. by R.H.H. Libermann Butterworths, New York, (1993)
11. I.Yamauchi, M.Ohmori, I.Ohnaka, *J. of Alloys and Compounds,* **299**, 269 (2000)
12. I.Yamauchi, M.Ohmori, I.Ohnaka, *J. of Alloys and Compounds,* **299**, 276 (2000)
13. I.Yamauchi, H.Kawamura, *J. of Alloys and Compounds ,* **370**, 137 (2004)
14. I.Yamauchi, H.Kawamura, K.Nakano, T.Takahara, *J. of Alloys and Compounds,* **387**, 187 (2005)

20.   Competitive Hydrogenation of Nitrobenzene, Nitrosobenzene and Azobenzene

Elaine A. Gelder[1], S. David Jackson[1], C. Martin Lok[2]

*[1] WestCHEM, Department of Chemistry, The University,
Glasgow, G12 8QQ Scotland*
*[2] Johnson Matthey Catalysts, Belasis Avenue,
Billingham, Cleveland, TS23 1LB, U.K.*

sdj@chem.gla.ac.uk

Abstract

The hydrogenation of nitrobenzene, nitrosobenzene and azobenzene has been studied singly and competitively. A kinetic isotope effect was observed with nitrobenzene but not with nitrosobenzene. Nitrosobenzene inhibits nitrobenzene hydrogenation in a competitive reaction, whereas azobenzene and nitrobenzene co-react but at lower rates. Taken together a more detailed mechanistic understanding has been obtained.

Introduction

The catalytic hydrogenation of nitrobenzene is an industrially important reaction, utilized in the production of aniline for the plastics industry. Commercially, the reaction is carried out in the gas phase over a nickel or copper based catalyst (1). However, the transformation is extremely facile and is carried out under relatively mild conditions. For this reason, hydrogenation occurs rapidly over most metals and is often employed as a standard reference reaction for comparing the activities of other hydrogenation catalysts (2 – 5). Despite the large volume of literature available citing the use of this reaction, very little has been published regarding actual mechanistic detail. Haber's initial scheme was published in 1898 (6) and proposed that nitrobenzene (NB) was transformed to aniline (A) in a three-step process involving nitrosobenzene (NSB) and phenylhydroxylamine intermediates (Scheme 1). In addition, it was also proposed that azobenzene (AZO) and azoxybenzene (AZOXY) by-products could be formed via reaction of the two intermediate species. This mechanism has been widely accepted since and a number of studies have reported the identification of the suggested reaction intermediates during hydrogenation (7, 8). In addition, Figueras and Coq (9) have also described the hydrogenation behavior of these intermediates and by products and found azobenzene to hydrogenate through to aniline. While these studies have appeared to provide further supporting evidence for Haber's reaction scheme, the mechanism is still not well understood and has never been fully delineated.

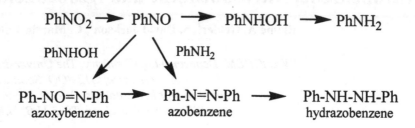

Scheme 1. Haber mechanism.

Although nitrobenzene, nitrosobenzene and azobenzene are often observed in nitrobenzene hydrogenation, we are aware of no studies of competitive reactions. In this paper we report on the competitive hydrogenations and their mechanistic implications.

Results and Discussion

The hydrogenation of nitrobenzene progressed to aniline without any significant by-product formation, only trace amounts of azobenzene were formed (< 1 %) as the reaction went to completion. NMR analysis showed no detectable phenyl hydroxylamine in solution. The hydrogen uptake displayed a smooth curve and the rate of hydrogen consumption coincided with the rate of aniline production. The rate of hydrogenation of nitrobenzene to aniline was 15.5 mmol.min^{-1}.g^{-1}.

In the hydrogenation of azobenzene, the rate of hydrogen consumption followed a smooth curve. The reaction profile showed a direct transformation to aniline with no by-product formation or intermediates detected. The rate of aniline production (8.3 mmol.min^{-1}.g^{-1}) was half the rate of nitrobenzene to aniline.

Unlike the hydrogenation of nitrobenzene, the hydrogen uptake curve for nitrosobenzene displayed two distinct stages. Both these stages proceeded at a significantly slower rate than the rate of hydrogen up-take during hydrogenation of nitrobenzene. This two-stage hydrogen up-take curve has also been reported by Smith et al. (10) over palladium/silica catalysts. Also, the rate of aniline production from nitrosobenzene was 0.35 mmol.min^{-1}.g^{-1}, which was much less than the rate of hydrogen consumption. Therefore aniline was produced 50 times slower from nitrosobenzene than nitrobenzene. The nitrosobenzene hydrogenation reaction profile also displays very different behavior to the nitrobenzene hydrogenation reaction profile (Figure 1).

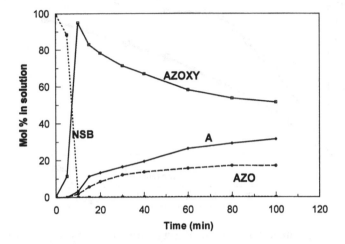

Figure 1. Reaction profile of nitrosobenzene hydrogenation.

In the reaction profile shown in Figure 1 (similar to that shown by Smith et al. (10)) the initial product was azoxybenzene. However this figure is deceptive; firstly azoxybenzene may be produced by a non-catalyzed reaction between nitrosobenzene and phenyl hydroxylamine (10), secondly the figure does not show the mass balance. Indeed at 10 min when all nitrosobenzene has been removed from the solution the amount of azoxybenzene formed was 18.6 mmol, equivalent to 37.2 mmol of reacted nitrosobenzene. Therefore, 42.8 mmol of the original 80 mmol of nitrosobenzene (53.5 %) were unaccounted for. It is possible that the missing mass is in the form of phenyl hydroxylamine in solution, which continues to disproportionate to produce aniline and nitrosobenzene and subsequently azoxybenzene and azobenzene. However as we shall subsequently discover this interpretation is unsustainable.

Azobenzene was also present during the hydrogenation of nitrosobenzene and a stepwise hydrogenation process of nitrosobenzene to azoxybenzene to azobenzene to aniline is a credible route. No azoxybenzene was observed during the hydrogenation of nitrobenzene.

The hydrogenation of nitrobenzene and nitrosobenzene were also studied using deuterium. The hydrogen/deuterium up-take graphs are shown in Figures 2 and 3.

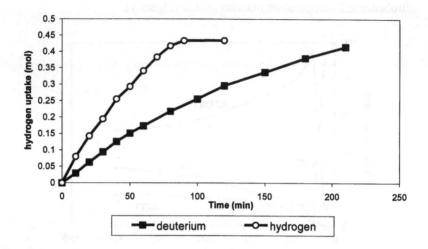

Figure 2. Hydrogen/deuterium up-take during nitrobenzene hydrogenation

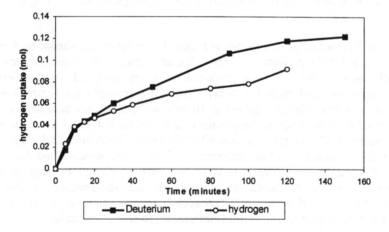

Figure 3. Hydrogen/deuterium uptake during nitrosobenzene hydrogenation

From figure 2 it can clearly be seen that there was a kinetic isotope effect with the reaction in H_2 occurring much more rapidly than the reaction in D_2. The value of the kinetic isotope coefficient (η) was calculated as 1.9. Therefore, it follows that the rate-determining step in the hydrogenation of nitrobenzene must involve a bond-breaking or bond-forming reaction with hydrogen. However when nitrosobenzene was reacted with deuterium (Figure 3) the uptake is considerably different. During the first 15 minutes of reaction, the rates of hydrogenation and deuteration were almost identical. After 15 minutes reaction time, an inverse kinetic isotope effect was evident as the reaction proceeded more rapidly with deuterium than with hydrogen. This was in direct contrast to the situation observed during nitrobenzene hydrogenation where a decrease in rate was observed on moving to deuterium gas. Two distinct processes were in operation; one over the first 15 minutes of reaction, that was not affected by the switch to deuterium gas which can be related to the production of azoxybenzene, and a second, the reaction of azoxybenzene to aniline, occurring after 15 minutes of reaction that was increased on moving to deuterium gas. A value for η, the kinetic isotope coefficient, was calculated as 0.7.

The competitive hydrogenation of azobenzene and nitrobenzene in a 0.5:1 molar mix was examined. A ratio of 0.5:1 was used as two nitrobenzene units are needed to produce a single azobenzene. The reaction profile is shown in Figure 4. In this reaction the concentrations of both nitrobenzene and azobenzene dropped simultaneously and coincided with an increase in aniline concentration. Aniline was produced at a rate of 3.5 mmol.min^{-1}g^{-1}, which is five times slower than nitrobenzene hydrogenation in the absence of azobenzene and two-and-a-half times slower than azobenzene in the absence of nitrobenzene. No other by-products were observed with this reaction.

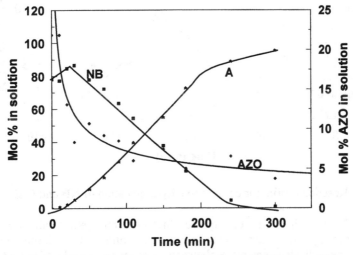

Figure 4. Reaction profile for the competitive reaction between NB and AZO.

In the competitive hydrogenation both azobenzene and nitrobenzene are reacting and the overall rate was reduced. In general, the kinetics of the nitrobenzene reaction have been reported as zero order in nitrobenzene and first order in hydrogen (2, 11, 12). Therefore any reduction in the surface concentration of hydrogen will result in a lowering of the rate. If azobenzene can access sites that would normally be occupied by hydrogen then the reduction in hydrogen concentration would result in a lowering of the reaction rates.

The competitive hydrogenation of nitrobenzene and nitrosobenzene in 1:1 molar mix was studied. The reaction profile is shown in Figure 5. In the early stages of the reaction (< 50 min) only nitrosobenzene reacted, the concentration of nitrobenzene remained almost constant. During this period the reaction profile was similar to that found for nitrosobenzene hydrogenation in the absence of nitrobenzene, however the rate of aniline production was significantly greater at 5.0 mmol.min^{-1}.g^{-1}. After 75 min nitrobenzene hydrogenation initiated and the rate of aniline production increased to 14.9 mmol.min^{-1}.g^{-1}, similar to that for nitrobenzene hydrogenation in the absence of nitrosobenzene.

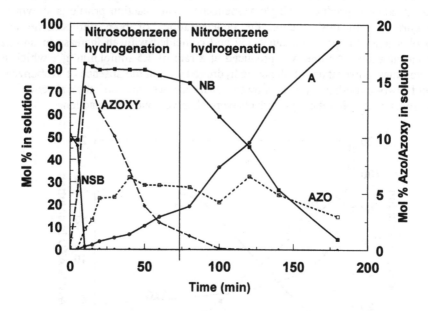

Figure 5. Reaction profile for competitive hydrogenation of NB and NSB.

From Figure 5 it can be clearly seen that nitrosobenzene totally inhibited nitrobenzene hydrogenation. The rapid adsorption and formation of azoxybenzene indicated that nitrosobenzene was more strongly adsorbed than nitrobenzene and that, given its high surface concentration, the principal surface reaction was coupling to form azoxybenzene with loss of water as shown in the reaction sequence;

Ph-NO → Ph-NO(a)
Ph-NO(a) + H(a) → Ph-N(OH)(a)
2Ph-N(OH)(a) → Ph-N=N(O)-Ph(a) + H$_2$O

Note that the rate of hydrogen up-take and the rate of formation of azoxybenzene were identical over the first 20 min of reaction (0.02 mol.min^{-1}g^{-1})

and that in this sequence it is probable that the final coupling reaction is the rate-determining step and such would not show a kinetic isotope effect when deuterium was used instead of hydrogen.

This behavior also suggests that during nitrosobenzene hydrogenation there is always a surface concentration of strongly adsorbed material that inhibits nitrobenzene adsorption on the palladium. There is rapid up-take of nitrosobenzene on the catalyst but most is adsorbed on the carbon support. Only when this strongly bound surface species is removed from the palladium does nitrobenzene hydrogenation initiate. This removes the possibility that the mass missing in nitrosobenzene hydrogenation is phenyl hydroxylamine in solution. The catalyst retains this mass during nitrosobenzene hydrogenation. We propose that this surface species is Ph-NOH(a), the reaction of ntrosobenzene with a surface saturated with adsorbed hydrogen, which can be considered an adsorbed form of phenyl hydroxylamine.

Nitrobenzene hydrogenation in deuterium revealed that there was a kinetic isotope effect indicating that a hydrogen bond making step was rate determining. The difference in the nitrobenzene and nitrosobenzene hydrogenation suggests that the rate-determining step in nitrobenzene hydrogenation must lie before the formation of Ph-NOH(a). In this way the surface concentration of Ph-NOH(a) is kept low. At low surface concentrations, this intermediate reacts with adsorbed hydrogen to continue to aniline. At high concentrations, i.e. during nitrosobenzene hydrogenation, Ph-NOH(a) dimerizes to produce azoxybenzene and water. This is shown in Scheme 2.

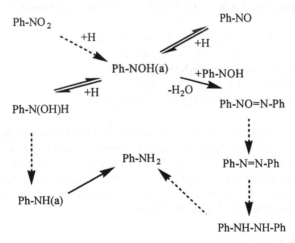

Scheme 2. Overview of NB and NSB hydrogenation (hashed lines denote multiple surface steps).

Hence the proposed reaction scheme has Ph-NOH(a) as the key surface intermediate. It can be formed from both nitrobenzene and nitrosobenzene. In this scheme nitrosobenzene is not an intermediate in nitrobenzene hydrogenation but a by-product formed under low surface hydrogen conditions. Proposed mechanisms are shown below.

Nitrobenzene hydrogenation:
1. $Ph-NO_2$ → $Ph-NO_2(a)$
2. $H_2(g)$ → $2H(a)$
3. $Ph-NO_2(a) + H(a)$ → $Ph-NO(OH)(a)$
4. $Ph-NO(OH)(a) + H(a)$ → $Ph-N(OH)_2(a)$
5. $Ph-N(OH)_2(a) + H(a)$ → $Ph-N(OH)(a) + H_2O$
6. $Ph-N(OH)(a) + H(a)$ → $Ph-N(OH)H(a)$ → phenyl hydroxylamine
7a. $2Ph-N(OH)H(a)$ → $Ph-NO(a) + Ph-NH_2(a) + H_2O$
7b. $Ph-NO(a) + Ph-NH_2(a)$ → $Ph-N=N-Ph(a) + H_2O$
7. $2Ph-N(OH)H(a)$ → $Ph-N=N-Ph(a) + 2H_2O$ → azobenzene
8. $Ph-N(OH)H(a) + H(a)$ → $Ph-NH(a) + H_2O$
9. $Ph-NH(a) + H(a)$ → $Ph-NH_2(a)$ → aniline

Reaction 7 is the sum of reactions 7a and 7b.

Nitrosobenzene hydrogenation:
1. $Ph-NO$ → $Ph-NO(a)$
2. $Ph-NO(a) + H(a)$ → $Ph-N(OH)(a)$
3. $2Ph-N(OH)(a)$ → $Ph-N=N(O)-Ph(a) + H_2O$ → azoxybenzene
4. $Ph-N=N(O)-Ph(a) + H(a)$ → $Ph-N=N(OH)-Ph(a)$
5. $Ph-N=N(OH)-Ph(a) + H(a)$ → $Ph-N=N-Ph(a) + H_2O$
6. $Ph-N=N-Ph(a) + H(a)$ → $Ph-N=NH-Ph(a)$
7. $Ph-N=N-Ph(a) + H(a)$ → $Ph-NH-NH-Ph(a)$
8. $Ph-NH-NH-Ph(a) + H(a)$ → $Ph-NH(a) + Ph-NH_2(a)$
9. $Ph-NH(a) + H(a)$ → $Ph-NH_2(a)$ → aniline

The hydrogenation of nitrobenzene and nitrosobenzene are complex and a range of factors can influence by-product reactions, e.g. hydrogen availability, support acid/base properties (13, 14). In this study we have examine competitive hydrogenation between nitrobenzene, nitrosobenzene and azobenzene. This methodology coupled with the use of deuterium has further elucidated the mechanism of these reactions.

Experimental Section

The catalyst was prepared by impregnation of the powdered activated carbon support, Norit SX Ultra, (surface area 1200 m^2g^{-1}) with sufficient palladium nitrate to produce a metal loading of 3 %. The resulting suspension was dried and calcined at 423 K for 3 hours. The dispersion of the catalyst was measured using CO chemisorption. Taking a CO: Pd ratio of 1:2 the dispersion was calculated as 42.0 % and an approximate metal particle size determined as 2.6 nm.

Hydrogenation was carried out in a Buchi stirred autoclave. The catalyst (0.1 g) was added to the reactor with the reaction solvent (280 mL methanol) and reduced in a flow of hydrogen (30 cm^3 min^{-1}) for 30 minutes with a stirring rate of 300 rpm. During reduction the reactor temperature was increased to 323 K. The reactant, nitrobenzene (8.5 mL, 0.08 mol), nitrosobenzene (8.85 g, 0.08 mol) or azobenzene (7.5 g, 0.04 mol) was then added and the hydrogen pressure set at 2 barg and the stirrer speed at 1000 rpm. In reactions involving the addition of one of the intermediates along with nitrobenzene, the intermediate was added first and mixed with the solvent under nitrogen before nitrobenzene was added and reaction commenced. No variation in hydrogenation rate was observed over a range of stirrer speeds (800 – 1500 rpm) indicating the absence of mass transfer control. Analysis of the reaction mixture was performed by GC-MS using a Varian CP-3800 GC with a Varian Saturn 2000 Trace MS detector fitted with a 25 m DB-5 column. The rate of hydrogenation was measured using the gas flow controller fitted to the autoclave and the rate of aniline production was calculated using the GC-MS data.

Acknowledgements

The authors would like to thank EPSRC and Johnson Matthey for funding this project and Dr. T. A. Johnson for his helpful suggestions. **Acknowledgment is made to the Donors of The American Chemical Society Petroleum Research Fund, for the support of this work through the award of a travel grant.**

References

1. K. Weissermel and H.-J Arpe, *Industrial Organic Chemistry,* 3rd ed. VCH, Weinheim, p. 377 1997
2. A. Metcalfe, M. W. Rowden, *J. Catal.,* **22,** 30 (1971)
3. D. J. Ostgard, M. Berweiler, S. Roder, K. Mobus, A. Freund and P. Panster, Chemical Industries (Dekker), **82,** *(Catal. Org. React.),* 75 (2001).
4. X. Yu, M. Wang, H. Li, *Appl.Catal. A,* **202,** 71 (2000)
5. S. R. de Miguel, J. I. Vilella, E. L. Jablonski, O. A. Scelza, C. Salinas-Martinez de Lecea, A. Linares-Solano, *Appl. Catal. A,* **232,** 237 (2002)
6. F. Haber, *Zeit. Electrochem.,* **4,** 506 (1898)
7. H. D. Burge, D. J. Collins, *Ind. Eng. Chem. Prod. Res. Dev.,* **19,** 389 (1980)
8. V. Höller, D. Wegricht, I. Yuranov, L. Kiwi-Minsker, A. Renken, *Chem. Eng. Technol.,* **23,** 251 (2000)
9. F. Figeras, B. Coq, *J. Mol. Catal. A.,* **173,** 223 (2001)

10. G. V. Smith, R. Song, M. Gasior, R. E. Malz Jr., Chemical Industries (Dekker), **53**, (*Catal. Org. React.*), 137 (1994).
11. D. J. Collins, A. D. Smith and B. H. Davis, *Ind. Eng. Chem. Prod. Res. Dev.*, **21**, 279 (1982).
12. G. J. K. Acres and G. C. Bond, *Platinum Metals. Rev.*, **10**, 122 (1966).
13. P. Zuman and B. Shah, *Chem. Rev.*, **94**, 1621 (1994).
14. D. Groskova, M. Stolcova and M. Hronec, *Catal. Lett.*, **69**, 113 (2000).

21. Selective Hydrogenation of Dehydrolinalool to Linalool Using Nanostructured Pd-Polymeric Composite Catalysts

**Linda Z. Nikoshvili[a] , Ester M. Sulman[a] , Galine N. Demidenko[a],
Valentina G. Matveeva[a] , Mikhail G. Sulman[a], Lumilla M. Bronstein[b],
Petr M. Valetskiy[c] and Irina B. Tsvetkova[c]**

[a] *Department of Biotechnology and Chemistry, Tver Technical University,
A. Nikitina str., 22, 170026, Tver, Russia*
[b] *Chemistry Department, Indiana University, Bloomington, IN 47405, USA*
[c] *Nesmeyanov Institute of Organoelement Compounds of RAS, Vavilov str., 28,
Moscow, Russia*

sulman@online.tver.ru

Abstract

Catalytic properties of the palladium nanoparticles impregnated into various synthetic polymeric matrixes (polydiallyldimethylammonium chloride (PDADMAC), poly (ethylene oxide)-block-polyvinylpyridine (PEO-b-PVP), polystyrene-block-poly-4-vinylpyridine (PS-b-P4VP), hypercrosslinked polystyrene (HPS) etc.) were studied in selective hydrogenation of long-chain acetylenic alcohols to olefinic ones, which are the intermediates in the production of vitamins A, K, E. We investigated both heterogeneous and homogeneous polymer containing catalysts. In the case of heterogeneous catalysts Pd nanoparticles formed in polyelectrolyte layers were deposited on γ-Al$_2$O$_3$. HPS was also investigated as a heterogeneous catalyst on the base of metal contained polymeric matrix. For all the catalysts, the optimal conditions (solvent, temperature, pH, catalyst and substrate amount, stirring) providing the high selectivity of hydrogenation up to 99.5% were found.

Introduction

The most important problem in industrial catalysis is the problem of achieving high selectivity, activity and technological performance of catalytic systems used in fine organic synthesis as it is connected with economical indices, quality of the target products and ecological situation.

In this study we report on the reaction of selective hydrogenation of DHL (3,7-dimethyl-6-octaene-1-yne-3-ol, dehydrolinalool) to olefin alcohol LN (3,7-dimethyl-octadiene-1, 6-ol-3, linalool). Fig. 1 shows the way of the DHL hydrogenation. In this case   DiHL   (3,7-dimethyl-6-octaene-3-ol,   dihydrolinalool)   is   an   unfavorable byproduct of the reaction.

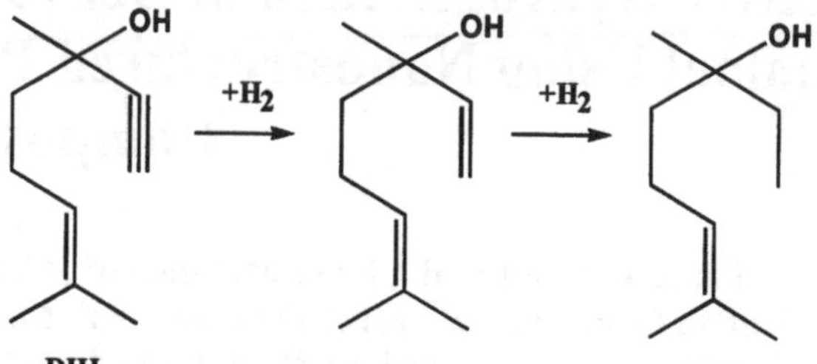

DHL   LN   DiHL

Figure 1. Hydrogenation of DHL to a completely saturated product DiHL

Linalool is one of the most widely used fragrant substances in cosmetic and pharmaceutical industry (as a composite of many cosmetics and perfumes and as an intermediate in synthesis of vitamins (A, E)) (1). But natural resources can't supply the growing needs, so the necessity in synthetic LN has been increasing. Therefore the reaction of selective catalytic hydrogenation of DHL to LN is one of the most significant reactions in the chemistry of fragrant substances (2), and the main problem of this research is the development of modern catalytic technology for preparation of linalool.

Catalytic hydrogenation of DHL by various catalysts such as bulk metals and metals deposited on the carbon, oxides and salts has been studied since the 70-s of the last century (3). The disadvantages of such catalysts are the low surface area and the high content of the noble metals, which result in their high cost price.

In contrast, modern catalysts on the base of metal nanoparticles have large surface area-to-volume ratio of the metal, which allows effectively utilizing expensive metals (4). However, without a suitable support the metal nanoparticles aggregate reducing the surface area and restricting the control over the particle size. To overcome this problem, catalytic nanoparticles have been immobilized on solid supports, e. g., carbon (5), metal oxides (6) and zeolites (7), or stabilized by capping ligands that ranged from small organic molecules to large polymers (8, 9). The encapsulation by polymers is advantageous because in addition to stabilizing and protecting the particles, polymers offer unique possibilities for modifying both the environment around catalytic sites and access to these sites (10-12). So, the protective polymer not only influences particle sizes but can also have a tremendous influence on catalytic activity and selectivity (13). Besides, polymers, which contain complexes or metal nanoparticles, combine the advantages both of homogeneous (high activity and selectivity) and heterogeneous catalysts (easy recovery from the reaction mixture and possibility of regeneration). Catalytic properties of such systems may be changed by varying the type of polymeric matrix and characteristics of metal nanoparticles (14).

Experimental Section

Catalysts on the Basis of Amphiphylic Block-Copolymers.

Amphiphylic PEO-b-P2VP and PS-b-P4VP synthesized by the living anion polymerization were purchased from Polymer Source Inc., Canada, and Max Planck Institute, Germany, respectively. PEO-b-P2VP has the following properties: M_n^{PEO} = 15400, M_n^{P2VP} = 14100, D = 1.04. The molecular structure of PEO-b-PVP is shown in Fig. 2 (a). The structure of PS-b-P4VP, which has the following properties: M_n = 22500, N_n^{P4VP} = 0,340, D = 1.02, is presented in Fig. 2 (b).

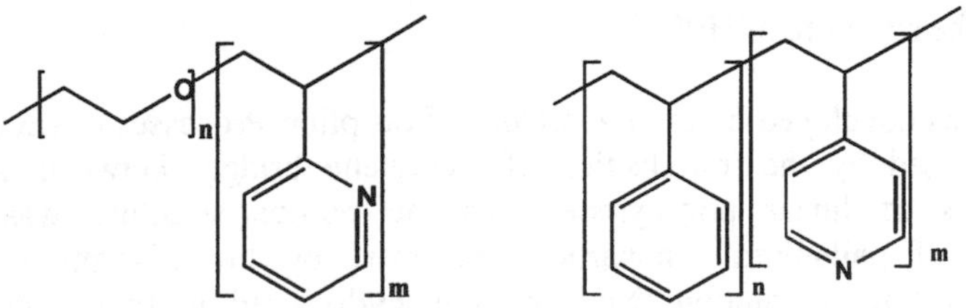

(a) (b)

Figure 2. Molecular structures of PEO-b-P2VP (a) and PS-P4VP (b)

Palladium containing homogeneous polymeric catalysts PEO-b-P2VP and PS-b-P4VP were synthesized by the immobilization of appropriate palladium salts into vinyl pyridine cores of PEO-b-P2VP and PS-b-P4VP micelles, respectively, followed by the reduction of the palladium.

Bimetallic colloids (Pd-Au, Pd-Pt and Pd-Zn) stabilized into the micelle cores of PS-b-P4VP were investigated to determine the influence of the second metal (18). In this case, during the preparation of the catalyst, the addition of both metals salts occurs simultaneously, followed by reduction of metal compounds.

Pd-PEO-P2VP/Al$_2$O$_3$ was studied as a heterogeneous catalyst. To obtain such a system palladium-containing PEO-P2VP was deposited on γ-Al$_2$O$_3$ by the impregnation method. Al$_2$O$_3$ was dried at 320°C for 2.5 h.; after cooling, the micellar solution of block-copolymer containing Pd colloids was added. The resulting mixture was stirred for 2-3 days at room temperature. The catalyst was filtered, washed with petroleum ether and dried in vacuum over P$_2$O$_5$ for 2 days.

Catalysts on the Basis of Polymeric Matrices.

Polymeric matrices such as HPS allow to control the nucleation and growth of the particles by the existence of nanosized voids with the high degree of monodispersion (6). The structure of HPS is presented in Fig. 3.

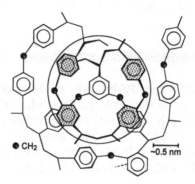

● CH₂   ⊢——⊣
                                             ~0.5 nm

Figure 3. The structure of HPS

HPS was developed by the Laboratory of Sorption Processes (INEOS RAS). It was synthesized by the introduction of methylene bridges between neighboring phenyl rings of linear polystyrene, thus the porous structure was obtained. Preparation of palladium nanoparticles occurred by the adsorption of freshly prepared $Na_2[PdCl_4]$ solution ($Na_2[PdCl_4]$ is hydrolyzed in time) into the HPS matrix. The reduction of palladium compound by hydrogen led to the formation of palladium nanoparticles.

Catalysts on the Basis of Polyelectrolytes.

Formation of metal nanoparticles in ultra thin polyelectrolyte systems was carried out using polydiallyldimethyl ammonium chloride (PDADMAC) by the "layer-by-layer" deposition method. PDADMAC was purchased from Aldrich (Mw = 400 000 – 500 000). The structure of PDADMAC is shown in Fig. 4.

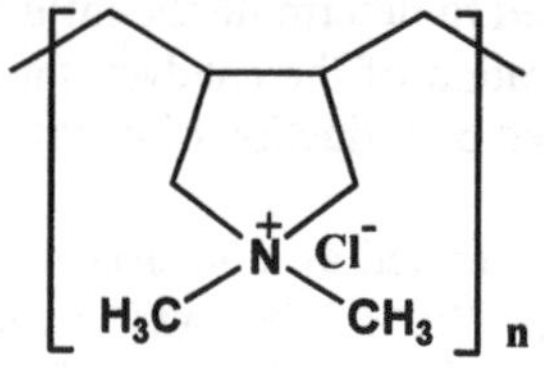

$$H_3C \quad \overset{+}{N} \quad Cl^- \quad CH_3$$

Figure 4. The structure of PDADMAC

CatalyticTests.

Catalytic reactions were carried out in the isothermal glass batch reactor installed in a shaker and connected to a gasometrical burette. The reactor was equipped with two inlets: one for catalyst, solvent, and substrate, and one for hydrogen feeding. The total volume of fluid phase is 32 mL. Hydrogen consumption was controlled by gasometrical burette. Before the test the catalysts were saturated with hydrogen. Both saturation and catalytic experiments were conducted at atmospheric pressure. Duration of the saturation was 1h. Samples (from 5 to 9 in each test) were taken depending on the hydrogen consumption. Most of the samples were collected in the beginning of the experiment (until the half of the hydrogen amount is consumed).

The influence of various solvents (water, methanol, ethanol, propanol, isopropyl alcohol (*i*-PrOH) and toluene) on the catalysts behaviour to provide the appropriate swelling of the above-mentioned polymeric matrixes and layers, their stability and catalytic activity during the experiments was investigated. PS-b-P4VP is soluble in organic solvents (e.g. toluene), while metal salt is not. So, the latter is solubilized solely in the micelle core due to interaction with pyridine units (15). In the case of water-soluble PEO-b-P2VP the probability of the existence of metal salts outside of the micelle cores is higher. Thus for Pd-PEO-b-P2VP and Pd-PEO-b-P2VP/Al$_2$O$_3$ catalysts, a 7:3 (vol.) ratio of isopropyl alcohol to water was found to be optimal (16).

Kinetic studies were performed at various of substrate-to-catalyst ratios and temperature. Physical-chemical investigations of the catalysts, substrate and product were conducted via gas chromatography (GC Chrom-5 equipped with FID), IR-, XP- and electron-spectroscopy. In order to find micelle characteristics providing a kinetic regime, TEM and AFM studies were performed. Micelle sizes in solutions were examined by spectra turbidity method using photoelectrical colorimeter FEK-56M (Russia) equipped with the high-pressure mercury–quartz lamp (DPK-120). The hypotheses of hydrogenation mechanisms were offered on the basis of the results of kinetic experiments, physical-chemical investigations and mathematical modeling.

Results and Discussion

For all investigated catalysts the optimal conditions (see Table 1) of the hydrogenation were found.

It can be seen from the Table 1 that in the case of Pd-PS-b-P4VP and Pd-HPS catalysts the optimal temperature is higher (90°C). The higher temperature of the reaction (which does not allow destructing the polymeric matrix) led to the higher selectivity due to the increase of the active sites accesibility because of the better swelling of the polymer.

Table 1. Optimal conditions

Catalyst	Tempe-rature, °C	C_c^a, mol Pd/l	C_o^b, mol/l	Solvent	Intensity of shaking, shaking/min
Pd-PEO-b-P2VP	70	$1.8 \cdot 10^{-5}$	0.44	*i*-PrOH + water	960

Pd-PEO-b-P2VP/A₂O₃	70	$1.8 \cdot 10^{-5}$	0.44	(7:3(vol.)) i-PrOH + water	960
Pd-PS-b-P4VP	90	$2.3 \cdot 10^{-5}$	0.44	(7:3(vol.)) toluene	480
Pd-PDADMAC	70	$2.8 \cdot 10^{-4}$	0.44	i-PrOH	960
Pd-HPS	90	$2.9 \cdot 10^{-4}$	0.55	toluene	480

[a] Catalyst concentration.
[b] Initial substrate concentration

Catalytic activity and selectivity of the catalysts developed are presented in Table 2.

Table 2. Catalytic activity and selectivity of the catalysts developed

Catalyst	Pd content, wt.%	W^a, m³(H₂)/(molPd ·molS·s)	TOF[b], l/s	Conversion, %	Selectivity, %
Pd-PEO-b-P2VP	0.060	9.3	3.4	100	98.4
Pd-PEO-b-P2VP/A₂O₃	0.060	10.0	7.9	100	98.5
Pd-PS-b-P4VP	0.036	31.8	18.5	100	99.8
Pd-PDADMAC	0.580	3.0	4.5	98.3	98.1
Pd-HPS	0.050	4.4	2.1	100	99.0

[a] Relative rate at 20% hydrogen uptake
[b] Turnover frequency (TOF = mol DHL/ (mol Pd · s))

It can be seen from Table 2 that the supported catalyst (Pd-PEO-b-P2VP/Al₂O₃) has almost twice the TOF compared to the unsupported one. It accounts for fact that in the case of heterogeneous catalyst the micellar aggregates are distributed on the support surface, and the accessibility of catalytic sites increases. Besides, all the investigated catalysts have rather high selectivity compared to the traditional catalysts (e.g. the commercial Lindlar catalyst (palladium deposited on CaCO₃ and modified with lead diacetate) provides the selectivity of 95% at 100% DHL conversion), which could be explained by the electronic structure of the catalysts. As discussed in the preceding papers (1, 18), modification of the metal nanoparticle surface, e.g. with aromatic rings or by the coordination of Pd atoms with nitrogen, causes the selectivity gain due to electron donation to Pd resulting in the decrease of LN adsorption strength (19).

DHL hydrogenation using Pd-HPS has never been reported. This catalyst will be discussed in more detail. Optimal conditions of the hydrogenation process were mentioned above (see Table 1). During the kinetic study we obtained the kinetic curves (see Figure 5 (a)) and the dependences of DHL content on the reaction time (see Figure 5 (b)). In Figure 5 q is the ratio of substrate to catalyst ($q=C_0/C_c$, C_0 – substrate [g], C_c – catalyst [g]). From the data presented in Figure 5 ((a) and (b)) one can see that increase of q leads to a decrease of the hydrogenation rate. The rate of hydrogenation was determined as a ratio of the volume of absorbed hydrogen per time unit (m^3H_2/s) to the total volume of hydrogen absorbed. These data show that the increase of the active phase amount leads to the change of the reaction rate

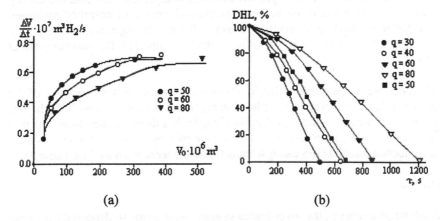

(a)   (b)

Figure 5. Kinetics of hydrogen consumption (a) and dependences of DHL content on the reaction time (b) (Pd/Zn-HPS)

The dependence of activity and selectivity of the DHL hydrogenation on the amount of the active phase (palladium) was investigated. The original catalyst contained 4.88% Pd, but the catalysts with the lower palladium content and bimetallic catalyst HPS-Pd-Zn were synthesized. The results of the investigations are presented in Table 3.

It is seen from Table 3 that the rate of the reaction increases almost three times with the decrease of the palladium content to 1%, and remains constant with the following decrease of the active phase amount. When Pd content is high (4.88%), most of active sites remain in the pores and inaccessible to DHL, so the selectivity is rather low. The selectivity of the process increases with the reduction of palladium content that is due to the increasing of the catalytic active sites amount.

Table 3. Dependences of activity and selectivity of the DHL hydrogenation on the active phase amount (all the experiments were conducted at kinetic conditions (960 shaking/min))

Catalyst	W, $m^3(H_2)/(molPd \cdot molS \cdot s)$	Selectivity, %
HPS(Pd=4.88%)	1.5	73.4
HPS(Pd=1%)	4.3	81.0
HPS(Pd=0.1%)	4.2	87.9
HPS(Pd=0.05%)	4.4	91.8
HPS(Pd/Zn Pd=0.05%)	4.9	96.0

According to the literature the second metal addition could produce the significant effect on the electronic properties of the catalyst (20). So both the energy of the metal-hydrogen and metal substrate bonds and amount of adsorbed hydrogen are expected to be drastically changed. Besides, the modifying metal addition may change the geometry of the catalytic surface, which is caused by the ensemble effect (21 – 23).

The decrease of the electron density favours the hydrogenation processes of the double bonds, but in the case of the triple bonds, use of the catalysts with the higher surface electron density is more preferable (24). Zn is a donor related to Pd, so it is able to increase the electron density at the surface palladium atoms. So the modifier (Zn) introduction into the HPS catalyst leads to the higher selectivity (see Table 3).

Conclusion

The catalytic properties of the investigated systems were found to depend on the type of the solvent and its composition and pH of the medium, as these parameters influence the micelle characteristics of the polymeric catalysts. The micellar parameters control diffusion and adsorption of DHL and hydrogen, and diffusion and desorption of LN within the micelles, thus, alter the reaction rate and selectivity. The denser the micelles are the higher the pyridine unit local concentration (which increases selectivity) and the slower the DHL diffusion within the micelles (which decreases the reaction rate) is. The micellar catalysts show an excellent colloidal stability (does not change micellar characteristics for months). For the better technological performance, PEO-b-P2VP–Pd can be deposited from the micellar solutions on various supports, e.g. Al_2O_3, to obtain heterogeneous catalysts. This would allow the recycling of the catalyst, although the excellent catalytic activity might be decreased.

While investigating the effect of the second metal addition on the catalytic properties of the PS-b-P4VP-Pd and Pd-HPS systems, the introduction of such metals as Au, Zn or Pt was found to allow modifying the structure of the bimetallic clusters, thus, influencing the accessibility and activity of the catalytic sites. It was determined that the higher the surface electron density of the catalyst is the higher the rate of the triple bonds hydrogenation is and the lower the activation energy is.

The nanocatalysts formed in polyelectrolyte layers and polymeric matrixes revealed high selectivity along with high activity and high stability in a hydrogenation process, so they may be recommended for industrial usage. Moreover, the use of polymeric catalysts allows solving some social problems, such as environmental issues and production of pure fine and special chemicals on a larger scale for the expected higher demands in the future.

Acknowledgements

We sincerely thank the Sixth Framework Program "Tailored nanosized metal catalysts for improving activity and selectivity *via* engineering of their structure and local environment" N 506621-1 and NATO Science for Peace program SfP - 974173.

References

1. E. Sulman, Y.Bodrova, V.Matveeva, N. Semagina, L. Cerveny, V. Kurtc, L. Bronstein, O. Platonova, P. Valetsky, *Appl. Catal., A*, **176**, 75 (1999).
2. E. Sulman, L. Bronstein, P. Valetsky, M. Sulman, V. Matveeva, D. Pyrog, Y. Kosivtsov, G. Demidenko, D. Chernishov, N. Businova, N. Semagina, Rus. Pat. №2144020. (2000).
3. A.M. Pak and D.V. Sokolsky, *Selective Hydrogenation of Unsaturated Oxycompounds,* Alma-Ata: Science, 1983.
4. J.H. Fendler, *Nanoparticles and Nanostructured Films: Preparation, Characterization and Application*; Wiley-VCH: Weinheim, Germany, 1998.
5. Z. Liu, X.Y. Ling, X. Su, J.Y. Lee, *J. Phys. Chem. B*, **108**, 8234 (2004).
6. K. Mallick, M.S. Scurrell, *Appl. Catal. A*, **253**, 527 (2003).
7. C. Sun, M.J. Peltre, M. Briend, J. Blanchard, K. Fajerwerg, J.M. Krafft, M. Breysse, M. Cattenot, M. Lacroix, *Appl. Catal. A*, **245**, 309 (2003).
8. D.E. Bergbreiter, C. Li, *Org. Lett*, **5**, 2445 (2003)
9. K.R. Gopidas, J.K. Whitesell, M.A. Fox, *Nano Lett.*, **3**, 1757 (2003).
10. M. Kralik, A. Biffis, *J. Mol. Catal. A*, **177**, 113 (2001).
11. S. Dante, R. Advincula, C.W. Frank, P. Stroeve, *Langmuir*, **15**, 193 (1999).
12. J.F. Ciebien, R.E. Cohen, A. Duran, *Mater. Sci. Eng. C*, 7, 45 (1999).
13. S. Kidambi, M. L. Bruening, *Chem. Mater.*, 17, 301 (2005).
14. E. Sulman, L. Bronstein, V. Matveeva, A. Eichhorn, J. Bargon, M. Sulman, A. Sidorov, *Chemie Ingenieur Technik.*, **6**, 693 (2001).
15. S. Sidorov, I. Volkov, V. Davankov, M. Tsyurupa, P. Valetsky, L. Bronstein, R. Karlinsey, J. Zwanziger, V. Matveeva, E. Sulman, N. Lakina, E. Wilder, R. Spontak, *J. Amer. Chem. Soc.*, **123**, 10502, (2001).

16. N. Semagina, A. Bykov, E. Sulman, V. Matveeva, S. Sidorov, L. Dubrovina, P. Valetsky, O. Kiselyova, A. Khokhlov, B. Stein, L. Bronstein, *J. Mol. Catal. A: Chemical*, **208**, 273 (2004).
17. N. Semagina, E. Joannet, S. Parra, E. Sulmana, A. Renken, L. Kiwi-Minsker, *Appl. Catal. A: General*, **280**, 141 (2005).
18. L.M. Bronstein, D.M. Chernyshov, I.O. Volkov, M.G. Ezernitskaya, P.M. Valetsky, V.G. Matveeva, E.M. Sulman, *J. Catal.*, **196**, 302 (2000).
19. A. Molnar, A. Sarkany, M. Varga, *J. Mol. Catal. A*, **173**, 185 (2001).
20. R. Adams, T. Barnard, *Organometallics*, **13**, 2885 (1998).
21. J. Candy, C. Santini, J. Basset, *J. Top. Catal.*, **3-4**, 211(1997).
22. A. Vieira, M. A. Tovar, C. Pfaff, B. Mendez, C.M. Lopez, F.J. Machado, J. Goldwasser, M.M.R. de Agudelo, *J. Catal.*, **1**, 60 (1998).
23. V. Sandoval, C. Gigola, *J.Appl.Catal.*, **14**, 81 (1996).
24. M. Russo, A. Furlani, P. Altamura, *Polymer*, **14**, 3677 (1997).

22. Modeling and Optimization of Complex Three-Phase Hydrogenations and Isomerizations under Mass-Transfer Limitation and Catalyst Deactivation

Tapio O. Salmi, Dmitry Yu. Murzin, Johan P. Wärnå, Jyri-Pekka Mikkola, Jeanette E. B. Aumo and Jyrki I. Kuusisto

Åbo Akademi, Process Chemistry Centre, FI-20500 Turku/Åbo, Finland
e-mail: Tapio.Salmi@abo.fi

Abstract

Internal and external mass transfer limitations in porous catalyst layers play a central role in three-phase processes. The governing phenomena are well-known since the days of Thiele (1) and Frank-Kamenetskii (2). Transport phenomena coupled to chemical reactions is not frequently used for complex organic systems. A systematic approach to the problem is presented.

Industrially relevant consecutive-competitive reaction schemes on metal catalysts were considered: hydrogenation of citral, xylose and lactose. The first case study is relevant for perfumery industry, while the latter ones are used for the production of sweeteners. The catalysts deactivate during the process. The yields of the desired products are steered by mass transfer conditions and the concentration fronts move inside the particles due to catalyst deactivation. The reaction-deactivation-diffusion model was solved and the model was used to predict the behaviours of semi-batch reactors. Depending on the hydrogen concentration level on the catalyst surface, the product distribution can be steered towards isomerization or hydrogenation products. The tool developed in this work can be used for simulation and optimization of stirred tanks in laboratory and industrial scale.

Introduction

Most parts of heterogeneously catalyzed reactions are carried out in porous catalyst layers, such as catalyst particles in fixed beds, fluidized beds, slurry reactors or in catalyst layers in structured reactors, such as catalytic monoliths, fibers or foams (4 - 6). In the manufacture of organic chemicals, three-phase processes are very frequent, catalytic three-phase hydrogenation being the most typical example. Such manufacturing processes very often involve a strong interaction of kinetics, mass transfer limitations and catalyst deactivation. Mass transfer might play a central role in various stages of a catalytic three-phase process: at the gas-liquid interface as well as inside the porous catalyst layer. In the presence of organic components, catalyst deactivation due to poisoning and fouling retard the activity and suppress the selectivity with time. At elevated reaction temperatures, sintering of the catalyst can take place. Production of organic fine and specialty chemicals is often carried out in batch and semibatch reactors, which imply that time-dependent, dynamic models are required to obtain a realistic description of the process.

Even though the governing phenomena of coupled reaction and mass transfer in porous media are principally known since the days of Thiele (1) and Frank-Kamenetskii (2), they are still not frequently used in the modeling of complex organic systems, involving sequences of parallel and consecutive reactions. Simple ad hoc methods, such as evaluation of Thiele modulus and Biot number for first-order reactions are not sufficient for such a network comprising slow and rapid steps with non-linear reaction kinetics.

In the current work, we present a comprehensive approach to the problem: dynamic mathematical models for simultaneous reactions media, numerical methodology as well as model verification with experimental data. Design and optimization of industrially operating reactors can be based on this approach.

Porous Catalyst Particles

Reaction, diffusion and catalyst deactivation in porous particles is considered. A general model for mass transfer and reaction in a porous particle with an arbitrary geometry can be written as follows:

$$\frac{dc_i}{dt} = \varepsilon_p^{-1}\left(r_i\rho_p - r^{-S}\frac{d(N_i r^S)}{dr}\right)$$

(1)

where the component generation rate (r_i) is calculated from the reaction stoichiometry,

$$r_i = \sum_j v_{ij} R_j a_j$$

(2)

where R_j is the initial rate of reaction j and a_j is the corresponding activity factor.

For the diffusion flux (N_i) various approaches are possible, ranging from the complete Stefan-Maxwell set of equations to the simple law of Fick (7). The symbols of eqs. (1)-(2) are defined in Notation.

The catalyst activity factor (a_j) is time-dependent. Several models have been proposed in the literature, depending on the origin of catalyst deactivation, i.e. sintering, fouling or poisoning (8). The following differential equation can represent semi-empirically different kinds of separable deactivation functions,

$$\frac{da_j}{dt} = -k_j'\left(a_j - a_j^*\right)^n$$

(3)

where k_j' is the deactivation parameter and a_j^* is the asymptotic value of the activity factor – for irreversible deactivation, $a_j^*=0$. Depending on the value of the exponent (n), the solution of eq. (3) becomes

$$a_j = a_j^* + \left(a_{0j} - a_j^*\right)e^{-k_j't}$$

(4a)

and

$$a_j = a_j^* + \left((a_0 - a_*)^{n-1} k_j'(n-1)t\right)^{\frac{1}{n-1}} , \ n \neq 1 \qquad (4b)$$

Some special cases of eq. (4) are of interest: for irreversible first (n=1) and second (n=2) order deactivation kinetics we get $a_j = a_{0j} e^{-k_j' t}$ and $a_j = a_{0j} / \left(1 + a_{0j} k_j' t\right)$ respectively.

In case that the effective diffusion coefficient approach is used for the molar flux, it is given by $N_i = -D_{ei}\left(\dfrac{dc_i}{dr}\right)$, where $D_{ei} = (\varepsilon_p / \tau_p) D_{mi}$ according to the random pore model. A further development of eq. (1) yields

$$\frac{dc_i}{dt} \varepsilon_p^{-1}\left(\rho_p \sum v_{ij} R_j a_j + D_{ei}\left(\frac{d^2 c_i}{dr^2} + \frac{s}{r}\frac{dc_i}{dr}\right)\right) \qquad (5)$$

The boundary conditions of eq. (5) are

$$\frac{dc_i}{dr} = 0 \qquad \text{at} \qquad r = 0 \qquad (6a)$$

and

$$D_{ei}\left(\frac{dc_i}{dt}\right)_{r=R} = k_{Li}(c_i - c_i(R)) \qquad \text{at} \qquad r = R \quad (6b)$$

where c_i is the bulk-phase concentration of the component. The initial condition at t=0 is $c_i(r)$=ci, i.e. equal to the bulk-phase concentration.

The molecular diffusion coefficients in liquid phase were estimated from the correlations Wilke and Chang (9) for organic solutions and Hayduk and Minhas (10) for aqueous solutions, respectively. The solvent viscosities needed in the correlations were obtained from the empirical equation based on the experimental data, $\ln\left(\dfrac{\eta}{cP}\right) = A + B/T$, where the parameters A and B are given in (11). Spherical (s=2) and trilobic s=0.49, were used in the experiments. Hydrogen solubility in the organic reaction mixture was obtained from correlation of Fogg and Gerrard (12), $\ln\left(x_{H_2}^*\right) = A + B/T$, where $x_{H_2}^*$ is the mole fraction of dissolved hydrogen in equilibrium, and A and B are experimentally determined constants. For aqueous sugar solutions, a correlation equation proposed by Mikkola et al. (13) was used.

The liquid-solid mass transfer coefficient was estimated from the correlation provided by Temkin et al. (14). The method is based on the estimation of Sherwood number (Sh), starting from Reynolds (Re) and Schmidt (Sc) numbers,

$$Sh_i = a \, Re^\alpha \, Sc_i^\beta \qquad (7)$$

where Sh=k_{Li}.d_p/D Reynolds number is calculated by using the turbulence theory of Kolmogoroff,

$$Re_d = \left(\frac{\varepsilon d}{v^3}\right)^{1/3}$$

and Schmidt number is defined as Sc= v/D

The critical issue in the calculation of the Reynolds number is the energy dissipated. The experiments of Hajek and Murzin (15) have shown that the dissipated energy can in many cases be clearly less than the energy imposed by the stirrer of a slurry reactor.

An analogous approach was applied on the calculation of the gas-liquid mass transfer coefficient of hydrogen (k_{LH}). The physical parameters needed for the mass transfer coefficients and the dimensionless parameters (Re, Sc) are listed in the Notation section near the end of this chapter.

Catalytic Reactor

For the semi-batch stirred tank reactor, the model was based on the following assumptions: the reactor is well agitated, so no concentration differences appear in the bulk of the liquid; gas-liquid and liquid-solid mass transfer resistances can prevail; and finally, the liquid phase is in batch, while hydrogen is continuously fed into the reactor. The hydrogen pressure is maintained constant. The liquid and gas volumes inside the reactor vessel can be regarded as constant, since the changes of the fluid properties due to reaction are minor. The total pressure of the gas phase (P) as well as the reactor temperature were continuously monitored and stored on a PC. The partial pressure of hydrogen (p_{H2}) was calculated from the vapour pressure of the solvent (p_{VP}) obtained from Antoine's equation (p_{VP0}) and Raoult's law:

$$p^{VP} = p^{Vp0} x_{solvent} \tag{10}$$

where the Antoine parameters are listed in the Notation section. The hydrogen pressure in gas phase thus becomes

$$p_{H_2} = p - p^{VPO} x_{solvent} \tag{11}$$

The volatilities of the reactive organic components (citral, its reaction products as well as the sugars) were negligible under the actual experimental conditions.

A general mass balance for an arbitrary liquid-phase component in the stirred tank reactor is thus written as follows,

$$N_i A_p = N_{GLi} A_{GL} + \frac{dn_i}{dt} \tag{12}$$

i.e. the fluxes (N_i, N_{GLi}) negative for reactants but positive for products. Introduction of the liquid-phase concentration (n_i=$c_i V_L$), assuming a constant liquid-phase volume and introducing the quantities a_P=A_P/V_L and a_{GL}=A_{GL}/V_L gives after some rearrangement:

$$\frac{dc_i}{dt} = N_i a_p - N_{GLi} a_{GL} \tag{13}$$

The initial condition is $c_i = c_i(0)$ at $t=0$. The flux at the particle surface is given by (see eq. 6b),

$$N_i = k_{Li}\left(c_i - c_i(R)\right) \tag{14}$$

while the flux at the gas-liquid interface is estimated from the two-film theory resulting in the expression

$$N_{GLi} = \frac{c_{Gi}^b - K_i c_{Li}^b}{\dfrac{K_i}{k_{Li}} - \dfrac{1}{k_{Gi}}} \tag{15}$$

In the actual case, this expression is used for hydrogen only, since the volatilities of organic reactants and products were negligible, and thus $N_{GLi}=0$ for the organics.

The concentration of hydrogen in gas phase was obtained from the ideal gas law, $c_{H2}=p_{H2}/(RT)$ (eq. 11), and the equilibrium ratio of hydrogen (K_{H2}) was calculated from the equilibrium solubility (x^*_{H2}, Fogg and Gerrard (12)) and Henry's law:

$$K_{H_2} = \frac{p_{H_2}}{x^*_{H_2} c_{TOT,L} RT} \tag{16}$$

where $C_{TOT,L}=\rho_L/M_L$ (Table 1).

Numerical Approach

The partial differential equations describing the catalyst particle were discretized with central finite difference formulae with respect to the spatial coordinate. Typically about 9 discretization points were used for the particle. The ordinary differential equations (ODEs) created were solved with respect to time together with the ODEs of the bulk phase. Since the system is stiff, the computer code of Hindmarsh (3) was used as the ODE solver. In general, the simulations progressed without numerical troubles.

Case Studies

Many organic hydrogenation and isomerization reactions have a parallel-consecutive reaction scheme of the type

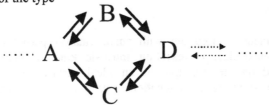

The hydrogenation steps can be preceded or succeeded by isomerization steps, which typically require the presence of hydrogen in the liquid phase, because hydrogen plays a role for the reaction mechanism, even though it is not consumed in the isomerisation itself. Lack of hydrogen on the catalyst surface typically leads to severe deactivation.

The model systems considered in the present work are hydrogenation of citral and xylose on nickel catalysts. The complete reaction schemes are displayed in Fig. 1. In an abbreviated form, the schemes can be presented as follows:

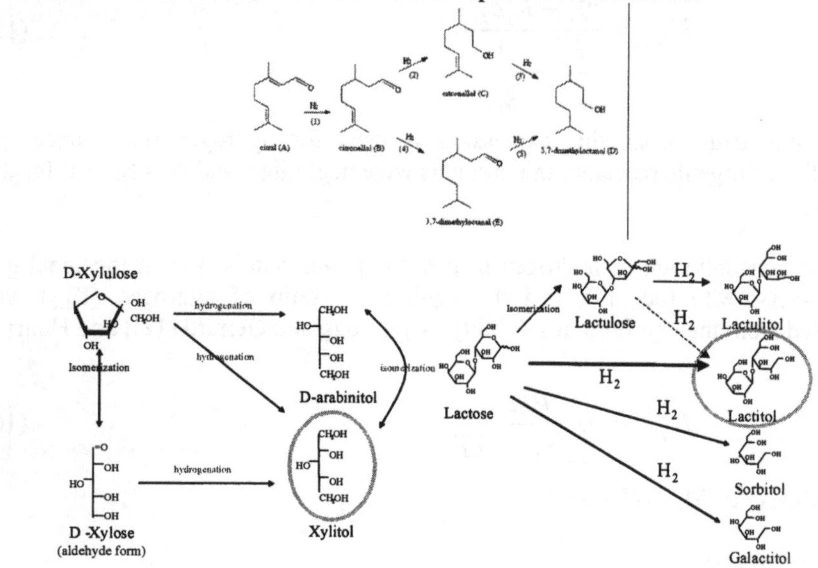

Figure 1. Reaction scheme for hydrogenation of citral, xylose and lactose.

Kinetic experiments were carried out isothermally in autoclave reactors of sizes 500 ml and 600 ml. The stirring rate was typically 1500 rpm. In most cases, the reactors operated as slurry reactors with small catalyst particles (45-90 micrometer), but comparative experiments were carried out with a static basket using large trilobic catalyst pellets (citral hydrogenation). Samples were withdrawn for analysis (GC for citral hydrogenation and HPLC for lactose hydrogenation). The experimental details as well as qualitative kinetics are reported in previous papers of our group Kuusisto et al. (17), Aumo et al. (5).

Modelling results

Based on experiments carried out with small catalyst particles under vigorous stirring, experimental data representing intrinsic kinetics were obtained. Rate expressions based on the principle of an ideal surface, rapid adsorption and desorption, but rate-limiting hydrogenation steps were derived. The competiveness of adsorption and the state of hydrogen on the catalyst surface has been the subject of intensive discussions in the literature (17), but in this case we remained with the simplest possible model, namely non-competitive adsorption of hydrogen and organics along with molecular adsorption of hydrogen. Previous comparisons (17) have shown that this simple approach gives an adequate description of the kinetics, simultaneously having the minimum number of adjustable parameters. Sample fits are provided by Fig. 2. A thorough comparison of experimental data with modeling results indicated that the model for intrinsic kinetics is sufficient and can be used for simulation of the reaction-diffusion phenomena in larger catalyst particles.

Some simulation results for trilobic particles (citral hydrogenation) are provided by Fig. 2. As the figure reveals, the process is heavily diffusion-limited, not only by hydrogen diffusion but also that of the organic educts and products. The effectiveness factor is typically within the range 0.03-1. In case of lower stirrer rates, the role of external diffusion limitation becomes more profound. Furthermore, the quasi-stationary concentration fronts move inside the catalyst pellet, as the catalyst deactivation proceeds.

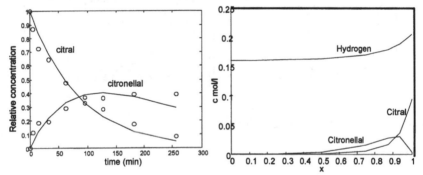

Figure 2. Fit of model to experimental data and concentration profiles inside a trilobic catalyst particle.

For "small" catalyst particles used in sugar hydrogenation (slurry reactors), one would intuitively conclude that the system is safely on the kinetic regime. The simulation results obtained for sugar hydrogenation strongly suggest that this is not the case. The organic molecules are large, having molecular diffusion coefficients of the magnitude $4.0 \cdot 10^{-9}$ m^2/s. The porosity-to-tortuosity factor can be rather small ($<<0.5$), thus the effective diffusion coefficient becomes small, and internal diffusion resistance starts paying a role for particles, whose diameters exceed 100 micrometer.

In addition, external diffusion resistance is often a crucial factor in sugar hydrogenation: due to inappropriate stirrer design and high viscosity, the risk for lack of hydrogen in the liquid phase is high. A simulations of the hydrogen content in the liquid phase indicated that a complete saturation is achieved at high conversions, while the hydrogen concentration is below saturation in the beginning of the experiment. By increasing the stirring efficiency, the k_L-values can be improved and the role of external mass transfer resistance is suppressed. The hydrogen concentration in the liquid phase plays a crucially important role for the production: in the case of external mass transfer limitation of hydrogen, the isomerisation reaction, which does not need any hydrogen, is favoured (Fig. 3).

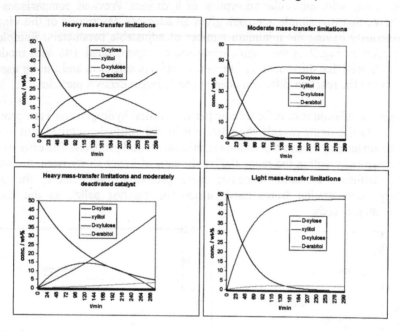

Figure 3. Simulation of hydrogenation of xylose.

Conclusions

We have presented a general reaction-diffusion model for porous catalyst particles in stirred semibatch reactors applied to three-phase processes. The model was solved numerically for small and large catalyst particles to elucidate the role of internal and external mass transfer limitations. The case studies (citral and sugar hydrogenation) revealed that both internal and external resistances can considerably affect the rate and selectivity of the process. In order to obtain the best possible performance of industrial reactors, it is necessary to use this kind of simulation approach, which helps to optimize the process parameters, such as temperature, hydrogen pressure, catalyst particle size and the stirring conditions.

Notation

a	catalyst activity factor
a_p	catalyst particle mass transfer area to reactor volume relation
a_{GL}	gas-liquid mass transfer area to reactor volume relation
A	mass transfer area
c	concentration
d	diameter
D_e	effective diffusion coefficient
k	rate constant,
k_L, k_G	mass transfer coefficient
K	gas-liquid equilibrium constant
n	molar amount
N	flux, mol/m^{2}s
p	partial pressure
P	total pressure
r	component reaction rate
r	radius, m
R	reaction rate
R	gas constant 8.3143 J/molK
s	catalyst particle shape factor
t	time, min
x	mole fraction
ε_p	porosity
ε	specific mixing power
ρ_p	Catalyst particle density
ν	stoichiometric coefficient, dimensionless
ν	kinematic viscosity
η	viscosity cP
Re	Reynolds number
Sc	Schmidt number
Sh	Sherwood number

Subscripts and superscripts

b	bulk property
G	gas phase
i	component index
j	reaction index
L	liquid phase
p	catalyst particle propery
*	equilibrium conditions

Acknowledgement
This work is part of the activities at the Åbo Akademi Process Chemistry Centre
within the Finnish Centre of Excellence Programme (2000-2011) by the Academy of
Finland

References

1. Thiele E.W., *Ind. Eng. Chem.*, 31, 916, (1939)
2. D.A.. Frank Kamenetskii, (1955), *Diffusion and Heat-transfer in Chemical Kinetics*, Princton Univ. Press, N.J., USA
3. Hindmarsh, A. C., 1983, *ODEPACK-A Systematized Collection of ODE-Solvers*, R. Stepleman *et al.*, Scientific Computing, IMACS/North Holland Publishing Company, p. 55-64
4. Aumo, J., Mikkola, J.-P., Bernechea, J., Salmi, T. and Murzin, D. Yu., Hydrogenation of Citral Over Ni on Monolith, *Int. J. Chem. Eng.*, (2005), Vol. 3, A25
5. Aumo, J., Oksanen, S., Mikkola, J.-P., Salmi, T. and Murzin, D. Yu., Novel Woven Active Carbon Fiber Catalyst in the Hydrogenation of Citral, *Cat. Today*, 102-103 (2005), 128-132
6. Aartun, I., Silberova, B., Venvik, H., Pfeifer, P., Görke, O., Schubert, K. and Holmen, A., Hydrogen Production from Propane in Rh-impregnated Metallic Microchannel Reactors and Aumina Foams, *Cat. Tod.*, (2005) 105, 469-478
7. A. Fick, Phil. Mag. (1855), 10, 30 and W.F. Smith, *Foundations of Materials Science and Eng.* 3rd ed., McGraw Hill (2004)
8. Murzin D. and Salmi T., *Catalytic Kinetics*, ISBN: 0-444-51605-0, 2005, Elsevier
9. Wilke, C.R., Chang, P., *Am. Inst. Chem. Engrs J.* 1, 264, 1955
10. Hayduk, W., Minhas, B.S., *Can. J. Chem. Eng.* (1982), 60, 295
11. Reid R.C., Prausnitz J.M., Poling B.E. (1988), *The Properties of Gases and Liquids* 5th ed., Mc graw Hill Book Company, New York
12. Fogg P.G.T and Gerrard W. (1991), *Solubility of Gases in Liquids*, Wiley, Chichester
13. Mikkola, J-P. Salmi, T., and Sjöholm, R., *J. Chem. Techn. and Biotechn.* (1999), 74, 655-662
14. A.I. Burschtein, A.A., Zcharikov and S.I., Temkin, *J. Phys.*, (1988), B 21,1907
15. Hajek, Jan; Murzin, Dmitry Yu. Liquid-Phase Hydrogenation of Cinnamaldehyde over a Ru-Sn Sol-Gel Catalyst. 1. Evaluation of Mass Transfer via a Combined Experimental/Theoretical Approach, *Industrial & Engineering Chemistry Research* (2004), 43(9), 2030-2038
16. Kuusisto, J., Mikkola, J-P., Sparv, M., Heikkilä, H., Perälä, R., Väyrynen, J., and Salmi, T., Hydrogenation of Lactose over Raney Nickel type Catalyst – Kinetics and Modeling, submitted 2005
17. Salmi, T., Murzin, D. Yu., Mikkola, J.-P., Wärnå, J., Mäki-Arvela, P., Toukoniitty, E., Toppinen, S., Advanced kinetic concepts and experimental methods for catalytic three-phase processes, *Indust. & Engin. Chem. Res.*, 43 (2004), 4540-4550

23.

The Effects of Various Catalyst Modifiers on the Hydrogenation of Fructose to Sorbitol and Mannitol

Daniel J. Ostgard, Virginie Duprez, Monika Berweiler, Stefan Röder and Thomas Tacke

Degussa AG, Rodenbacher Chaussee 4, 63457 Hanau, Germany

dan.ostgard@degussa.com

Abstract

Fructose hydrogenations have been performed with various sponge-type Ni and Cu catalysts and the reaction data have been correlated to the catalysts' properties. Cu is less active than Ni and it favors the production of mannitol over sorbitol by a ~2:1 ratio, while Ni generates them on a ~1:1 basis. Promoting Ni with Mo increased the rate of hydrogenation, and decreased the mannitol selectivity to 44.1%. The data show that the least active catalysts adsorb fructose weaker, have less zero order behavior and higher activation energies leading to higher mannitol selectivities from the preferred hydrogenation of the sparser less sterically hindered α-furanose with retention at the anomeric carbon. Highly active catalysts adsorb fructose stronger for a more competitive, but still not favored, adsorption of the more abundant β-furanose leading to more sorbitol. Depositing carbonaceous residues onto the catalyst prior to the reaction showed that smaller ensembles favor mannitol, thereby confirming its source to be the readily adsorbed α-furanose.

Introduction

Mannitol is widely used as a dusting, bulking and anticaking agent in the food and pharmaceutical industries due to its nonhygroscopicity (1). It is made by reducing invert sugar (1:1 glucose:fructose) over sponge-type Ni (2) to give a ~25%:~75% mannitol:sorbitol mix that is separated by fractional crystallization (3). Glucose gives almost exclusively sorbitol, and fructose generates roughly a 50:50 mannitol:sorbitol mix. Although mannitol's market is much smaller than sorbitol's, its price is 3.9 (1) to 4.4 (4) times higher meaning that an in-depth understanding of fructose hydrogenation could help to optimize this process for the marketplace.

Unlike most organic compounds, aqueous reducing sugars can form up to six different tautomers (α-pyranose, β-pyranose, α-furanose, β-furanose, acyclic and hydrated forms) (5). It is the relative concentrations, activities and adsorption strengths of these tautomers on the catalyst that determine the activity and selectivity of the sugar's hydrogenation (4,6,7). Figure 1 shows the mutarotation of fructose with the expected hydrogenation products when assuming retention at the anomeric carbon. The rate of mutarotation is much faster than the rate of hydrogenation for

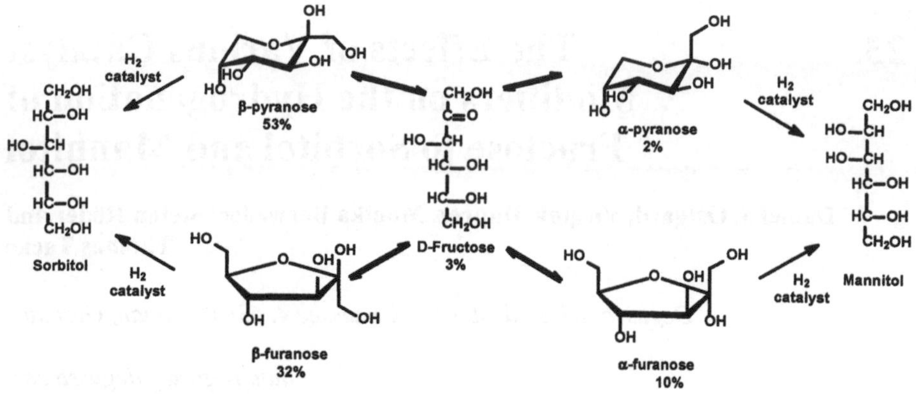

Figure 1. Fructose mutarotation (at 80°C) and hydrogenation (assuming retention).

fructose (4) meaning that a less abundant but more active species can influence the reaction's selectivity more than its equilibrium concentration would predict. The above mentioned 1:1 mannitol:sorbitol ratio found with Ni confirms this finding and suggests that not all of the tautomeric forms are equally active. This and other findings have led to the proposal of various fructose hydrogenation mechanisms. Ruddlesden et al. (7) have shown that fructose hydrogenation on Ni and Cu does not occur by either the open ketose or the 1,2-enediol forms. They proposed that it proceeds via the hydrogenolysis of the bond between the ring oxygen and the anomeric carbon. They also showed that a hydroxyl group must be on the anomeric carbon for this reaction to take place indicating that this is not a typical hydrogenolysis reaction. Although they admittedly could not identify the adsorbed active species, they did propose that ketal hydrogenolysis occurs with retention at the anomeric carbon. Makkee (6) and colleagues (8) studied the hydrogenation of fructose over Cu, Ni and other metals. They found that the reaction is first order to hydrogen and initially zero order to fructose with a shift to first order at concentrations lower than 0.3 M implying that the rate determining step at higher concentrations is the attack of hydrogen on the adsorbed fructose. They also stated that the pyranose rings are preferentially adsorbed over the furanose rings and that the furanose species were far more active, meaning that most of the adsorbed fructose reacts slowly. A similar trend was found for Ru/C by Heinen et al. (4), where they claimed that pyranose and furanose rings adsorb with equal strengths and that only furanose is active. While the products did not inhibit fructose adsorption on Ni or Cu (6), they were found to do so on Ru (4). Makkee et al. (6) went on to say that adsorbed fructose forms an ionized ketal species where the anomeric carbon to ring oxygen bond becomes less than single and the hydroxyl oxygen to anomeric carbon bond becomes more than single. This ionized adsorbed species is then thought to be attacked by a hydride-like species in a S_N2 fashion to give the polyol with inversion at the anomeric carbon. The term hydride-like species is not clear, because hydrogen dissociates on these metals to form chemisorbed hydrogen atoms that are active for hydrogenation (9). While these metals may form metallic hydrides (10), they will not form active surface hydrides (11) under the conditions used here (10,12). Moreover, these metals form the metallic type of hydrides and not the ionic saltlike hydrides (13) that one would need for an S_N2 reaction. Hence, the goal of this work was to clear up the inconsistencies of the currently proposed mechanisms.

Results and Discussion

Since the tautomeric species' concentrations at different temperatures can influence fructose hydrogenation, we reviewed the literature to find the data in Table 1 (5). Although the listed maximum temperature was 80°C and our reactions were at 100°C (for Ni and Mo doped Ni) and 110°C (for Cu and its doped varieties), these data suggest that our reaction solution had > 42% furanose forms, the overall β-to-α ratio was < 7 and the furanose β-to-α ratio was close to ~3. None of our catalysts were 100% selective for either sorbitol or mannitol and their mannitol selectivities ranged from 44.1% for the Mo doped Ni to 67.7% for the 80°C formaldehyde treated (FT) Cu catalyst. The catalyst's properties clearly affect the interactions of the tautomeric forms with its surface to give it a very specific activity and selectivity profile. The catalysts used here were chosen to observe the affects of coordinating, blocking and chemisorption assisting promoters as well as the type of active metal and its ensemble size. We used the deposition of formaldehyde to decrease the ensemble size of the active metal as described in the literature (14). It is known that formaldehyde disproportionates over Ni (110) as low as 95 K to give methanol and CO (15). This CO adsorbs strongly on the metal as a site-blocker and conceivably as an electronic modifier that can form bridged and linear species whose relative amounts depend on the surface coverage and temperature. This strongly held species doesn't desorb from Ni until 170°C (16,17,18,19), and this agrees with the temperature programmed oxidation (TPO) of this FT catalyst (14) that gave off a measured amount of CO_2 from 200 to 370°C. The TPO data also indicated that FT did not change the catalyst's other attributes (20). The Ni adsorbed CO is clearly stable enough to survive the conditions used here for fructose hydrogenation and this technique helped us to evaluate the influence of the adsorbed species on the reaction.

As seen in Figure 2, the Mo doped Ni is clearly the most active of the not FT catalysts followed closely by Ni, and all the Cu catalysts are more than an order of magnitude less active (albeit at a higher reaction temperature) than the Ni ones. Fe and Pt dopants do not change the activity of the Cu catalysts very much and this is in stark contrast to the overwhelmingly positive affects these promoters exert on the dehydrogenation of aminoalcohols (21). Increasing the level of FT steadily decreases the activity of the Ni and Cu catalysts in a predictable fashion and the activity of the lowest level FT Mo doped catalyst is very similar to the Ni catalyst without FT. The drop in activity is strongly dependent on treatment temperature for the Cu catalyst, where 80°C is more effective than 25°C for the decomposition of formaldehyde on this surface. As seen in Figure 3, the not FT Cu catalyst produced a ~2:1 mannitol:sorbitol ratio where promotion with Pt and Fe led to slightly less mannitol. The FT Cu catalysts show a very shallow maximum at 32.3 mmol formaldehyde per mol of Cu for the 25°C treatment, and the FT of Cu at 80°C led to the most mannitol selective surface. The FT also improved the mannitol selectivity

Fructose Hydrogenation

Table 1. The affect of temperature on the percentages of fructose's tautomers (5).

°C	% α-Pyr	% β-Pyr	% α-Fur	% β -Fur	% Open Keto	%Fur	Fur/Pyr	Fur β / α	Overall β / α
0	0	85	4	11	0	15	0.18	2.75	24.0
30	2	70	5	23	0	28	0.39	4.6	13.3
80	2	53	10	32	3	42	0.76	3.2	7.08

Pyr => Pyranose.   Fur => Furanose

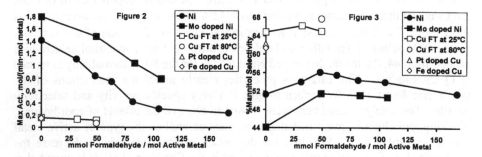

Figure 2. The effects of the FT level on the catalysts' activity.

Figure 3. The effects of the FT level on the catalysts' % mannitol selectivity.

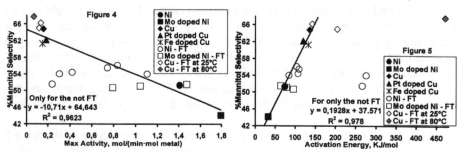

Figure 4. The relationship between maximum activity and % mannitol.

Figure 5. The relationship between activation energy and % mannitol.

of the Ni catalyst as it went through an optimum at ~48.2 mmol of formaldehyde per mol of Ni implying that there is an optimal ensemble size for this polyol. The initial FT of the Mo doped Ni had the largest mannitol selectivity increase of all of the treated catalysts and it was just as selective as the not FT Ni catalyst. This suggests that formaldehyde selectively decomposes on Mo to give a surface with a similar fructose activity and selectivity profile as one without Mo. There could also be an optimum FT level for the Mo doped Ni and this behavior mirrors that of the Ni catalysts. The trend seen in Figure 4 for the not FT catalysts shows that mannitol selectivity increases with lower activity and this is also true for the lower levels of

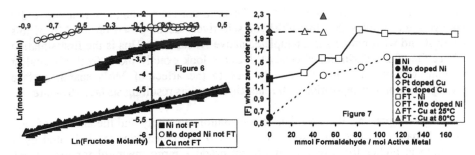

Figure 6. Dependence of fructose hydrogenation activity on its concentration.

Figure 7. The affect of FT on the concentration in which zero order stops.

FT, where steadily increasing its amount past the optimum leads to more notable deviations from this tendency. This is also the same relationship between mannitol selectivity and activation energy for the FT and not FT catalysts shown in Figure 5.

The relationships between fructose concentration and activity shown in Figure 6 display the different levels of zero order behavior for the Ni, Mo doped Ni and Cu catalysts. The Mo doped Ni catalyst exhibits zero order until 0.60 M fructose and then it becomes first order. The Ni catalyst without promoters follows a similar pattern where zero order stops earlier at 1.23 M and then the reaction becomes 1.38 order. Zero order activity in fructose is an indication that it is strongly adsorbed on the surface as the most abundant surface intermediate (MASI) and that the reaction rate is determined by the ability of hydrogen to competitively adsorb and react with this MASI. The Cu catalyst shows very little if any zero order behavior and it is first order with respect to fructose. The fructose concentration dependent activity of Cu shows that fructose adsorbs more weakly on Cu than Ni and this would favor the hydrogenation of the least sterically hindered tautomeric form for this mannitol selective catalyst. The data of Figure 7 show that the reaction rate changes sooner from zero order to a concentration dependent one for Ni with and without Mo as the level of FT is increased. It is interesting to note that the 49.2 FT Mo doped Ni catalyst behaves once more like the Ni one in that both of them stop their zero order dependence at ~1.26. As mentioned before, the Cu catalyst shows very little if any zero order behavior and this doesn't change by either FT or the addition of Fe or Pt. There is a small, but measurable, difference between the 25°C and the 80°C FT Cu catalyst and this is attributed to the more effective FT of the catalyst at 80°C. Figure 8 shows the affect of FT on the reaction order after the fructose zero order stops. Even though there are similar trends for all of the catalysts at the higher FT levels, each type of catalyst seems to react differently at the lower FT amounts. The lower levels of FT for the Ni catalyst result in a constant reaction order until 48.2 mmol formaldehyde per mol of Ni and from this point on, the reaction order drops as more formaldehyde is added. This is also the FT level with the highest mannitol selectivity for the Ni catalyst indicating that the reaction's selectivity is clearly influenced by this change of reaction order. The most surprising result was the drastic increase in the reaction order for the Mo doped Ni catalyst at the level of 49.2

mmol formaldehyde per mol of Ni. This extraordinary jump in reaction order confirms the concept that formaldehyde selectively decomposed on the Mo of this catalyst and what remains is a templated active Ni surface. This is the first situation where the 49.2 FT Mo doped catalyst does not look exactly like a fresh Ni catalyst without Mo or FT, and this may be due to the affect of Mo's spatial surface arrangement on the deposition of formaldehyde and its resulting surface structure.

Figure 9 presents the trend between mannitol selectivity and the amount of zero order behavior, where the lower levels of zero order behavior (as indicated by it stopping at higher fructose concentrations) produce more mannitol for the not FT catalysts. It is obvious that the catalysts with the lower FT levels also follow the least squares fit of Figure 9 and that the higher FT amounts produce catalysts that no longer fit this trend. The reaction order data demonstrate unambiguously that zero order activity resulting from the strong adsorption of fructose will enhance the hydrogenation of the most abundant active species in the reaction solution (β-furanose) to form more sorbitol, while less zero order activity created from the weak adsorption of fructose will favor the least sterically hindered species (α-furanose) to give more mannitol.

The kinetics of catalytic reactions are usually treated with the Langmuir-Hinshelwood model (6,9), however this is not as effective as the Michaelis-Menten equation for the description of catalytic surfaces where the desorption of the substrate can influence the reaction rate as portrayed in Figure 10. These data can be analyzed graphically with Lineweaver-Burk (LB) double reciprocal plots (please see Figure 11, Figure 12, Figure 13 and Figure 14) and the details of these analyses are described thoroughly in the literature (22). The changes in the slopes and the y-intercepts of these graphs can visually show a change from one type of surface behavior to another. These values can also be used to calculate the maximum activity (V_{max}) under zero order conditions without diffusional constraints and the Michaelis constant (K_m) as seen with the equations in Figure 12. K_m can also be described as the apparent dissociation constant of the adsorbed fructose (K_{-1} / K_1 of Figure 10). Although this is a bisubstrate reaction, the hydrogen pressure was kept constant and we will simply consider the effects of formaldehyde deposition (the source of the irreversibly adsorbed inhibitor) on the conversion of fructose in our interpretation of the data. The visual inspection of Figure 11 shows that the addition of low levels of formaldehyde (from 31.5 to 62.9 mmol formaldehyde per mol metal) results in an incrementally directly proportional drop in activity as the slopes of these lines gradually increase with higher FT levels. At these levels the residues of formaldehyde act as non-competitive inhibitors (i.e., they don't occupy the active sites) that don't impede adsorption but they do slow down the conversion of the adsorbed species into product via electronic and/or steric factors. There is a large difference in both the slopes and the y-intercepts of the plots as the FT level increases from 62.9 to 81.8 mmol of formaldehyde per mol of Ni and this correlates to both the departure from the optimal FT area for mannitol selectivity and the transformation from non-competitive to something similar to uncompetitive inhibition for the FT. Uncompetitive inhibition occurs when both the inhibitor and

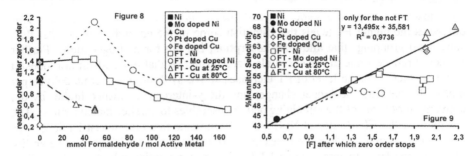

Figure 8. The effect of FT on the reaction order after zero order behavior.

Figure 9. The effect of zero order behavior on the % mannitol.

$$\text{Fructose}_{(aq)} \underset{k_{-1}}{\overset{k_1}{\rightleftharpoons}} \text{Fructose}_{(ads)} \xrightarrow{k_{Hyd.}} \text{Sorbitol or Mannitol}$$

Figure 10. Adsorption/desorption of fructose-catalyst species and its hydrogenation.

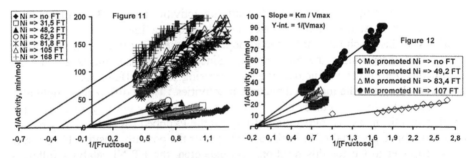

Figure 11. The Lineweaver-Burk double reciprocal plots of the FT Ni.

Figure 12. The Lineweaver-Burk double reciprocal plots of the FT Mo promoted Ni.

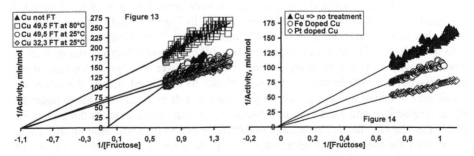

Figure 13. The Lineweaver-Burk double reciprocal plots of the FT Cu.

Figure 14. The Lineweaver-Burk double reciprocal plots of the promoted Cu.

the substrate occupy the same site and the resulting complex is inactive. This describes the higher FT levels that diminish some of the ensembles to smaller than subcritical groupings that allow for adsorption, but not for hydrogenation. Further FT could reduce some of the ensemble sizes further so that even adsorption doesn't occur on them. As the level of FT increases above 81.8 mmol of formaldehyde per mol of Ni the slope no longer changes and the y-intercept is pushed higher by the conversion of some of the remaining active ensembles to smaller inactive ones. The LB plots of the FT Mo doped Ni (Figure 12) indicate that the highest FT level was not enough to turn the deposited residues into uncompetitive inhibitors. The ability of Mo doped Ni to undergo higher FT levels while maintaining higher activity than Ni is due to the selective deposition of formaldehyde on Mo and the positive structural effects this has on these residues (*vide supra*). The LB plots of the FT Cu catalysts (Figure 13) show that the FT of this catalyst changes its surface by both non-competitive and uncompetitive inhibition and that FT at 80°C has the greatest effect. The addition of Pt to Cu increases the average activity of the adsorbed species more than Fe (reciprocal plots of the FT Cu Figure 14) but neither one seems to have much impact. Comparing the LB plots to the selectivity data indicates that optimally smaller ensembles favor mannitol formation for all of the catalysts and that there is an optimum ensemble size for mannitol formation that depends on the catalyst type.

Figure 15 compares the measured maximum activities to the LB calculated V_{max} for the catalysts studied here. If one mentally draws a line through the not FT catalysts of this graph (except for the Pt doped Cu), there is a linear correlation that shows that these values are comparable. Since these measurements are not all completely zero order in fructose, it isn't surprising to see differences between the calculated V_{max} and the measured maximum activities. These differences could also be associated with the accuracy of the LB plots. Surprisingly the V_{max} value is much greater than the measured maximum activity for the Pt doped Cu and this may be due to the untapped potential of Pt itself as a catalyst or the possibly positive H-spillover effect from Pt to Cu for this reaction. As expected, the FT Ni catalysts follow a linear drop in both activity values as the level of FT increases. A similar drop in both activities is also seen for the increasing levels of FT for the Mo doped Ni, whereas the measured activity is much greater than that predicted by the LB plot and this could be from the aforementioned effect that Mo has on the FT. The correlation of V_{max} to mannitol selectivity (Figure 16) is very similar to that of the measured maximum activity to mannitol selectivity shown in Figure 4 and that confirms the trend that more active catalysts produce more sorbitol. The trend between K_m and activity (Figure 17) shows, as expected, that higher dissociation constants lead to lower activity. It is interesting to note that the lower FT levels lead to lower activities but not to a decrease in the K_m. Hence as stipulated before, the lower FT levels do not inhibit adsorption, but they do lead to a drop in activity due to potential steric interactions and electronic influences. It isn't until one reaches the higher FT levels that the K_m also decreases, and this corresponds to the same amount of FT that is required for the mannitol selectivity to drop from its optimum level. Figure 18 shows that the mannitol selectivity increases as the K_m value does for the not FT catalysts, and this confirms the previously mentioned theory that a weaker fructose adsorption strength favors the formation of mannitol. Treating the catalysts with

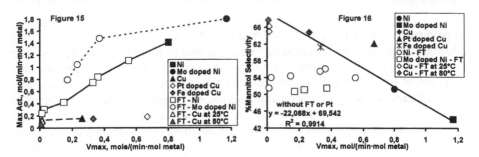

Figure 15. The correlation between the experimental max activity and V_{max}.

Figure 16. The correlation between %mannitol and V_{max}.

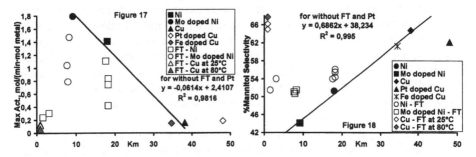

Figure 17. The correlation between the experimental max activity and K_m.

Figure 18. The correlation between %mannitol and K_m.

formaldehyde does push this relationship to higher mannitol levels for the same K_m values at the lower FT levels. However the higher FT amounts inhibit the adsorption so strongly that the K_m is actually reduced, thereby implying that if fructose does manage to adsorb on this heavily treated surface it will actually have a longer residence time than even on a clean surface. On possibility would be the creation of 3-dimentional carbonaceous residues on the surface that not only inhibit fructose adsorption, but it also inhibits its desorption via potential steric effects.

The results show us that mannitol formation is favored by a less active catalyst with a higher activation energy that has as little as possible zero order behavior, a weak fructose adsorption strength (i.e., with a higher K_m value) and an optimally somewhat smaller ensemble size. The production of sorbitol is logically favored by just the opposite trends. These tendencies allow us to formulate a fructose hydrogenation mechanism with the possible intermediates shown in Figure 19. It has been shown that one needs an abstractable hydrogen atom on the hydroxyl group attached to the anomeric carbon for this reaction to occur (7) and we propose that the adsorbed reaction intermediate must have both the ring oxygen and the hydroxyl group of the anomeric carbon chemisorbed onto the catalytic metal surface in order for this reaction to occur. It will then be possible for the catalyst to homolytically cleave the oxygen to hydrogen bond of the hydroxyl group on the anomeric carbon to produce an adsorbed cyclic alkoxy species and an adsorbed hydrogen atom. This adsorbed hydrogen atom will then be well positioned to either return to the adsorbed alkoxy moiety to reproduce the hydroxyl group or to attack the ring oxygen resulting in a shift of electron density to the adsorbed alkoxy group for the development of a π-bond thereby creating an acyclic adsorbed ketose. In other words, this is a surface induced adsorbed ketal to adsorbed hydroxy ketose equilibrium via a 1,3 intramolecular hydrogen atom shift. It has already been proven unambiguously with deuterium labeling that olefin isomerization can occur via the perpendicular adsorption of the π-bond onto the metal, the abstraction of the lower allylic hydrogen atom (due to its closer proximity to the surface) to give an adsorbed hydrogen atom and a π-allyl species and the attack of that adsorbed hydrogen atom to the bottom side of the opposite end of the adsorbed π-allyl to produce the isomer (23). This work also showed that there was no measurable mixing between the hydrogen involved in this isomerization and the nearby chemisorbed hydrogen. Although there are clearly differences between olefin isomerization and the proposed equilibrium of the adsorbed ketal and hydroxy ketose species, there is no mechanistic reason to exclude the occurrence of a 1,3 intramolecular hydrogen atom shift during this equilibrium. Even if neighboring chemisorbed hydrogen atoms do participate, this would not change the conclusions of the mechanism proposed here.

If the developing π-bond from this equilibrium is able to become perpendicular to the metal surface with very little or no strain and if the metal is able to stabilize this developing π-character, there will be a shift to a more ketoselike species that can be hydrogenated at a lower activation energy. If a particular tautomeric form is not able to accommodate this metal-stabilization of the developing π-bond, it will have less keto character and this will undergo a reaction that resembles more the hydrogenolysis of the ketal at considerably higher activation energy. The reaction will also have a higher activation energy and the adsorbed intermediate will have less keto-character if the catalytic metal is not very effective at stabilizing the π-bond character as it is being developed. Thus, the preferred reaction intermediate will be adsorbed by both the ring oxygen and the hydroxyl oxygen on the anomeric carbon (C2), where the bond between the ring oxygen and C2 is less than single and the bond between C2 and its hydroxyl oxygen is more than single. The partially adsorbed hydrogen atom from the C2 hydroxyl group will then reside somewhere between the C2 hydroxyl oxygen and the ring oxygen. A nearby chemisorbed hydrogen atom could then attack the anomeric carbon to give a "half-hydrogenated" adsorbed alkoxy species that can easily react with another chemisorbed hydrogen atom to afford the polyol. Figure 19 displays the possible flat and edge adsorbed intermediates proposed here and since the above mentioned developing π-bond will need to be perpendicular to the C1-C2-C3 plane, it is obvious that the orientation of this π-bond will be parallel to the metal surface for all of the edge adsorbed species. This lack of metal-stabilization for the edge adsorbed intermediates, means that they will not be the major contributors to the resulting product distribution under normal reaction conditions and this also refutes the preferred adsorbed surface intermediates proposed by Makkee et al (6).

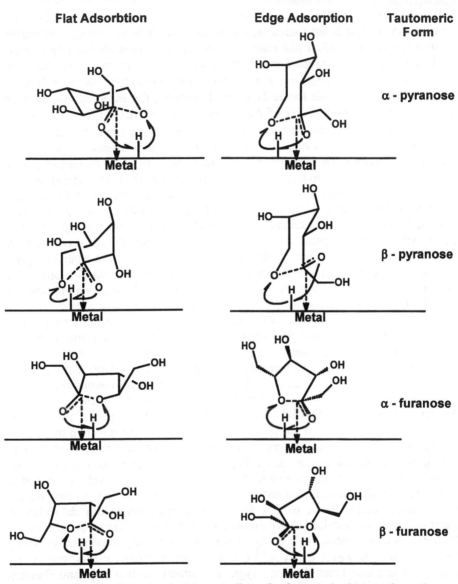

Figure 19. The possible adsorbed intermediates for fructose hydrogenation, where the more active flat adsorption is better positioned for the metal stabilization of the developing π-bond character between the anomeric carbon and its hydroxyl oxygen.

The only flat adsorbed β-pyranose ring that can accommodate the adsorption of both the ring oxygen and the hydroxyl oxygen on the anomeric carbon also suffers from a 1,3 steric interaction between the methylene hydroxyl group on the C2 position and the hydroxyl group on C4 resulting in it being less stable and less preferred to the remaining flat adsorbed species. The flat adsorbed α-pyranose ring fulfills the requirements of an active intermediate for this mechanism and its lower concentration (~2%) is not something that would limit it from making a noticeable contribution to the product distribution. However this 6-membered ring is more stable and less active than the 5-membered α-furanose and its lower contribution to the mannitol/sorbitol product mix should have a similar selectivity as the α-furanose flat adsorbed species. This leaves the α- and β-furanose flat adsorbed species and it is obvious due to the position of the C5 methylene hydroxyl group that the α-furanose ring will approach the surface with far less steric hindrance than the β-furanose ring. Thus a metal, such as Cu, that weakly adsorbs fructose will preferentially adsorb and hydrogenate the α-furanose over its β counterpart so that its contribution to the product mix will be far greater than its concentration in solution predicts. It is known that hydrogen dissociation is activated on Cu and non-activated on Ni (10, 24, 25) meaning that the Cu surface will contain less hydrogen than the Ni one at reaction conditions. This could explain why Cu is typically used under higher hydrogen pressures (10). This also increases the required residence time of the active intermediate on the Cu surface as it waits for the relatively sparse surface hydrogen to react with it and this too favors the α- over the β-furanose ring even more. Seeing how Cu is less effective than Ni for the dissociation of hydrogen, it could also be less efficacious for the abstraction of the hydrogen atom from the hydroxyl group on the anomeric carbon as needed for the above mentioned preferred reaction intermediate. Once this intermediate is formed, Cu may not be able to stabilize the developing π-bond character as well as Ni, resulting in a higher activation energy for this hydrogenation in comparison to Ni. During the reaction, the adsorbed hydrogen will add to the bottom of the adsorbed intermediate and this may result in retention or so-called inversion at the anomeric carbon. This will depend on how the C2-C3 bond rotates to stabilize the developing π-bond. If the C2-C3 bond of the adsorbed flat species slightly rotates in the direction that pushes the hydroxyl group on the anomeric carbon away from the ring, there will be stabilization of the forming π-bond with a minimum of strain and energy, and the addition of hydrogen to the bottom to this species will give a polyol with retention. This rotation will also maintain some degree of bonding between the ring oxygen and anomeric carbon bond so that the adsorbed intermediate can delocalize its electron density as described above for the preferred reaction intermediate. The C2-C3 bond could also rotate the hydroxyl group towards the ring to stabilize the newly formed π-bond, however that would not only require more rotation, energy and strain; it would also break the ring oxygen to anomeric carbon bond and eliminate any stabilization this may bring to the adsorbed intermediate. In agreement with the reaction data and as explained above, Cu clearly favors the hydrogenation of the least sterically hindered α-furanose with retention at the anomeric carbon to give a mannitol-rich product mixture and its less effective stabilization of the developing π-bond is the reason that it has a higher activation energy in comparison to Ni.

Because mutarotation (Figure 1) is faster than hydrogenation, it is possible for a less concentrated but more active tautomer to determine the overall selectivity of the reaction (4). This was the case for Cu where the α-furanose was ~6 times more active than the more concentrated β-furanose to generate the ~ 2:1 mannitol:sorbitol product mixture. The Ni catalyst produced a ~1:1 mannitol:sorbitol ratio indicating that the α-furanose is only ~3 times more active than the β analog. The β-furanose is more competitive, albeit still not preferred, on the Ni surface because its stronger adsorption strength increases the likelihood that this more concentrated tautomer will find an adsorption site and the higher hydrogen content of Ni will decrease its required residence time before it is hydrogenated. Doping the Ni catalyst with Mo increased both its activity and sorbitol selectivity. The literature suggests that molybdate species readily form complexes with the open keto and hydrated forms of fructose (26). This means that the surface molybdate species on the Ni catalyst assisted in the formation of reactive intermediates that have the most keto character of all the catalysts used here resulting in the highest hydrogenation activity and the lowest activation energy. This increased activity led to the mannitol:sorbitol ratio of ~0.8:1 meaning that α-furanose is now only ~2.4 times more active than β-furanose for this reaction. This is attributed once more to the improved competitive adsorption of the β-furanose on the Mo doped Ni surface.

As discussed earlier, the FT deposits carbonaceous residues onto the surface and splits it up into smaller ensembles. The smaller ensembles favor the adsorption of the least sterically hindered α-furanose leading to an increase in mannitol selectivity. This measurable increase in mannitol selectivity continues until the ensemble size becomes so small that it starts to inhibit adsorption and this leads to lower mannitol selectivity. This should force fructose to start adsorbing in more of a tilted fashion and then eventually to almost the edge adsorption modes shown in Figure 19. As this adsorption becomes more tilted, the steric interaction of the C5 methylene hydroxyl group should have a lower impact on the selectivity of the reaction and this could provide for the limited drop in the mannitol selectivity at the higher FT levels seen here. One would also expect that the flat adsorbed α-pyranose will stop adsorbing and reacting at lower FT levels than the flat adsorbed α-furanose and β-furanose rings leading to fewer sources for mannitol and contributing to the drop in mannitol selectivity at the higher FT levels. The fastest increase in mannitol selectivity was after the FT of the Mo doped Ni, were the formaldehyde selectively blocks the Mo promoters and changes the activity/mannitol selectivity profile of this catalyst to what would be expected for a fresh Ni catalyst without Mo or FT. The highest mannitol selectivity was obtained with the 80°C FT Cu catalyst where the combination of weak fructose adsorption and smaller ensemble sizes pushed the preference for α-furanose adsorption high enough to give 67.7% mannitol.

An increasingly tilted adsorption will also increase the ketal hydrogenolysis character of this reaction leading to the observed higher activation energies. It is generally thought that there is an optimal ensemble size with the right combination of atoms for hydrogenolysis on Ni (27). Thus it is interesting that the catalyst's activity doesn't decrease more as the reaction is forced further into a hydrogenolysis mode while the average ensemble size becomes smaller. The Ni catalyst used here typically has an average Ni crystal size of about 10 nm (28) and this should have been completely covered at the level of ~120 mmol formaldehyde per mole of active metal when one assumes a formaldehyde:surface metal ratio of 1:1. Obviously this is not the case, and this could indicate that there is 3-dimensional carbonaceous structure being created that leaves a certain amount of the surface clean for reactions.

In conclusion, fructose preferably adsorbs in a flat manner where the adsorption and hydrogenation of the least sterically hindered α-furanose is favored over the more abundant β-furanose. This adsorption involves both the ring oxygen and the hydroxyl oxygen on the anomeric carbon that gives up its hydrogen atom to form an adsorbed ketal to hydroxy ketose equilibrium via a 1,3 intramolecular hydrogen atom shift. This equilibrium intermediate could have more ketose hydrogenation or ketal hydrogenolysis behavior as determined by the metal's ability to stabilize the developing π-bond character where the shift towards more ketal hydrogenolysis leads to lower activity and higher activation energies. This adsorbed species is hydrogenated with retention at the anomeric carbon so that α-furanose produces mannitol and β-furanose gives sorbitol. All of our catalysts preferentially adsorbed and hydrogenated the sparser less sterically hindered α-furanose over the β-furanose fructose and this tendency was enhanced by weaker fructose adsorption, less first order behavior, a higher activation energy (more ketal hydrogenolysis behavior) and a lower hydrogen content. For these reasons, Cu was found to be less active and more mannitol selective than Ni. The addition of Mo to Ni not only enhanced the adsorption strength of fructose to increase the amount of adsorbed β-furanose, but it also stabilized the ketose form of the equilibrium to give the highest hydrogenation activity and sorbitol selectivity at the lowest activation energy measured here.

Experimental Section

The hydrogenation of 500 grams of a 40 wt.% aqueous fructose solution was carried out in a 1L steel autoclave stirred at 1015 rpm under 50 bars of hydrogen with either 2.4 wt.% of an activated Ni catalyst (a.k.a., sponge-type) at 100°C or 7.2 wt.% of an activated Cu catalyst at 110°C. The hydrogen uptake was monitored during the reaction and the samples of the product mixture were analyzed by HPLC. The catalysts used here were products of Degussa AG and their formaldehyde treatments were performed according to the patent literature (14).

References

1. N. Borgeson, B. Brunner and K. Sakota, Food Additives, A Specialty Chemicals / SRI International Report, Dec. 1999.
2. M. Makkee, A.P.G. Kieboom, H. van Bekkum, *Starch/Stärke*, **37**, 136-141 (1985).
3. A.A. Unver and T. Turkey, U.S. Patent 3,632,656 to Atlas Chemical Industries, Inc. (1972).
4. A. Heinen, J. Peters, and H. van Bekkum, *Carbohydr. Res.*, **328**, 449-457 (2000).
5. S.J. Angyal, *Advances in Carbohydrate Chemistry and Biochemistry*, **42** (1984) 15-68.

6. M. Makkee, A.P.G. Kieboom and H. van Bekkum, *Carbohydr. Res.*, **138**, 225-236 (1985).
7. J.F. Ruddlesden A. Stewart, D.J. Thompson and R. Whelan, *Faraday Discuss. Chem. Soc.*, **72**, 397-411(1981).
8. A.P.G. Kieboom, *Faraday Discuss. Chem. Soc.*, **72**, 397-411(1981).
9. N. Chang, S. Aldrett, M. Holtzapple, R. R. Davidson, *Chem. Eng. Sci.*, **55**, 5721-5732 (2000).
10. J.W. Geus, *Hydrogen Effects in Catalysis*, Zoltán Paál and P.G. Menon eds., Marcel Dekker, Inc., New York, 1988, p 85–115.
11. P. Fouilloux, *App. Catal.*, **8**, 1-42 (1983).
12. W. Palczewska, *Hydrogen Effects in Catalysis*, Zoltán Paál and P.G. Menon eds., Marcel Dekker, Inc., New York, 1988, p 373–395.
13. K.R. Christmann, *Hydrogen Effects in Catalysis*, Zoltán Paál and P.G. Menon eds., Marcel Dekker, Inc., New York, 1988, p 3–55.
14. D.J. Ostgard, R. Olindo, V. Duprez, M. Berweiler and S. Röder, European Patent 2004012862 to Degussa AG (2004).
15. L.J. Richter and W. Ho, *J. Chem. Phys*, **83**, 5, 2165-2169 (1985).
16. J.T. Yates and D.W. Goodman, *J.Chem. Phys.*, **73**, 10, 5371-5375 (1980).
17. A. Bandara, S. Katano, J. Kubota, K. Onda, A. Wada, K. Domen and C. Hirose, *Chem. Phys Lett.*, **290**, 261-267 (1998).
18. A. Bandara, S. Dobashi, J. Kubota, K. Onda, A. Wada, K. Domen, C. Hirose and S.S. Kano, *Surf. Sci.*, **387**, 312-319 (1997).
19. A. Bandara, S.S. Kano, K. Onda, S. Katano, J. Kubota, K. Domen, C. Hirose and A. Wada, *Bull. Chem. Soc. Jpn.*, **75**, 1125-1132 (2002).
20. D.J. Ostgard, M. Berweiler and S. Laporte, Degussa AG, unpublished results.
21. D.J. Ostgard, J. Sauer, A. Freund, M. Berweiler, M. Hopp, R. Vanheertum and W. Girke, US Patent 6794331 to Degussa AG (2002).
22. R.L. Augustine, *"Heterogeneous Catalysis for the Synthetic Chemist"*, Marcel Dekker, New York, 1996, 119–146.
23. G.V. Smith, *Chemical Industries* (Dekker), **68**, *(Catal. Org. React.)*, 1–14 (1996).
24. B. Hammer and J.K. Norskov, *Nature*, **376**, 238-240 (1995).
25. J.K. Norskov, *Cattech*, June, 14–15 (1998).
26. J.P. Sauvage, J.F. Vershére and S. Chapelle, *Carbohydr. Res.*, **286**, 67–76 (1996).
27. M. Che and C.O. Bennett, *Advances in Catalysis*, **36**, 55–172 (1989).
28. S. Knies, G. Miehe, M. Rettenmayr and D.J. Ostgard, *Z. Metallkd.*, **92**, 6, 596–599 (2001).

24. Cavitating Ultrasound Hydrogenation of Water-Soluble Olefins Employing Inert Dopants: Studies of Activity, Selectivity and Reaction Mechanisms

Robert S. Disselkamp, Sarah M. Chajkowski, Kelly R. Boyles, Todd R. Hart, and Charles H.F. Peden

Institute for Interfacial Catalysis, Pacific Northwest National Laboratory, Richland, WA 99352 USA

robert.disselkamp@pnl.gov

Abstract

Here we discuss results obtained as part of a three-year investigation at Pacific Northwest National Laboratory of ultrasound processing to effect selectivity and activity in the hydrogenation of water-soluble olefins on *trans*ition metal catalysts. We have shown previously that of the two regimes for ultrasound processing, high-power cavitating and high-power non-cavitating, only the former can effect product selectivity dramatically (>1000%) whereas the selectivity of the latter was comparable with those obtained in stirred/silent control experiments [R.S. Disselkamp, Y.-H. Chin, C.H.F. Peden, *J. Catal.*, **227**, 552 (2005)]. As a means of ensuring the benefits of cavitating ultrasound processing, we introduced the concept of employing *inert dopants* into the reacting solution. These *inert dopants* do not partake in solution chemistry but enable a more facile *trans*ition from high-power non-cavitating to cavitating conditions during sonication treatment. With cavitation processing conditions ensured, we discuss here results of isotopic H/D substitution for a variety of substrates and illustrate how such isotope dependent chemistries during substrate hydrogenation elucidate detailed mechanistic information about these reaction systems.

Introduction

Using ultrasound to enhance activity, and to a lesser extent to alter selectivity, in heterogeneous condensed phase reactions is well known [1-7], with the first paper on sonocatalysis having been published over 30 years ago [8]. In principle, there exists two separate domains for sonochemistry, these are non-cavitating and cavitating ultrasound regimes. For commercially available instruments, bath systems by virtue of their lower acoustic intensity are usually non-cavitating whereas probe (e.g., horn) systems can be either (high power) non-cavitating or cavitating. One objective of this paper is to contrast differences in a heterogeneous catalytic reaction for non-cavitating and cavitating ultrasound compared to a control (stirred and silent) system. Only through "doping" our solution were we able to initiate the rapid onset of cavitation during ultrasound treatment, and enable the chemical effects arising from cavitating conditions to be studied. Since not all reaction liquid mixtures readily cavitate, this technique decreases the power threshold for cavitation making it of general use.

The first system we investigate will be the *inert dopant* study just introduced. Three additional studies will also be discussed. The first examines *cis* to *trans* isomerization of *cis*-2-buten-1-ol and *cis*-2-penten-1-ol on Pd black. Isomerization is important in edible oil partial hydrogenation, where it is desirable to partially hydrogenate a C18 *cis* multiple-olefin without isomerizing its unconverted double bonds. We will see here that cavitating ultrasound processing reduces the amount of isomerization of these *cis* olefins to their *trans* form. The second system we discuss pertains to the use of H/D isotope substitution in hydrogenation to yield information about the mechanisms of reaction. Here the two substrates 3-buten-2-ol and 1,4-pentadien-3-ol were employed. We compared D-atom substitution to control experiments (e.g., H-atom processes) using D_2O instead of water for alcohol substitution and/or D_2 instead of H_2 during hydrogenation. The final investigation made is that of using *trans*ition state theory to explain the origin of olefin exchange.

Experimental Section

Materials and Apparatus

The experimental details of our approach have been given previously [9-12], therefore only the salient features are outlined here. The reagents were obtained commercially from the following vendors at the stated purities: *cis*-2-buten-1-ol, ChemSampCo, 97%; *cis*-2-penten-1-ol, ChemSampCo, 95%; 3-buten-2-ol and 1,4-pentadien-3-ol, Aldrich Chem. Co., 97+%). Unless otherwise stated, a substrate concentration of 100 mM was employed. All experiments except for the last (non-equilibrium system) employed a Pd-black catalyst (Aldrich, 99.9% purity metals basis) with a N_2 BET surface area of 42 m^2/g. The non-equilibrium investigation employed Raney Nickel (W.R. Grace Co. type 2800). Each system discussed below will present the substrate and catalyst mass employed. Deionized water (18 MΩ-cm) was used as the solvent. In some systems D_2O was used as the solvent (Aldrich Chem. Co., 99.9% atom purity). Hydrogenations were performed with H_2 or D_2 gas (A&L specialty gas, 99.99% purity) at a pressure of 6.5 atm (80 psig). All components used for the reaction apparatus are commercially available and have been described in detail previously [9,10,13]. Experiments were conducted isothermally at a temperature typically of 298 K with an uncertainty of ±2.5 K controlled using a water-bath circulation unit. All the studies used a 20 kHz Branson Ultrasonics digital model 450 sonifier II unit capable of delivering up to 420 W. A jacketed Branson Ultrasonics reactor that screwed onto the horn assembly was used to contain the reacting systems. Samples collected during an experiment were analyzed on a Hewlett-Packard GC/MS (5890 GC and 5972 MSD). Authentic standards were employed in the calibration of mass area counts when available. The column selected for separation was typically a 30 meter, 0.5 micron film, DB-5MS column. In a prior study we have shown that only synthetic chemistry occurs, namely that conservation of initial substrate concentration into well-defined products results, so that no sample "atomization" or other undesirable loss processes due to ultrasound treatment occurs [13].

Experimental Procedure

For all experiments, 50 mL of water and catalyst were added to the reaction cell. For ultrasound-assisted, as well as stirred (blank) experiments, the catalyst was reduced with hydrogen (80 psig) in water using non-cavitating ultrasound at an average power of 360 W (electrical; 90% amplitude) for at least one minute prior to reaction. The first sample for each experiment was taken for time equal to zero minutes and filtered through a 0.45 μm hydrophilic Millipore filter to remove catalyst powder into a capped vial for subsequent GC/MS analyses. During control (magnetically stirred/silent-MS) experiments, the reactor was connected to the sonifier probe assembly, purged and then pressurized with hydrogen (60 psig). Stirring was commenced and after a defined time interval, a (filtered) sample was collected. This process was repeated throughout the course of reaction. Conversely, during ultrasound-assisted experiments (US), the reactor was connected, purged and then pressurized with hydrogen (60 psig) as just described. The pressure chosen was optimal for cavitation. This is because the ability of a solution to achieve cavitation increases with greater acoustic power delivered which is favored to a degree at larger static pressures. However, a competing effect is that cavitation ability is hindered by large static pressures. Hence only a narrow pressure range ensured cavitation, often only realizable with inert dopants, occurred around 60 psig. The solution in the cell was irradiated with ultrasound and samples collected after similarly defined time intervals. In principle the hydrogenation pressure employed may affect chemical selectivity, however for the sake of experimental brevity such investigations were not made here. During non-cavitating sonication, an amplitude of 90% was employed, resulting in 360 ± 15 W delivered from the power supply. For cavitating ultrasound, a dopant was used (see below) to cause cavitation within 7 seconds of turning on the sonifier resulting in 190-280 W delivered to the convertor, here at >90% sonifier amplitude.

Results and Discussion

Inert Dopants

The hydrogenation of 3-buten-2-ol by our group [9] is an excellent choice for probing the chemical selectivity and activity of ultrasound processing because it undergoes competing reaction pathways yielding two products. For all experiments, 3.0 ± 0.2 mg of Pd-black catalyst were used. The concentration of substrate was 100 mM (33 M/g-catalyst based on initial concentrations). The four experimental regimes of stirred (without dopant), non-cavitating high-power ultrasound, stirred (with 1-pentanol dopant), and cavitating ultrasound (with 1-pentanol) were employed. In the latter experiments, 50 mM of 1-pentanol was used as an inert dopant to facilitate the onset of cavitation. The exact mechanism by which the onset of cavitation onset is enhanced through the use of these inert dopants is unclear, but likely involves solvent (or solution) modification such as by viscosity or surface tension effects. Experiments were performed at 298 K.

The full reaction process is summarized in scheme 1.

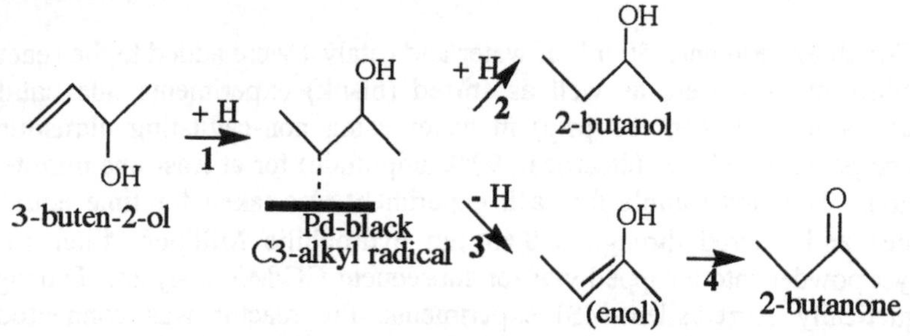

Scheme 1.

For example, in one pathway H-atom elimination reactions generate the enol intermediate, which eventually rearranges to 2-butanone. In a second competing reaction pathway, H-atom addition results in direct hydrogenation to the saturated alcohol 2-butanol.

A kinetic analyses of the data was performed by noting the pseudo-first order loss of substrate together with selectivity. This enabled a pseudo-first order kinetic description of the two pathways to be obtained. Table 1 lists the lifetimes of 2-butanone and 2-butanol production for the various experiments. Here the lifetimes refers to the inverse of the pseudo-first order reaction rate coefficients.

Table 1. Results of kinetics for stirred, non-cavitating ultrasound, and cavitating ultrasound experiments.

Experiment	t_1 (min.) (to 2-butanone)	t_2 (min.) (to 2-butanol)	t_{total} (min.)
Stirred	82.0	60.7	34.9
Non-cavitating ultrasound	19.8	11.1	7.1
Stirred (with 1-pentanol)	125.2	125.2	62.6
Cavitating ultrasound (with 1-pentanol)	8.9	0.71	0.66

Four primary conclusions can be put forth regarding the data of Table 1. First, a small difference in selectivity is seen for the non-cavitating ultrasound compared to the control experiment (obtained from the inverse of the ratio of lifetimes). For example, the ratio of 2-butanone to 2-butanol products for the stirred without 1-pentanol is 0.74 (equal to 60.7/82.0). Second, comparing these values to the selectivity estimate for the cavitating ultrasound run shows the latter having a ketone to alcohol ratio of 0.080. Therefore there is factor of ~12-fold enhanced alcohol selectivity for cavitating ultrasound compared to stirred with dopant. Third, the total activity can be examined via the substrate (total) lifetimes. It is seen that there is a 8.8-fold enhancement in reaction rate for the non-cavitating ultrasound compared to stirred with dopant and that the cavitating ultrasound activity enhancement is a factor of 10.8 greater than the non-cavitating run. Fourth, the greatest activity enhancement is seen for cavitation compared to stirred with dopant where a 14-fold enhancement in ketone formation rate is seen, but a 176-fold alcohol formation rate increase occurs.

It is concluded from this investigation that cavitating ultrasound can alter selectivity by ~1200% and activity by a factor of 94-fold. Thus cavitating ultrasound should be viewed as enabling novel chemical syntheses to be performed to complement traditional processing approaches. It should be emphasized that the use of *inert dopants* to ensure cavitation during ultrasound treatment, where it otherwise would not have occurred, is a key finding here.

Cis to trans Isomerization

We have studied the effect of cavitating ultrasound on the heterogeneous aqueous phase hydrogenation of *cis*-2-buten-1-ol (C4 olefin) and *cis*-2-penten-1-ol (C5 olefin) on Pd-black (1.5±0.1 mg catalyst) to form the *trans*-olefins (*trans*-2-buten-1-ol and *trans*-2-penten-1-ol) and saturated alcohols (1-butanol and 1-pentanol, respectively). These chemistries are illustrated in Scheme 2. A full analysis of this study has been recently presented [10].

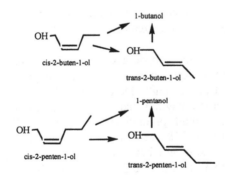

Scheme 2.

Again, silent (and magnetically stirred) experiments served as control experiments. In these studies, we employed 200 mM of the *inert dopant* 1-propanol in the reaction mixture to ensure the rapid onset of cavitation in the ultrasound-assisted reactions. The motivation for this study is to examine whether cavitating ultrasound can increase the [saturated alcohol/*trans* olefin] molar ratio during the course of the reaction. This could have practical application in that it may offer an alternative processing methodology for synthesizing healthier edible seed oils by reducing *trans*-fat content of, for example, C18 oils.

Results for this system are given in Figure 1 as the [saturated alcohol/*trans*] molar ratio versus extent of reaction (e.g., conversion). We have observed that cavitating ultrasound results in a [(saturated alcohol/*trans* olefin)$_{ultrasound}$/(saturated alcohol/*trans* olefin)$_{silent}$] ratio quantity greater than 2.0 at the reaction mid-point for both the C4 and C5 olefin systems. This indicates that ultrasound reduces *trans*-olefin production compared to the silent control experiment. Furthermore, there is an added 30% reduction for the C5 versus C4 olefin compounds again at reaction mid-point. We attribute differences in the ratio quantity as a moment of inertia effect. In principle, C5 olefins have a ~52% increase in moment of inertia about C2=C3 double bond relative to C4 olefins, thereby slowing isomerization. Since seed oils are C18 multiple *cis* olefins and have a moment of inertia even greater than our C5 olefin here, our study suggests that even a greater reduction in *trans*-olefin content may occur for partial hydrogenation of C18 seed oils.

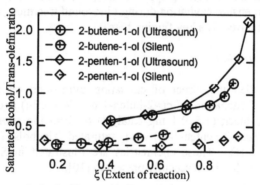

Figure 1. The saturated alcohol/*trans*-olefin ratio for the 2-buten-1-ol and 2-penten-1-ol systems are shown versus extent of reaction (ξ).

The following question can be asked in light of our study here: Can partial hydrogenation of edible seed oils as an industrial process ever be viable using cavitating ultrasound processing? Despite this study that suggests that cavitating ultrasound hydrogenation may have advantages over traditional processing methods, we can only make broad predictions extrapolating this work to the partial hydrogenation of actual C18 seed oils. For example, based on the cavitating ultrasound C5-olefin result here that ~50% hydrogenation occurred at ~80 seconds, it can be computed that it cost ~$0.35USD/lb for 50% hydrogenation (assuming electricity cost is $0.05USD/ kW-h and power usage was 280 W during treatment). Therefore, assuming similar costs for C18 seed oils, from an economic perspective the cost is likely not prohibitive. A second issue is the catalytic selectivity towards hydrogenation relative to isomerization (to the *trans*-olefin). Here, as an example, if we consider isomerization about the C9 position in linoleic acid there would be a roughly ~6.1-fold increase in moment of inertia for this particular double bond versus our C5-olefin [6-8], which from our result comparing our C4 and C5 olefins here suggests there would be an approximate 3.5-fold decrease in *trans*-olefin production in linoleic acid using cavitating ultrasound. For comparable amounts of hydrogenated (e.g., saturated alcohol) and *trans*-olefin content in our C5 experiment, this indicates that for a C18 seed oil there would still be ~18% [*trans*-olefin/hydrogenated species] content at reaction mid-point. This is likely still too much isomerization, thus improvements beyond the effects seen here would need to be seen in ongoing research of cavitating ultrasound chemical processing.

H/D Isotope Substitution

An H/D isotope effect study of the (H_2 versus D_2) hydrogenation of the aqueous substrates 3-buten-2-ol (3B2OL) and 1,4-pentadien-3-ol (14PD3OL) was performed using a Pd-black catalyst. Either H_2O or D_2O solvents were employed (for alcohol H/D isotope substitution). Despite the fact that this study suggests cavitation-only ultrasound processing is employed, again two experimental conditions of cavitating ultrasound (US) and stirred/silent (SS) methods were used. Products formed include 2-butanol and 2-butanone for the former, and 3-pentanol and 3-pentanone for the latter. The observed selectivity, and the pseudo-first order reaction rate coefficients (i.e., activity) to these products enabled a mechanistic interpretation for the various reaction conditions to be proposed.

Temperature was maintained at 298 K. All substrate concentrations were 100 mM and either 1.0, 1.5, or 5.0 mg ($\pm$0.2 mg) of Pd-black catalyst were utilized. For the 3B2OL substrate, 100 mM of 1-propanol inert dopant was employed. Conversely, due to the difficulty in cavitating solutions containing the di-olefin substrate (14PD3OL), 275 mM of 1-propanol was used in this case. The concentrations chosen, in part, were based on a cavitation time onset of less than ca. 6 seconds for each substrate. For 3B2OL and SS processing, it has been noted that the difference in activity between 50 mM of 1-pentanol inert dopant and no inert dopant was a 50% enhancement in activity and only a 10% difference in selectivity [9]. In our study here for 14PD3OL and CUS processing yielded no difference in selectivity when 1-propanol versus 50 mM 1-pentanol inert dopant was employed, therefore we will not consider the different amounts of *inert dopants* used significant.

Numerous experimental combinations of process conditions (SS or US), hydrogenation gas (H_2 or D_2), and solvent (H_2O or D_2O) have been explored. A summary of combinations we have chosen for study is presented in Table 2. In this table it is seen that the experiments are labeled B1-B7 for 3B2OL and P1-P6 for 14PD3OL. The second column lists the experimental conditions, whereas the third column lists the initial system concentration based on 100 mM of substrate and the amount of catalyst used. The penultimate column lists the final (extent of reaction > 95%) selectivity to ketone (2-butanone or 3-pentanone) and the final column lists the pseudo-first order substrate loss rate coefficient. The dataset contained in Table 2 enables numerous conclusions to be made regarding the reaction systems. The differences in initial concentrations (e.g., 67 versus 100 M/g-cat.) arise from the chosen convenience of having similar activities and therefore comparable reaction times.

Substrate	Conditions	M/g-cat.	% Sel. ketone	k (min.$^{-1}$)
3B2OL – B1	H$_2$/H$_2$O – US	67.	16.	2.7
3B2OL – B2	H$_2$/D$_2$O - US	67.	9.	4.3
3B2OL – B3	H$_2$/H$_2$O – SS	67.	42.	0.0038
3B2OL – B4	H$_2$/D$_2$O - SS	67.	32.	0.010
3B2OL – B5	D$_2$/H$_2$O – US	67.	17.	3.3
3B2OL – B6	D$_2$/H$_2$O – SS	67.	36.	0.012
3B2OL – B7	D$_2$/H$_2$O – SS	20.	68.	0.030
14PD3OL – P1	H$_2$/H$_2$O – US	100.	17.	3.5
14PD3OL – P2	H$_2$/D$_2$O - US	100.	16.	2.3
14PD3OL – P3	H$_2$/H$_2$O – SS	100.	28.	0.0068
14PD3OL – P4	D$_2$/H$_2$O – US	100.	18.	2.3
14PD3OL – P5	D$_2$/H$_2$O – SS	100.	32.	0.0070
14PD3OL – P6	D$_2$/H$_2$O – SS	20.	42.	0.042

Table 2. Summary of 3-buten-2-ol (3B2OL) and 1,4-pentadien-3-ol (14PD3OL) experiments are given. The abbreviations US and SS are defined as cavitating ultrasound and stirred/silent processing, respectively. The percent selectivity to final ketone plus saturated alcohol sum to 100%.

Based on the generally accepted model of olefin hydrogenation of Horiuti and Polayni [14], it is straightforward to postulate the reaction mechanism leading to formation of ketone and saturated alcohol for our two substrates. The proposed reaction mechanism for 3B2OL is given as scheme 1 in the Inert Dopant section. In scheme 1 it is seen that initially a single H-atom addition to the substrate occurs leading to the C3 alkyl radical intermediate. This intermediate, in turn, serves as a source of the two products (i.e., a branching point) in that either a second H-atom adds to the surface bound alkyl radical leading to 2-butanol (via reaction 2), or that a unimolecular C2 H-atom elimination occurs from the substrate to yield the enol that then tautomerizes to 2-butanone (via reaction 4). Based on scheme 1 it is straightforward to predict, at least qualitatively, the effect of H/D isotope substitution on the various reaction steps. For example, employing D$_2$O instead of water as a solvent will result in deuteration of the (substrate) alcohol throughout the reaction sequence. Thus, aside from general solvent effects (which are assumed to affect activity more than selectivity) for all reaction steps, this scheme predicts that primarily the enol tautomerization step will be most effected as the olefin/O-H versus olefin/O-D bond rearrangement processes can be expected to be most different. Similarly, when comparing H$_2$ versus D$_2$ hydrogenation, the selectivity is expected to be different since the H-addition step (via reaction 2) may differ between H-addition and D-addition, however the unimolecular H-elimination step (via reaction 3) is identical as it remains unchanged upon differing hydrogenation gases employed. So here it can be concluded that differences will arise solely from differences in H-addition (via reaction 2).

A somewhat more involved reaction scheme is required for 14PD3OL hydrogenation, and is illustrated in Scheme 3 below.

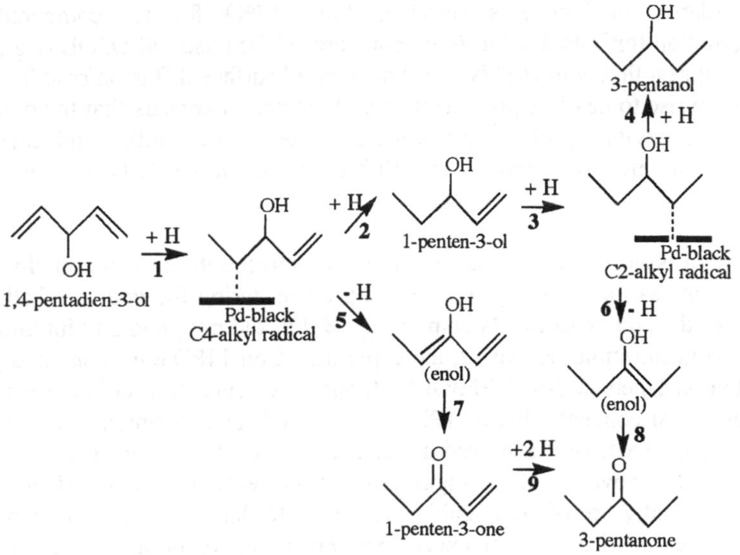

OH
3-pentanol
4 ↑ + H

OH
1-penten-3-ol
+ H
3 →

OH

OH
+ H
2

OH
1,4-pentadien-3-ol
+ H
1 →

Pd-black
C4-alkyl radical

Pd-black
C2-alkyl radical

6 ↓ - H

- H
5

OH
(enol)

OH
(enol)

8 ↓

↓ 7

O
1-penten-3-one

+2 H
9 →

O
3-pentanone

Scheme 3.

In Scheme 3 for 14PD3OL, since this substrate is a di-olefin, there is hydrogenation (H-atom addition via reaction **1**) to the first C4 alkyl radical intermediate that serves as the first branching point for either further hydrogenation (reaction **2**) or enol formation (reaction **5**). The latter enol, again, is expected to undergo tautomerization to the potentially stable intermediate 1-penten-3-one (reaction **7**) and from here eventually hydrogenate to 3-pentanone (reaction **9**). Alternatively, the intermediate 1-penten-3-ol can gain a second H-atom (reaction **3**) to a second C2 surface-bound alkyl radical and serve as a second branching point for either additional hydrogenation to the saturated alcohol (3-pentanol, reaction **4**), or another unimolecular H-elimination process (reaction **6**) to the enol and (via reaction **8**) form 3-pentanone. Thus because there are two surface bound alkyl radicals (i.e., two branching points), there are two opportunities for 3-pentanone formation, whereas scheme 1 above for 3B2OL only had one opportunity for ketone formation.

Some general observations and tentative conclusions can be made of the results of Table 2. First, the activity of the US compared to SS processing were at least 250-fold larger (e.g., compare for example B1/B3, B2/B4, or B5/B6). Second, variable catalyst loading experiments for stirred/silent D_2 hydrogenation processing (e.g., compare B6/B7 or P5/P6) indicated that mass *trans*fer of deuterium gas to the Pd-surface may have played a role such that higher catalyst loading reduced surface D-atom concentrations and reduced saturated alcohol formation (e.g., via perhaps reduced D-atom addition to surface alkyl radicals). Third, for US processing the ketone selectivities for experiments employing water compared to D_2O indicated that 3B2OL were twice as large (e.g., compare B1/B2), whereas for 14PD3OL they were comparable (e.g., compare P1/P2). According to scheme **1** of the previous section, this suggests, somewhat surprisingly, that for 3B2OL enol tautomerization to ketone is a slow, and possibly rate-controlling, process. Finally, again for US processing, the similarity in ketone selectivities (all ~17%) for H_2 compared to D_2, hydrogenation for both 3B2OL (e.g., compare B1/B5) and 14PD3OL (e.g., compare P1/P4) suggest that both H/D isotopes have rapid surface diffusion/reaction rates and hence give rise to nearly equal selectives. Restated, it appears that the control (e.g., thermal) or cavitating ultrasound kinetic processes are similar and each provide sufficient energies to surmount any differences in surface H/D diffusional energy barriers.

A final issue worthy of discussion is the concentration of intermediates during the reactions as measured by the sampling taken during the course of all reactions investigated. The experiments employing 3B2OL did not yield any intermediates of notable concentration. However, the experiments on 14PD3OL, conversely, formed a substantial 1-penten-3-ol (1PE3OL) intermediate concentration in the experiments. The maximum concentration of 1PE3OL occurred for experiment P4 (e.g., D_2/H_2O US) of Table 2 where it achieved a concentration of 42%. According to scheme **3** for 14PD3OL above, it can be suggested that the reaction velocity of the combined reactions **1** and **2** are of comparable magnitude to that of reaction **3**. Furthermore, data for the identical system except SS processing (experiment P5 of Table 2) yielded an intermediate 1PE3OL concentration of only 18% (data not shown). The reason for the observed differences between the 1PE3OL intermediate concentrations for the US (42%, experiment P4) compared to SS processing (18%, experiment P5) is not known, but may be related to the possibility that 1PE3OL for the SS system desorbs less from the Pd surface following reaction **2** compared to US processing. The cause of this phenomenon would be the enhanced energy gained by the 1PE3OL intermediate at the Pd surface arising from cavitating ultrasound processing.

Application of Transition State Theory to Traditional and Cavitating Ultrasound Olefin Exchange

An olefin exchange process for *cis*-2-buten-1-ol is illustrated in scheme **4**. Olefin exchange occurs when a simple *cis* olefin undergoes a D-atom addition yielding surface alkyl radicals that can eliminate either the substrate H-atom or newly gained D-atom in an activated unimolecular process to reform the olefin. The chemical mechanism we adopt is illustrated in Scheme 4, which clearly displays the addition-subtraction process, as summarized by Bond and Wells [15]. Upon deuteration of substrate to form the +D intermediate (I), two pathways can occur via the intermediate to form the *trans*-olefin, these include the C3(-H) or C3(-D) elimination reaction u_3 resulting in isomerization. These processes are essentially C-H or C-D bond activation processes, respectively. The process whereby *cis*-olefins isomerize to their *trans* form is generally understood as occurring through C-H activation of surface bound alkyl radical species. Here we present aqueous phase deuteration results of *cis*-2-buten-1-ol (100 mM) on Raney Nickel (125 mg).

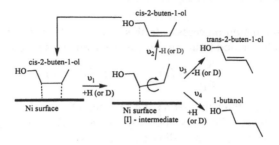

Scheme 4.

The results of our olefin exchange experiments are summarized in Figure 2. The deuterium number is given as a function of extent of reaction. The deuterium number is defined as the fractional amount of a given species that has gained +1 amu relative to the hydrogenation-only expected mass. The extent of reaction is simply the progress of the reaction where at c=0 all substrate exists, whereas at c=1 all product(s) exist. In our discussion here, substrate refers to the *cis*-olefin with zero deuterium number, and the (only) product is the saturated alcohol (e.g., 1-butanol) not of general interest here. A beneficial feature of our GC/MS analysis approach of the *cis* and *trans* olefins was the presence of a strong parent ion signal and weak parent-1 mass feature. Thus it was straightforward to quantify parent+1 olefin exchange. We did not observe parent+2, or larger, mass signals indicative of higher order exchange processes. It is seen that the substrate (*cis*) deuterium values are much less than those of the newly formed *trans*-olefins. This is expected as initially (at c=0) only H-containing *cis*-olefins are present. Significantly, the cavitating ultrasound *trans* values (*trans*-US) are much larger than the conventional processing values (*trans*-MS). This data will be analyzed further, after a model of the deuteration process is presented.

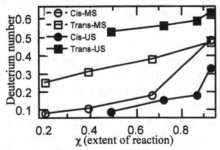

Figure 2. Deuterium number is shown versus extent of reaction for control (MS) as well as ultrasound (US) experiments.

Two limiting cases of this C3(-H) or C3(-D) elimination chemistry are possible. In one mechanism, rapid energy input into the C-H/D reaction coordinate suggests that an excess of energy, beyond the barrier height, can occur. Here one expects a nearly statistical distribution of reaction pathways. Thus, the deuterium number (deuterium exchange fraction) should be 0.5. At more moderate vibrational temperatures, we can rigorously apply *trans*ition-state theory to describe the u_3 C3(-H) and C3(-D) elimination processes. In this scenario, the ratio of rate coefficients for C-D and C-H elimination from the surface intermediate is given by [16],

$$\frac{k_D}{k_H} = (Q_{t,D}/Q_{t,H})\,(Q^\dagger_D / Q^\dagger_H)$$

(1)

where $Q_{t,D}$ and $Q_{t,H}$ are tunneling terms and $Q^\dagger_D$ and $Q^\dagger_H$ are the partition functions of the *trans*ition state. For a sufficiently large barrier to reaction, which is reasonable for our system, the tunneling correction is close to unity [16-18]. Here we approximate the $Q^\dagger_D$ and $Q^\dagger_H$ terms of equation (1) by the two bending modes of both C-C-H and C-C-D, respectively. In particular, we estimate these terms using data for ethane and deutero-ethane [19] where: $w_1(D)=503$, $w_2(D)=661$, $w_1(H)=671$, and $w_2(H)=849$ cm^{-1}, where w_1 is the out-of-plane CH bond bending mode. Using expressions for the vibrational partition function results in $k_D/k_H=0.39$, or a *trans*ition-state theory predicted deuterium number (via $k_D/(k_D+k_H)$) of 0.28 at 298 K.

These predictions can now be compared to experiment. Examining our dataset of Figure 2 leads to extrapolated c→0 *trans*-olefin deuterium numbers of 0.20 (MS) and 0.46 (US). The former value is close to that predicted by *trans*ition state theory just described. We propose, therefore, that the less energetic conventional (thermal) processing approach yields *trans*-olefin primarily through the C3-H/D elimination given by *trans*ition-state theory employing the reaction temperature of 298 K. Conversely, the results for the more energetic cavitating ultrasound system is also explainable using *trans*ition-state theory except that a much higher, non-equilibrium (compared to the thermal bath), vibrational temperature is required to model the deuterium number of 0.5. This is because *trans*ition-state theory predicts that the deuterium number asymptotically approaches 0.5 as the temperature approaches infinity. A temperature of 800 K results in a computed deuterium number of 0.40, thus this is likely the minimum vibrational temperature that describes the cavitating ultrasound system.

This assignment is supported, indirectly, by the measured activities of these experiments. For example, the activities were measured to be (in M/g-catalyst hour): 0.98 (MS-H$_2$); 0.61 (MS-D$_2$); 180 (US-H$_2$); and 190 (US-D$_2$). Hence, the ~250-fold greater activities of the ultrasound systems is consistent with the expected, more rapid, statistical C-H/D dissociation process as compared to the conventional (e.g., stirred/silent) mediated systems. Additional support for this model arises from a study of gas phase *cis*-2-butene isomerization to *trans*-2-butene [15] at 291 K. Here the c→0 extrapolated *trans* deuterium number of ~0.27 is supportive of C3-H/D elimination predicted by *trans*ition-state theory in this system at thermal equilibrium (e.g., vibrational temperature equal to *trans*lational temperature).

In the context of the accepted olefin isomerization mechanism, our results illustrate that *trans*ition-state theory can accurately model the competition between C-H and C-D activation for olefin exchange (isomerization) for the case of conventional catalytic processing. This is the case also for a catalytic process that includes cavitating ultrasound, although the model then requires a much higher vibrational temperature (at least ~800 K) in order to simulate the selectivity of the deuterium exchange process. Thus, cavitating ultrasound likely incorporates a high level of molecular vibrational excitation, suggesting that the vibrational temperature is not in equilibrium with the thermal (e.g., *trans*lational) temperature as the chemistry proceeds along a traditional reaction path. These results suggest that cavitating ultrasound chemistries can only be explained by a surface energy of >800 K which is much greater than the solution (e.g., bath) temperature of 300 K employed.

Acknowledgements

We would like to thank Dr. James F. White (PNNL) for fruitful discussions during the course of this project. SMC and KRB were supported from DOE Pre-service Teacher (PST) and Community College Initiative (CCI) Office of Fellowship programs for undergraduates, respectively, during the summer of 2005. This project was performed in the Environmental Molecular Sciences Laboratory (EMSL) and funded from a Laboratory Directed Research and Development (LDRD) grant administered by Pacific Northwest National Laboratory (PNNL). The EMSL is a national scientific user facility located at PNNL and supported by the U.S. DOE Office of Biological and Environmental Research. PNNL is operated by Battelle Memorial Institute for the U.S. Department of Energy.

References

1. J.L. Luche (ed.), *Synthetic Organic Sonochemistry* (Plenum Press) New York (1998).
2. T.J. Mason, *Sonochemistry* (Oxford University Press) Oxford, United Kingdom (2000).
3. K.S. Suslick, *Handbook of Heterogeneous Catalysis*, vol.3, G.Ertl, H. Knozinger, J. Weitkamp, eds., (Wiley-VCH), Weinheim, Germany (1997).
4. T.J. Mason, *Ultrasonics Sonochem.*, **10**, 175 (2003).
5. B. Torok, K. Balazsik, M. Torok, Gy. Szollosi, M. Bartok, *Ultrasonics Sonochem.*, **7**, 151 (2000).
6. M.M. Mdleleni, T. Hyeon, K.S. Suslick, *J. Am. Chem. Soc.*, **120**, 6189 (1998).
7. R.S. Disselkamp, T.R. Hart, A.M. Williams, J.F. White, C.H.F. Peden, *Ultrasonics Sonochem.*, **12**, 319 (2004).
8. J.W. Chen, J.A. Chang, G.V. Smith, *Chem. .Eng. Progress, Symp. Ser.*, **67**, 18 (1971).
9. R.S. Disselkamp, Ya-Huei Chin, C.H.F. Peden, *J. Catal.*, **227**, 552 (2004).
10. R.S. Disselkamp, K.M. Denslow, T.R. Hart, J.F. White, C.H.F. Peden, *Appl. Catal. A: General*, **288**, 62 (2005).
11. K.R. Boyles, S.M. Chajkowski, R.S. Disselkamp, C.H.F. Peden A Cavitating Ultrasound Study of the H/D Isotope Effect in the Hydrogenation of Aqueous

3-buten-2-ol and 1,4-pentadien-3-ol on Pd-black, *Ind. Eng. Chem. Res..*, submitted (2005).

12. R.S. Disselkamp, K.M. Denslow, T.R. Hart, C.H.F. Peden Non-equilibrium Effects in the Hydrogenation-mediated Isomerization Mechanism of Olefins during Cavitating Ultrasound Processing, , *Catal. Commun.*, in press (2006).

13. R.S. Disselkamp, K.M. Judd, T.R. Hart, C.H.F. Peden, G.J. Posakony, L.J. Bond, *J. Catal.*, **221**, 347 (2004).

14. J. Horiuti, M. Polayni, *Trans. Faraday Soc.*, **30**, 1164 (1934).

15. G.C. Bond, P.B. Wells, *Adv. Catal.*,, **15**, 91-226 (1964).

16. L. Melander, W.H. Saunders Jr., *Reaction Rates of Isotopic Molecules*, (Robert E. Krieger Publ. Co.) Malabar, Florida (1987).

17. R.P. Bell, *The Tunnel Effect in Chemistry* (Chapman and Hall Publ. Co.) London (1980).

18. N. Moiseyev, J. Rucker, M.H. Glickman, *J. Am. Chem. .Soc.*, **119**(17), 3853 (1997).

19. E.B. Wilson Jr., J.C. Decius and P.C. Cross, *Molecular Vibrations*, Dover Publ. Inc., New York, p.253 (1955).

25.   The Treatment of Activated Nickel Catalysts for the Selective Hydrogenation of Pynitrile

Daniel J. Ostgard,[1] Felix Roessler,[2] Reinhard Karge[2] and Thomas Tacke[1]

[1]Degussa AG, Exclusive Synthesis & Catalysts, Rodenbacher Chaussee 4
D-63457 Hanau, Germany

[2]DSM Nutritional Products Ltd, Postfach 3255, CH-4002 Basel, Switzerland

dan.ostgard@degussa.com

Abstract

Vitamin B1 can, among various other synthetic routes, be prepared via the key intermediate 5-cyano-pyrimidine (698-29-3) followed by the selective hydrogenation of its nitrile function with a base metal catalyst to the corresponding primary amine "Grewe diamine". The selectivity of this hydrogenation is typically controlled by the addition of ammonia and this usually increases the Grewe diamine selectivity up to 96.4%. However the current commercial conditions demand that one improves this selectivity to over 99%. It was found that one could improve the selectivity of the activated nickel catalyst by treating it with formaldehyde, carbon monoxide, acetone or acetaldehyde prior to its use. The most productive treatment was with formaldehyde leading to a Grewe diamine selectivity of 99.7% and carbon monoxide was the next best modifier giving a selectivity of 98.8% for this primary amine. The other modifiers were clearly less effective. The comparison of the various modifiers and their performances for nitriles other than pynitrile, have indicated that the efficacy of the modifier is dependent on its ability to restructure and decompose on the catalytic surface leading to the formation of more selective active sites.

Introduction

Vitamin B1 (a.k.a., thiamin and aneurin) serves a number of essential metabolic functions such as the conversion of fats and carbohydrates to energy as well as for the maintenance of healthy nerves and muscles (1,2). Vitamin B1 occurs naturally in small amounts in many foods, however it is not stable under conditions of heat and in the presence of alkali, oxygen and radiation (e.g., sunlight). Hence, a considerable amount of the naturally occurring vitamin B1 is destroyed during food preparation (2,3,4). Since the body has a high vitamin B1 turnover and is not able to store it very well, the use of vitamin B1 supplements (especially for diets high in carbohydrates and physically active people) is advisable for the avoidance of disorders such as Beriberi and the Wernicke-Korsakoff syndrome. Vitamin B1 is produced on an industrial scale in multiple step syntheses (1) meaning that even the smallest improvement in the yield of one of the steps can have a large impact on the overall economics of this product. This is particularly true for the mid-to-later steps such as the hydrogenation of pynitrile to the Grewe diamine that is processed further to vitamin B1 (5). The yield of this hydrogenation could be improved from 96.4% to 99.7% by the use of newly developed modified activated Ni catalysts (6). This work explores the reasons for this improvement and use of this new catalyst for the hydrogenation of other nitriles.

Results and Discussion

Table 1 shows the reaction conditions of the tests performed here and Table 2 describes the modifications of these catalysts with different modifying agents. The mechanism of pynitrile hydrogenation and the formation of the secondary amine is depicted in Figure 1 and Figure 2 displays the reaction data for the hydrogenation of pynitrile over activated Ni (B 113 W), activated Co (B 2112 Z) and Ni/SiO$_2$ catalysts with and without the formaldehyde treatment. The reaction results seem to be very similar for all of the catalysts before formaldehyde treatment and even though all of the catalysts are considerably better after this treatment, the activated Ni catalyst observed the greatest improvement and overall result.

It was found earlier that formaldehyde disproportionates over Ni (110) surfaces at low temperature (95 K) to form methanol and carbon monoxide (7). The freshly formed carbon monoxide adsorbs strongly on the metal to function as a site blocker and possibly as an electronic modifier of the nearby active sites. Studies on the chemisorption of CO over Ni have shown that both the bridged and linear species are formed and their amounts are dependent on the surface coverage and temperature, where desorption occurs around 170°C (8,9,10,11). This agrees with our temperature programmed oxidation data, where the formaldehyde treated catalyst formed a measured amount of CO$_2$ in the range of 200 to 370°C, while maintaining the other structural characteristics of the catalyst (12). Clearly the chemisorbed CO on Ni is strong enough to survive the conditions used here for both the treatments and the hydrogenations. Other reactions which could take place are: (a) the reverse reaction, i.e., the hydrogenation of CO with H$_2$ to CH$_2$O, however Newton and Dodge (13) found that Ni at even at 200°C is far more likely to decompose formaldehyde to CO and H$_2$ ($K_{decomposition}$ = 1800) than it is to hydrogenate CO to formaldehyde ($K_{hydrogenation}$ = 2.3 x 10^{-5}). Hence the adsorbed CO will not be reduced during the hydrogenation; (b) the *in situ* generated methanol could also readsorb to form chemisorbed CO and hydrogen, or even possibly form a hydrogen deficient polymeric species on the catalyst. The reaction data further suggest (as from the evolution of CO and formation of methanol) that the activated Ni catalyst utilize the formaldehyde treatment more effectively than the activated Co and the Ni/SiO$_2$. From this it can be expected that the activated Ni performs better than the activated Co in this respect, however it is surprising that the supported Ni/SiO$_2$ catalyst is not enhanced as much as the activated Co by this treatment. This unexpected difference may have to do with a possible interaction between the support and Ni to influence parameters such as the level of Ni reduction and the resulting Ni crystal size. Due to these results and the unpredictable price swings of the more expensive Co, it was decided to focus on the activated Ni catalyst for the rest of these studies.

Table 1. The reaction conditions for the various nitrile hydrogenations

Reaction	bar	°C	hours	grams nitrile	grams NH_3	ml of solvent[a]
Benzonitrile	40	100	3	100	0	2000
Benzonitrile	40	100	3	100	15	2000
Pynitrile	40	110	5	40	300	1475
Pyridine-3-carbonitrile	40	100	3	40	31	2000
Valeronitrile	20	120	5	80	15	2000

[a] the solvent was methanol in all cases

Table 2. The treatment conditions with the aqueous slurries of the modifiers.

Catalyst	Metal	Modifier	Mod. amt.[a]	°C	hours	Intended reaction
B 113 W	Ni	formaldehyde	0.839	25	1	Benzonitrile
B 113 W	Ni	carbon dioxide	1.487	25	1	Benzonitrile
B 113 W	Ni	benzaldehyde	0.591	25	1	Benzonitrile
B 113 W	Ni	formaldehyde	0.840	25	0.5	Pynitrile
B 2112 Z	Co	formaldehyde	0.840	25	1	Pynitrile
Ni/SiO$_2$	Ni	formaldehyde	0.831	25	1	Pynitrile
B 113 W	Ni	carbon monoxide	0.268	25	0,5	Pynitrile
B 113 W	Ni	acetone	2.296	25	1	Pynitrile
B 113 W	Ni	acetaldehyde	0.908	60	1	Pynitrile
B 113 W	Ni	formaldehyde	0.908	25	1	py-3-cn[b]
B 113 W	Ni	formaldehyde	3.090	25	1	Valeronitrile

[a] Amount of modifier in mmoles of modifier per gram of catalyst.
[b] Pyridine-3-carbonitrile

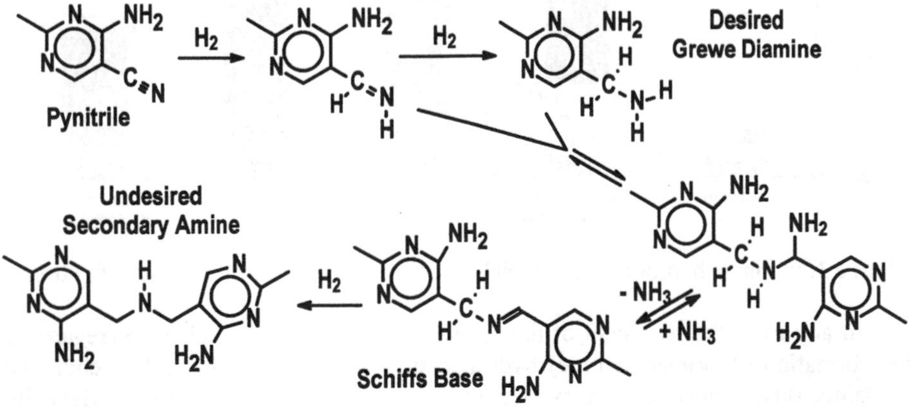

Figure 1. The hydrogenation mechanism of pynitrile and the possible formation of the corresponding secondary amine.

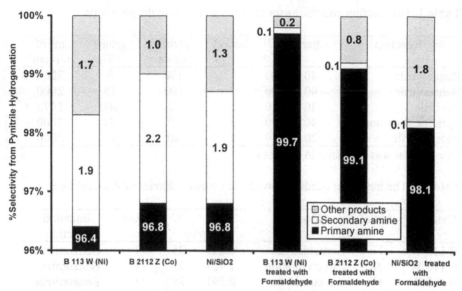

Figure 2. Pynitrile hydrogenation over fresh and treated activated Ni (B 113 W), activated Co (B 2112 Z) and Ni/SiO$_2$ catalysts in the presence of NH$_3$.

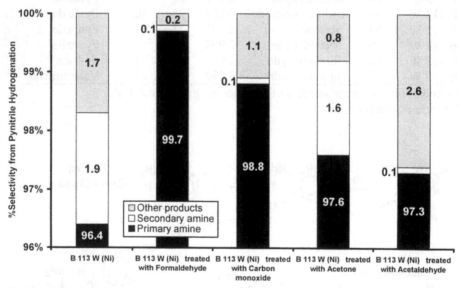

Figure 3. Pynitrile hydrogenation in NH$_3$ over B 113W after different modifications.

In any case, the treatment of the catalyst results in a more selective (regarding the formation of primary amine) hydrogenation of nitriles. This effect could be explained on a molecular basis by electronic (work function) and/or steric effects by strongly adsorbed species that function to partition the wide-open active surface of Ni and Co catalysts into smaller active sites composed of a controlled number of metal atoms. These newly formed ensembles would then prefer reactions like the hydrogenation of the nitrile to the primary amine that can take place on smaller sites, while disallowing the formation and adsorption of larger molecules such as the alpha amino secondary amine and the Schiff's base that are depicted in Figure 1.

Figure 3 displays the effectiveness of other modifiers for Grewe diamine formation and interestingly CO itself is not as effective as formaldehyde. Apparently the presence of hydrogen in the modifying molecule improves its ability to generate more selective sites, meaning that the formaldehyde generated carbonaceous layer on the catalyst is more than just strongly adsorbed CO. However the modification with CO is still better than that with acetone and acetaldehyde. While the acetone and acetaldehyde modified catalysts are similarly slightly more selective than the untreated catalyst, the acetone modified one produces far more secondary amines and less of the other side products than the acetaldehyde modified catalyst. Clearly acetaldehyde forms a different type of residue on the metal than acetone and that is to be expected by the structures of these molecules.

Figure 4 shows the benzonitrile hydrogenation results over fresh and formaldehyde treated catalysts both with and without ammonia. In the absence of ammonia, the formaldehyde treated catalyst is distinctly better than the fresh one and its primary amine selectivity is comparable to that of the fresh catalyst with ammonia. Logically, using the treated catalyst with ammonia gave the best results. Since less ammonia was used for the hydrogenation of benzonitrile than pynitrile, its primary amine selectivity was also lower, however the selectivity trends for both of these nitriles were the same for the formaldehyde treated catalyst. Treating the catalyst with carbon dioxide also increased its benzylamine selectivity to plainly show that carbon dioxide is not as inert in the presence of an activated Ni catalyst as it is sometimes thought to be. The data suggest that carbon dioxide can decompose on the catalyst to form an effective carbonaceous layer that enhances benzylamine formation. Benzaldehyde treatment also improved benzylamine selectivity in a similar fashion. It is known that benzaldehyde decarbonylates on Pd and Pt catalysts (14) and it is not surprising that it could also do so on Ni to generate *in-situ* CO for the effective modification of the catalyst.

Figure 5 displays the structures of the other nitriles studied here and Figure 6 shows their hydrogenation data. Due to the different amounts of ammonia (Table 1), it is difficult to directly compare the absolute primary amine selectivities of these tests, however the formaldehyde treatment clearly improves the primary amine selectivity more for aromatic nitriles than for aliphatic ones. Unlike the aliphatic nitriles, the aromatic ones adsorb stronger and longer on clean catalytic surfaces via both the aromatic ring and the nitrile moiety leading to higher levels of secondary amines. Hence weakening the adsorption of aromatic nitriles with smaller ensembles via the deposition of carbonaceous residues inhibits the formation and adsorption of larger molecules leading to lower secondary amine levels. Nonetheless, the valeronitrile data distinctly show that this treatment is also useful for improving the already high primary amine selectivity of aliphatic nitrile hydrogenation.

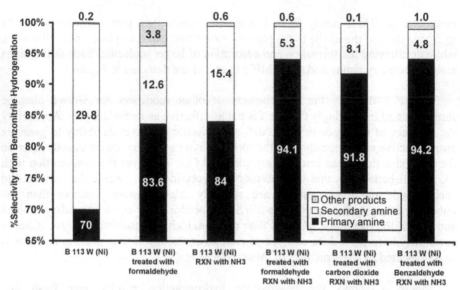

Figure 4. The effects of NH_3 and modifiers on Benzonitrile hydrogenation over Ni.

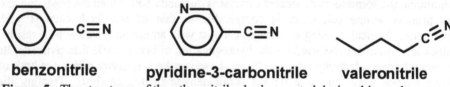

benzonitrile   **pyridine-3-carbonitrile**   **valeronitrile**

Figure 5. The structures of the other nitriles hydrogenated during this work.

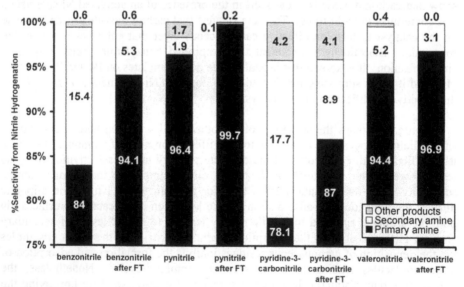

Figure 6. The hydrogenation of various nitriles with NH_3 over fresh and formaldehyde treated (FT) Ni catalysts (B 113 W). See Table 1 for details.

In conclusion, the treatment of activated Ni, activated Co and Ni/SiO$_2$ catalysts with formaldehyde improves the primary amine selectivity of nitrile hydrogenation. This treatment works better for the activated Ni catalyst than it did for the activated Co and these sponge-type catalysts respond better to it than the Ni/SiO$_2$. The best modifiers (e.g., formaldehyde) are those that decompose into strongly adsorbing compounds (e.g., CO) and the presence of hydrogen in the modifying compound is beneficial. The improved primary amine selectivity appears to be caused by electronic and/or steric effects such as the formation of smaller ensembles via the deposition of the appropriate carbonaceous residues on the catalyst. While this treatment improves the primary amine selectivity for aliphatic nitriles, it enhances the primary amine selectivity for aromatic nitriles even more so as seen by the 99.7% yield of Grewe diamine from pynitrile.

Experimental Section

The hydrogenation of pynitrile was carried out in an autoclave with either 5 grams of an activated Ni (B 113 W), 10 grams of an activated Co (B 2112 Z) or 8 grams of a Ni/SiO$_2$ catalyst under the conditions described in Table 1. Only in the case of the acetone treated activated Ni catalyst was the amount increased to 6 grams for the hydrogenation of pynitrile. Benzaldehyde, pyridine-3-carbonitrile and valeronitrile were hydrogenated with 5.7 to 6.0 grams, 5.3 grams and from 20 to 21.9 grams of an activated Ni catalyst respectively according to the conditions in Table 1. The modification treatments were carried out as the aqueous suspensions of the catalyst with the modifier under the conditions described in Table 2.

Acknowledgments

The authors thank DSM Nutritional Products Ltd. and Degussa AG for the permission to present this work.

References

1. http://www.dsm.com/en_US/html/dnp/prod_vit_b1.htm
2. http://www.chm.bris.ac.uk./webprojects2002/schnepp/viatminb1.html
3. V. Tanphaichitr, Thiamin, in *Modern Nutrition in Health and Disease*, 9th ed., William & Wilkins, Baltimore, 1999, 391-399.
4. M. Kimura, Y. Itokawa and M. Fujiwara, *J.Nutr.Sci.Vitaminol*, **36**, 17-24 (1990)
5. J.H. Hui, *Encyclopedia of Food Science and Technology*, John Wiley, New York, 1992.
6. Degischer, O.G. and Roessler, F. EP 1108469 to Hoffmann-La Roche AG (2000).
7. L.J. Richter and W. Ho, *J. Chem. Phys*, **83**, 5, 2165-2169 (1985).
8. J.T. Yates and D.W. Goodman, *J.Chem. Phys.*, **73**, 10, 5371-5375 (1980).
9. A. Bandara, S. Katano, J. Kubota, K. Onda, A. Wada, K. Domen and C. Hirose, *Chem. Phys Lett.*, **290**, 261-267 (1998).

10. A. Bandara, S. Dobashi, J. Kubota, K. Onda, A. Wada, K. Domen, C. Hirose and S.S. Kano, *Surf. Sci.*, **387**, 312-319 (1997).
11. A. Bandara, S.S. Kano, K. Onda, S. Katano, J. Kubota, K. Domen, C. Hirose and A. Wada, *Bull. Chem. Soc. Jpn.*, **75**, 1125-1132 (2002).
12. D. Ostgard, M. Berweiler and S. Laporte, Degussa AG, unpublished results.
13. R.H. Newton, B.F. Dodge, *J.Am. Chem. Soc.*, **55**, 4747-4759 (1933).
14. S. Ruozhi, D. Ostgard and G.V. Smith, Chemical Industries (Dekker), **47** (*Catal. Org. React.*), 337-349 (1992).

26. Deactivation of Sponge Nickel and Ru/C Catalysts in Lactose and Xylose Hydrogenations

Jyrki I. Kuusisto, Jyri-Pekka Mikkola and Tapio Salmi

Laboratory of Industrial Chemistry, Process Chemistry Centre,
Åbo Akademi University, Biskopsgatan 8, FIN-20500 Turku, Finland

Abstract

Catalyst deactivation during consecutive lactose and xylose hydrogenation batches over Mo promoted sponge nickel (Activated Metals) and Ru(5%)/C (Johnson Matthey) catalysts were studied. Deactivation over sponge nickel occurred faster than on Ru/C in both cases. Product selectivities were high (between 97 and 100%) over both catalysts. However, related to the amount of active metal on the catalyst, ruthenium had a substantially higher catalytic activity compared to nickel.

Introduction

The importance of catalyst stability is often underestimated not only in academia but also in many sectors of industry, notably in the fine chemicals industry, where high selectivities are the main objective (1). Catalyst deactivation is inevitable, but it can be retarded and some of its consequences avoided (2). Deactivation itself is a complex phenomenon. For instance, active sites might be poisoned by feed impurities, reactants, intermediates and products (3). Other causes of catalyst deactivation are particle sintering, metal and support leaching, attrition and deposition of inactive materials on the catalyst surface (4). Catalyst poisons are usually substances, whose interaction with the active surface sites is very strong and irreversible, whereas inhibitors generally weakly and reversibly adsorb on the catalyst surface. Selective poisons are sometimes used intentionally to adjust the selectivity of a particular reaction (2).

Catalyst deactivation often plays a central role in manufacturing of various alimentary products. Sugar alcohols, such as xylitol, sorbitol and lactitol, are industrially most commonly prepared by catalytic hydrogenation of corresponding sugar aldehydes over sponge nickel and ruthenium on carbon catalysts (5-10). However, catalyst deactivation may be severe under non-optimized process conditions.

Experimental Section

Aqueous lactose (40 wt-% in water) and xylose (50 wt-%) solutions were hydrogenated batchwise in a three-phase laboratory reactor (Parr Co.). Reactions with lactose were carried out at 120 °C and 5.0 MPa H_2. Xylose hydrogenations were performed at 110 °C and 5.0 MPa. The stirring rate was 1800 rpm in all of the experiments to operate at the kinetically controlled regime.

For lactose hydrogenations were used 5 wt-% (dry weight) sponge nickel and 2 wt-% (dry weight) Ru/C catalyst of lactose amount. In case of xylose, 2.5 wt-% (dry weight) sponge nickel and 1.5 wt-% (dry weight) Ru/C catalyst of xylose amount were used. Prior to the first hydrogenation batch, the Ru/C catalyst was reduced in the reactor under hydrogen flow at 200 °C for 2 h (1.0 MPa H_2, heating and cooling rate 5 °C/min). The reactor contents were analysed off-line with an HPLC, equipped with a Biorad Aminex HPX-87C carbohydrate column.

Results and discussion

Xylose hydrogenation gave xylitol as a main product (selectivity typically over 99 %) and arabinitol, xylulose and xylonic acid as by-products. In lactose hydrogenation, the main product was lactitol (selectivity typically between 97 and 99 %) and lactulitol, galactitol, sorbitol and lactobionic acid were obtained as by-products.

Studies about xylose hydrogenation to xylitol suggested that the main reasons for the sponge nickel deactivation were the decay of accessible active sites through the accumulation of organic species in the catalyst pores and by poisoning of the nickel surface. Deactivation during consecutive xylose hydrogenation batches over Ru/C catalyst was insignificant (Fig. 1A). Catalyst deactivation during consecutive lactose hydrogenation batches occurred faster than during the xylitol manufacture (Fig. 1B).

One of the problems encountered in the catalytic hydrogenation of aldose sugars is the formation of harmful by-products, such as aldonic acids. E.g. formation of D-gluconic acid is known to deactivate glucose hydrogenation catalysts by blocking the active sites (9,10). Furthermore, aldonic acid formation increases leaching of metals, since they are strong chelating agents. Under non-optimized conditions, lactobionic acid is formed as a by-product during lactose hydrogenation and xylonic acid in xylose hydrogenation.

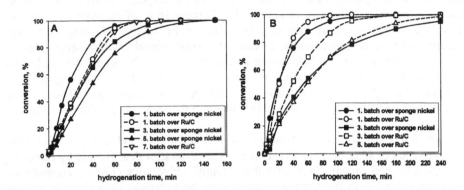

Figure 1. A. Consecutive xylose hydrogenation batches over 2.5 wt-% sponge nickel and 1.5 wt-% Ru/C catalyst. **B.** Catalyst deactivation during consecutive lactose hydrogenation batches over 5 wt-% sponge nickel and 2 wt-% Ru/C catalyst.

Influence of aldonic acids (lactobionic acid and xylonic acid) to reaction rate and catalyst deactivation were tested in consecutive hydrogenations of lactose and xylose over the nickel catalyst, with and without aldonic acid addition in the aqueous reactant solutions (Fig. 2A and 2B). Lactobionic acid seemed to retard more lactose hydrogenation than xylonic acid influenced xylose hydrogenation. Moreover, the abilities of different sugar aldehydes to get dehydrogenated varies (11). In general, sugars containing glucose moiety (such as glucose, lactose and maltose) are dehydrogenated easier than xylose.

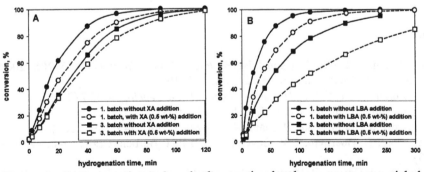

Figure 2. A. Consecutive xylose hydrogenation batches over sponge nickel catalyst (XA=xylonic acid). **B.** Influence of lactobionic acid (LBA) on lactose hydrogenation rate.

Moreover, in consecutive lactose hydrogenation batches, recycled sponge nickel catalyst was able to adsorb substantially less hydrogen compared to a fresh catalyst according to our hydrogen temperature programmed desorption (TPD) measurements. Also, this indicates poisoning of active sites. Regeneration of catalysts poisoned by strongly adsorbed acidic species may be achieved by catalyst washing in a basic medium, such as caustic solution. In case of sponge nickel catalyst deactivated by lactobionic acid, we were able to desorb by alkali wash 2 wt-% lactobionic acid of the total catalyst amount, thus returning the catalyst activity almost to the original level. However, too high pH during alkali wash of sponge nickel catalyst should be avoided, since at higher alkali concentrations, aluminium and molybdenum leaching increases.

An integrated ultrasound treatment of the sponge nickel catalyst during the hydrogenation of sugar species has shown to retard catalyst deactivation by removing strongly adsorbed organic impurities that occupy active sites (12,13). Furthermore, hydrogen solubility in the reaction mixture may be improved by adding an organic solvent into the aqueous sugar solution and thus suppressing the formation of aldonic acids and retarding catalyst deactivation. For instance, xylose hydrogenation over sponge nickel catalyst proceeded much faster in ethanol-water solutions and catalyst deactivation was retarded compared to hydrogenations in pure water (14). However, various sugar species have very limited solubilities in organic solvents, which limits the use of this method.

Acknowledgements

This work is part of the activities at the Åbo Akademi Process Chemistry Centre within the Finnish Centre of Excellence Programme (2000-2011) by the Academy of Finland. Financial support from the National Technology Agency (Tekes), Danisco Sweeteners and Swedish Academy of Engineering Sciences in Finland is gratefully acknowledged.

References

1. D. Yu. Murzin, E. Toukoniitty and J. Hájek, *React. Kinet. Catal. Lett.,* **83**, 205 (2004).
2. P. Forzatti and L. Lietti, *Catalysis Today,* **52**, 165 (1999).
3. D. Yu. Murzin and T. Salmi, *Trends in Chemical Engineering,* **8**, 137 (2003).
4. M. Besson and P. Gallezot, *Catal. Tod.,* **81**, 547 (2003).
5. B. Kusserow, S. Schimpf and P. Claus, *Adv. Synth. Catal.,* **345**, 289 (2003).
6. K. van Gorp, E. Boerman, C.V. Cavenaghi and P.H. Berben, *Catal. Tod.,* **52**, 349 (1999).
7. P. Gallezot, P.J. Cerino, B. Blanc, G. Flèche and P. Fuertes, *J. Catal.,* **146**, 93 (1994).
8. J.-P. Mikkola, T. Salmi and R. Sjöholm, *J. Chem. Technol. Biotechnol.,* **74**, 655 (1999).

9. B.W. Hoffer, E. Crezee, F. Devred, P.R.M. Mooijman, W.G. Sloof, P.J. Kooyman, A. D. van Langeveld, F. Kapteijn and J.A. Moulijn, *Appl. Catal. A*, **253**, 437 (2003).
10. B. Arena, *Appl. Catal. A*, **87**, 219 (1992).
11. G. de Wit, J.J. de Vlieger, A.C. Kock-van Dalen, R. Heus, R. Laroy, A.J. van Hengstum, A.P.G. Kieboom and H. van Bekkum, *Carbohydrate Research*, **91**, 125 (1981).
12. B. Toukoniitty, J. Kuusisto, J.-P. Mikkola, T. Salmi and D. Yu. Murzin, *Indus.& Engin. Chem. Res.*, **44**, 9370 (2005).
13. J.-P. Mikkola and T. Salmi, *Chem.Eng.Sci.*, **54**, (10), 1583 (1999).
14. J.-P. Mikkola, T. Salmi and R. Sjöholm, *J. Chem. Technol. Biotechnol.*, **76**, 90 (2001).

27. Selectivity Control in 1-Phenyl-1-Propyne Hydrogenation: Effect of Modifiers

S. David Jackson and Ron R. Spence

WestCHEM, Department of Chemistry, The University, Glasgow, G12 8QQ, Scotland

sdj@chem.gla.ac.uk

Abstract

The use of modifiers in controlling the selectivity in liquid phase alkyne hydrogenation has been studied. Modifiers that are more strongly bound than the intermediate alkene inhibit hydrogenation and isomerization giving high stereo- and chemo-selectivity. One modifier increased the rate of alkyne hydrogenation

Introduction

Control of selectivity in alkyne hydrogenation is a constant industrial problem. In large scale acetylene hydrogenation plants carbon monoxide is often used as a modifier to enhance the selectivity to ethylene (1, 2, 3). This option is not well suited to liquid phase hydrogenations. However the basic concept where a more strongly bonding agent is added to the feedstream is one that can be developed for liquid phase systems. Ideally the system should achieve 100 % conversion of the alkyne with no conversion of the alkene. Typically alkene selectivity is high until most of the alkyne has reacted. This is usually considered to be due to the stronger adsorption of the alkyne inhibiting re-adsorption of the alkene, so that only when there is insufficient alkyne to maintain high coverage does alkene hydrogenation begin in earnest.

In this paper we will report on using a series of modifiers to enhance selectivity during 1-phenyl-1-propyne over a Pd/alumina catalyst. The modifiers, trans-cinnamaldehyde, trans-cinnamonitrile, 3-phenylpropionitrile, and 3-phenylpropylamine, were chosen to have a functionality that potentially could adsorb more strongly than an alkene and to be unreactive under the reaction conditions.

Results and Discussion

The reaction profile of 1-phenyl-1-propyne over Pd/alumina is shown in Figure 1. 1-Phenyl-1-propyne was hydrogenated to cis-β-methylstyrene and phenylpropane. Only trace levels of trans-β-methylstyrene were detected and further hydrogenation of cis-β-methylstyrene to phenylpropane occurred once the majority of the 1-pheny-1-propyne had reacted.

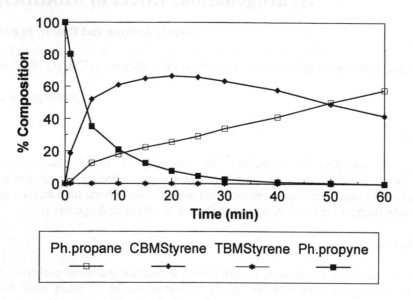

Figure 1. Reaction profile for 1-phenyl-1-propyne hydrogenation.

The reaction of 1-phenyl-1-propyne (1PP) was then studied after modifying the catalyst with trans-cinnamaldehyde (TCA), trans-cinnamonitrile (TCN), 3-phenylpropionitrile (3PPN), and 3-phenylpropylamine (3PPA). The first-order rate constant calculated for the loss of 1-phenyl-1-propyne in each of the systems is reported in Table 1. All the modifiers were unreactive under the conditions used.

Table 1. First-order rate constants (min⁻¹) for the hydrogenation of 1-phenyl-1-propyne in the presence of modifiers.

NM, 1PP[a]	TCA[b]	TCN[c]	3PPN[d]	3PPA[e]
0.10	0.03	0.07	0.13	0.09

[a] NM, not modified, 1PP, 1-phenyl-1-propyne.
[b] TCA, trans-cinnamaldehyde.
[c] TCN, trans-cinnamonitrile.
[d] 3PPN, 3-phenylpropionitrile.
[e] 3PPA, 3-phenylpropylamine

Rather surprisingly the 3PPN modifier increased the rate of 1PP hydrogenation. This type of effect has been observed before with unsaturated systems (4, 5) and has been related to enhancing surface hydrogen transfer due to changes in retained species (5).

The alkene selectivity as a function of conversion is shown in Figure 2.

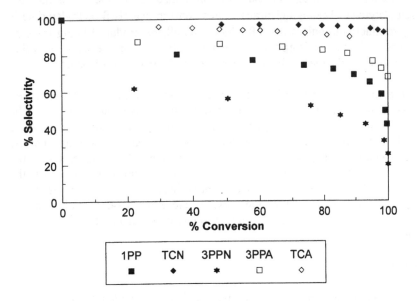

Figure 2. Plot of conversion Vs selectivity

The average cis:trans ratio for methylstyrene was determined and is reported in Table 2.

Table 2. Average cis:trans ratio of alkene intermediate in 1PP hydrogenation.

1PP	TCA	TCN	3PPN	3PPA
519	1164	1031	19	905

It is clear from Figure 2 that the modifiers have an effect on the selectivity to the alkene. TCN and TCA modified systems maintain alkene selectivity at ≥ 90 % over the full range of 1PP conversion. 3PPA modified system shows an enhanced selectivity but as conversion increases to 100 % there is a slight drop off. These three modified systems all show higher alkene selectivity than the unmodified 1PP, which has a declining selectivity with conversion, but lies mainly between 80 – 60 %. The TCA, TCN, 3PPA modifiers also affect the cis:trans ratio of the methylstyrene. Very high cis:trans ratios are observed in these systems. These results show that the TCA, TCN, and 3PPA modifiers do inhibit re-adsorption of the alkene. When there is little or no re-adsorption, the production of phenylpropane is inhibited. At 55 % conversion, 1PP hydrogenation modified with TCA produces only 13 % of the amount of phenyl-propane that is produced in the absence of the modifier. At 100 % conversion of 1PP a similar reduction is seen when TCN is the modifier. This inhibition of alkene re-adsorption is also observable in the cis:trans ratio of methylstyrene. The high cis:trans values confirm that cis-β-methylstyrene is the primary product with only around 0.1 % of the alkene produced being trans-β-methyl-styrene. When re-adsorption and isomerisation occur the cis:trans value drops to 19, reflecting the thermodynamic drive to produce the trans isomer.

Therefore the unsaturated aldehyde, the unsaturated nitrile, and the saturated amine all bond more strongly than methylstyrene to the palladium surface. The saturated amine has adsorption energetics that are close to those of methylstyrene showing only a limited effect. The saturated nitrile however enhances not just the alkyne hydrogenation but also the alkene hydrogenation. This suggests that the saturated nitrile affects the rate determining hydrogen transfer aspect of the hydrogenation. Rate enhancement effects have been previously observed in competitive reaction systems containing alkynes and alkenes (4, 5) and a similar conclusion drawn. The exact mode of operation is however unclear and requires further investigation.

Experimental Section

The catalyst used throughout this study was a 1% w/w palladium on alumina supplied by Johnson Matthey. The support consisted of θ-alumina trilobes (S.A. ~100 m^2g^{-1}) and the catalyst was sized to <250μ for all catalytic studies. The reactants and modifiers (all Aldrich >99 %) were used without further purification. No significant impurities were detected by GC. The gases (BOC, >99.99 %) were used as received.

The reaction was carried out in a 0.5 L Buchi stirred autoclave equipped with an oil jacket and a hydrogen-on-demand delivery system. 0.05 g of catalyst was added to 278 mL of degassed solvent, propan-2-ol. Reduction of the catalyst was performed *in situ* by sparging the solution with H_2 (300 cm^3min^{-1}) for 30 minutes at 343 K at a stirring speed of 300 rpm. After reduction, the autoclave was adjusted to 313 K under a nitrogen atmosphere and 1 mL of modifier was added in 10 mL of degassed propan-2-ol. The system was pressurized to 3 bar hydrogen and the stirrer set to 1000 rpm. After 1 h, 1 mL of 1-phenyl-1-propyne was added to the reactor in 10 mL of degassed solvent. Samples were taken at defined time intervals and analyzed by GC. Specific reactions were repeated at different stirrer speeds and equivalent rates and selectivities were observed indicating and absence of mass transfer within the system.

Acknowledgements

The authors would like to acknowledge the ATHENA project, which is funded by the Engineering & Physical Sciences Research Council (EPSRC) of the U.K. and Johnson Matthey plc.

References

1. Li Zon Gva and Kim En Kho, *Kinet. Katal.*, **29**, 381 (1988).
2. A. N. R. Bos and K. R. Westerterp, *Chem. Eng. Process.*, **32**, 1 (1993).
3. F. King, S. D. Jackson and F. E. Hancock, Chemical Industries (Dekker), **68**, (*Catal. Org. React.*), 53-64 (1996).
4. C. A. Hamilton, S. D. Jackson, G. J. Kelly, R. R. Spence and D. de Bruin, *Appl. Catalysis A*, **237**, 201 (2002)
5. S. D. Jackson and G. J. Kelly, *Curr. Top. Catal.* **1**, 47 (1997).

28. Hydrogenation and Isomerization Reactions of Olefinic Alcohols Catalyzed in Homogeneous Phase by Rh(I) Complexes

Maria G. Musolino, Giuseppe Apa, Andrea Donato and Rosario Pietropaolo

Department of Mechanics and Materials, Faculty of Engineering, University of Reggio Calabria, Loc. Feo di Vito, I-89060 Reggio Calabria, Italy

pietropaolo@ing.unirc.it

Abstract

Homogeneous hydrogenation and isomerization reactions of α,β-unsaturated alcohols have been investigated in ethanol by using tris(triphenylphosphine) chlororhodium(I), $RhCl(PPh_3)_3$, and triethylamine at 303 K and 0.1 MPa hydrogen pressure. The results can be interpreted on the basis of a proposed mechanism in which $RhH(PPh_3)_3$ is the active species. A comparison is also reported with analogous reactions carried out in the absence of NEt_3.

Introduction

In a previous paper we have investigated homogeneous hydrogenation and isomerization reactions of *(Z)*-2-butene-1,4-diol in ethanol at 303 K by using the Wilkinson catalyst ($RhCl(PPh_3)_3$) in the presence of triethylamine. Although $RhCl(PPh_3)_3$ was, so far, largely used for hydrogenation reactions and mainly affords fully hydrogenated compounds, it is worth noting that such a catalyst, in the presence of triethylamine and *(Z)*-2-butene-1,4-diol, is more selective towards the geometric isomerization product, *(E)*-2-butene-1,4-diol (1).

In this work we extend our study to the hydrogenation and isomerization of a series of α,β-unsaturated alcohols, such as 2-propen-1-ol (A2), *(E)*-2-buten-1-ol (EB2), *(Z)*-2-penten-1-ol (ZP2), *(E)*-2-penten-1-ol (EP2), *(Z)*–2-hexen-1-ol (ZH2), *(E)*–2-hexen-1-ol (EH2), carried out in the presence of $RhCl(PPh_3)_3$, with and without triethylamine (NEt_3), at 303 K, using ethanol as solvent. The major targets of our research are to investigate the influence of the unsaturated alcohol structure on the product distribution and to verify the possibility of extending the results, previously obtained with *(Z)*-2-butene-1,4-diol, to other analogous substrates.

Results

Reactions of α,β-unsaturated alcohols with hydrogen were carried out in the presence of $RhCl(PPh_3)_3$, with and without NEt_3, at 303 K, 0.1 MPa hydrogen pressure and using ethanol as solvent. Under the experimental conditions adopted, the reaction, in the presence of NEt_3, proceeds according to the following scheme (in the case reported below we consider a *Z* geometric isomer as starting material):

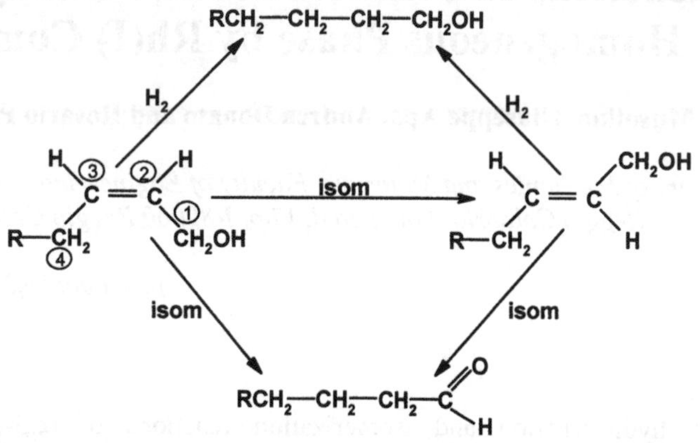

$$RCH_2\text{---}CH_2\text{---}CH_2\text{---}CH_2OH$$

(reaction scheme showing isomerization and hydrogenation pathways with intermediate Z and E alkene structures and final aldehyde)

$$RCH_2\text{---}CH_2\text{---}CH_2\text{---}\overset{O}{\underset{H}{C}}$$

Three main reaction routes are operating. In addition to the fully hydrogenated product, double bond and geometric isomerization derived compounds were also detected. Indeed, the double bond migration process from 2-3 carbons to the 2-1 (see the above scheme), involving, in this case, the carbon atom bearing the -OH group, affords the aldehydes through a vinyl alcohol intermediate. No hydrogenolysis products were observed and only a small amount of compounds, where the double bond moves from the 2-3 to the 3-4 carbons, was detected.

The selectivity and the yield of the process depend mainly on three different factors: i) the catalytic species in solution; ii) the geometric isomer structure of the organic substrate; iii) the chain length of the reacting compound. Taking into account some of our previous results (1), showing the formation of an 1:1 electrolyte by conductometric measurements on the same rhodium(I) coordinated system, used in this paper, and including also an interesting observation by Schrock and Osborn (2), the following equilibria may be considered in our system:

$$RhCl(PPh_3)_3 + H_2 + solv \rightleftharpoons H_2RhCl(PPh_3)_2(solv) + PPh_3$$

$$H_2RhCl(PPh_3)_2(solv) + NEt_3 + PPh_3 \rightleftharpoons RhH(PPh_3)_3 + NEt_3H^+ + Cl^- + solv$$

Consequently, in the absence of NEt_3, the main catalytic species should be $H_2RhCl(PPh_3)_2(solv)$, whereas, in the presence of NEt_3, $RhH(PPh_3)_3$ should be formed. We are aware that such a conclusion is somewhat speculative; however, it seems the most likely if we look at all the experimental data reported on the subject. Table 1 reports experimental results concerning both activity and product distribution, determined in different conditions. Since the Rh(I) mono hydride complex is a catalytic species, it is reformed in every step of the reaction and its concentration remains constant. Therefore, rate data are calculated by the ratio of slopes of plots of organic substrate concentration, divided by the Rh(I) concentration, versus reaction time. The slope of these curves is obtained at about 70 % conversion of the substrate.

Figure 1 shows a typical product composition-time plot of hydrogenation and isomerization reactions at the reported experimental parameters.

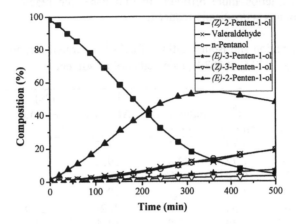

Figure 1. Composition - time profile for the hydrogenation and isomerization reaction of *(Z)*-2-penten-1ol in ethanol at 303 K and at 0.1 MPa H_2 pressure, in the presence of $RhCl(PPh_3)_3$ and NEt_3.

Discussion

Data in table 1 allow a direct comparison of both activity values and selectivity, chosen at 80% conversion of the substrate, and clearly demonstrate that the reaction, in the presence of NEt_3, is faster than that carried out in its absence. Indeed, whereas the dihydride species mainly behaves as a hydrogenation catalyst, the mono hydride complex behaves both as an isomerization and hydrogenation species (2). All our results may be interpreted on the basis of these fundamental concepts. In fact, when a Z substrate reacts with H_2, in the presence of $RhCl(PPh_3)_3$ and NEt_3 in ethanol, the geometric E derivatives are formed with a good selectivity. Conversely, *(E)*-2-buten-1-ol is mainly converted to the corresponding hydrogenated compound.

As we expect, the reactivity of E isomers, both in the presence and in the absence of NEt_3, is slower than that of the analogous Z derivatives. Such behavior is mainly due to an enhanced steric hindrance of the E derivatives in comparison with the analogous Z isomers and the reactivity slows down as the chain length increases, at least when we consider the reactivity of *(E)*-2-penten-1-ol and *(E)*-2-hexen-1-ol, for which no selectivity values were detected owing to the very low reaction rate obtained.

With analogous substrates, the heterogeneous hydrogenation vs. isomerization reactions show behavior that is worth considering. Comparison of the results reported here, with those previously obtained for analogous organic molecules hydrogenated on Pd/TiO_2 systems (3), reveals a reduced reactivity variation between Z and E geometric isomers in the heterogeneous catalysis. A possible explanation is that a metal surface is more open to interactions with olefinic substrates while, in the presence of a homogeneous catalyst, the crowding of ligands around a metal may make such an interaction more difficult. In this case, the reactivity of the more hindered *E* compounds is drastically reduced.

Table 1. Hydrogenation and isomerization of α,β-unsaturated alcohols catalyzed by $RhCl(PPh_3)_3$ (~ 3 x 10^{-4} M) at 303 K, using ethanol as solvent.

Substrate	[NEt$_3$] (M)	r^b (min^{-1})	Saturated alcohol	Selectivity (%)[a]		Other product
				Aldeyde	*E*-isomer	
A2	0	1.12	81.5	18.5		
A2	0.03	2.72	80.8	19.2		
EB2	0	0.18	n. d.[c]	n. d.		
EB2	0.03	1.47	79.1	19.4		1.4
ZP2	0	0.54	67.8	8.2	24.0	
ZP2	0.03	2.49	13.8	14.4	64.1	7.7
EP2	0	0.16	n.d.	n.d.		
EP2	0.03	0.32	n.d.	n.d.		
ZH2	0	0.31	n.d.	n.d.	n.d.	
ZH2	0.03	1.94	37.8	10.6	48.4	3.15

[a] At 80% conversion of the unsaturated alcohol; [b] Rate data (see text); [c] n.d. = non-determined.

Taking into account all these considerations, the following general mechanism may be operating, when the hydrogenation of a *Z* geometric isomer is carried out in the presence of NEt$_3$:

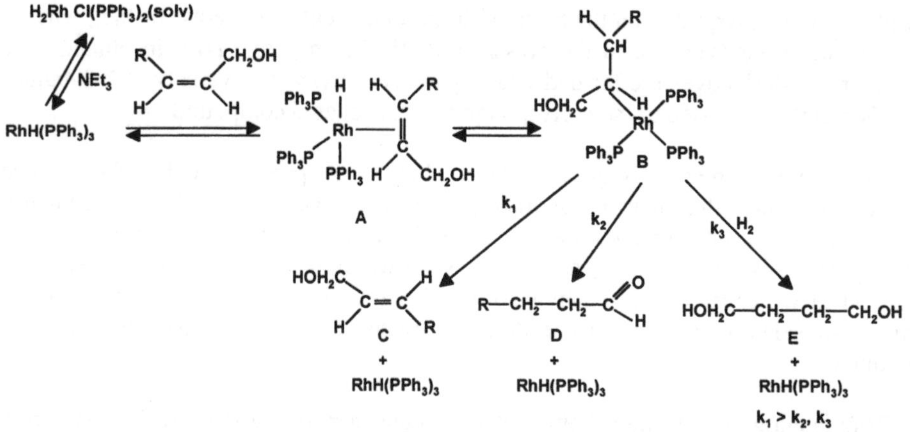

We infer that final products, C, D and E, stem from the same metal-carbon σ-bonded intermediate B, and their relative amounts are due to kinetic factors. Carrying out the same reaction starting from an *E* geometric derivative, the *E/Z* isomerization is, as observed, practically suppressed. The *E* starting compound is, in fact, thermodynamically more stable than the corresponding *Z* one.

Conclusions

The main aim of the results concerns the comparison between *E* and *Z* isomers, both towards hydrogenation and/or isomerization processes. In particular, the less stable *Z* isomers favor the geometric isomerization products, whereas the most stable *(E)*-2-buten-1-ol mainly affords a hydrogenated compound. This agrees with earlier findings, where *Z* isomers generally favor isomerized (both geometric and double bond) derivatives (3).

When 2-propen-1-ol reacts in the presence of NEt₃ and Rh(I), the reduced steric hindrance and the lack of a possible geometric isomerization are to be expected. So, we explain the large amount of the hydrogenated compound formed (Table 1). In the absence of NEt₃, as a rule, hydrogenation of organic substrates generally occurs. However, additionally in this specific case, the reduced reactivity of the system suggests that a different catalytic species is operating.

Considering all these data, we conclude that the steric hindrance of coordinated phosphines is the most important factor in determining the behavior of the reactions studied.

Experimental Section

$RhCl(PPh_3)_3$ was prepared according to the literature (4). The hydrogenation of different α,β-unsaturated alcohols, 2-propen-1-ol, *(E)*-2-buten-1-ol, *(Z)*-2-penten-1-ol, *(E)*-2-penten-1-ol, *(Z)*–2-hexen-1-ol, *(E)*–2-hexen-1-ol, was carried out in liquid phase at 303 K and 0.1 MPa hydrogen pressure, using ethanol as solvent in a batch reactor. A stirring rate of 500 rpm was used. The experimental setup was thoroughly purged with nitrogen before the beginning of the reaction. The rhodium complex (1.08×10^{-5} moles) was dissolved, under stirring, in the solvent used (25 mL) in atmosphere of H_2 at 303 K; then, NEt₃ was added in the molar ratio $RhCl(PPh_3)_3$/NEt₃ 1:100, and the system was allowed to equilibrate for one hour. Finally 15 mL of a 0.6 M solution of the α,β-unsaturated alcohol in ethanol, containing an internal standard, was added through one arm of the flask.

The progress of the reaction was followed by analyzing a sufficient number of samples withdrawn from the reaction mixture. Product analysis was performed with a gas chromatograph (Agilent Technologies model 6890N) equipped with a flame ionization detector. The product separation was obtained by a capillary column (J&W DB-Waxetr, 50 m, i.d. = 0.32 mm). Quantitative analyses were carried out by calculating the area of the chromatographic peaks with an electronic integrator.

References

1. M. G. Musolino, G. Apa, A. Donato and R. Pietropaolo, *Catal. Tod*, **100**, 467 (2005).

2. R. R. Schrock and J. A. Osborn, *J. Am. Chem. Soc.*; **98**, 2134 (1976).
3. M. G. Musolino, P. De Maio, A. Donato and R. Pietropaolo, *J. Mol. Catal.: A Chem.*; **208**, 219 (2004).
4. J. A. Osborn, F. H. Jardine, J. F. Young and G. Wilkinson, *J. Chem. Soc. A*; 1711 (1966).

29. Reductive Amination of Isobutanol to Diisobutylamine on Vanadium Modified Raney® Nickel Catalyst

Sándor Gőbölös and József L. Margitfalvi

Chemical Research Center, Institute of Surface Chemistry and Catalysis, H-1025 Budapest, Pusztaszeri út 59-67, Hungary

gobolos@chemres.hu

Abstract

Amination of i-butanol to diisobutylamine was investigated on vanadium modified granulated Raney® nickel catalyst in a fixed bed reactor. The addition of 0.5 wt.% V to Raney® nickel improved the yield of amines and the stability of catalyst. Factorial experimental design was used to describe the conversion of alcohol, the yield and the selectivity of secondary amine as a function of "strong" parameters, i.e. the reaction temperature, space velocity and NH_3/i-butanol molar ratio. Diisobutylamine was obtained with 72% yield at 92% conversion and reaction parameters: P=13 bar, T=240°C, WHSV=1 g/g h, and molar ratios NH_3/iBuOH= 1.7, H_2/NH_3= 1.9.

Introduction

Amines find application as intermediates in many fields of industry and agriculture (1). Lower alkylamines are usually produced by the amination of the corresponding alcohols with ammonia. However, only scare data are available in the literature on the preparation of isobutylamines. In the alkylation of NH_3 or amines with alcohols, Co-, Ni- or Cu-containing catalysts are mainly used (1). Nickel catalysts produce mainly primary amine, however, in many applications secondary amines are needed. Diisobutylamine ($(iBu)_2NH$) was obtained from i-butanol (iBuOH) and i-butylamine with 60% yield on 20wt.%Co-5wt.%Ni/Al_2O_3 catalyst at 200°C (2). Supported nickel catalysts containing 25-45 wt.% metal were active in producing $(iBu)_2NH$ from iBuOH and NH_3 with 65-72% yield at 86-92% conversion, and also at 200°C (3).
In this work the effect of process parameters on the amination of iBuOH to $(iBu)_2NH$ was studied over V-modified Raney® nickel. V is known to increase the yield of amines and the stability of catalyst (4,5). Factorial experimental design was used to describe the conversion of alcohol, and the yield and selectivity of secondary amine as a function of reaction temperature, space velocity and NH_3/iBuOH molar ratio.

Results and Discussion

The V-modified Raney® nickel catalyst showed noticeable stability in time-on-stream experiment for 55 h, and tolerated several heating-cooling cycles. It is noteworthy that the conversion of i-butanol was ca. 86%, and yield of mono-secondary and tertiary amine was ca. 18, 65 and 3%, respectively (see Figure 1). The conversion and the amine yield listed in Table 1 indicate that the H_2 pressure has little effect on catalytic performance in the range investigated.

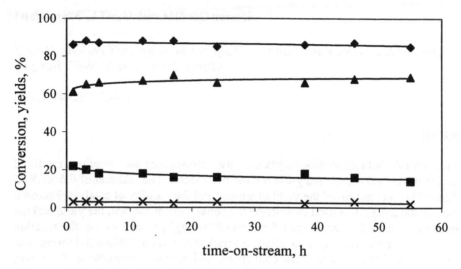

Figure 1. Amination of i-butanol. Time-on-stream experiment.
Reaction condition: T=225°C; WHSV=1.0g/(g_{catalyst} h); P=13bar; NH_3/iBuOH=1.4; H_2/NH_3=1.6. Abbreviations: ◆-conversion; yields of $(iBu)_2NH$, $iBuNH_2$, and $(iBu)_3N$, ▲, ■, and ✕, respectively.

Table 1. Amination of isobutanol. Effect of hydrogen pressure

P, bar	Conversion, %	Yield, %		
		$iBuNH_2$	$(iBu)_2NH$	$(iBu)_3N$
7	89	19	66	4
11	88	20	65	3
12	90	19	68	3
13	89	18	67	4
14	91	19	69	3
21	88	19	67	2
31	89	19	67	3

Reaction condition: T=225°C; WHSV=1 g/(g_{catalyst} h); NH_3/iBuOH=1.6; H_2/NH_3=1.6

Catalytic activity data summarized in Table 2 indicate that both the reaction temperature and the NH_3/iBuOH ratio strongly affect the conversion of iBuOH and the selectivity and thus the yield of secondary amine. Upon increasing the reaction temperature and the NH_3/iBuOH ratio the conversion of alcohol significantly increased. Because of the interaction of the two parameters, i.e. the temperature and the NH_3/iBuOH ratio, the yield and the selectivity of secondary amine shows up maximums in the function of NH$_3$/iBuOH ratio (see also Fig. 2). Under the same condition (t=220°C, NH$_3$/iBuOH=1.4) the yield of amine is lower on the unmodified Raney® nickel than on the V-modified one (compare 5th and 6th rows in Table 2).

Table 2. Amination of isobutanol. Effect of temperature and NH$_3$/iBuOH ratio

T, °C	NH$_3$/iBuOH	Conversion, %	Yield, % (iBu)$_2$NH	Selectivity, % (iBu)$_2$NH
200	1.1	72	53	74
200	1.4	78	59	76
200	2.3	82	59	72
220	0.9	69	51	73
220^a	1.4	88	60	68
220	1.4	87	65	75
220	1.7	88	67	76
240	1.2	88	66	75
240	1.7	92	72	78
240	2.5	95	60	63

Reaction condition: WHSV=1.0g/g$_{catalyst}$ h; P=13bar; H$_2$/NH$_3$=1.9
^{a}Experiment was carried out on unmodified Raney nickel.

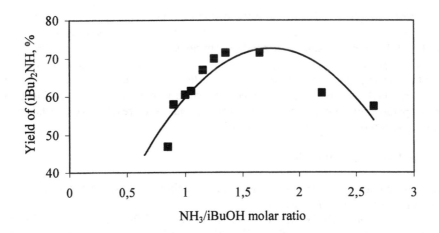

Figure 2. Amination of i-butanol. Effect of NH$_3$/i-BuOH ratio on (iBu)$_2$NH yield. Reaction condition: T=225°C;WHSV=1.0g/8$_{catalyst}$ h9; H$_2$/NH$_3$=1.6

Low NH$_3$/iBuOH molar ratio is not sufficient to achieve high conversion and degree of amination, whereas at high ratios, e.g. NH$_3$/iBuOH=2.5 the selectivity of primary amine increases at the expense of secondary one. It has been found that the temperature, space velocity and NH$_3$/iBuOH ratio strongly affect both the conversion and (iBu)$_2$NH selectivity. The lower the space velocity the higher the conversion, however upon increasing the space velocity the secondary amine yield slightly increases (see Table 3). Upon increasing the H$_2$/NH$_3$ ratio between values of 0.7 and 1.9 the selectivity and the yield of (iBu)$_2$NH increases (see Table 3). Neither aldehyde nor nitrile was identified in the reaction product. All these results suggests that the maximum of (iBu)$_2$NH yield can be expected at reaction parameters as follows: T= 220-240°C, WHSV=1.0-1.5 g/g h, NH$_3$/iBuOH and H$_2$/NH$_3$ molar ratios of 1.7 and 1.9, respectively. Under this condition the yield of (iBu)$_2$NH is (67-72 %), whereas the yield of primary and tertiary amines is in the range of 14-16 % and 1-3 %, respectively.

Table 3. Amination of isobutanol. Effect of H$_2$/NH$_3$ ratio and space velocity

H$_2$/NH$_3$	WHSV g/g$_{catalyst}$ h	Conversion, %	Yield, % (iBu)$_2$NH	Selectivity, % (iBu)$_2$NH
0.7	1.0	93	59	63
1.4	1.0	91	64	70
1.9	1.0	89	67	75
0.7	1.5	90	63	70
1.4	1.5	88	65	74
1.9	1.5	87	67	77

Reaction condition: T=240°C, P=13bar; NH$_3$/iBuOH=1.2

The highest (iBu)$_2$NH yield (72 %) was obtained a conversion level of 92% and at reaction parameters P=13 bar, T=240°C, WHSV=1.0 g/g h, NH$_3$/iBuOH= 1.7, H$_2$/NH$_3$= 1.9. In conclusion, a secondary amine yield above 70 % can be was obtained in fixed bed reactor using vanadium promoted Raney nickel catalyst without recycling unconverted alcohol. In order to describe the conversion of alcohol, as well as the yield and selectivity of diisobutylamine in the function process parameters, experiments were carried out and results were evaluated according to orthogonal factorial design (6,7).

Factorial Experimental Design

In the factorial design the parameter ranges and the abbreviations are as follows:
Parameter ranges and abbreviations:
T= 200-240°C, NH$_3$/iBuOH(n)= 0.8-2.0, WHSV(w)= 1.0-2.2g/g h, P= 1.3MPa, H$_2$/NH$_3$= 1.9, C=conversion of iBuOH (%), S and Y selectivity and yield of (iBu)$_2$NH in %, respectively.
t= confidence level of coefficients (%), F=fit of calculated equation (%)

$$C = 198.50 - 0.541T - 76.63n - 107.7w + 0.5067Tn + 0.3192Tw - 8.806n^2 + 7.417w^2$$
(Maximum of C=99.7% at T=240°C, n=2.0, w=1.0g/g h) (t=90%, F=95%)

$$S = 72.34 - 0.2015T + 70.99n - 76.04w + 0.3925Tw - 12.17nw - 22.75n^2$$
(Maximum of S=80.2% at T=240°C, n=1.3, w=1.0g/g h) (t=90%, F=95%)

Y=421.79-1.676T-268.1n-245.1w+1.265Tn+1.099Tw+106.1nw-0.511Tnw
(Maximum of Y=76.1%, T-240°C, n=2.0, w=1.0g/g h) (t=95%, F=95%)

The t-test for individual parameters showed that all three parameters were significantly different from zero within 90 or 95% confidence level. F-statistic at 95% significance indicates the fit between calculated and experimental data. The coefficients of the equations describing the conversion, end the yield and selectivity of secondary amine indicate strong interaction between the temperature and the space velocity and the temperature and NH_3/iBuOH ratio. With respect to the conversion the strength of parameters is roughly equal. The strength of parameters in determining $(iBu)_2NH$ selectivity decreases in the order: n > w > T. The yield of secondary amine is strongly influenced by all three parameters. The maximum values of conversion, selectivity and yield calculated by using derived equations, as well as related reaction parameters are given in parentheses. With respect to the maximum selectivity and yield of $(iBu)_2NH$ the measured and calculated values are in quite good agreement.

Experimental Section

The granulated Raney® nickel catalyst was prepared by leaching a 50wt.% Ni-Al alloy with 20wt.% NaOH-H_2O solution at 50°C as described elsewhere (8). Modification of Raney® nickel catalyst with 0.5 wt.% V was carried out by adsorption using an aqueous solution of NH_4VO_3 (3). Catalytic test was carried out in a continuous-flow fixed bed reactor charged with 20g granulated Raney® nickel catalyst. Prior to activity test the catalyst was heated to 250°C at a rate of 2°C/min in a flow of 75%H_2-25%N_2 mixture and kept at 250°C for 3 hours. Liquid organic reaction product was analyzed by gas chromatography using FID and a glass column (3m x 3mm) filled with 60/80 mesh Chromosorb P NAW containing 5wt.% KOH and 18wt.% Carbowax 20M. Factorial experimental design (6,7) was applied to explore the effect of temperature (T), weight hour space velocity (WHSV) and NH_3/iBuOH molar ratio on the catalytic performance of V-modified Raney® nickel.

References

1. K. S. Hayes, *Appl. Catal. A*, **221**, 187 (2001).
2. A. Buzas, C. Fogarasan, N. Ticusan, S. Serban, I. Krezsek and L. Ilies, Ger. Offen. 2,937,325 (1981).
3. M. Deeba, U.S. Patent 4,918,234 to Air Products (1990).
4. J. Antal, J. Margitfalvi, S. Gőbölös et al.., Hung. Patent 206,667 to Nitrogen Works Pét, Hungary (1987).
5. S. Gőbölös, M. Hegedűs, E. Tálas, J.L. Margitfalvi, *Stud. Surf. Sci. Catal.* Vol. **108**, Elsevier, Amsterdam, pp. 131-138 1997
6. G.E.P. Box and K.B. Wilson, *J. Royal Statist. Soc.*, Ser.B **13**, 1-45 (1951).S.
7. A. Dey and R. Mukerjee, *Fractional Factorial Planes*, Wiley, N.Y. 1999.
8. Gőbölös, E. Tálas, M. Hegedűs, J.L. Margitfalvi, J.Ryczkowski, *Stud. Surf. Sci. Catal.* Vol. **59**, Elsevier, Amsterdam, 1991, pp. 335-342

IV. Symposium on Novel Methods in Catalysis of Organic Reactions

30. How to Find the Best Homogeneous Catalyst

Gadi Rothenberg[1], Jos A. Hageman[2], Frédéric Clerc[3], Hans-Werner Frühauf[1] and Johan A. Westerhuis[2]

[1] Van't Hoff Institute for Molecular Sciences and [2] Swammerdam Institute of Life Sciences, University of Amsterdam, Nieuwe Achtergracht 166, 1018 WV Amsterdam, The Netherlands. [3] Institut de Recherches sur la Catalyse, 2 Avenue Albert Einstein, 69626 Villeurbanne Cedex, France

gadi@science.uva.nl

Abstract

We present in this communication an alternative concept for optimizing homogeneous catalysts. With this method, one extracts and screens virtual catalyst libraries, indicating regions in the catalyst space where good catalysts are likely to be found. This automated screening is done in two stages: A 'rough screening' using 2D descriptors, followed by a "fine screening" using 3D descriptors. The model is demonstrated for a set of Pd-catalyzed Heck reactions using bidentate ligands.

Introduction

One of the biggest challenges in chemistry is discovering new chemical reactions that will enable society to function in a sustainable manner. Atom economy and waste minimization are at the heart of industrial policy, driven both by governmental incentives and by market considerations. Homogeneous catalysis is a most promising area in this respect (1). The Nobel prizes (2, 3) awarded in 2001 (to Sharpless, Noyori and Knowles for their discovery of chirally-catalyzed oxidation and hydrogenation reactions), and in 2005 (to Chauvin, Grubbs, and Schrock for their discovery of metathesis catalysts) exemplify how a new catalyst can cause a paradigm shift in the chemical industry (4).

The last two decades have seen enormous developments in catalyst discovery and optimization tools, notably in the area of high-throughput experimentation (HTE) and process optimization (5). However, the basic concept used for exploring the catalyst space in homogeneous catalysis has not changed: Once an active catalyst complex is discovered, small modifications are made on the structure to try and screen the activity of neighboring complexes, covering the space much like an ink drop spreads on a sheet of paper. This is not a bad method, but can we do better with the new tools that are available today?

Here we present an alternative concept for optimizing homogeneous catalysts. Using a "virtual synthesis" platform, we assemble large catalyst libraries (10^{15}–10^{17} candidates) *in silico*, and use statistical models, molecular descriptors, and quantitative structure-activity and structure-property relationship (QSAR/QSPR) models to extract subsets from these libraries and predict their catalytic performance.

A full discussion of all the issues related to this concept is out of the scope of this preliminary communication. Instead, we present here the basics of the general approach, as well as a specific example illustrating one iteration in the optimization of Pd-catalyzed Heck reactions using bidentate ligands, which demonstrate higher catalytic activities and lifetimes than monodentates (6). The full technical details of the algorithms and the theoretical treatment of the catalyst library diversity will be published elsewhere (7).

Results and Discussion

Let us first consider the problem of homogeneous catalyst optimization. In this situation, one has some initial data on the figures of merit (e.g. that such-and-such metal-ligand complexes are "good catalysts" because they display a high turnover frequency). The objective then is to pinpoint "active regions" in the catalyst space. This space, defined here as space **A**, is not continuous. Rather, it is a grid containing all the possible metal-ligand complexes (each point in space **A** denotes a unique catalyst). It is important to realize that **A** is multi-dimensional and very large. Therefore, any method used for exploring it must *(i)* reduce the dimensionality of the problem and *(ii)* employ fully automated screening techniques.

To do this, we define two additional multi-dimensional spaces: **B** and **C** (Figure 1). Space **B** contains the values of the catalyst descriptors that pertain to these catalysts (*e.g.* backbone flexibility, partial charge on the metal atom, lipophilicity) as well as the reaction conditions (temperature, pressure, solvent type, and so on). Finally, space **C** contains the catalyst figures of merit (i.e., the TON, TOF, product selectivity, price, and so forth). Spaces **B** and **C** are continuous, and are arranged such that each dimension in each space represents one property.

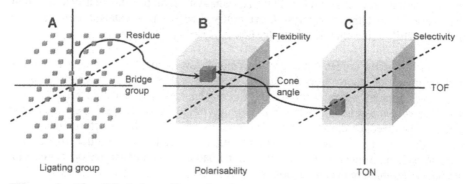

Figure 1. Simplified three-dimensional representation of the multi-dimensional spaces containing the catalysts, the descriptor values, and the figures of merit.

By dividing the problem this way, we translate it from an abstract problem in catalysis to one of relating one multi-dimensional space to another. This is still an abstract problem, but the advantage is that we can now quantify the relationship between spaces **B** and **C** using QSAR and QSPR models. Note that space **B** contains molecular descriptor values, rather than structures. These values, however, are directly related to the structures (8).

Of course, the real catalyst space is infinite, and it is not possible to study all of it. Instead, we generate a very large space **A** (10^{15}–10^{17} catalysts) *in silico*, using a virtual synthesis platform, developed in our group and based on a 'building block' synthesis concept (9). One basic assumption that we make here is that the 'good catalyst' we are seeking is somewhere on this grid. Note that less than 50 building blocks are needed to create a space of 10^{17} catalysts, even when using only simple species that are joined selectively using a number of well defined reactions.

To optimize the catalyst, we use an iterative approach (Figure 2), with consecutive modelling, synthesis, and analysis steps. First, we consider all of the available data (from earlier experiments or from literature), and build a regression model that connects the catalyst descriptors and the figures of merit (10). The screening is done in two stages. In the first, 'rough screening', we use 2D descriptors to examine relatively large areas of space **A**. We select random subsets from this space (typically 10,000–50,000 catalysts). The program calculates the 2D catalyst descriptor values and uses the above model to predict the figures of merit for these new catalysts. Depending on the data available, one can also apply genetic algorithms (GAs) at this stage to try and optimize the catalyst structure based on the 2D descriptors using meta-modelling (11). The best catalysts (typically 200–500 structures) are then selected for the next stage.

In the second, 'fine screening' stage, the program computes the 3D descriptors for this new subset, and again projects the results on the model and predicts the figures of merit. Basically, 3D descriptor models are more costly than 2D ones, but they give better results (12). As we showed earlier (10, 13), nonlinear models that combine chemical and topological descriptors are well suited for predicting activity/selectivity trends in homogeneous catalyst libraries, with typical correlation coefficients of $R^2 = 0.8$–0.9.

The result is a small subset of 20–50 new catalysts. These are then synthesized and tested experimentally. The model is then updated and the cycle repeats. In theory this process can repeat indefinitely, but our results on industrial data show that the figures of merit usually converge after 5–6 cycles. This means that in principle it is possible to indicate an optimal region in a space of a million catalysts after testing less than 300 ligand-metal complexes!

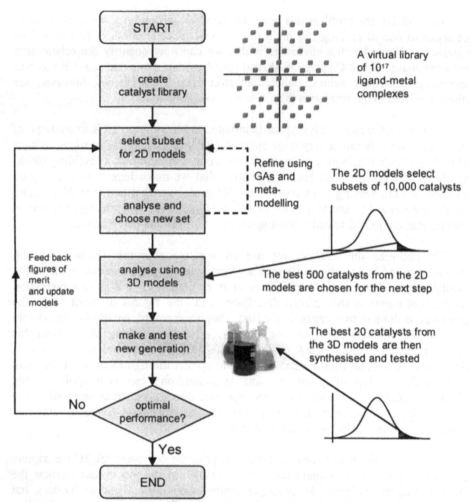

Figure 2. Iterative approach flowchart for homogeneous catalyst optimization.

An important feature is that, with the exception of the catalyst synthesis, this is a fully automated process. Indeed, we envisage that in industry the entire process, including the synthesis and testing, can be automated using commercially available synthesis robots. If robotic synthesis and screening is available, the size of the test sets can be increased. Moreover, it is precisely the automated modelling and analysis aspect that is lacking in the robotic laboratory workflow.

Another important feature of GAs is that they are tunable. This means that we can define the algorithm's fitness function to reflect the actual requirements from the catalyst. An optimal catalyst exhibits high activity, high stability, and high selectivity. These three figures of merit are directly related to the product yield, the turnover number (TON) and the turnover frequency (TOF), respectively. Often, however, an increase in one comes at the expense of another. Using GAs you can pre-define the weight of each figure of merit. For example, if you know beforehand that the catalytic activity is the most important parameter, you can assign a heavier weight to the TOF. In this way, the computer searches for the most suitable catalyst.

To demonstrate this concept, let us consider the first optimization cycle for the palladium-catalyzed Heck reaction in the presence of bidentate ligands. The literature shows us some promising leads (14-18), with yields > 95% and TOFs > 1000, but much of the catalyst space is unexplored territory.

In this example, we will assume that each catalyst consists of one Pd atom and one bidentate ligand. The ligand includes two ligating groups L_1 and L_2, a backbone group B, and three residue groups R_1, R_2, and R_3. To simplify, we will limit the ligating groups to **1–14**, the backbone groups to **15–21**, and the residue groups to **22–29** (Figure 3). Further, we will constrain the R groups to one per ligating or backbone group. There is no restriction on group similarity, *i.e.* it is possible that $L_1 \equiv L_2$ and so forth. Each ligand has a unique $\{L_1(R_1)\text{-}B(R_2)\text{-}L_2(R_3)\}$ identifier. The connection points for the R groups and between the L and B groups are predefined for each building block (for example, the tetrahydrothiophene ligating group **9** connects to the Pd *via* the S atom, and to the backbone and the residue group on positions 2, 3, or 4). The total number of ligand-Pd complexes one can assemble from the above 29 building blocks (connecting only *via* the specified connection points and limiting ourselves to the $L_1(R_1)\text{-}B(R_2)\text{-}L_2(R_3)$ form) is 2.61×10^{17}. This is a huge number, well beyond the combined synthetic capabilities of all of the laboratories in the world. Note that these building blocks were chosen specifically for this example, mimicking some of the ligand types in the following dataset to enable good intrapolation. The precise relationship between building block structure and connectivity and the resulting catalyst diversity is very complicated and will be discussed in a separate paper (19).

To simplify things, we will show here only one iteration (the meta-modelling is not necessary for demonstrating the two-stage screening). As a starting point, we assemble a dataset containing 253 published Heck reactions performed using 58 different catalysts and/or under different reaction conditions (Table 1 shows a partial representation of this dataset). For each reaction, we include the substrates, catalyst, reaction conditions, and three figures of merit: Product yield, TON, and TOF. We then calculate a set of thirty-one 2D descriptors (20). This gives a 253×31 matrix. As we showed earlier, both linear regression models (such as partial least squares, PLS), and nonlinear ones (*e.g.* artificial neural networks) can be used (10). In this case, however, there are not enough initial data to 'feed' a neural network, so we use a PLS model, which is also more robust (21). This model is used for correlating the 2D descriptors and the reaction conditions (temperature, Pd concentration, and solvent) with the above three figures of merit.

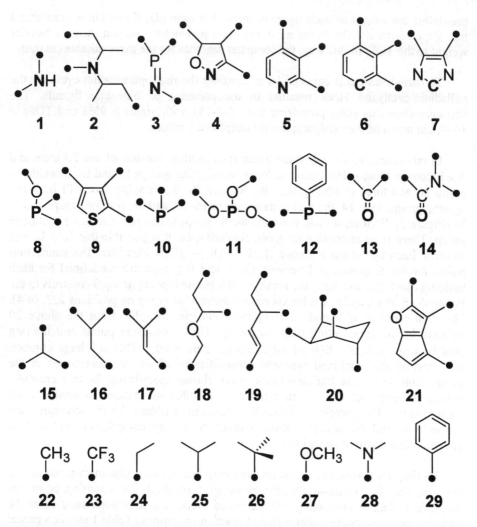

Figure 3. Building blocks used in assembling the virtual libraries. Ligating groups (structures **1–14**; the ligating atom is indicated in boldface type), backbone or 'bridging' groups (**15–21**), and residue groups (**22–29**). The '•' symbols indicate the possible connection points. After assembling the ligand, the program assigns H atoms to any unused connection points.

Table 1. Partial representation of the initial Heck reaction dataset.

Cat	t °C	ArX[a]	Alkene[a]	Solv.[a]	Yield%	TON	TOF h^{-1}	Ref.
1	50	PhI	NBA	DMA	100	41	1000	(18)
2	30	CNBr	NBA	DMA	50	1	50	(18)
3	100	PhI	TBA	Et$_3$N	98	57	981	(17)
4	100	NO$_2$Br	TBA	Et$_3$N	100	14	100	(17)
5	85	PhI	EtA	DMF	100	222	1000	(15)
6	85	PhI	EtA	DMF	48	209	480	(15)
7	85	PhI	EtA	DMF	63	191	630	(15)
8	120	PhBr	NBA	DMA	91	113	910	(22)
9	120	NO$_2$Br	NBA	DMA	95	237	950	(22)
10	60	PhOTf	DHF	THF	85	4.25	1	(23)
11	120	MeOBr	styrene	TBAB	99	2.78	50	(24)

[a] NBA = *n*-butyl acrylate; TBA = *tert*-butyl acrylate; EtA = ethyl acrylate; CNBr = *p*-CN(C$_6$H$_4$)Br; NO$_2$Br = *p*- NO$_2$(C$_6$H$_4$)Br; MeOBr = *p*- CH$_3$O(C$_6$H$_4$)Br; TBAB = tetrabutylammonium bromide.

Already at this early stage, using simple 2D descriptors, the model yields important mechanistic information: The correlation for the TON and the TOF depends strongly on the reaction temperature, with a cut-off point at 120 °C (Table 2). The chemical reason for this is that Pd nanoclusters form much faster above 120 °C (25), and the reaction follows a pathway that is independent of the ligand.

Table 2. PLS model prediction quality as a function of reaction temperature

Figure of merit[a]	Full dataset (253 reactions)	<120 °C (62 react.)	=< 120 °C (105 react.)	>=120 °C (191 react.)
TON	0.23	0.79	0.57	0.27
TOF	0.18	0.79	0.47	0.32
Product yield	0.86	0.92	0.90	0.87

We now select 10,000 bidentate catalysts at random (because this is the first iteration) from the large space generated using building blocks **1–29**, and calculate the 2D descriptor values for these ligands. Projecting the results for these 10,000 structures on the PLS model gives the predicted figures of merit (Figure 4). From this 'rough screening' we choose the best 206 structures, by combining the 100 best-performing structures for each figure of merit (the TON and TOF show a high overlap, which is to be expected).

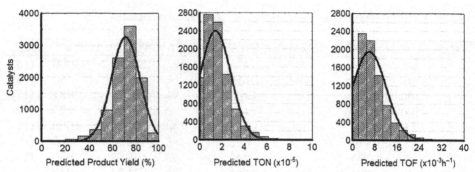

Figure 4. Predicted distribution of the 10,000 catalysts in the first screening.

In the second stage, we compute the 3D descriptors for these 206 structures. This entails geometry optimization, and is much more costly. The advantage is that we have already discarded 9,794 less-promising cases, and 3D descriptors can tell us much more about the catalyst. The results are then projected on the 3D PLS model for the original dataset, and the final results indicate the best-performing structures with regard to the figures of merit. In this case, 12 of the original 10,000 structures show a predicted yield of > 85%, TON > 10,000, and TOF > 2000. Three of the most promising computer-generated structures (**30–32**, all with TON > 10^5, TOF > 5000, and yield > 80%) are shown below. The proof of the pudding, of course, will be to synthesize and test these structures (for synthesis purposes different building blocks will be used, but the principle remains unchanged).

In conclusion, we present in this communication an alternative method for optimizing homogeneous catalysts. This concept is based on iterative modelling and synthesis steps. We do not claim to replace serendipity in catalyst discovery and optimization using this approach. Rather, we believe that this approach can complement serendipity by steering synthetic chemists clear from null regions of the catalyst space.

Experimental Section

The technical details of the descriptor selection algorithms will be published elsewhere (7). Libraries and subsets of structure strings were generated using Matlab (26). Geometry optimization (for calculating the 3D descriptors) was performed in Hyperchem (27), using the MM+ force field in combination with a conjugate gradient optimization method (Polak-Ribiere). Structures were optimized in batch mode using dedicated Hyperchem/Matlab scripts. The 2D descriptors were computed using Matlab and OptiCat (9), and analyzed with Statistica (28). The 3D QSAR parameter were calculated using Codessa (29). We used topological, geometrical and electrostatic descriptors (the CPU time costs of quantum-chemical and thermodynamic QSAR parameters are too high for such large sets). The PLS models were built using the NIPALS algorithm (30), including the intercept. Leave-one-out validation was performed for all models. All three figures of merit were predicted simultaneously. Models with an average $R^2 > 0.8$ were accepted. The 2D and the 3D models contained 9 components each. The most important variables for the 2D model were the number of H and C atoms. For the 3D model, the most relevant descriptors were related to electronic charges per surface area.

References

1. P. W. N. M. van Leeuwen, *Homogeneous Catalysis: Understanding the Art*, Kluwer Academic Press, Amsterdam, 2004.
2. C. P. Casey, *J. Chem. Educ.*, **83**, 192 (2006).
3. A. Ault, *J. Chem. Educ.*, **79**, 572 (2002).
4. Another good example is the Metolachlor process, see H.-U. Blaser, *Adv. Synth. Catal.*, **344**, 17 (2002).
5. G. Li, Chemical Industries (CRC Press), **104**, (*Catal. Org. React.*), 177-184 (2005).
6. N. D. Jones and B. R. James, *Adv. Synth. Catal.*, **344**, 1126 (2002).
7. For a description of some of the optimization methods discussed here see J. A. Hageman, J. A. Westerhuis, H.-W. Frühauf and G. Rothenberg, *Adv. Synth. Catal.*, **348**, 361 (2006).
8. See, for example P. W. N. M. van Leeuwen, P. C. J. Kamer and J. N. H. Reek, *Pure Appl. Chem.*, **71**, 1443 (1999).
9. F. Clerc, *OptiCat - A Combinatorial Optimisation Software*. OptiCat is available free of charge from the author.
10. E. Burello, D. Farrusseng and G. Rothenberg, *Adv. Synth. Catal.*, **346**, 1845 (2004).
11. For a recent review see Y. Jin, *Soft Computing*, **9**, 3 (2005).
12. For a discussion on using 2D and 3D descriptors see E. Burello and G. Rothenberg, *Adv. Synth. Catal.*, **347**, 1969 (2005).
13. E. Burello and G. Rothenberg, *Adv. Synth. Catal.*, **345**, 1334 (2003).

14. For a comprehensive review on Pd-catalyzed Heck reactions see I. P. Beletskaya and A. V. Cheprakov, *Chem. Rev.*, **100**, 3009 (2000).

15. I. P. Beletskaya, A. N. Kashin, N. B. Karlstedt, A. V. Mitin, A. V. Cheprakov and G. M. Kazankov, *J. Organomet. Chem.*, **622**, 89 (2001).

16. K. R. Reddy, K. Surekha, G.-H. Lee, S.-M. Peng and S.-T. Liu, *Organometallics*, **19**, 2637 (2000).

17. T. Kawano, T. Shinomaru and I. Ueda, *Org. Lett.*, **4**, 2545 (2002).

18. C. S. Consorti, M. L. Zanini, S. Leal, G. Ebeling and J. Dupont, *Org. Lett.*, **5**, 983 (2003).

19. For an excellent review on measuring chemical similarity/diversity see N. Nikolova and J. Jaworska, *QSAR Comb. Sci.*, **22**, 1006 (2003).

20. For a discussion on the choice of descriptors for homogeneous catalysts see E. Burello, P. Marion, J.-C. Galland, A. Chamard and G. Rothenberg, *Adv. Synth. Catal.*, **347**, 803 (2005).

21. S. Wold, M. Sjostrom and L. Eriksson, *Chemom. Intell. Lab. Sys.*, **58**, 109 (2001).

22. Z. Xiong, N. Wang, M. Dai, A. Li, J. Chen and Z. Yang, *Org. Lett.*, **6**, 3337 (2004).

23. W.-M. Wei-Min Dai, K. K. Y. Yeung and Y. Wang, *Tetrahedron*, **60**, 4425 (2004).

24. W. A. Herrmann and V. P. W. Böhm, *J. Organomet. Chem.*, **572**, 141 (1999).

25. M. B. Thathagar, J. Beckers and G. Rothenberg, Chemical Industries (CRC Press), **104**, (*Catal. Org. React.*), 211-215 (2005).

26. MATLAB is commercially available from MathWorks, Natick USA, version 6.1, 2001.

27. HyperChem™ Professional 7.51, Hypercube, Inc., 1115 NW 4th Street, Gainesville, Florida 32601, USA.

28. Statistica is distributed by StatSoft, Inc., 2300 East 14th Street, Tulsa, OK 74104, USA.

29. Codessa version 2.642, University of Florida (1994).

30. S. Rännar, F. Lindgren, P. Geladi and S. Wold, *J. Chemom.*, **8**, 111 (1994).

31. Novel Chloroaluminate Ionic Liquids for Arene Carbonylation

Ernesto J. Angueira and Mark G. White

School of Chemical and Biomolecular Engineering,
Georgia Institute of Technology, Atlanta, GA 30332-0100

mark.white@chbe.gatech.edu

Abstract

Novel ionic liquid formylating agents were synthesized from alkyl-methylimidazolium (RMIM$^+$) chlorides and AlCl$_3$. These intrinsically Lewis acids containing excess AlCl$_3$ absorbed HCl to develop strong Brønsted acidity. Modeling by semi-empirical methods showed that HCl experienced three, local energy minimum positions in the IL. *Ab initio* predictions of the ^{1}H-NMR spectra using Hartree-Fock methods showed that the three sites for the absorbed HCl had different chemical shifts (CSs). Additional modeling of the Al-species (^{27}Al-NMR) showed that two distinct dinuclear aluminum chloride anions, having different CS's could be stabilized in the IL. Indirect acidity measurements of the HCl protons by ^{13}C-NMR of 2-^{13}C-acetone confirmed the predictions of multiple, strong acid sites in the IL. Subsequent measurements of the ^{27}Al-NMR confirmed the existence of two Al species in the intrinsically-Lewis acidic IL's and most remarkable, the concentrations of these two species depended upon the length of the alkyl groups, R, present in RMIM$^+$. The pseudo, first-order rate constants observed for the toluene carbonylation reaction decreased systematically in ILs having increasing chain length of R in a way that could be easily correlated by the mole fraction of the dinuclear Al chloride species: [Cl$_3$AlClAlCl$_3$]$^-$ observed by ^{27}Al-NMR.

Introduction

Ionic liquids (IL's) are substances that form liquids at room temperature and lower at atmospheric pressure (1). The very hydrophilic, chloroaluminate Ils, O$^+$[Al$_n$Cl$_{3n+1}$], are said to exhibit Lewis acidity when n > 1 (2, 3) where O$^+$ is the organic cation. These IL's may be prepared using organic cations such as the pyridinium or the substituted imidazolium cations. Recently, it was reported that strong, Brønsted acidity could be created in EMIM$^+$(Al$_2$Cl$_7$)$^-$ upon exposure to dry HCl (4). Moreover, it was shown (5) that arene carbocations were formed when combined with IL's derived from trimethylsulfonium bromide-AlCl$_3$/AlBr$_3$ and exposed to HBr gas. We showed how a combination of HCl with EMIM$^+$(Al$_2$Cl$_7$)$^-$ resulted in a potent conversion agent for toluene carbonylation (6).

Acidic Ionic Liquids

The effect upon the ^{27}Al-NMR spectra for changing the Al/RMIM$^+$ ratio and adding HCl to the system was attributed to either one (7) or more (8) equilibria involving anionic aluminum species.

$$O^+[Al_2Cl_7]^- + O^+Cl^- \Leftrightarrow 2O^+[AlCl_4]^- \text{ and}$$
$$O^+[Al_2Cl_7]^- + HCl \Leftrightarrow O^+[AlCl_4]^- + H^+[AlCl_4]^-$$

We (9) reported predictions, after the manner of Chandler and Johnson (10), to show the effect of HCl upon the Al-speciation in the IL and these results suggested that two structures could exist for the anion having the following stoichiometry: $[Al_2Cl_8H]^-$. Subsequent molecular modeling suggested that the HCl molecule could reside in three different environments and these sites showed different free energies of formation (11). The decreasing reactivity of these IL's as the structure of the cation was changed to show increasing chain length of R [RMIM]$^+$, R = C_nH_{2n+1}; n = 2, 4, 6, 8, and 12, could not be related to the amount of HCl adsorbed in the IL's. It became apparent that the molecular environment of these IL's was more complicated than what had been reported in the literature.

The aim of this manuscript is to examine the molecular structure of the IL as the ratio of Al/RMIM$^+$ was changed with and without HCl added. The working hypothesis is that the aluminum speciation determines the reactivity of the IL and this speciation depends upon the structure of RMIM$^+$.

Results and Discussion
^{27}Al-NMR data of the [n-BuMIM]$^+$/[Al$_n$Cl$_{3n+1}$]$^-$ IL's for n = 1, 3/2, and 2 are shown in Fig. 1 without any HCl added to the IL's. The observed resonances downfield from Al(NO$_3$)$_3$ (aq) are shown in Table 1. These observations of peak shape are similar to those reported earlier (7, 8); however, neither of these earlier studies used an internal standard, nor did they employ quantum mechanical methods to assist in the peak assignment. The association of the organic halide with the AlCl$_3$ was modeled by transition state methods to give the species shown in Figure 2. The products of this transition state were two dinuclear Al-species showing different structures but similar free energies of formation. Figure 2-a shows the species before the transition state optimization and Figure 2-b is the

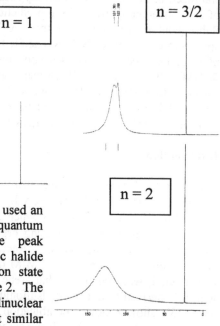

Figure 1. ^{27}Al-NMR of IL's with R = butyl and n = 1; 3/2; and 2 mol/mol (clockwise)

optimized geometry at the transition state. Product 1 (Figure 2-c) was formed by cutting two Al-μ–Cl bonds indicated by vertical lines; whereas, product 2 (Figure 2-d) was formed by cutting one Al-μ–Cl bond as shown by the horizontal line. The predicted ^{27}Al-NMR chemical shifts from Al(NO$_3$)$_3$ (aq) are also shown in the same figures.

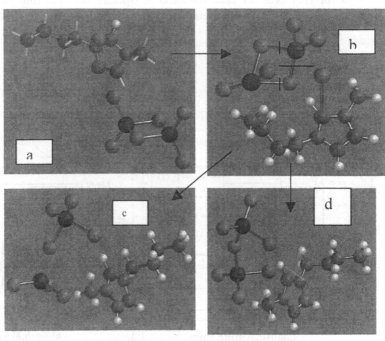

Figure 2: a-reactants, b-optimized transition state geometry, c-product 1, d-product 2

The free energies for forming products 1 and 2 were 99.68 and 99.69 kcal/mol, which are indistinguishable, considering that the uncertainty in these predictions is 4-5 kcal/mol (12). The predicted ^{27}Al-NMR chemical shifts downfield from Al(NO$_3$)$_3$ (aq) were 69.8, 69.7 ppm for product 1 and 62.7, 60.0 ppm for product 2. While the absolute values of the predicted chemical shifts are ~30 ppm smaller than the observed chemical shifts, these calculations permit the assignment shown in Table 1.

The success of this modeling can be ascertained by the ability to replicate the observed peak shapes using Gaussian peaks centered at peak positions suggested by the modeling (Fig. 3). For the case where n = 3/2, five peaks were needed: 1 for the monomeric Al and 2 peaks each for each of the two dinuclear Al species. These peaks were combined to successfully replicate the observed NMR peaks recorded for this sample. When n = 2, the data were reproduced using only 4 peaks, two each for each of the two dinuclear Al species. Our earlier predictions (11) showed that HCl could combine with either of the dinuclear Al-species in three different positions, which showed different acid strengths.

We attempted to model the ^{1}H-NMR of HCl in these three positions to show chemical shifts of 15, 14, and 2-3 ppm (Fig. 4). These predictions suggest that three sites for HCl exist for each dinuclear Al-species and that four of the six sites are much more acidic than the remaining two sites. Subsequent ^{13}C-NMR measurements

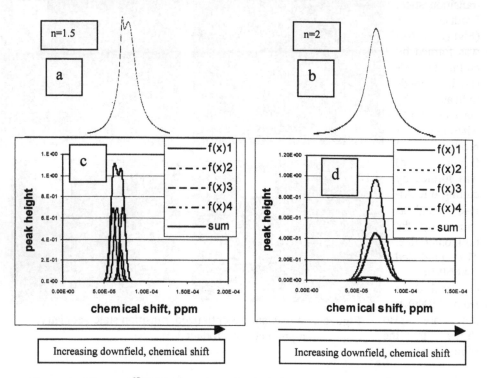

Figure 3. Observed ^{27}Al-NMR spectrum for IL derived from [n-BuMIM]$^+$ [Al$_n$Cl$_{3n+1}$]$^-$, a: n = 3/2; b: n = 2; Predicted ^{27}Al-NMR spectrum for IL derived from [n-BuMIM]$^+$ [Al$_n$Cl$_{3n+1}$]$^-$. c: n = 3/2: b: n = 2.

of 2-^{13}C-acetone in the BMIM-IL confirmed the presence of six acidic protons (Fig. 4) showing chemical shifts of 246.7 to 239.8 ppm. The chemical shifts of 2-^{13}C-acetone in contact with sulfuric acid and triflic acid confirmed that labeled acetone chemical shifts greater than 245 ppm are super acidic. (13)

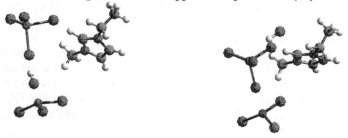

Figure 4 – a: 15.3 ppm(245)   **Figure 4 – b:** 14.3 ppm(242)

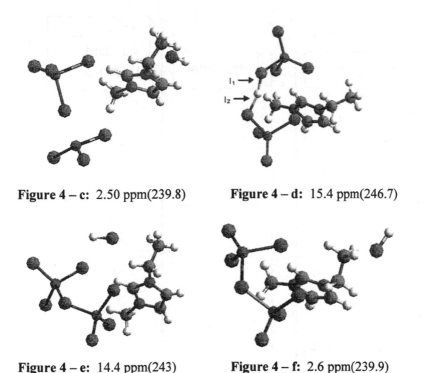

Figure 4 – c: 2.50 ppm(239.8)   **Figure 4 – d:** 15.4 ppm(246.7)

Figure 4 – e: 14.4 ppm(243)   **Figure 4 – f:** 2.6 ppm(239.9)

Figure 4. Models of IL's showing optimized positions of HCl sited in the structures and the calculated ^{1}H-NMR chemical shifts from TMS along with ^{13}C-NMR chemical shifts observed in 2-^{13}C-Acetone in IL reported in parentheses

The ^{27}Al-NMR of the IL's derived from [RMIMCl][Al$_2$Cl$_6$] showed systematic changes in the shapes of the spectra as the chain length of R increased. The NMR spectra for IL's derived from ethyl- and dodecyl-MIM$^+$ are shown in Fig. 5. Vertical lines show the chemical shifts of the two, dinuclear species [AlCl$_3$-AlCl$_4$]$^-$ at 98.2 ppm and [Al$_2$Cl$_7$]$^-$ at 102.2 ppm. These spectra and the others were deconvoluted to determine the mole fractions of these two Al-containing species (Table 2). The observed rate constants are included in the same table for toluene carbonylation over these same IL's at room temperature.

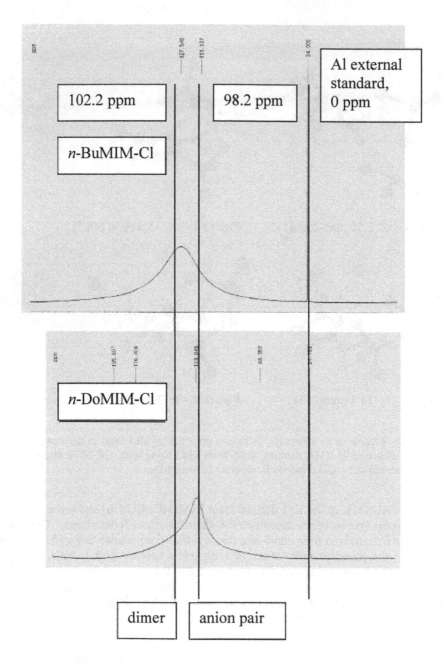

Figure 5. ^{27}Al-NMR of ILs derived from [RMIM]$^+$[Al$_2$Cl$_7$]$^-$ where R = *n*-butyl and *n*-dodecyl

Table 1. ^{27}Al-NMR chemical shifts in $[n\text{-BuMIM}]^+[\text{Al}_n\text{Cl}_{3n+1}]^-$; n = 1, 3/2, & 2

n	Peak 1	Peak 2	Peak 3
1	103.4		
3/2		102.2	98.2
2		102.2	98.2
Assignment	$[\text{AlCl}_4]^-$	$[\text{Al}_2\text{Cl}_7]^-$	$[\text{AlCl}_3\text{-AlCl}_4]^-$
Predicted CS's	72.4	69.8; 69.7	62.7; 60.0

Apparently, the mole fraction of the species $[\text{Al}_2\text{Cl}_7]^-$ was governed by the structure of the cation where the concentration of this anion was highest when the chain length of R was shortest (ethyl) and it was smallest when the chain length of R was longest (dodecyl). The rate constants could be correlated by a linear function of the mole fraction of the $[\text{Al}_2\text{Cl}_7]^-$, Figure 6. In the proposed mechanism, Scheme 1, the $(\text{RMIM})^+$ structure determines the equilibrium distribution, K, between species **I** & **II**. HCl reacts with **I** to form a super acidic species (**III**), which combines with CO to give the formyl cation (**IV**). This formyl cation reacts with substrate to form tolualdehyde that is stabilized as an adduct with AlCl_3 in species **V**. The aldehyde product can be recovered upon addition of water to the system. Without water addition, the aldehyde adduct cannot be used again either 1) for lack of sufficient acidity, or 2) there is no other aluminum species to stabilize the aldehyde product. These data can be explained by the reaction mechanism, Scheme 1.

Table 2. Effect of cation structure upon the mole fraction $[\text{Al}_2\text{Cl}_7]^-$ and the observed toluene carbonylation rate constant.

Cation	x$[\text{Al}_2\text{Cl}_7]^-$	k, kmin^{-1}
EMIM$^+$	0.812	4.67
BMIM$^+$	0.702	3.92
HMIM$^+$	0.599	2.98
OMIM$^+$	0.512	2.34
DoMIM$^+$	0.377	1.35

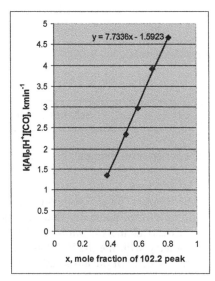

Figure 6. Correlation of observed rate constants with x$[\text{Al}_2\text{Cl}_7]^-$

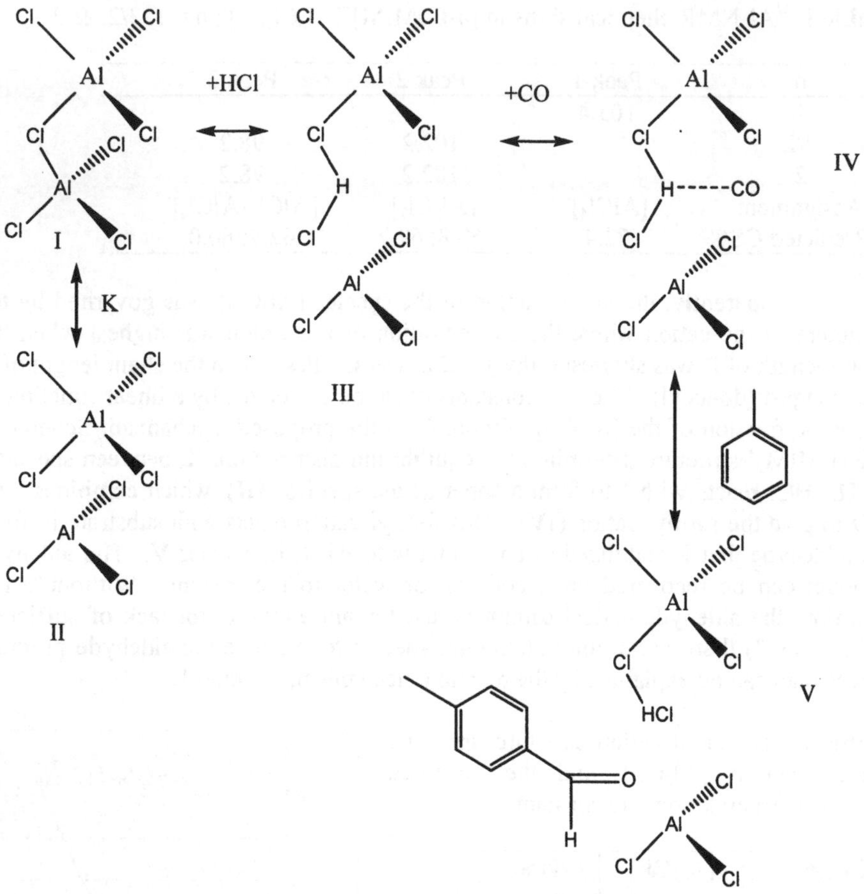

Scheme 1: Mechanism for arene formylation

Experimental Section

Calculations

The Spartan '02 & '04 molecular modeling software packages were used to predict the optimized geometry and the NMR spectra. The ^{27}Al-, ^{1}H-NMR spectra were predicted at the *ab initio* level using the Hartree-Fock to obtain the equilibrium geometry with the 3-21G* basis set. In a similar manner, we predicted the proton NMR spectra of the absorbed HCl.

Chemicals

The imidazolium compounds 1-ethyl-3-methyl-imidazolium-chloride (EMIM-Cl), 1-butyl-3-methyl-imidazolium-chloride (n-BuMIM-Cl), and 1-hexyl-3-methyl-imidazolium-chloride (n-HeMIM-Cl), toluene (anhydrous, 99.99%), and HCl were obtained from Sigma Aldrich; whereas, 1-octyl-3-methyl-imidazolium-chloride (n-OcMIM-Cl), 1-Dodecyl-3-methyl-imidazolium-chloride (n-DoMIM-Cl) were obtained from Merck Chemicals and used without further purification. Aluminum chloride (99.99%), obtained from Sigma Aldrich, was sublimed under a vacuum before use. Carbon monoxide, CP grade, was obtained from Airgas. Acetone, enriched in ^{13}C in the 2-position (99%), was purchased from Sigma-Aldrich and used without further purification.

^{27}Al-NMR and ^{13}C-NMR

The external standard for ^1H, and ^{13}C was dimethyl sulfoxide (99%). The aluminum external reference was aqueous, aluminum nitrate (Fisher) at a concentration of 1 M. IL's under HCl partial pressures (~ 3 bar) were examined in NMR tubes from NEW ERA Enterprises, Inc; NE-PCAV-5-130). The capillary tube was approximately 76.2 - 88.9 mm long 1.5 mm diameter and was filled to approximately 50.8 - 63.5 mm of the tube's length with the standard. The capillary tube containing the standard was placed inside the NMR tube and the IL or acid (sulfuric or triflic acid) was added. The NMR tube was sealed, evacuated and filled up with the desired gas to the desired pressure, and contacted with this gas for 30 minutes. During the gas addition, the sample was shaken to promote mixing of the gas with the liquid, and bubbling of the liquid was observed to indicate that good mixing had been achieved. All data were recorded on a Bruker 300X. Typically, 256 and 64 scans were obtained for ^{13}C, ^{27}Al, respectively. The spectrometer settings for these analyses were standard settings already established by the scientists at the Georgia Tech NMR Center.

Preparation of IL's

The detailed synthesis of the IL's used here was described earlier (6). Two samples were prepared having R = butyl but the Al/cation ratio was 3/2 and 2 mol/mol. One sample each was prepared having R = ethyl, hexyl, octyl, and dodecyl for which the Al/cation = 2 mol/mol.

Acknowledgements

We gratefully acknowledge the support from the U.S. Department of Education for a GAANN Award to E.A.

References

1. P. Wasserscheid and W. Keim, *Angew. Chemie (Int. Ed.)*, **2000** *39*, 3772-89.
2. H. A. Øye, M. Jagtoyen, T. Oksefjell, & J. S. Wilkes, *Mater. Sci. Forum*, (1991), **73-75**, 183-9.
3. Z. J. Karpinski, and R. A. Osteryoung, *Inorg. Chem.* (1984), **23**, 1491-4; Kbdul-Sada, A. A. K., K. R. Greeway, K. R. Seddon, & T. Welton, *Org. Mass Spectrom.* (1993), **28**, 759-65.
4. P. Smith, A. S. Dworkin, R. M. Pagni and S. P. Zingg, *J. Am. Chem. Soc.,* 1989, **111**, 525.
5. Minhui Ma and K. E. Johnson, *J. Am. Chem. Soc.* (1995), **117**, 1508-13.
6. E. J. Angueira and M. G. White, *J. Mol. Catal. A.* **227**/1-2, 51-58 (2005).
7. J. L. Gray and G. E. Maciel, *J. Am. Chem. Soc.,* 1981, **103**, 7147.
8. J. S. Wilkes, J. S. Frye and G. F. Reynolds, *Inorg. Chem.* (1983) **22**, 3870.

9.   E. J. Angueira, and M. G. White, *A. I. Ch. E. J.* **51**, No. 10, 2778-85 (2005).
10. W. D. Chandler and K. E. Johnson, *Inorg. Chem.* **38**, 2050-6 (1999)
11. E. J. Angueira and M. G. White, *J. Mol. Catal. A.*, **238,** 163-74 (2005).
12. MOPAC 2002 Manual:
     http://www.cachesoftware.com/mopac/Mopac2002manual/node650.html
13. E.J. Angueira, Ph. D. thesis, Georgia Institute of Technology (2005).

32. Chemoselective Hydrogenation of Nitro Compounds to the Corresponding Amines on Raney® Copper Catalysts

Simon Robitaille, Geneviève Clément, Jean Marc Chapuzet and Jean Lessard

*Laboratoire de Chimie et Electrochimie Organiques, Département de Chimie,
Université de Sherbrooke, Sherbrooke, Québec, J1K 2R1 Canada*

Jean.Lessard@USherbrooke.ca

Abstract

The selectivity of the electrocatalytic hydrogenation (ECH) method for the reduction of nitro compounds to the corresponding amines is compared with that of reduction by Raney copper (RCu) alloy powder in alkaline aqueous ethanol. In the former method, chemisorbed hydrogen is generated *in situ* by electrochemical reduction of water. In the latter method (termed "chemical catalytic hydrogenation" (CCH)), chemisorbed hydrogen is also generated *in situ* but by reduction of water by aluminium (by leaching of the alloy). Finally, the selectivity and efficiency of the electrochemical reduction of 5-nitro-indoles, -benzofurane, and -benzothiophene at RCu electrodes in neutral and alkaline aqueous ethanol is compared with that of the classical reduction with zinc in acidic medium.

Introduction

The electrocatalytic hydrogenation (ECH) of a nitro group to the corresponding amine in neutral or basic aqueous or mixed aqueous-organic media is described by equations [1] to [7] where M represents an adsorption site of the catalyst and M(H) chemisorbed hydrogen, and $M(RNO_2)$, $M(RNH(OH)_2)$, and $M(RNHOH)$ represent

[1] $6H_2O + 6e^- + M \longrightarrow 6M(H) + 6OH^-$

[2] $RNO_2 + M \rightleftharpoons M(RNO_2)$

[3] $M(RNO_2) + 2 M(H) \longrightarrow M(RNH(OH)_2)$

[4] $M(RNH(OH)_2) + 2 M(H) \longrightarrow M(RNHOH) + H_2O$

or/and $\begin{cases} [5] \ M(RNH(OH)_2) \longrightarrow M(RNO) + H_2O \\ [6] \ M(RNO) + 2 M(H) \longrightarrow M(RNHOH) + H_2O \end{cases}$

[7] $M(RNHOH) + 2 M(H) \longrightarrow M(RNH_2) + H_2O$

$$M + RNH_2$$

the adsorbed organic substrate and adsorbed reduction intermediates. The stoichiometry applies to the adsorbed species only, not to the adsorption site M (1-3). ECH involves the same hydrogenation steps (steps [2] to [7]) as those of classical catalytic hydrogenation (CH). In both ECH and CH, the hydrogenolysis of the adsorbed dihydroxylamine, $M(RNH(OH)_2)$ (step [4]), could be faster than its dehydration to the adsorbed nitroso derivative M(RNO) (step [5]) (or alternatively, than its desorption followed by dehydration to RNO (not shown)).

On Ni, Co, and Cu electrodes, two mechanisms are competing for the electrochemical reduction of nitro groups to hydroxylamines: 1) the ECH mechanism (equations [1] to [4]); and 2) electron transfer to the nitro group (and to the intermediates formed) followed by protonation by water (electronation-protonation (EP) mechanism) (equations [8] to [11]) (3, 4). For the hydro-genolysis of the hydroxylamine to the amine, the sole mechanism operating is the ECH mechanism (equation [4]). This is because unprotonated hydoxylamines (pH > 5) are not reducible by electron transfer (EP mechanism) (5, 6). For that same reason, the electrohydrogenation of nitro compounds at a high hydrogen overvoltage cathode (Hg, glassy carbon) in neutral (pH > 5) and basic medium stops at the hydroxylamine (5, 6).

[8] $RNO_2 + e^- \rightleftharpoons RNO_2^{\bullet-}$

[9] $RNO_2^{\bullet-} + H_2O \longrightarrow RNO_2H^{\bullet} + OH^-$

[10] $RNO_2H^{\bullet} + e^- \longrightarrow RNO + OH^-$

[11] $RNO + 2e^- + 2H_2O \longrightarrow RNHOH + 2OH^-$

In this paper, the selectivity of the ECH method for the reduction of nitro compounds to the corresponding amines on RCu electrodes will be compared with that of reduction by RCu alloy powder in alkaline aqueous ethanol. In the latter method (termed chemical catalytic hydrogenation (CCH)), chemisorbed hydrogen is generated *in situ* but by reduction of water by aluminium (by leaching of the alloy) (equation [12]). The reductions by *in situ* leaching must be carried out in a basic medium in order to ensure the conversion of insoluble $Al(OH)_3$ into soluble aluminate (equation [12]). The selectivity and efficiency of the electrochemical reduction of 5-nitro-indoles, -benzofurane, and -benzothiophene at RCu electrodes in neutral and alkaline aqueous ethanol will also be compared with that of the classical reduction with zinc in acidic medium.

[12] $CuAl + 3H_2O + HO^- \longrightarrow 3Cu(H) + Al(OH)_4^-$

Results and Discussion

Reduction of nitroaryl groups

Nitrobenzene (1) was used as a model to determine the best conditions for the "chemical catalytic hydrogenation" (CCH) using the Raney Cu alloy. In this work, we used ethanol as organic co-solvent because it is greener than methanol. The best conditions for CCH were 0.05 M KOH in ethanol-water (45:55 v/v) (pH ≈ 12.5), at room temperature and under a nitrogen atmosphere, using an amount of alloy corresponding to 3 equation of Al. The conversion was complete after 5 h and aniline was obtained in a quantitative yield (100% by vapor phase chromatography). The ECH of nitrobenzene at Raney Cu electrodes in alkaline aqueous methanol gives aniline in a nearly quantitative yield (99%) (7).

When these CCH conditions were applied to *o*-iodonitrobenzene (1), *o*-iodoaniline (2) together with iodoaniline (4) and nitrobenzene (1) were obtained in the yields indicated in Scheme 1 (88% mass balance), which corresponds to a 62% selectivity for the formation of *o*-iodoaniline (3). Under the same conditions, the ECH of *o*-iodonitrobenzene (2) gave aniline (90% yield) as the sole product. There was complete hydrogenolysis of the C−I bond (no selecti-vity). Thus, in basic medium (pH ≈ 12.5), CCH method is much more selective than ECH. However, in weakly acidic medium (pH 3, pyridine.HCl buffer), it has been reported that ECH at a RCu cathode in methanol-water 95:5 (v/v) gave *o*-iodoaniline in a 97% yield (1, 3).

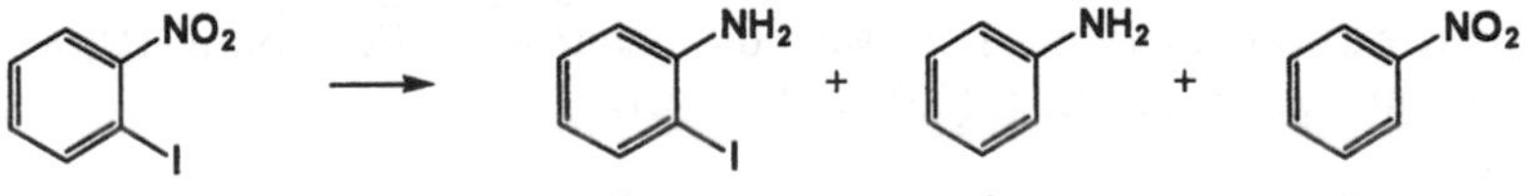

	3	4	1
CCH (pH 12.5)	55%	28%	5%
ECH (pH 12.5)	0%	90%	0%
ECH (pH 3) (1, 3)	97%	n.d	n.d

Scheme 1

The CCH of *p*-nitroacetophenone (5) (Scheme 2) was inefficient and unselective giving *p*-aminoacetophenone (6) in about 20% yield together with some 10-12% of unidentified compounds (30% mass balance). Attempts to recover more material were unsuccessful. The ECH at a RCu cathode in an alkaline (0.28 M KOH, pH ≈ 13.5) MeOH-H$_2$O (1.5% of H$_2$O) solution was reported to give exclusively *p*-amino-acetophenone (6) (79% yield of isolated product) (3,8).

5	CCH (pH 12.5)	20% 6
	ECH (pH 13.5) (3,8)	79%

Scheme 2

The CCH and ECH of *p*-cyanonitrobenzene (**7**) at pH 12-13 gave the three products shown in Scheme 3 with low selectivities for the formation of *p*-cyanoaniline (**8**) respectively of 8-13% (82% mass balance) and 7% (65% mass balance). In neutral medium, the ECH of **7** has been reported to be very efficient, giving only *p*-cyanoaniline (**8**) in a quantitative yield (1, 3).

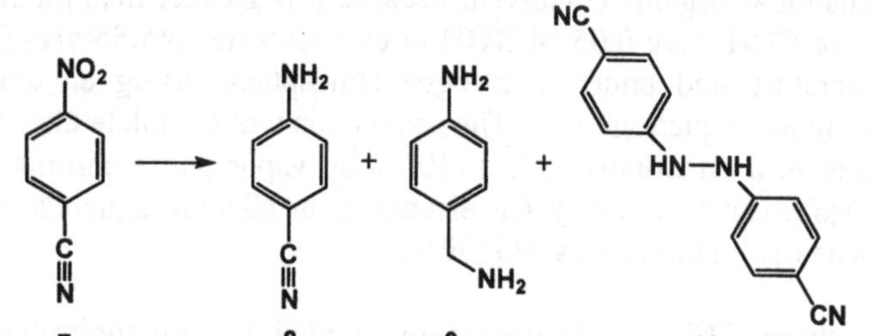

	7	**8**	**9**
CCH (pH 12.5)		36-47%	17-8%
ECH (pH 12.5)		27%	10%
ECH (pH 6) (1, 3)		100%	n.d

(with **10**: CCH 29-27%, ECH 38%, ECH (pH 6) n.d)

Scheme 3

The CCH and ECH of the *o*-nitrodioxolane **11** was very selective giving only the corresponding *o*-aminodioxolane **12** (Scheme 4). No cleavage of the dioxolane ring giving the aminodiol **13** and cyclohexanone by hydrogenolysis of the benzylic C–O bond was observed. Such cleavage did occur upon ECH on a RNi electrode under the same basic conditions (60% yield of **13**).

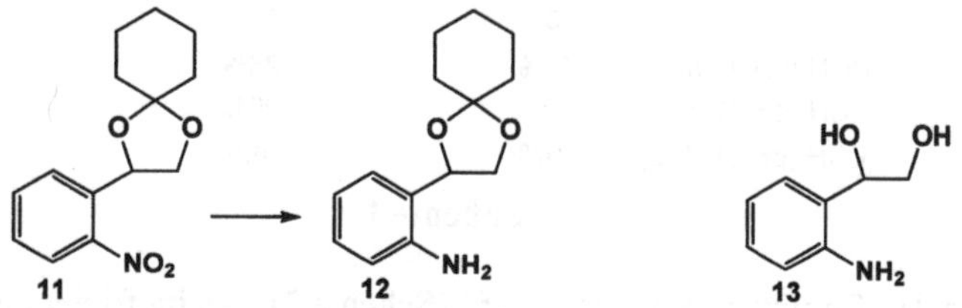

	11	**12**	**13**
CCH (pH 12.5)		100%	
ECH (pH 12.5)		92%	

Scheme 4

The CCH of 2-(*p*-nitrobenzyl)-2-nitropropane (**14**) allowed the selective reduction of the nitroaryl group, without reduction of the tertiary nitroalkyl group, giving 85-93% of 2-(*p*-aminobenzyl)-2-nitropropane (**15**) (Scheme 5). No diamino

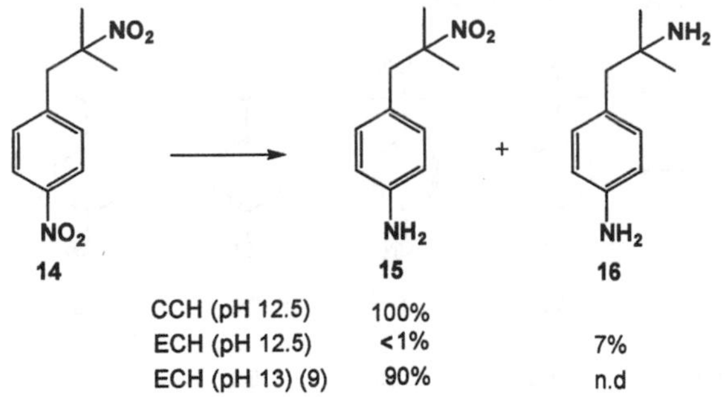

Scheme 5

derivative **16** was detected. The ECH under the same conditions (0.05 M KOH) gave very poor results as shown in Scheme 5. However, ECH in a neutral medium (acetate buffer) in MeOH-H$_2$O (93:7 v/v)) has been reported to be 100% selective giving nitroamine **15** as the sole product in a 100% yield (9).

Nitroalkyl groups

The CCH of nitroacetylenic acetal **17** gave the corresponding amine **18** in a low yield (23%) (Scheme 6). The reduction stopped at the hydroxylamine which underwent decomposition to acetylenic alcohol **19** (11%) and an ene-yne acetal which was further reduced to diene acetal **20** (67%). The ECH in neutral medium (acetate buffer in CH$_3$OH-H$_2$O 93:7 v/v) has been reported to give 72% of acelylenic amine **18** (3, 10). The products resulting from the decomposition of the intermediate acetylenic hydroxylamine were not identified (3, 10).

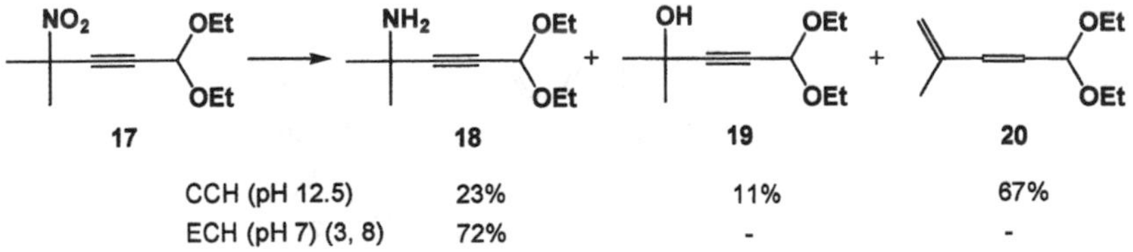

Scheme 6

The CCH of *p*-cyano-nitrocumene (**21**) under the standard conditions gave *p*-cyano-aminocumene (**22**) as the main product (84%) (Scheme 7), the alcohol **23** (14%) and the styrene **24** (5%) both resulting from the decomposition of the intermediate hydroxylamine. The ECH under the same conditions is much less selective giving only 26% of *p*-cyano-aminocumene (**22**) together with alcohol **23** (36%, the main product), the styrene **24** (17%), *p*-cyanocumene (**25**) (8%), and *p,p'*-dicyanobicumyl (**26**) (18%) (Scheme 7). The ECH in neutral medium (pH 5, NaCl 0.1 M in methanol-water 95:5 (v/v)) has been reported to be highly selective giving 91% of isolated *p*-cyano-aminocumene (**22**) (1, 3). The bicumyl **26** results from elec-

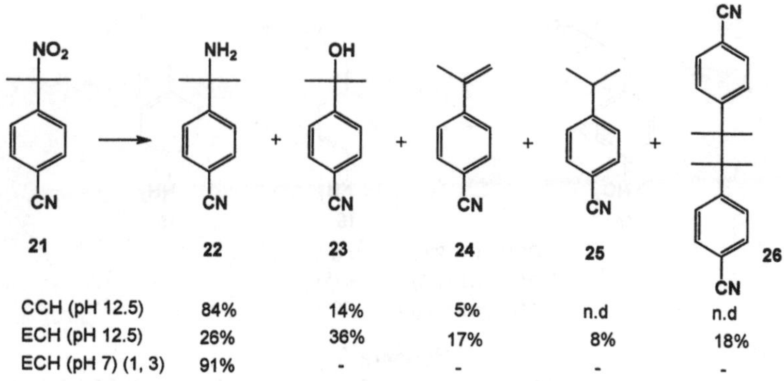
NO2
NH2
OH
CN
21
22
23
24
25
26
CCH (pH 12.5) 84% 14% 5% n.d n.d
ECH (pH 12.5) 26% 36% 17% 8% 18%
ECH (pH 7) (1, 3) 91% - - - -

Scheme 7

tron transfer to **21** (equation [8]) followed by fast cleavage of the resulting radical anion (11) (Scheme 8) and the electron transfer does compete with reaction with chemisorbed hydrogen (equations [2] and [3]) as previously shown (12).

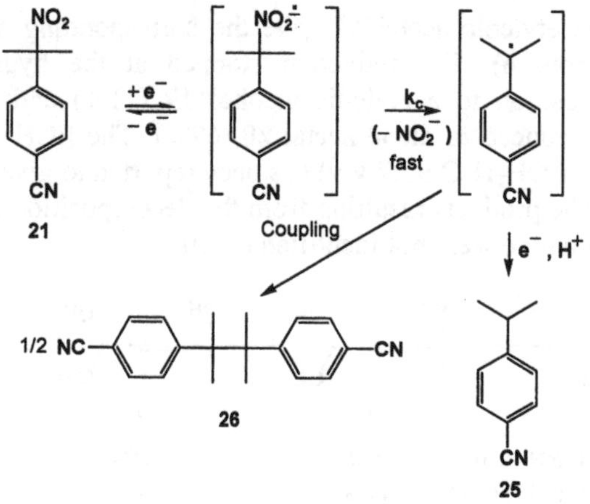
NO2
CN
21
+e⁻
−e
kc
(− NO2⁻)
fast
Coupling
e⁻, H⁺
1/2 NC
CN
26
25

Scheme 8

ECH of 5-nitroindoles, 5-nitrobenzofurane, and 5-nitrobenzothiophene

As shown in Scheme 9, the ECH of the 5-nitro derivatives **26** of the indole family in EtOH-H_2O 50:50 (v/v) gives higher yields of 5-amino derivatives **27** (range over three experiments) in basic medium (0.15 M KOH, pH $\approx$ 13) than in neutral medium (acetate buffer, pH $\approx$ 6) (13, 14). By comparison, reduction under the classical acidic conditions (Zn/HCl in the same solvent) gave a mixture of 5-amino derivative **27** in the yields indicated in Scheme 9 and of 4-chloro-5 amino derivatives **28** (see inset of Scheme 9) in 15%, 17%, 8% and 23% yield for X = NH, NCH_3, O, and S respectively (13, 14).

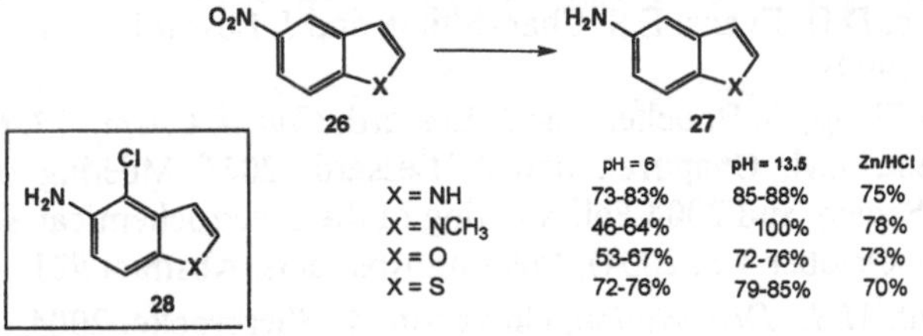

	pH = 6	pH = 13.5	Zn/HCl
X = NH	73-83%	85-88%	75%
X = NCH₃	46-64%	100%	78%
X = O	53-67%	72-76%	73%
X = S	72-76%	79-85%	70%

Scheme 9

Conclusions

The reduction of polyfunctional nitro compounds, nitroaryl as well as nitroalkyl compounds, to the corresponding amines in basic aqueous alcoholic solutions on Raney copper (RCu) is more selective if carried out by generating chemisorbed hydrogen by electroreduction of water (ECH method) than by generating it by leaching of the alloy *in situ* (CCH method) except in the case of *o*-iodonitrobenzene (2) for which the CCH method is more selective. However, the most selective method in all cases studied is ECH in neutral medium (pH 3-7).

Acknowledgements

Financial support from NSERC of Canada, the Fonds FCAR of Quebec, the Ministry of Energy and Natural Resources of Québec, and the Université de Sherbrooke is gratefully acknowledged.

References

1. J.M. Chapuzet, R. Labrecque, M. Lavoie, E. Martel, and J. Lessard. *J. Chim. Phys.*, **93**, 601 (1996).
2. J.M. Chapuzet, A. Lasia and J. Lessard, *Electrocatalysis*, Frontiers of Electrochemistry Series, J. Lipkowski and P.R. Ross, Eds, Wiley-VCH, Inc., New York, 1998, p. 155-196.
3. J. Lessard, Chemical Industries (CRC Press), **104**, (*Catal. Org. React.*) 3-12 (2005).
4. E. S. Chan-Shing, D. Boucher and J. Lessard, *Can. J. Chem.*, 77, 687 (1999).
5. H. Lund, *Organic Electrochemistry*, H. Lund and O. Hammerich, Eds, Marcel Dekker, Inc., New York-Basel, 2001, p. 379.
6. A. Cyr, E. Laviron, and J. Lessard, *J. Electroanal. Chem.*, **263**, 69 (1989).
7. A. Cyr. P. Huot, G. Belot, and J. Lessard, *Electrochim. Acta*, **35**, 147 (1990).
8. G. Belot, S. Desjardins, and J. Lessard, *Tetrahedron Lett.* 25, 5347 (1984).
9. B. Côté. *M.Sc. Dissertation*, Université de Sherbrooke, 1993.
10. J.M. Chapuzet, B. Côté, M. Lavoie, E. Martel, C. Raffin, and J. Lessard. "The Selective Reduction of Aliphatic and Aromatic Nitro Compounds" in *Novel Trends in Electroorganic Synthesis*, S. Torii, Ed., Kodansha, Tokyo, 1995, p. 321.

11. Z-R. Zheng, D.H. Evans, E.S. Chan-Shing, and J. Lessard, *J. Am. Chem. Soc.*, **40**, 9429 (1999).
12. E.S. Chan-Shing, D. Boucher, and J. Lessard, *Can. J. Chem.*, **77**, 687 (1999).
13. G. Clément, J.M. Chapuzet, and J. Lessard, 206th Meeting of the Electrochemical Society and 2004 Fall Meeting of the Electrochemical Society of Japan (Honolulu, October 3-8, 2004), Meeting Abstracts, Abstract #2130.
14. G. Clément. *M.Sc. Dissertation*, Université de Sherbrooke, 2004.

33.   Selective Hydrogenolysis of Sugar Alcohols over Structured Catalysts

Chunshe (James) Cao, James F. White, Yong Wang and John G. Frye

Pacific Northwest National Laboratory, 902 Battelle Blvd, MS K8-93, Richland, WA 99352
chunshecao@yahoo.com

Abstract

A novel gas-liquid-solid reactor based on monolith catalyst structure was developed for converting sugar alcohols to value-added chemicals such as propylene glycol. The structured catalyst was used intending to improve product selectivity. Testing at the pressure of 1,200 psig and 210°C with H_2 to sorbitol molar ratio of 8.9 and a space velocity range from 0.15 to 5 hr^{-1} demonstrated that as high as 41 wt% of propylene glycol selectivity and 13 wt% ethylene glycol selectivity can be obtained. In addition, monolith catalysts gave higher C3/C2 ratio than that in the conventional trickle bed reactor with similar liquid hourly space velocities.

Introduction

Bio-based feedstocks such as glucose, sorbitol etc. can be converted into value-added chemicals such as ethylene glycol, 1,2-propylene glycol and glycerol by reacting with hydrogen over the catalysts (1-4). Such catalytic hydrogenolysis of sugar alcohols occurs in gas-liquid-solid three phase reaction systems. Conventional reactors used in this process are slurry or trickle bed reactors. In the scale-up and commercial demonstration, selectivity to desired products may be limited due to severe mass transfer limitation in the three-phase system. The selectivity to desired products are limited by the resistance in the interfaces of gas-liquid, liquid solid, and gas-solid, preventing the reactant liquid molecules from contacting the catalytic sites. Consequently, the reduced overall reaction rate requires low space velocity operation to achieve high conversion. However, long residence time may cause unwanted secondary reactions so that the selectivity is compromised.

This study uses novel monolith structured catalysts in aiming to improve process productivity and selectivity. Apart from the advanced characteristics of low pressure drop, less backmixing, convenient change out of catalysts, the monolith reactor structure reduces the mass transfer limitation and potentially improves the selectivity.

Results and Discussion

The present investigation provides a hydrogenolysis method in which sorbitol is reacted with hydrogen, at a temperature of 210°C and 1200psig. The solid catalyst is present as a new form of a structured monolith containing Nickel and Rhenium as the multimetallic catalyst. The study provides a method of improving the catalytic selectivity of sorbitol hydrogenolysis to valued added products. It has been demonstrated that the total selectivity to glycerol, ethylene glycol, and propylene glycol can be improved by 20% using such a structured mass transfer reduced catalyst compared to conventional trickle bed format catalyst. This method can be extendedly applied to hydrogenolysis of other sugar alcohols such as glycerol, xylitol and potentially also to glucose.

As shown in Table 1, at the same temperature and pressure, sorbitol conversion and selectivity to value-added products are listed, in which the monolith reactor covers the liquid hourly space velocity (LHSV) condition of the trickle bed. It was surprisingly observed that total selectivity to glycerol, ethylene glycol (EG), and propylene glycol(PG) from the monolith reactor was as 12.5% higher than that of conventional trickle bed. The PG selectivity is particularly high in the monolith reactor, which gives high C3/C2 ratio in the product mixtures. Such selectivity increase does sacrifice sorbitol conversion to a certain extent, but the lower conversion is caused by higher weight hourly space velocity(WHSV) due to less active material loading. It is noted that the comparison was made at the similar volume based LHSV. In the meantime, the monolith reactor was operated at a much higher WHSV, as the catalyst loading on the monolith substrate is much less than that in the packed bed. This demonstrated the high efficiency of catalyst utilization with reduced mass transfer resistance. The reduction of mass transfer limitation is in part attributed to the decreased mass transfer distances with thin coating as well as the unique Taylor flow pattern within the tunnel structure of the monolith catalyst.

Table 1. Performance comparison between the monolith reactor and the conventional trickle bed reactor.

Catalyst	Ni/Re/TiO2	T = 210°C	P = 1200 psig				
				Product Polyol Selectivity			
		Sorbitol			Ethylene	Propylene	Total EG +PG	Carbon
Monolith # 1	LHSV	Conversion	Glycerol	Glycol	Glycol	+ Glycerol	Balance
	0.38	65.40%	17.90%	11.10%	30.70%	59.70%	96.50%
	0.76	52.20%	19.30%	12.70%	41.30%	73.30%	97.90%
	1.52	44.80%	20.20%	13.40%	38.90%	72.50%	96.90%
Trickle Bed							
Ni/Re/Carbon							
	0.83	89.00%	16.70%	12.60%	31.50%	60.80%	

Experimental Section

Cordierite monolith (400cpi) (Corning, Inc) was used as a substrate. The opening of each channel has 1x1 mm of dimension. The monolith was pre-machined to fit into a stainless steel tubular reactor with 1.27cm of diameter and 52.8cm of length. The monolith substrate was first treated with 10wt% HNO_3 for 1 hour at 80°C followed by washing with DI H_2O. This served to both clean the monolith and rehydrate the surfaces. The monolith was then dried at 50°C overnight. TiO_2 contained Tyzor LA material (Dupont) was applied as catalyst support which was coated onto the monolith, The TiO_2 concentration in Tyzor was 2.2M. Tyzor coating was applied sequentially with inter-drying steps and calcinations at 350°C. The total TiO_2 loading is 5.2g. The coating thickness is 25µm. $Ni(NO_3)_2$. $6H_2O$ and $HReO_4$ were the precursors of the active metals. Aqueous phase solution containing 8.6wt% Ni and 1.22wt% Re was impregnated into the TiO_2 coating on the monolith. The catalyst was then calcined at 300°C for 3 hours. The final composition was 18% Ni/ 2.6% Re/TiO_2.

The monolith catalyst was snugly fit into a tubular reactor with a jacked heat exchanger. The catalyst was activated by flowing hydrogen across the bed at atmospheric pressure, and the bed was heated to 285°C and held for 8 hours. The reactor was then cooled under hydrogen. The reactor was then raised to 1200 psig and 210°C. 25wt% sorbitol, 2.1wt% NaOH aqueous solution was pumped into the system with flowrates of 25-100 ml/hr. Hydrogen flow was regulated by a mass flow controller to keep the H_2/ Sorbitol molar ratio of 8.9 The reactor pressures was maintained at 1200 psig by a dome-loaded back pressure regulator.

Acknowledgements

This work was funded by Laboratory Directed Research and Development (LDRD) Program at Pacific Northwest National Laboratory. Pacific Northwest National Laboratory is operated by Battelle Memorial Institute for US Department of Energy.

References

1. T. A. Werpy, J. G. Frye, A. H. Zacher and D. J. Miller, US Pat. 6,841,085, to Battelle Memorial Institute and Michigan State University (2005).
2. D. C. Elliott, U.S. Pat. Appl. 2002,169,344, Battelle Memorial Institute (2002).
3. T. A. Werpy, J. G. Frye, A. H. Zacher and D. J. Miller, US Pat. 6,479,713, to Battelle Memorial Institute and Michigan State University (2002).
4. S. P. Chopade, D. J. Miller, J. E. Jackson, T. A. Werpy, J. G. Frye, A. H. Zacher, US Pat. 6,291,725, to Michigan State University and Battelle Memorial Institute (2000).

34. Twinphos: A New Family of Chiral Ferrocene *Tetra*-Phosphine Ligands for Asymmetric Catalytic Transformations

Benoit Pugin, Heidi Landert, Martin Kesselgruber, Hans Meier, Richard Sachleben, Felix Spindler and Marc Thommen

Solvias AG, Klybeckstrasse 19, Postfach CH 4002, Basel, Switzerland

Marc.Thommen@Solvias.com

Abstract

A new family of chiral ligands for asymmetric homogeneous hydrogenation has been developed. The performance of mono- and bis-rhodium complexes of these chiral ferrocene tetraphosphine ligands in the hydrogenation of model substrates was surveyed in comparison to their ferrocene bis-phosphine analogs.

Introduction

Over the past 1-2 decades, asymmetric homogeneous hydrogenation has developed into a reliable and practical technology widely used in the pharmaceutical and chemical industries to synthesize chiral molecules.[1] This is due in no small part on the availability of a wide selection of chiral ligands and efficient techniques for rapid screening of catalyst systems for specific prochiral substrates.[2] Bisphosphine ligands, in particular, have shown broad applicability for enantioselective hydrogenation,[3] with those constructed on a ferrocene nucleus showing particular utility.[4]

Two basic modes exist for assembling two coordinating functions (e.g. two tertiary phosphines) on a ferrocene nucleus: (i) attachment of both groups onto the same cyclopentadienyl ring and (ii) attachment of a single donor function to each cyclopentadienyl ring. Among the many ferrocene bisphosphines developed to date, these two motifs can be exemplified by the ligand families Josiphos[5] and Mandyphos[6] (Figure 1).

In the course of our research on new ligand families for homogeneous catalysis, we became intrigued with the idea of combining both of these motifs in a single chiral ferrocene tetraphosphine ligand. One example of a ligand system exemplifying this concept, that we have named "Twinphos", is shown in Figure 1. This ferrocene tetraphosphine ligand structure offers a number of potential advantages over previously studied phosphine ligands, including the possibility of unique metal binding modes, improved performance for select prochiral substrates, the possibility of forming mixed metal complexes, easier synthesis, and better performance relative to total catalyst mass. Consequently, we have prepared three new Twinphos ligands, examined their rhodium complexes by NMR, and evaluated their performance relative to their 'parent' Josiphos and BoPhoz ligands in the homogeneous hydrogenation of model substrates.

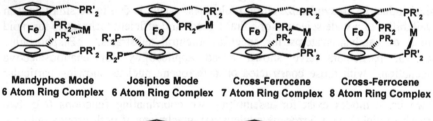

Figure 1. Structures of chiral ferrocene bisphosphine and tetraphosphine ligands.

Results and Discussion

In theory, tetradentate ferrocene ligands can form a variety of unimolecular 1:1 metal complexes, including two monodentate (assuming the two ferrocenes bear the same phosphine substituents), four bidentate (shown in Figure 2), two tridentate (Fig. 2), and one tetradentate coordination mode, in contrast to the two monodentate and one bidentate unimolecular modes for ferrocene bisphosphines. For 1:2 ligand-metal complexes, only two unimolecular complexes are reasonable (Fig. 2), since the relative orientation of the phosphines preclude bidentate complexation of a second metal in the Mandyphos and 8-atom ring Cross-Ferrocene binding modes

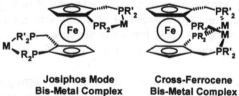

Figure 2. Binding modes for mono-Rh and bis-Rh complexes of ferrocene tetraphosphine ligands.

Thus, while the behaviour of the mono-rhodium tetraphosphine complexes in homogeneous hydrogenation may be expected to correspond to that previously observed for Josiphos or Mandyphos if the additional phosphine act purely in a spectator mode, new behaviour may be observed if either of the two new cross-ferrocene binding modes dominate or if the one or both of the additional phosphine groups interact with the complexed metal. While a mixture of complexes may be present, in any catalysis reaction it may be expected that each complex will behave independently and therefore the observed product distribution will represent the sum of the contributions of the individual catalytically active species. Given that activity, selectivity, and productivity of the various species will differ, the observed product distribution will not necessarily reflect the population distribution of the different species. In fact, such an outcome would be unexpected.

The six ligands studied in these investigations are shown in Figure 3. For each new ferrocene tetraphosphine prepared, we evaluated the hydrogenation behavior relative to the corresponding diphosphine of the "Josiphos"[7] or "Bophoz"[8] type.

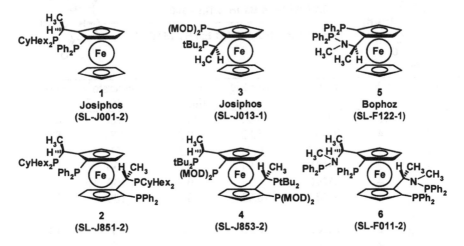

Figure 3. Structures of ligands used in this study (Cyh = cyclohexyl, Ph = phenyl, MOD = 3,5-dimethyl-4-methoxyphenyl, tBu = 2-methyl-2-propyl).

For our initial studies we chose to evaluate the hydrogenation of two unsaturated carbonyl model prochiral substrates with rhodium complexes of chiral ferrocene diphosphine and tetraphosphine ligands using a standard set of conditions. The substrates screened were methyl α-acetamido cinnamate (MAC) and dimethyl iticonate (DIMI). The substrates, catalysts, conditions, and experimental results are shown in Table 1.

Table 1. Hydrogenation of methyl α-acetamido cinnamate (MAC) and dimethyl itaconate (DMI) in MeOH.

Ligand	# P	Substrate	Rh / Lig	reaction time [h]	Conversio (%)	ee (%)
1	2	MAC	0.91	0.4	98	81
2	4	MAC	1.82	0.3	98	87
2	4	MAC	0.91	0.3	72	85
1	2	DMI	0.91	0.4	100	99
2	4	DMI	2	0.3	100	99
2	4	DMI	0.91	0.3	100	92
3	2	MAC	0.95	1	27	14
4	4	MAC	2	1	67	64
4	4	MAC	1	1	66	74
3	2	DMI	0.91	1	49	13
4	4	DMI	1.82	1	100	53
4	4	DMI	0.91	1	80	45
5	2	MAC	1	1	63	98
6	4	MAC	1.9	1	80	97
5	2	DMI	0.95	1	100	95
6	4	DMI	2	1	100	62
6	4	DMI	0.91	1	100	52

S/Rh = substrate to rhodium ratio (200 corresponds to 0.5% Rh loading)
NBD = norbornadiene
# P = number of phosphines in the ligand.
% ee =   100% x (moles major enantiomer − moles minor enantiomer) /
(moles major enantiomer + moles minor enantiomer)

The hydrogenation reactions were all performed at the same substrate-to-rhodium ratio and, since the tetraphosphine ligands **2**, **4**, and **6** can form bidentate complexes with one and two rhodium ions, were performed at both 1:1 and 2:1 rhodium-to-ligand ratios.

As seen in Table 1, the mono- and bis-rhodium complexes of tetraphosphine **2** provide similar enantioselectivities in the chiral hydrogenation of both substrates as the rhodium complex of the diphosphine (Josiphos) ligand **1** does. The bis-rhodium complex of **6** provides higher conversion but similar enantioselectivity as the rhodium complex of the diphosphine (Bophoz) ligand **5** in the chiral hydrogenation of **MAC**.

In contrast, the mono- and bis-rhodium complexes of tetraphosphine **4** provide higher conversion and enantioselectivity in the chiral hydrogenation of **MAC** compared to the rhodium complex of diphosphine (Josiphos) ligand **3**. For **DMI**, the mono-rhodium complex of **4** provides much higher conversion and enantioselectivity compared to the rhodium complex of **3**, while the bis-rhodium complex of **4** is intermediate in both conversion and enantioselectivity to **3•Rh** and **4•1Rh**. Finally, both the mono- and bis-rhodium complexes of **6** provide similar conversion with lower enantioselectivities in the chiral hydrogenation of **DMI** compared to **5•Rh**.

From these initial survey results, we postulate that the dominant species formed by the rhodium complexes of tetraphosphine ligand **2** are the same as the rhodium complex formed by Josiphos ligand **1** and the two catalytic sites in ligand **2** act essentially independently.

However, because the rhodium complexes of tetraphosphine ligand **4** behave differently than the rhodium complex of Josiphos ligand **3** for both substrates, we postulate that either a) different rhodium complexes are being formed (different binding modes as discussed above, see Fig. 2) by **3** and **4**, or b) the substituents on the second cyclopentadiene (cp) ring influence the catalytically active site in the rhodium complexes of **4**.

The unique performance of **4•Rh** and **4•2Rh** relative to **3•Rh** led us to initiate investigations to better understand rhodium complexation by these ligands. As a first step we prepared catalyst precursors by mixing ligands **3** and **4** with [Rh(NBD)$_2$]BF$_4$. We obtained ^{31}P NMR spectra of **3**, **3 + 1Rh**, **4**, **4 + 1Rh**, and **4 + 2Rh** which are shown in Figure 4.

The spectrum of **3** exhibits two phosphorus resonances at -26.6 (sidearm PMOD$_2$) and 50.1 ppm (PPh$_2$ attached to ferrocene), each showing ^{31}P coupling of 50 Hz, while the spectum of **4** exhibits a corresponding pair of resonances at -27.2 and 52.0 ppm and a ^{31}P coupling of 63 Hz.

Addition of one equivalent of [Rh(NBD)$_2$]BF$_4$ to **3** or two equivalents of [Rh(NBD)$_2$]BF$_4$ to **4** produces similar spectra with two strong resonances at 22.5 and 76.3 ppm for **3•Rh** and 22.2 and 78.2 ppm for **4•2Rh**, each resonance appearing as four lines due to combined ^{31}P (30 Hz) and ^{103}Rh (155 Hz) coupling. Both spectra show minor resonances around -25 to -35 ppm, indicative of non-complexed phosphorus, while **4•2Rh** also exhibits a weak resonance at 97.8 ppm split by Rh but not by ^{31}P.

The major resonances suggest that complexation of two Rhodium ions by the tetraphosphine ligand **4** occurs in the same manner as binding of one rhodium by the diphosphine Josiphos ligand **3** (see Fig. 2). While the minor resonances observed prevents ruling out the presence of species representing alternative binding modes, the lack of ^{31}P coupling on the resonance at 97.8 ppm in the **4•2Rh** spectrum suggests this may be due to a monophosphine impurity rather than alternative species of **4•nRh**.

Addition of one equivalent of [Rh(NBD)$_2$]BF$_4$ to **4** produces a spectrum with multiple ^{31}P resonances. Comparison to the spectra of **4** and **4•2Rh** allows tentative assignment of one pair of the observed resonances to **4** and a second pair of resonances to **4•2Rh**. In addition, two resonances exhibiting ^{31}P coupling at -28.6 and -30.6 ppm, two resonances exhibiting ^{31}P coupling at 52.2 and 54.1 ppm, and one resonance at 76.8 ppm exhibiting both ^{31}P and ^{103}Rh coupling are observed

If the mono-rhodium complex of **4** adopts the "Josiphos" binding mode as shown in Figure 2, the ^{31}P NMR would be expected to exhibit four resonances corresponding to one pair for the complexed phosphines, similar to **4•2Rh** and one pair for the non-complexed phosphines, as in **4**. Therefore, we have tentatively assigned to **4•1Rh** the resonances at -30.6, 54.1, and 76.8 ppm, along with a resonance at 22.2 ppm overlapping that of **4•2Rh**. If these assignments are correct, then a 1:1 mixture of **4** and [Rh(NBD)$_2$]BF$_4$ generates a mixture that is composed of mainly **4**, **4•1Rh**, and **4•2Rh**.

There remains a pair of resonances at -28.6 and 52.2 exhibiting ^{31}P coupling that are not assigned. In the absence of a corresponding pair of resonance exhibiting ^{31}P and ^{103}Rh coupling, it is reasonable to assume that these do not represent a rhodium complex of **4**, however in the absence of more detailed analysis the identity of the species producing these resonances cannot be established.

Thus, the ^{31}P NMR spectra for **3**, **3•Rh**, **4**, **4•1Rh**, and **4•2Rh** can be interpreted as being consistent with the "Josiphos" binding mode for both the mono- and bis-rhodium complexes of ferrocene tetraphosphine ligand **4**. Unfortunately, this is not sufficient to explain why tetraphosphine ligand **4** exhibits behavior different from Josiphos ligand **3** in the hydrogenation experiments, since we cannot rule out that the catalytic behavior results from highly active minor species not identified in the ^{31}P NMR spectra. On the other hand, the apparent predominance of the "Josiphos" binding mode in the ^{31}P NMR of the ferrocene tetraphosphine rhodium complexes coupled with the observation that **1** and **2** exhibit similar hydrogenation behavior while **3** and **4** differ, suggests that interaction of the phosphine substituents on the opposing ferrocene rings in **4** may be influencing the catalytic sites in the rhodium complexes of **4**, warranting further studies.

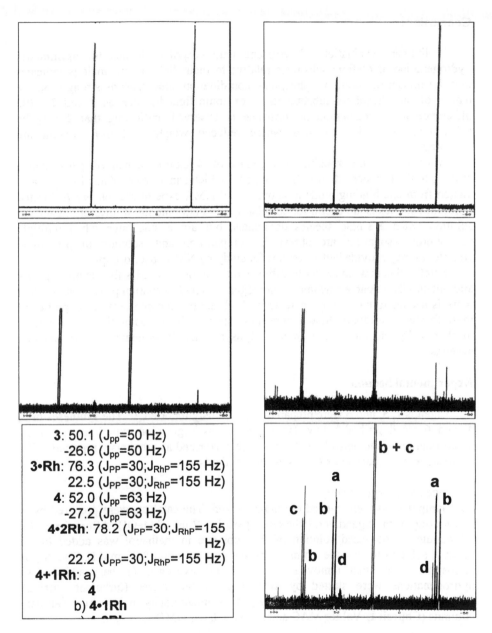

3: 50.1 (J_{pp}=50 Hz)
   -26.6 (J_{pp}=50 Hz)
3•Rh: 76.3 (J_{PP}=30;J_{RhP}=155 Hz)
   22.5 (J_{PP}=30;J_{RhP}=155 Hz)
4: 52.0 (J_{pp}=63 Hz)
   -27.2 (J_{pp}=63 Hz)
4•2Rh: 78.2 (J_{PP}=30;J_{RhP}=155 Hz)
   22.2 (J_{PP}=30;J_{RhP}=155 Hz)
4+1Rh: a)
   4
   b) 4•1Rh

Figure 4. ^{31}P NMR spectra of **3**, **3 + 1Rh**, **4**, **4 + 2Rh**, and **4 + 1Rh** in CD$_3$OD / THF (2/1 V/V) with putative assignments of resonances for **4 + 1Rh**.

Conclusions

Preliminary evaluation of ferrocene tetraphosphine ligands for asymmetric hydrogenation of olefinic substrates relative to their diphosphine analogs indicates that the impact of additional phosphine coordination sites depends strongly on the nature of the phosphine substituents. For some ligands, such as **1** and **2**, little difference in hydrogenation performance is observed, indicating that **2** may be substituted for **1** in processes where reduced weight and iron content are advantageous.

Alternatively, the rhodium complexes of **4** exhibit enhanced performance relative to the ferrocene diphosphine analog **3** which cannot be explained simply as a switch from one binding mode to another. ^{31}P NMR experiments on *in situ* formed rhodium complexes are useful for correlating rhodium binding of different ligands in solution when a single species dominates, but are inconclusive when multiple phosphorus resonances are observed. Preparation and isolation of preformed complexes may provide better systems for study by NMR spectroscopy.

Finally, this new ligand motif offers additional opportunities for identifying new and potentially valuable asymmetric homogeneous hydrogenation processes. Further work is needed to more fully characterize both the metal complexation properties of these ferrocene tetraphosphine ligands and opportunities for using these new ligands in chemically and industrially interesting asymmetric homogeneous hydrogenation systems.

Experimental Section

Ligands **1**, **3**, and **5** were prepared as previously reported.[7,8] The syntheses of **2**, **4**, and **6** are described below. Homogeneous hydrogenation experiments and the determination of the optical yields (ee) were performed as previously reported.[9] The catalysts were prepared *in situ* by standard methods.

Hydrogenation experiments

All manipulations were carried out under Argon. The catalysts were prepared by *in situ* mixing of the ligand and [Rh(NBD)$_2$]BF$_4$ in ~2 ml of MeOH. After stirring for 10 minutes, a degassed solution of the substrate in methanol was added to the catalyst solution to provide a final substrate concentration 0.25 mol/L. The Argon atmosphere was then removed *in vacuo* and hydrogen (1 bar) added. The hydrogenations were started by switching on the stirrer (turbulent stirring). Conversion and ee were determined by gas chromatography, using chiral capillary columns (Chirasil-L-Val for MAC and Lipodex-E for DMI).

Preparation of the NMR-samples

Approximately 0.01 mmol of ligand and the appropriate molar equivalents of [Rh(NBD)$_2$]BF$_4$ were weighed into a Schlenk tube. A magnetic stir bar was added, the Schlenk tube was evacuated, filled with Argon, and 1.2 ml of degassed CD$_3$OD / THF (2:1 V/V) mixture was added. The solution was stirred for 10 min at room temperature, then transferred by syringe into an Argon filled NMR tube. The NMR spectra were recorded on a Bruker DPX spectrometer operating at 300.13 MHz proton resonance.

Synthesis of 2

Dicyclohexylphosphine (0.34 ml, 1.7 mmol) was added to 1,1'-bis(diphenylphospino)-2,2'-bis(α-dimethylaminoethyl)ferrocene (536 mg, 0.77 mmol) in 5 ml of acetic acid and the red solution is stirred overnight at 105°C. After cooling, the reaction mixture was partitioned between toluene and water. The aqueous phase was saturated with sodium chloride and extracted with toluene. The organic phase was dried over anhydrous Na_2SO_4, evaporated *in vacuo*, and chromatographed on silica gel (Merck Si60, heptane/TBME 50:1 eluent) to provide 424 mg (55%) of **2** as a yellow crystalline solid. ^{1}H-NMR (C_6D_6): δ 0.8-2.0 (m, 50H), 3.11 (s, 2H), 3.56 (m, 2H), 4.40-4.55 (m, 4H), 6.85-7.55 (m, 20H). ^{31}P-NMR (C_6D_6): δ +16.3 (d); -25.5 (d).

Synthesis of 4

s-BuLi (30.45 ml of 1.3 molar in cyclohexane, 39.6 mmol) was added dropwise over 30 minutes to a stirred solution of (S,S)-1,1'-bis[1-(dimethylamino)ethyl]ferrocene[10] (5.144 g, 39.6 mmol) in 25 ml of diethyl ether cooled in an ice-water bath. After stirring with cooling for 3.5 hours, $(MOD)_2PCl$ (14.44 g, 42.9 mmol) was added, the cooling bath was removed, and the reaction mixture was stirred overnight. The mixture was quenched with with water and extracted with TBME. The organic phase was dried over anhydrous Na_2SO_4, evaporated *in vacuo*, and purified by chromatography on silica gel (Merck Si60, ethanol). Recrystallization from ethanol provided 7.03 g (46%) of a yellow, crystalline solid[11] which was used in the next reaction.

Di-t-butylphosphine (18.9 g of 10% solution in acetic acid, 12.02 mmol) was added to a solution of the product (4.0 g, 4.31 mmol) from the previous step in 20 ml of acetic acid and this reaction mixture was stirred overnight at 105°C. After cooling, the mixture is partitioned between dichloromethane and water. The organic phase was dried over anhydrous Na_2SO_4, evaporated *in vacuo*, and purified by chromatography on silica gel (Merck Si60, 10/1/0.1 heptane/TBME/triethylamine) to provide **4** as orange, crystals compound (yield: 50%). ^{1}H-NMR (C_6D_6): δ 7.73 (s, 2H), 7.70 (s, 2H), 7.23 (s, 2H), 7.21 (s, 2H), 4.18 (m, 2H), 3.93 (m, 2H), 3.70 (q, 2H), 3.65 (m, 2H), 3.36 (s, 6H), 3.26 (s, 6H), 2.33 (m, 6H), 2.24 (s, 12H), 2.12 (s, 12H), 1.42 (d, 18H), 1.15 (d, 18H). ^{31}P-NMR (C_6D_6): δ +52.2 (d), -26.5 (d).

Synthesis of 6

40% Aqueous methylamine (64 ml, 33.2 mmol) was added to 1,1'-bis(diphenylphospino)-2,2'-bis(α-acetoxyethyl)ferrocene[12] (403 mg, 0.55 mmol) in isopropanol (5 ml) and the reaction mixture was stirred in a closed pressure ampoule at 90°C for 66 hours. After evaporation *in vacuo*, the residue was dissolved in ethyl acetate/heptane 1:1 and extracted with 10% aqueous citric acid. The aqueous phase was washed with ethyl acetate/heptane 1:1, basified with 2N NaOH basic and extracted with dichlromethane. The organic phase was dried over anhydrous Na_2SO_4, evaporated *in vacuo*, and purified by chromatography on silica gel (Merck Si60, 1% cyclohexane in ethonal) to provide a yellow solid[13] (58%) of that was used in the next reaction.

A solution of the product from the previous step (209 mg, 0.313 mmol), triethylamine (0.2 mL, 1.4 mmol), and diphenylphosphine chloride (0.15 ml, 0.81 mmol) in toluene (2 mL) was stirred at 50 °C overnight. After cooling, heptane (10 mL) was added, the suspension was filtered, the filtrate was evaporated *in vacuo*, and chromatographed on silica gel (Merck Si60 80/20/2.5 heptaneethyl acetate/triethylamine) to provide **6** as solid orange foam (yield: 96%). ^{1}H-NMR (C_6D_6), δ7.5 – 6.8 (m, 40H), 5.24 – 5.12 (m, 2H), 4.52 (m, 2H), 4.25 (m, 2H), 3.01 (m, 2H), 2.19 (d, 6H), 1.60 (d, 6H). ^{31}P-NMR (C_6D_6): δ +59.1 (d); -24.6 (d).

References

1. H.U. Blaser, F. Spindler, and M. Studer, *Applied Catalysis A: General* **221** 119 (2001). H.U. Blaser, E. Schmidt eds., *"Large Scale Asymmetric Catalysis,"* Wiley-VCH, Weinheim, 2003.
2. *Chim. Oggi / Chem. Today* **22**(5) (2004), *Supplement on Chiral Catalysis*
3. H.U. Blaser, Ch. Malan, B. Pugin, F. Spindler, H. Steiner, and M. Studer, *Adv. Synth. Catal.* **345**, 103 (2003). W. Tang and X. Zhang, *Chem. Rev.* **103**, 3029 (2003).
4. H.U. Blaser, M. Lotz, F. Spindler, "Asymmetric Catalytic Hydrogenation Reactions with Ferrocene Based Diphosphine Ligands" in *Handbook of Chiral Chemicals*, 2nd Edition, D. J. Ager (Ed.), CRC Press, Boca Raton 2005
5. H. U. Blaser, W. Brieden, B. Pugin, F. Spindler, M. Studer, and A. Togni, *Topics in Catalysis* **19**, 3 (2002)
6. J. J. Almena Perea, A. Borner, P. Knochel, *Tetrahedron Lett.* **39**, 8073 (1998); J. J. Almena Perea, M. Lotz, P. Knochel, *Tetrahedron: Asymmetry* **10**, 375 (1999); M. Lotz, *et al.*, *ibid.* **10**, 1839 (1999).
7. A. Togni, *et al.*, *J. Am Chem Soc.* **116**, 4062 (1994).
8. N. W. Boaz, *et al.*, *Org. Lett.* **4**, 2421 (2002).
9. W. Weissensteiner, *et al. Organometallics* **21**, 1766 (2002).
10. L. Schwink, P. Knochel, *Chem. Eur. J.* **4**, 950 (1998).
11. ^{1}H-NMR (C_6D_6): δ 7.52 (s, 2H), 7.50 (s, 2H), 7.14 (s, 2H), 7.11 (s, 2H), 4.37-4.28 (m, 6H), 3.86 (m, 2H), 3.30 (two s, 12H), 2.1 (s, 12H), 2.09 (s, 12H), 1.90 (s, 12H), 1.40 (d, 6H). ^{31}P-NMR (C_6D_6): δ -23.7 (s).
12. Hayashi *et al.*, *Organometal. Chem.* **370**, 129 (1989).
13. ^{1}H-NMR (C_6D_6): δ 7.43 – 7.37 (m, 4H), 7.3 – 7.25 (m, 4H), 6.99 – 6.86 (m, 12H), 4.55 (s, 2H), 4.39 (m, 2H), 4.10 – 4.03 (m, 2H), 3.21 (m, 2H), 2.06 (s, 12H), 1.51 (d, 6H). ^{31}P-NMR (C_6D_6): δ - 24.2 (s).

35.   Catalyst Library Design for Fine Chemistry Applications

József L. Margitfalvi, András Tompos, Sándor Gőbölös, Emila Tálas and Mihály Hegedűs

Chemical Research Center, Institute of Surface Chemistry and Catalysis, Department of Organic Catalysis, 1025 Budapest, Pusztaszeri út 59-67 Hungary

joemarg@chemres.hu

Abstract

In this study general principles of catalyst library design for a 16-well high-throughput high-pressure reactor system (SPR 16) used for combinatorial catalytic experiments in the field of fine chemistry are described and discussed. The focus is laid on heterogeneous catalytic selective hydrogenation reactions. It has been shown that Holographic Research Strategy (HRS) combined with Artificial Neural Networks (ANNs) is an excellent tool both for catalyst library design and the visualization of multidimensional experimental space.

Introduction

In the last decade methods of combinatorial catalysis and high throughput experimentation has obtained great interest [1-4]. In the field of heterogeneous catalysis most of the efforts are devoted to the investigation of gas phase reactions, where several hundreds catalysts can simultaneously be tested [5,6]. Contrary to that, in high-pressure liquid phase catalytic reactions in a single reactor module only 8-16 parallel experiments can be performed. There are reports to use up to six modules as a parallel setup [7].

In combinatorial heterogeneous catalysis two general approaches are in practice: experiments without [8-10] and experiments with the use of a given library optimization method [11,12]. The use of an optimization tool strongly reduces the number of experiments required to find catalysts with optimum performance.

The methods used for catalyst library design are quite divers. Industrial companies, like Symix, Avantium, hte GmbH, Bayer AG are using their own proprietary methods. In academic research the Genetic Algorithm (GA) is widely applied [11,12]. Recently Artificial Neural Networks (ANNs) and its combination with GA has been reported [13,14]. In these studies ANNs have been used for the establishment of composition-activity relationships.

In gas phase reactions the size of catalyst libraries can be over couple of thousands. For instance, in the synthesis of aniline by direct amination of benzene around 25000 samples were screened in about a year [15], however, the optimization method used was not discussed. In contrast, in liquid phase reactions taking place at elevated pressure and temperature, due to technical difficulties the rational approach does not allow testing libraries containing more than 200-250 catalysts. Consequently, the informatic platform and the strategy used to design catalyst libraries for high-pressure liquid phase reactions should have very unique optimization tools.

Although combinatorial and high throughput methods are often used in the field of fine chemistry [16-17], there are only scarce data in the literature for the use of high-throughput or combinatorial approaches in liquid phase selective hydrogenation in the presence of heterogeneous catalysts. Simons has reported [18] the use of high-throughput method for both the preparation and testing of Pt based supported hydrogenation catalysts. However no optimization tools were used. A split-plot experimental design has been applied to investigate the type of catalyst, catalyst concentration, the pressure and the temperature [7]. Recently, continuous-flow micro-reactors were used for high throughput application in the area of hydrogenation and debenzylation [19,20]. However, no optimization tools were applied in these studies.

The lack of the use of catalyst library design tools in the field of heterogeneous catalytic hydrogenation inspired us to describe our approach used in this area. In this presentation we shall depict our complex approach to design, optimize and mapping catalyst libraries.

The aim of the present study is to show the strength of our optimization tools for fine chemistry applications. We shall discuss the peculiarities of library design for selective hydrogenation reactions. The basis of this approach is the availability of a high-throughput automatic reactor set-up allowing to perform 16 parallel hydrogenation experiments at different temperatures, hydrogen pressure, stirring rate, concentrations and using different solvents.

Results and Discussion

General Considerations

The Basis of the Optimization

The design and optimization of a catalyst library for selective hydrogenation is based on the knowledge accumulated in the patent and open literature. In this study we shall focus on catalysts libraries related to supported metals. The catalyst library optimization is performed in an iterative way. First a "rough experimental space" is created, tested and optimized by HRS.

In the "rough experimental space" the distance between discrete levels of the experimental variables is relatively large. After testing three – four catalyst generations different Data Processing methods, such as general statistical approaches or Artificial Neural Networks (ANNs), can be applied to determine the contribution of each variable into the overall performance or establish the Activity - Composition Relationship (ACR).

Based on the ACR "virtual" catalytic tests and optimization can be performed using HRS (further details are given in the experimental part). Alternatively the whole experimental space can be mapped (see Fig. 1). After subsequent verification step a "high resolution experimental space" is created and further optimization takes place by HRS creating 1-3 more catalyst generations. After the last verification procedure the

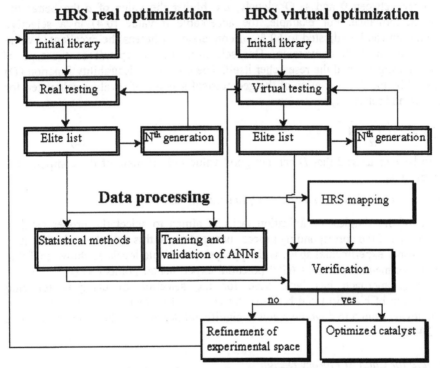

composition of optimized catalyst can be obtained. This strategy is represented by Scheme 1.

Scheme 1. Flowchart of catalyst library optimization.

The Specificity of Optimization in Selective Hydrogenation

In the field of selective hydrogenation two important properties are used to describe the catalytic performance: the activity and the selectivity of the catalysts. Their values have to be optimized. The simplest approach is to fix the desired conversion level and ranking the catalysts according to their selectivity data. An alternative way for catalyst optimization is the use of the so called "desirability function" (*d*). Upon using this function different optimization parameters can be combined in a common function (*D*). In the combination different optimization parameters (often called as objective functions) can be taken into account with different weights [21]. The single desirability function for the conversion (α) can be described by the following formula:

$$d_\alpha = e^{-(e^{-(b_0\alpha + b_1\alpha \cdot \alpha)})}$$

(1)

Similar function can also be applied for the selectivity as well. In these formulas the b_0 and b_1 parameters can be determined if two corresponding d and α values are available. These values are usually arbitrary selected by the researcher. *d* can have values only between 0 and 1. Obviously, the higher the value of *d* the better the catalyst performance. For example, the acceptable *d* value (0.4) in a selective hydrogenation can be adjusted to 60 % of conversion, whereas the excellent *d* value (0.9) belongs to 80 % conversion. This selection always depends on the type of reaction investigated and the researcher itself. The combined desirability function (D) is obtained by the determination of the geometrical average of *d* values calculated for conversion and selectivity:

$$D = \sqrt{d_\alpha \cdot d_s}$$

(2)

It has to be emphasized that *D* can get good value only if none of the component *d* values are small.

Variables and their Levels in a Simple Optimization Task

One of the simplest optimization tasks is aimed to select the proper catalyst combination and the corresponding process parameters. In this case the main task is to create a proper experimental space with appropriate variable levels as shown in Table 1. This experimental space has 6250 potential experimental points (N) (N = 2 x 5^5= 6250). This approach has been used for the selection of catalysts for ring hydrogenation of bi-substituted benzene derivatives. The decrease of the number of variable levels from 5 to 4 would result in significant decrease in the value of N (N= 2 x 4^5 = 2048).

Input Data for Catalyst Library Design

The first step in the library design is the definition of the key metal or the combination of key metals involved in the hydrogenation of the given functional group. The second step is the selection of the support, what is followed by choosing modifiers. Both the active site and the support can be modified. In both types of modifiers the determination of their proper concentration levels is the most important task. In general, the modification of active sites requires less amount of modifier than that of the support. The modifiers can be added to the catalyst during its preparation or during the catalytic reaction ("compositional" and "process" modifiers, respectively).

In the selection of modifiers the key issues are as follows:
(i) which functional group has to be hydrogenated, (ii) what group has to be preserved, and (iii) what type of undesired side reactions should be suppressed. In selective hydrogenation the following differentiation was done between components used to prepare multi-component catalysts:
(i) active metals, (ii) modifiers of the active metals, (iii) modifiers of the support, and (iv) selective poisons. The list of modifiers used in different hydrogenation or related reactions are given in Table 2.

Table 1. Variables and their levels in a simple optimization task.

Variables	Variable levels				
Catalyst 1, amount, g	0.05	0.10	0.15	0.20	0.25
Catalyst 2, amount, g	0.00	0.05	0.10	0.15	0.20
Hydrogen pressure, bar	25	50	75	100	125
Temperature , °C	80	90	100	115	130
Substrate concentration, M	0.08	0.15	0.30	0.60	0.90
Stirring rate, rpm	LOW	HIGH			

Table 2. Modifiers used in selective hydrogenation or related reactions.

Modifier	Reaction	Type
Transition metals (Pb)	Triple bond	M – C
Transition metal salts	Various	M - C
Transition metal oxides	Various	S – C
Quinoline and analogues	Dienes	M – P
Tertiary amines	Keto groups	M – P
NH_3	Reductive amination	P
Trialkyl and triaryl phosphines	hydrofromylation	M – P
Alkali hydroxides and carbonates	Nitrile hydrogenation	S – C
Long chain amides	For unsaturated nitriles	M-P
H_3PO_3 and H_3PO_4	Reductive amination of unsaturated compounds	M-P
Sulfur containing compounds (CS_2, tiophene)	Selective hydrogenation	M-C
Metal acetyl-acetonates	Selective hydrogenation	M-P
Metal alkyls (SnR_2Cl_2,)	Nitrile hydrogenation	M-C
Metal (Sn, Ge) tetraalkyls	Unsaturated aldehydes	M-C
CO	Triple bond	M-P
Chiral moieties	Prochiral ketones, imines	M-C, M-P
Alkali and alkali earth metals	Selective hydrogenation of phenol	M-C
VO_x	Reductive amination of ketones	M-P

M - modifier for the metal, **S** - modifier for the support, **C** - Compositional, **P** - Process.

The above list of modifiers unambiguously shows their large diversity. There are examples indicating that the best results can be achieved by combination of several modifiers [22]. Consequently, catalyst modification is an excellent field for combinatorial research provided a library optimization method is available.

Basic Principle of HRS

The Holographic Research Strategy (HRS) developed for catalyst library design and optimization can be considered as a deterministic approach, i.e. the route of optimization is unequivocally determined by the applied setting parameters [23,24]. In HRS similarly to other methods, such as the genetic algorithm (GA) [11]], the test results of the (n-1)th generation is used to design the nth generation. During the optimisation, the Holographic Research Strategy uses the rank between catalysts tested. This rank is usually called as an "elite list" which is used to design the next catalyst generation.

In HRS the key setting parameters are as follows [23,24]: (i) the total number of experimental variables, (ii) the selected levels of experimental variables, (iii) initial arrangement of variables along the axes (see Fig. 1) and the way of variable position changes, (iv) the number of the best hits around which the new catalyst generation is created, and (v) the size and the form of the experimental regions used to design the next catalyst generation.

Undoubtedly all of these parameters can influence both the rate and the certainty of optimum search. In this respect the importance of the size of the experimental region has already been discussed in our previous studies [23,24].

When all compositional variables have been selected the next task is the definition of the levels of these variables. The levels can be given either in absolute concentration or as a relative ratio. The necessary number of levels for a variable is the arbitrary decision of the experimenter. It depends on the range that has to be explored. Special attention has to be devoted to possible non-smooth areas. In a reasonably narrow range the effect of a variable are usually investigated using 4 – 5 levels, which according to our results proved to be sufficient. In an eight-dimensional experimental space containing 78,000 – 125,000 possible compositions less than 150-200 measurements were sufficient to find or approach the optimum [23,24].

The levels of the compositional variables strongly depend on the role of the given component in the catalyst composition. For key components steps in 0.5 w % is very common. For compositional modifiers of the active metal small steps has to be used in 0.05 or 0.1 w % interval. For the determination of the required amount of process modifiers (quinoline, amides, sulfur) the dispersion (D) of the key metal has to be determined or estimated. The amount of this type of modifiers (M) is usually in the range of 0.05-0.2 M_{mod}/A_{metal} where M_{mod} = amount of the modifier in moles and A $_{metal}$ is the total amount metal in gram atoms. The amount of modifiers for the support can be one order higher than the metal content of the catalyst.

According to the above considerations if the optimization is performed under fixed process parameters the initial step in library design is finished, i.e. the catalysts of the initial library can be introduced into the experimental hologram. However, it is strongly recommended to include one or two process parameters into the library design procedure. Reaction temperature and hydrogen pressure is the two most important process parameters influencing both the activity and the reactivity.

Combination of HRS with ANNs

In catalysis research upon using high-throughput experimentation the most crucial problem is to discover patterns related to the composition and the catalytic performance (activity, selectivity and lifetime). For this purpose "information mining" methods are applied [25]. Artificial Neural Networks (ANNs) is one of the commonly used "information mining" tools. Due to their strong pattern recognition abilities ANNs has been used by different authors for information mining [13,26]. We are using ANNs to establish the relationship between the composition of catalysts and their catalytic activity. Due to the black-box character of ANNs they can be used to perform "virtual catalytic" experiments or to map the whole experimental space (see Scheme 1.).

Visualization Ability of HRS

Since the visualisation ability of the experimental space is an inherent property of HRS it can be exploited in data mining as well. In this case upon using ANNs the activity for each composition in the hologram can be determined. Figure 1. shows the mapping of the experimental space after combined use of HRS and ANNs.

In this small experimental space four variables (A, B, C, D) were optimized. The concentration waves of components A and B are placed along the X axis, while that of the components C and D along the Y axis. As far as there are only two variables along each axis there are only four combinations for the visualization of the experimental hologram. Two selected holograms are shown in Figure 1.

As emerges from Figures 1 all variables have six concentration levels, i.e. the total number of combinations in this experimental space is 1296. The activity of samples above 89 % conversion is shown by a white color, while that of below 50 % is shown by black. The maximum value of conversion is shown by a cross. The analysis of these holograms shows the following activity composition relationship:

- There are distinct light and dark areas corresponding to compositions with high and low activity, respectively;
- The activity crescents upon increasing the concentration of components **B, C** and **D,** however the optimum does not requires the highest concentration in components **B** and **D**. Low activity area can be found when the concentrations of both **C** and **D** are low. However, the contribution of component **C** is more pronounced than that of component **D** as in its absent the activity is the lowest (see right-side hologram);
- The coordinates of the maximum activity show also that there is an optimum in the concentration of **A**.
- High amount of **C** results in high activity (see lines 1a,b,c).
- Both figures clearly show the formation of islands with high activity. It is an indication for the strong synergism between components **C** and **B** (see left-side hologram). There is also a synergism between component **A** and **D** (see right-side hologram.
- The higher the amount of **B** and **C** the broader the area with high activity. It means also that the increase of the amount of B allows decreasing the amount of

D in compositions with high activity (see the area between the two lines in the right-side hologram).

- When the concentration of component **D** is low **A** has a definite negative effect on the activity (see left-side hologram).
- When the amount of **D** is low the activity is diminishing upon increasing the amount of **A** (see line 1a). The circle shows the area of high activity at low **A** content.
- At high concentration of **C** upon increasing the concentration of **B** the high activity areas are shifted to areas with low **D** content (see lines at the right-side hologram). It means that the addition of component **C** replaces component **D**.

These results show that HRS possesses an excellent visualization ability. The visualization allows us to elucidate the influence of all components on the activity. The results show that similar activity ranges can be obtained in different compositional areas. Consequently, in this case the difference in the price of components can be taken into account before using a given catalyst in the production.

Catalyst and Process Optimization

Catalyst library design is considered as an optimization procedure in a multi-dimensional experimental space. The variables in the multi-dimensional space can be differentiated as follows: (i) compositional variables, and (ii) process variables. The term compositional variables have already been discussed.

The most common process variables are as follows: temperature, pressure, concentrations, pH, catalyst pore size, flow rate, stirring rate, etc. In the process of creating a compositional catalyst library the initial steps are as follows:

1. Selection of the components,
2. Choosing the concentration levels of components,
3. Introduction of a limit for a given component (Pt_{max} = 3 wt%),
4. Introduction of a limit for the total amount of components (total metal content = 5 wt. %).

For catalysts used in fine chemistry the following approaches can be combined in the catalyst library design and process optimization:

1. Preparation of multi-component modified catalysts;
2. Creation of an optimum composition of modifiers (promoters, inhibitors, selective poisons, etc.);
3. Creation of a multi-dimensional experimental space containing also process variables (pressure, temperature, pH, flow rate, etc.)

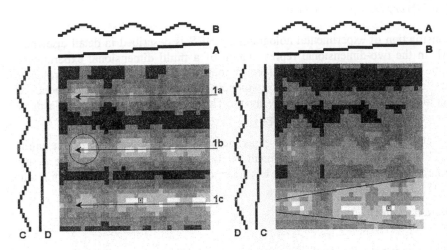

Figure 1. Mapping of catalyst library. Coordinates optimum activity (shown by ☒) :
D - fifth level,
C - last level, A - fourth level, B - fifth level. Conversion values in %: □ - > 89-100,
▨ - 85-89, ▥ - 75-85, ▦ - 50-75, ■ - 0-50.

The main steps in catalyst optimization are as follows (see Scheme 1.):

1. Preparation and testing 3-4 catalyst generations by HRS (80-100 catalytic experiments);
2. Refinement, i.e., establishment of the activity - composition relationship by using ANNs as an information mining tool;
3. Performing "virtual experiments" using ANNs and HRS;
4. Testing the hits of "virtual experiments" (16 experiments);
5. Final tuning of the catalysts accomplishing additional 64-80 "real experiments" using HRS.

Conclusions

In this contribution a short description of HRS and its combination with ANNs aimed to design catalyst libraries for selective hydrogenation was given. The approach developed is based on the use of both "real" and "virtual" optimization algorithms. A set of results using HRS optimization represents the "real" optimization process. The use of ANNs as an information mining tool provides a possible to perform "virtual" experiments. In this "virtual" experiments the objective function provided by ANNs is used to move into the direction of global optimum by performing "virtual" optimization. The combination of "real" and "virtual" experiments strongly accelerates the process of catalyst library optimization. The success in catalyst library design by using the above tools has been verified both by the rate and the certainty of optimum search.

Experimental Section

Catalytic Reactions
Reactions were carried out in a multi-reactor system (AMTEC GmbH, SPR-16) having 16 mini autoclaves working in a parallel way. Product analysis was done by GC.

Catalyst Library Design Procedures

The construction of experimental holograms by HRS is described in detail elsewhere [23,24]. In the two-dimensional representation of a multi dimensional experimental space the discrete concentration levels of components and modifiers are represented by lines (see Fig.1). The level of each component increases gradually till it reaches its maximum then it decreases gradually again. This mode of representation leads to wavelike arrangement of levels (see Fig. 1).

The elements of symmetry of the experimental space is used to create the initial catalyst library resulting in 16-48 different catalyst compositions [23,24]. The design of forthcoming generations by HRS has been described in detail in our previous studies [23,24].

Information Mining

Artificial neural networks have been used for information mining ANNs provide the quantitative relationship between composition and catalytic performance. ANNs describing the objective functions in the given experimental space were previously trained with data obtained during HRS optimization. For appropriate formation of ANNs and for evaluation of their predictive ability the available data of each catalyst library have been divided into three sets: (i) training, (ii) validation and (iii) testing in the following ratios: 70:15:15, respectively. The networks are trained with resilient back-propagation algorithm [26]. Training is stopped if the validation error increases for more then two consecutive epochs. Nineteen different network architectures were investigated to achieve acceptable model accuracy [26]. Every neural network architecture has been trained 1000 times (each training has been initialized with different, random node-to-node weights) [26]. According to the average mean square errors (MSE) the resulted 19000 networks were ranked. The best 100 networks have been involved into <u>O</u>ptimal <u>L</u>inear <u>C</u>ombination [27], during which so called OLC-network has been created. The resulting OLC-network has been applied in this study for "virtual" catalytic tests.

"Virtual" catalytic tests

Two optimization tools can be used for "virtual" catalytic experiments: (i) HRS and Genetic Algorithm (GA). We have recently demonstrated [28] that HRS is a faster optimization tool than the GA. The only advantage of GA with respect to HRS is that GA uses a continuous experimental space, while HRS makes use of levels.

In "virtual" catalytic experiments the objective function determined by ANNs is used for optimization, i.e. for finding compositions or experimental parameters with optimum performance. In "virtual" catalytic experiments "virtual" catalyst libraries are created and just using computational methods several catalyst libraries can be virtually tested, while in the virtual experimental space we are moving towards the virtual optimum determined by the given objective function. Having found the virtual optimum one new "real" catalyst library is created in the neighborhood of virtual optimum. In this way it is possible to accelerate the process of optimization of a given catalyst library.

Acknowledgements

Thanks to Lajos Végvári (Meditor Bt., Hungary) for his help using HRS and Dr. Ernő Tfrist for his contribution to create ANNs.

References

1. B. Jandeleit, H.W. Turner, T. Uno, J.A.M. van Beek, and W.H. Weinberg, *CATTECH*, **2**, 101, (1998).
2. P.P. Pescarmona, J.C. van der Waal, I.E., Maxwell, and T. Meschmayer, *Catal. Lett.,* **63**, 1 (1999).
3. W.F. Maier, G. Kirsten, M. Orschel, and P.A. Weiss, *Chemica Oggi-Chemistry Today,* **18**, 15 (2000).
4. S. Senkan, *Angew. Chemie, Int. Ed.,* **40**, 312 (2001).
5. S. Bergh, S. Guan, A. Hagemeyer, C. Lugmair, H. Turner. A.F.Volpe, Jr., W.H. Wienberg, and G. Mott, *Appl. Catal.,* **254**, 67 (2003).
6. G. Kirsten, and W.F. Maier, *Appl. Surf. Science*, 223, 87 (2004).
7. F. A. Castello, J. Swerney, P. Margl, and W. Zirk, *QSAR and Comb. Sci.*, **24**, 38 (2005).
8. S.Thomson, Ch. Hoffman, S. Ruthe, H.-W. Schmidt, and F. Schuth, *Appl. Catal. A: General,* 220, 253 (2001).
9. S. Duan, and S. Senkan, *Ind. & Eng. Chem. Res.*, **44**, 6381 (2005).
10. Ch.M. Snively, G. Oskarsdottir, and J. Lauterbach, *Catalysis Today*, **67**, 357 (2001).
11. D. Wolf, O.V. Buevskaya, and M. Baerns, *Appl. Catal. A: Gen.,* **200**, 63 (2000).
12. U. Rodemerck, D. Wolf. O.V. Buyevskaya, P. Claus, S. Senkan, and M. Baerns, *Chem. Eng. Journal*, **82**, 3 (2001).
13. A. Corma, J.M. Serra, E. Argente, V. Botti, and S. Valero, *Chem. Phys. Chem.*, **3**, 939 (2002).
14. U. Rodemerk, M. Baerns, H. Holene, and D. Wolf., *Appl. Surf. Sci.,* **223**, 168 (2004).
15. A.Hagemeyer, R. Borade, P. Desrosiers, Sh. Guan, D.M. Love, D.M. Poojary, H. Turner, H. Wienberg, X. Zhou, R. Armbrust, G. Fengler, and U. Notheis, *Appl. Catal. A: General*, **227**, 43, (2002).
16. V.Aranujatesan, ShaRee L. MacIntosh, M. Cruz, R.F.Renneke, and B. Chen, Chemical Industries (Dekker), **104**, (*Catal. Org. React.*), 195-200. (2005).
17. G.Y.Li, Chemical Industries (Dekker), **104**, (*Catal. Org. React.*), 177-194 (2005).
18. K.E. Simons, *Topics in Catal.*, **13**, 201 (2000).
19. B. Desai, and C.O. Kappe, *J. Comb. Chem.,* **7**, 641 (2005).
20. S. Saaby, K. R. Knudsen, M. Ladlow, and S. V. Ley, *Chem. Comm.*, 2909 (2005).
21. E.C. Harrington, *Industrial Quality Control*, **21**, 494 (1965).
22. H.U. Blaser, personal communications.
23. L. Végvári, A. Tompos, S. Gőbölös, and J. Margitfalvi, *Catal. Today,* **81**, 517 (2003).
24. A. Tompos, J.L. , Margitfalvi, E. Tfirst, and L. Végvári, *Appl. Catal.* A: General, **254**, 160 (2003).
25. K. Huang, F.Q. Chen, and D.W Lu, *Appl. Catal. A: General*, **219**, 61 (2001).
26. T.R. Cundari, J. Deng, and Y. Zhao, *Ind. Eng. Chem. Res.* **40**, 5475 (2001).

27. S. Hashem, *Neural Networks,* **10**, 599 (1997).
28. A. Tompos, J.L. Margitfalvi, E. Tfirst and L. Végvári, *Appl. Catal. A: General,* accepted for publication

36.

Dendrimer Templates for Pt and Au Catalysts

Bethany J. Auten, Christina J. Crump, Anil R. Singh and Bert D. Chandler

Department of Chemistry, Trinity University, 1 Trinity Place, San Antonio, TX, 78212-7200

bert.chandler@trinity.edu

Abstract

We are developing a new method for preparing heterogeneous catalysts utilizing polyamidoamine (PAMAM) dendrimers to template metal nanoparticles.(1) In this study, generation 4 PAMAM dendrimers were used to template Pt or Au Dendrimer Encapsulated Nanoparticles (DENs) in solution. For Au nanoparticles prepared by this route, particle sizes and distributions are particularly small and narrow, with average sizes of 1.3 ± 0.3 nm.(2) For Pt DENs, particle sizes were around 2 nm.(3) The DENs were deposited onto silica and Degussa P-25 titania, and conditions for dendrimer removal were examined.

The focus of these studies has been on identifying mild activation conditions to prevent nanoparticle agglomeration. Infrared spectroscopy indicated that titania plays an active role in dendrimer adsorption and decomposition; in contrast, adsorption of DENs on silica is dominated by metal-support interactions. Relatively mild (150° C) activation conditions were identified and optimized for Pt and Au catalysts. Comparable conditions yield clean nanoparticles that are active CO oxidation catalysts. Supported Pt catalysts are also active in toluene hydrogenation test reactions.

Introduction

Industrial heterogeneous catalysts and laboratory-scale model catalysts are commonly prepared by first impregnating a support with simple transition metal complexes. Catalytically active metal nanoparticles (NPs) are subsequently prepared through a series of high temperature calcination and / or reduction steps. These methods are relatively inexpensive and can be readily applied to numerous metals and supports; however, the NPs are prepared *in-situ* on the support via processes that are not necessarily well understood. These inherent problems with standard catalyst preparation techniques are considerable drawbacks to studying and understanding complex organic reaction mechanisms over supported catalysts.(4)

We are developing a new route for preparing model catalysts that uses Polyamidoamine (PAMAM) Dendrimer Encapsulated Nanoparticles (DENs) as NPs templates and stabilizers. PAMAM dendrimers, which bind transition metal cations in defined stoichiometries, can be used to template mono- and bimetallic NPs in solution. DENs have been used as homogeneous catalysts for a variety of organic reactions, including hydrogenations (particularly size-selective hydrogenations), and Suzuki, Heck, and Stille coupling reactions.(5) Bimetallic DENs are also active hydrogenation catalysts in solution, and their catalytic activity can be tuned by controlling NP composition.(5)

The potential to ultimately control NP properties makes DENs extremely attractive as precursors to supported catalysts. The possibility of controlling particle size and composition makes DENs uniquely suited to exploring the relative importance of these effects on catalytic reactions. Because the nanoparticles are prepared *ex situ* and can be deposited onto almost any substrate or support, DENs offer the opportunity to examine tailored NPs using materials comparable to those employed as industrial catalysts. However, before DENs can be utilized as heterogeneous catalyst precursors, appropriate methods must be developed to remove the organic template. If activation conditions are too harsh, particle agglomeration may suppress the potential advantages of the dendrimer method. If activation conditions are too mild, incomplete removal of the dendrimer may leave organic residues on the particles and poison the catalyst.

Background

Previous work has focused largely on dendrimer removal from Pt-DENs supported on silica. Our studies have shown the amide bonds that comprise the dendrimer backbone are relatively unstable (they begin decomposing at mildly elevated temperatures, ca. 100 °C) and that the Pt nanoparticles help to catalyzed the dendrimer decomposition.(3,6,7) However, extended higher temperature oxidation and/or reduction treatments (several hours at 300 °C) are required to completely remove organic material from Pt DENs. For Pt/SiO_2 catalysts, dendrimer oxidation appears to lead to the formation of surface carboxylates, which partially poison catalytic activity.(6,8) A recent surface science study by Chen and coworkers supports these general findings and provides convincing evidence that Pt plays an important role in catalyzing dendrimer decomposition.(9)

Crooks and coworkers, who studied Pd and Au DENs immobilized in sol-gel titania, similarly reported the onset of dendrimer mass loss at relatively low temperatures (ca. 150 °C). Pd helped to catalyze dendrimer decomposition in their system, as well. Temperatures of 500 °C or greater were required to completely remove organic residues from their materials.(10) This treatment resulted in substantial particle agglomeration, particularly for Au-based materials, although pore templating by the dendrimer mitigated the particle growth.

Using Pt-DENs/SiO$_2$, we showed that activation temperatures could be reduced to as low as 150 °C by using CO oxidation reaction conditions. Supported Pt catalysts pretreated in 1% CO & 25% O$_2$ at 150 °C for 16 hours had essentially the same CO oxidation activity as catalysts oxidized at 300 °C for 16 hours. In this activation protocol, CO essentially acts as a protecting group: strong CO adsorption prevents Pt from participating in dendrimer oxidation and prevents dendrimer fragments from fouling the nanoparticle catalyst.

Results and Discussion

Based on our previous work developing activation conditions for supported Pt DENs, we began investigating dendrimer removal from supported Au DENs under CO oxidation catalysis conditions (1% CO, 25% O$_2$). Au based catalysts are extremely sensitive to preparation techniques and sintering, so minimizing activation temperatures is in important consideration for this system. Additionally, in spite of the potential for high CO oxidation activity, it is unclear whether Au NPs will be capable of participating in dendrimer oxidation. Consequently, we began our study by directly monitoring CO oxidation activity during catalyst pretreatment.

In these experiments, shown in Figure 1, the CO oxidation reactor system was charged with supported, in tact Au-DENs/TiO$_2$, and CO oxidation activity was monitored as a function of time on stream at various temperatures (100, 125, and 150 °C). The time required to reach maximum activity varied from 8 hours at 150 °C to 24 hours at 100 °C. We sought to keep activation temperatures as low as

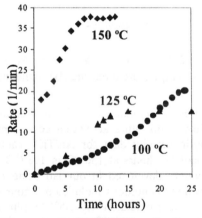

Figure 1. Au-DENs/TiO$_2$ activation under CO/O$_2$ atmosphere.

possible to minimize sintering, so higher temperatures were not investigated. Additionally, activation under CO oxidation conditions relies on adsorbed CO to "protect" the metal particle surface from poisoning by dendrimer decomposition products. The "protective" properties of CO diminish as adsorption equilibira favor free CO at higher temperatures.

The differences between catalysts treated at 125 and 100 °C suggest that lower activation temperatures might produce catalysts that are more active. This data is collected with different conversions, at different temperatures, and with small masses of catalyst, so observed rates in these experiments are only qualitative activity measures. Variances in surface water content, which might be substantial at these temperatures, also can dramatically affect the activity of supported Au catalysts.(11) Consequently, it is difficult to draw meaningful conclusions regarding catalyst activity from the low temperature activation data.

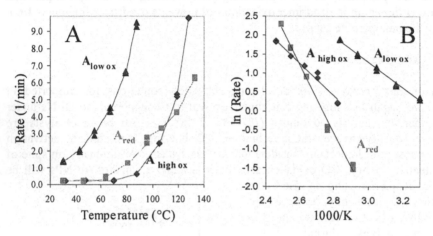

Figure 2. CO oxidation catalysis by Au/TiO_2 for various pretreatments. (A) rate vs. temperature plots and (B) Arrhenius plots.

For further studies, we focused on catalysts activated at 150 °C as a model for the low temperature activation conditions. Figure 2a shows carefully determined catalytic activity data for Au/TiO_2 catalysts activated with the low temperature protocol (16 hours at 150 °C in 1% CO and 25% O_2) versus activation treatments previously developed in our lab using other supported DENs. These previously determined conditions included a 6 hour oxidation with 25% O_2 at 300 °C ($A_{high-ox}$) and a 4 hour oxidation with 25% O_2 plus two hour reduction with 25% H_2 at 300 °C (A_{red}). For the Au/TiO_2 catalysts, activated with these protocols, only minimal differences in reactivity were observed.

Catalyst A_{low-ox} was substantially more reactive than catalysts prepared with shorter, high temperature treatments (rates were roughly 6 times faster with the A_{low-ox} pretreatment at about 80 °C). Although the Au/TiO_2 catalysts prepared from DENs are more active than the supported Pt catalysts we have previously examined,(6)[1] they are not as active as the best literature reports for CO oxidation by supported Au catalysts prepared by other means. Arrhenius plots using rate data between 2 and 12% conversion for the Au/TiO_2 catalysts activated with these three protocols (A_{low-ox}, A_{red}, and $A_{high-ox}$) are in Figure 2b. Extracted apparent activation energies (E_{app}) from this data, found in Table 1, indicate a substantial change to the catalyst upon high temperature reduction (*vida infra*).

Table 1. Apparent activation energies (E_{app}) from Arrhenius plots.

Catalyst Designation	Pretreatment	E_{app} (kJ/mole)
$A_{high-ox}$	25% O_2 @ 300 °C for 6 hrs	36
A_{red}	25% O_2 @ 300 °C for 4 hrs + 25% H_2 @ 300 °C for 2 hrs	75
A_{low-ox}	1% CO + 25% O_2 @ 150 °C for 16 hrs. • activation & testing immediately after preparation	32
B_{low-ox}	1% CO + 25% O_2 @ 150 °C for 16 hrs. • activation & testing immediately after preparation	31
$B_{low-ox+1mo.}$	1% CO + 25% O_2 @ 150 °C for 16 hrs. • activation immediately after preparation, testing 1 month later	32
$B_{1mo+low-ox}$	1% CO + 25% O_2 @ 150 °C for 16 hrs. • activation & testing 1 month after preparation	40

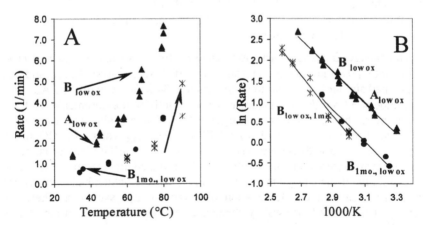

Figure 3. CO oxidation by batch B catalysts activated under CO + O_2 at 150 °C. (A) rate vs. temperature plots and (B) Arrhenius plots.

We encountered substantial difficulties in reproducing the A_{low-ox} activity data over the course of several weeks using the original batch of catalyst. Subsequent activations yielded catalysts with lower activities than the freshly prepared and activated material.

Although supported Au catalysts are notorious for their problems with reproducibility, these results were surprising, since our previous work with supported Pt catalysts prepared from DENs was highly reproducible among different members of the group.

To examine the peculiarities associated with the Au/TiO_2 catalysts activated at lower temperatures, a second batch (B) was prepared by a second research student using identical procedures as the first. Beyond evaluating the general reproducibility of the preparation method, the goal of these experiments was to determine if it is necessary to activate supported Au DENs immediately after catalyst preparation. Additionally, we hoped to begin investigating the cause of the apparent catalyst deactivation.

Catalyst batch B was separated into three parts: B_{low-ox} was activated and tested immediately after preparation, $B_{low-ox+1mo}$ was activated immediately after preparation but activity tests were performed one month later, and $B_{1mo+low-ox}$ was activated and tested one month after preparation. During the waiting period, catalysts were stored in foil wrapped containers under air in a desiccator. The data for catalysts A_{low-ox} and B_{low-ox} in Figure 3 and Table 1 show that both the CO oxidation rates and E_{app} values were highly reproducible between batches, provided that the catalysts are activated and tested soon after preparation. Catalyst $B_{low-ox+1mo}$ was a factor of two less active than the freshly prepared and activated B_{low-ox}; catalyst $B_{1mo+low-ox}$ showed similar decreases in activity.

The changes in apparent activation energies for the various samples offer initial insights into possible changes in catalytic active sites. The A_{low-ox} and $A_{high-ox}$ catalysts had similar E_{app} values of approximately 34 kJ/mole, while the A_{red} catalyst, which underwent a reduction treatment prior to activity measurements, had an E_{app} that was more than a factor of two larger at 75 kJ/mole (Table 1). In our previous work with Pt catalysts prepared from DENs, lower E_{app} values for CO oxidation generally correlated with cleaner surfaces.(3,7) For the Au/TiO_2 catalysts, however, the higher temperature reduction treatments also reduced the overall catalyst activity. The magnitude of the change in E_{app} for the A_{red} sample, which had been reduced at 300 °C, suggests that the deactivation observed for this sample is likely due to sintering. This catalyst also visibly changes color, consistent with an increase in Au nanoparticle size.

The apparent activation energies for the unreduced samples are all very similar. The samples pretreated with low temperature oxidation in the presence of CO all have E_{app} values within experimental error of 32 kJ/mole. The high temperature oxidation and delayed low temperature oxidation treatments are only slightly higher (36 and 40 kJ/mole, respectively). The observation that the E_{app} values from the less active B catalysts do not change substantially from the freshly prepared and activated catalyst, and are similar to the catalyst treated with high temperature (300 °C) oxidation, indicates that the active sites on all these catalysts are at least qualitatively similar. This, in turn, suggests that the primary activity differences from sample to sample are not due to intrinsic activity difference associated with different types of active sites. Rather, the differences observed in catalytic activity are likely to be due to differences in the number of active sites present on each catalyst.

In order to better understand changes to the catalyst surface during activation, infrared spectroscopy was used to monitor the dendrimer decomposition process. IR spectra of intact Au-DENs/TiO_2 are

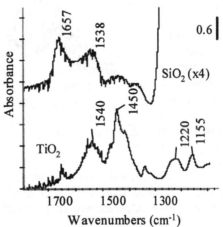

Figure 4. IR spectra of (top) Pt-DENs/SiO_2 and (bottom) Au-DENs/TiO_2

compared with a spectrum of Pt-DENs/SiO_2 in Figure 4. The spectrum of Pt-DENs/SiO_2 shows the amide I and II bands at 1657 and 1538 cm^{-1}. These values are consistent with free PAMAM dendrimers in solution and with previous reports for unactivated dendrimer templated catalysts on SiO_2.(3,7,9) The TiO_2 supported DENs are strikingly different, with 2 prominent peaks from the dendrimer visible at 1540cm^{-1} and 1440 cm^{-1}. Two additional smaller peaks at 1220 cm^{-1} and 1155 cm^{-1} - are also apparent on the TiO_2 support.

The 100 cm^{-1} shift of the two most prominent bands indicates strong interactions between the dendrimer and TiO$_2$; these interactions are not present when SiO$_2$ is used as the support. A previous study comparing FTIR adsorption of benzamide and acetamide on TiO$_2$ shows that the amide bonds bind strongly to TiO$_2$, shifting the amide IR vibrations about 100 wavenumbers.(12) This study concluded that amides lose the amino H+ to TiO$_2$ upon adsorption, and the resulting IR band is due to the symmetric and antisymmetric stretch of the -CON- group.(12) Results with titania supported Au-DENs are consistent with this explanation, and suggest that the adsorption of the DENs onto TiO$_2$ is driven primarily by the interaction between the dendrimer amide bonds and the support. This conclusion is somewhat different than for other supports (SiO$_2$, Al$_2$O$_3$) where adsorption appears to be driven by NP-support interactions (3).

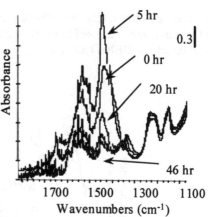

Figure 5. IR spectroscopy of Au-DENs/TiO$_2$ during low-ox pretreatment.

Dendrimer decomposition was also monitored during activation. In this experiment, *in situ* IR spectra were collected as the catalyst was heated to 150 °C in the presence of CO, O$_2$ and He, simulating the "low-ox" activation conditions. Figure 5 shows that the amide bonds at 1540 cm^{-1} and 1450 cm^{-1} do not fully decompose after 20+ hours of treatment at 150 °C, and the two smaller peaks at 1150 cm^{-1} and 1230 cm^{-1} do not change at all. Since typical activation conditions consists of heating the catalyst at 150 °C for only 16 hours, it is clear that substantial dendrimer fragments and oxidation byproducts remain on the catalyst surface after activation.

The persistence of the dendrimer decomposition products is the likely cause of the catalyst deactivation over time. The presence of dendrimer and dendrimer byproducts indicates that even the more active catalysts are not particularly clean. It is difficult to distinguish between species adsorbed on the NPs from those primarily on the support; however, it is likely that the location of the dendrimer decomposition varies widely along the surface of the catalyst. The dendrimer fragments present on the support could migrate over time and poison the metal active sites, resulting in the lower catalytic activity over time. It is also possible that the residual dendrimer undergoes some slower oxidation processes that result in a stronger, unobservable poison.

Experimental Section

Au-DENs were prepared via literature procedures (2)[2] and deposited onto Degussa P25 Titania by stirring at pH 6 overnight. In situ infrared spectroscopy, catalyst activation, and CO oxidation experiments were performed using previously described procedures.(3) Catalyst activation under CO oxidation conditions were used 23 mg catalyst samples diluted 10:1 with α-Al_2O_3. In CO oxidation activity measurements, the feed composition was 1.1% CO, 27% O_2 balance He, and the flow rate was kept constant at 20 mL/min. Conversion was measured as a function of temperature and rate data was determined only for conversions between 1 and 12%.

Acknowledgements

We gratefully acknowledge the Robert A. Welch Foundation (Grant number W-1552) for financial support of this work. BJA also thanks the Beckman Foundation for support during the 05-06 academic year. Acknowledgement is made to the donors of the American Chemical Society's Petroleum Research Fund supporting the CO oxidation reactor system construction.

References

1. R. M. Crooks, M. Zhao, L. Sun, V. Chechik, and L. K. Yeung, *Accts. Chem. Res.*, **34**, 181-190 (2001).
2. Y.-G. Kim, S.-K. Oh, and R. M. Crooks, *Chem. Mater.* **16**, 167-172 (2004).
3. H. Lang, R. A. May, B. L. Iversen, and B. D. Chandler, *J. Am. Chem. Soc.*, **125**, 14832-14836 (2003).
4. V. Ponec and G. C. Bond, *Catalysis by Metals and Alloys*, Elsevier, Amsterdam, 1995.
5. R. W. J. Scott, O. M. Wilson, and R. M. Crooks, *J. Phys. Chem. B,* **109**, 692-704 (2005).
6. A. Singh and B. D. Chandler, *Langmuir,* **21**, 10776-10782 (2005).
7. H. Lang, R. A. May, B. L. Iversen, and B. D. Chandler, Chemical Industries (Taylor & Francis Group / CRC Press), **68**, (*Catal. Org. React.*), 367-377 243-250 (2005).
8. L. Beakley, S. Yost, R. Cheng, and B. D. Chandler, *Appl. Catal. A: General*, **292**, 124-129 (2005).
9. O. Ozturk, T. J. Black, K. Perrine, K. Pizzolato, C. T. Williams, F. W. Parsons, J. S. Ratliff, J. Gao, C. J. Murphy, H. Xie, H. J. Ploehn, and D. A. Chen, *Langmuir* **21**, 3998-4006 (2005).
10. R. W. J. Scott, O. M. Wilson, and R. M. Crooks, *Chem. Mater* **16**, 5682 - 5688 (2004).
11. G. C. Bond and D. T. Thompson, *Catal. Rev. Sci. Eng.*, **41**, 319-388 (1999).
12. L.-F. Liao, C.-F. Lien, D.-L. Shieh, F.-C. Chen, and J.-L. Lin, *Phys. Chem. Chem. Phys.*, **4**, 4584-4589 (2002).

V. Symposium on Acid and Base Catalysis

37. Development of an Industrial Process for the Lewis Acid/Iodide Salt-Catalyzed Rearrangement of 3,4-Epoxy-1-Butene to 2,5-Dihydrofuran

Stephen N. Falling, John R. Monnier (1), Gerald W. Phillips and Jeffrey S. Kanel
Eastman Chemical Company, Kingsport, TN 37662

Stephen A. Godleski
Eastman Kodak Company, Rochester, NY 14650

sfalling@eastman.com

Abstract

2,5-Dihydrofuran (2,5-DHF) is typically produced by the catalytic rearrangement of 3,4-epoxy-1-butene (**1**) using an inorganic Lewis acid and inorganic or onium iodide in a polar, aprotic solvent. For years, commercial processes utilizing this chemistry were unattractive due to the high cost of **1** and side reactions in the rearrangement step. Recovery of the expensive catalysts and solvent was difficult and presented a serious problem for scale-up. Following the discovery of an economical process for **1**, the production of 2,5-DHF was once again of industrial interest. This paper describes the development of a continuous, liquid phase process utilizing a trialkyltin iodide and tetraalkylphosphonium iodide co-catalyst system which gives high selectivity for 2,5-DHF and provides an efficient means for catalyst recovery.

Introduction

3,4-Epoxy-1-butene (**1**) is a versatile intermediate for the production of commodity, specialty and fine chemicals (2). An important derivative of **1** is 2,5-dihydrofuran (2,5-DHF). This heterocycle is useful in the production of tetrahydrofuran (3), 2,3-dihydrofuran (4), 1,4-butanediol (5), and many fine chemicals (e.g., 3-formyltetrahydrofuran (6) and cyclopropanes (7)). The homogeneous, Lewis acid and iodide salt-catalyzed rearrangement (isomerization) of **1** to 2,5-DHF has been known since 1976 (8) and is the only practical method for 2,5-DHF synthesis.

$$\Delta H = -16.8 \text{ Kcal/mole}$$

1   **2,5-DHF**

Typically, this rearrangement process is catalyzed by an inorganic Lewis acid and inorganic or quaternary onium iodide in a polar, aprotic solvent (e.g., ZnI_2, KI, N-methyl pyrrolidone) (8,9). For years, however, commercial processes utilizing this chemistry were unattractive due to the high cost of **1** and poor reaction selectivity.

In 1986, a process to produce **1** by the continuous, vapor phase oxidation of 1,3-butadiene over a silver on alumina catalyst was discovered by Monnier and Muehlbauer of the Kodak Corporate Research Laboratories (10). The process was further developed and commercialized by Eastman Chemical Company at its Longview, Texas plant (11). Following this discovery of an economical process for **1**, the production of 2,5-DHF was once again of commercial interest.

Results and Discussion

In the development of a 3,4-epoxy-1-butene (**1**) rearrangement process suitable for industrial scale-up, a number of factors were evident. The product (2,5-DHF) and starting material (**1**) are both liquids with identical boiling points (66°C). No practical method is known by which to separate these isomers. This fact demands that the catalytic process be performed at high conversion for acceptable economics. The common practice of recycling unreacted starting material was not an option for this process.

The major side products in this homogeneous catalysis process are crotonaldehyde (also isomeric with starting material) and the polyether oligomer of **1**. Crotonaldehyde is a volatile liquid with a boiling point of 104°C so it is separable from the reaction mixture and product. However the oligomeric material is non-volatile and accumulates in the reaction mixture causing dilution of the catalysts and filling of the reaction vessel. Because Lewis acid and iodide catalysts are normally too expensive to discard, a rearrangement process was needed with high selectivity for product and a means for catalyst recovery (or oligomer removal). For the rearrangement process to be economically and environmentally acceptable, the recovery/reuse of the catalyst components is imperative.

Scheme 1

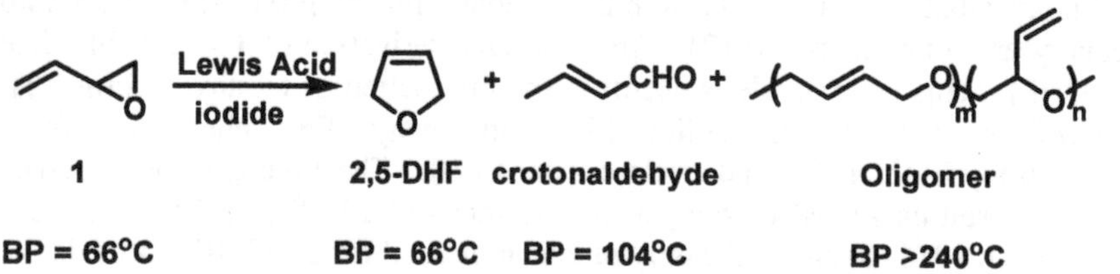

1		**2,5-DHF**	**crotonaldehyde**	**Oligomer**
BP = 66°C		BP = 66°C	BP = 104°C	BP >240°C

Due to 3,4-epoxy-1-butene's (**1**) conjugated vinyl and epoxide functional groups it can polymerize (oligomerize) with repeat units which are linear 1,4-substituted (m-units in Scheme 1) or branched 1,2-substituted (n-units). The m/n repeat unit ratio depends on the catalyst and conditions but is usually about 15/85 (12). The oligomer produced as a side product in the 2,5-DHF process is typically a viscous liquid with a broad molecular weight distribution averaging about 1100 and density greater than 1. An important property of this unavoidable side product is the poor solubility of the higher molecular weight oligomers in non-polar alkane solvents. Thus the possibility existed for separation of catalysts from oligomers using an alkane solvent in a liquid-liquid extraction.

With these process factors in mind, a continuous, homogeneous reaction process concept was developed in which the starting epoxide **1** is fed as a liquid to the reactor containing the soluble Lewis acid and iodide salt while continuously removing the 2,5-DHF and crotonaldehyde by distillation. Continuously, or as needed, the catalysts are recovered from the reaction mixture by extraction with an alkane and the undissolved oligomer is discarded.

Catalyst Requirements

The rearrangement reaction of **1** requires a two component catalyst system consisting of a Lewis acid and a soluble source of iodide anion. The process concept had a number of requirements for the catalysts in order to achieve commercially useful operation. These were:

1. High catalytic activity.
2. High selectivity (especially low oligomerization).
3. Chemical and thermal stability (for acceptable catalyst lifetime).
4. Soluble in 2,5-DHF.
5. Alkane solubility (for recovery/oligomer removal).
6. Low vapor pressure (to prevent losses during distillation steps).
7. Low melting point (for easy handling as liquids).
8. Low cost.
9. Low toxicity.

Iodide Salt Catalyst Component

For adequate reaction rates, a high concentration of iodide anion is necessary. The cation portion of the salt appears to have little or no effect on catalytic activity or reaction selectivity. Inorganic iodides (such as potassium iodide) are the obvious first choice based on availability and cost. Unfortunately these catalysts have very poor solubility in the reaction mixture without added solubilizers or polar, aprotic solvents. These solubilizers (e.g., crown ethers) and solvents are not compatible with the desired catalyst recovery system using an alkane solvent. Quaternary onium iodides however combine the best properties of solubility and reactivity.

Quaternary ammonium iodides are attractive choices because they generally have good activity, low cost, and solubility in the reaction and recovery processes. Simple quaternization of the wide variety of available tri(n-alkyl)amines with n-alkyl iodides allows optimization of the tetra(n-alkyl)ammonium iodide salt properties:

$$R_3N \ + \ R'I \ \rightarrow \ R_3R'N^+I^-$$

The long-term operation of a continuous reaction process with catalyst recovery requires a catalyst with very good thermal stability. Unfortunately the tetra(n-alkyl)ammonium iodides which have good solubility properties also have poor thermal stability due to breakdown by dealkylation. This was observed during operation of the continuous process and in thermal analysis. Thermogravimetric analysis (TGA) of tetra(n-heptyl)ammonium iodide shows decomposition beginning at 180°C and complete loss of weight by 285°C.

Quaternary phosphonium iodides are also good choices for the iodide salt catalyst component because they are highly active and, in some cases, soluble in the reaction and recovery processes. The simple quaternization of tri(n-alkyl)phosphines or triarylphosphines with n-alkyl iodides produces a wide variety of low cost phosphonium iodide salts:

$$R_3P + R'I \rightarrow R_3R'P^+I^-$$

Although the number of available tri(alkyl/aryl)phosphine starting materials may be somewhat lower than with the tri(n-alkyl)amines, the large number of commercially available n-alkyl iodides allows for the synthesis of many phosphonium iodide catalyst candidates. Triphenylphosphine is an attractive starting material due to its industrial availability and low cost. However it was found that the resulting (n-alkyl)(triphenyl)phosphonium iodides had poor solubility in alkane solvents. They also tended to have high melting points which is a disadvantage in process operation. Fortunately a number of trialkylphosphines are also produced in bulk industrially. Of particular interest is tri(n-octyl)phosphine which is manufactured in large volume for production of tri(n-octyl)phosphine oxide (TOPO) – a mining extraction solvent.

The critical advantage of the quaternary phosphonium iodides is their excellent temperature stability. For example, the selected tetra(alkyl)phosphonium iodide salt, tri(n-octyl)(n-octadecyl)phosphonium iodide [(n-Oct)$_3$(n-Octadecyl)P$^+$I$^-$, referred to as "**TOP18**"], shows no TGA weight loss up to 300°C.

Table 1. Tetra(n-alkyl)phosphonium iodides tested.

Tetra(n-alkyl) phosphonium iodide	Carbon Count	M.W.	M.P. (°C)	Solubility in Octane
(n-Oct)$_3$(n-Bu)P$^+$I$^-$	28	554.63	55-57	insol.
(n-Oct)$_3$(n-Hexyl)P$^+$I$^-$	30	582.68	semisolid	insol.
(n-Oct)$_3$(n-Heptyl)P$^+$I$^-$	31	596.71	77-79	insol.
(n-Oct)$_4$P$^+$I$^-$	32	610.73	86-88	sol. hot
(n-Oct)$_3$(n-Nonyl)P$^+$I$^-$	33	624.76	79-81	sol. hot
(n-Oct)$_3$(n-Decyl)P$^+$I$^-$	34	635.79	64-65	sol. hot
(n-Oct)$_3$(n-Dodecyl)P$^+$I$^-$	36	666.84	50-52	sol. warm
(n-Oct)$_3$(n-Hexadecyl)P$^+$I$^-$	40	722.94	56-58	sol. warm
(n-Oct)$_3$(n-Octadecyl)P$^+$I$^-$, **TOP18**	42	750.99	62-65	sol. warm
(n-Dodecyl)$_4$P$^+$I$^-$	48	835.15	77-79	sol. hot
(n-Hexadecyl)$_4$P$^+$I$^-$	64	1059.57	87-88	sol. hot

Thus the selected iodide catalyst, **TOP18**, has good catalytic activity, selectivity, and stability. It is readily soluble in 2,5-DHF and in warm alkane solvents. As a salt, **TOP18** has virtually no vapor pressure and it is reasonably low melting. It is easily synthesized by reaction of tri(n-octyl)phosphine with n-octadecyl iodide at a reasonable cost. Toxicity studies on **TOP18** showed it to be of low toxicity: Oral LD-50 (rat), >2000 mg/kg; Dermal LD-50 (rat), >2000 mg/kg.

Lewis Acid Catalyst Component

As in the case of the iodide salt component, a high concentration of Lewis acid is necessary for adequate reaction rate. Inorganic Lewis acids (such as zinc iodide) are the obvious first choice based on availability and cost. However these catalysts also have very poor solubility in the reaction mixture without the use of a polar, aprotic solvent. Fortunately a family of Lewis acids was found with improved reaction selectivity and good solubility—tri(organo)tin iodides. These Lewis acids were investigated due to their known activity in the formation of cyclic carbonates from epoxides and carbon dioxide (13). Tri(organo)tin iodides were found to be considerably better than di(organo)tin diiodides and mono(organo)tin triiodides for the desired **1** rearrangement. Tri(organo)tin bromides and chlorides (in conjunction with an onium bromide or chloride) were less active and less selective than the all-iodide systems.

Organotin compounds are produced commercially for use in PVC stabilization and as agricultural chemicals (14). Tri(organo)tin iodides are appropriate candidates for consideration due to their excellent catalytic activity, high selectivity, availability of starting materials and ease of preparation. Triphenyltin iodide is a crystalline solid (mp 121°C) and is soluble in hot alkanes. It is easily prepared from commercially-available triphenyltin chloride by reaction with sodium iodide. Although triphenyltin iodide had the highest activity for rearrangement with low oligomerization, it had less than acceptable stability. In extended continuous laboratory runs, the breakdown of triphenyltin iodide to diphenyltin diiodide and benzene was observed. This reaction can occur by reaction with low levels of hydrogen iodide that are present in the system.

$$Ph_3SnI \; + \; HI \; \rightarrow \; Ph_2SnI_2 \; + \; PhH$$

This loss of catalyst and contamination of product with benzene caused us to select the more stable tri(n-alkyl)tin iodides. These catalysts are not as active as the triaryltin iodides but exhibit very good stability under normal reaction conditions. They are also readily prepared from industrially-available bulk starting materials. Tri(n-octyl)tin iodide [(n-Oct)$_3$SnI, referred to as "**TOT**"] can be prepared from three different starting materials although the iodide displacement is preferred:

$$(n\text{-Oct})_3SnCl \; + \; NaI \; \rightarrow \; (n\text{-Oct})_3SnI \; + \; NaCl$$
$$\textbf{TOT}$$

Tri(n-butyl)tin iodide is also an effective catalyst and its starting materials are available but **TOT** was selected due to its lower volatility. The octyltin compounds are also generally much less toxic than their butyltin counterparts (15).

Table 2 Tri(organo)tin iodides tested.

Tri(organo)tin iodide	M.W.	M.P. (°C)	B.P. (°C (mm))
Ph₃SnI	476.91	121	253 (13.5)
(n-Bu)₃SnI	416.94		168 (8)
(n-Oct)₃SnI, **TOT**	585.26		215 (5)

Thus the selected Lewis acid catalyst, **TOT**, has good catalytic activity, selectivity and stability. It is a non-viscous liquid which is compatible with **TOP18** and is miscible with 2,5-DHF and non-polar alkane solvents. It has a very low vapor pressure so it is not lost during product distillation and catalyst recovery operations. **TOT** is easily synthesized at low cost and it has low toxicity (Oral LD-50 (rat), >2000 mg/kg; Dermal LD-50 (rat), >2000 mg/kg).

Final Catalyst System

The optimized catalyst system consisted of a 1:1 mole ratio of **TOP18** and **TOT** (16). Fresh **TOP18** is charged to the reactor as a melt or a THF solution (THF is stripped during start-up). No additional solvents are used in the reaction stage of the 2,5-DHF process. The reaction mixture therefore consists of only reactant **1**, 2,5-DHF product, catalysts and side products (crotonaldehyde and oligomer). Indeed because of the high loading of catalysts, the reaction mixture may be considered as an ionic liquid. Analysis of the reaction mixture is best performed using X-ray Fluorescence Spectrometry (XFS, for % P, Sn, I, and Cl) and by quantitative carbon-13 NMR (for % catalysts and oligomer). Carbon-13 NMR is especially useful for observing the condition of the organotin catalyst (17).

The use of organotin reagents and reactions is well entrenched in organic synthesis (19). Organotin alkoxides have been shown to be valuable in the preparation of ethers (20). Heating 4-haloalkoxytributyltin compounds gives quantitative yields of tetrahydrofurans (21). In view of this precedent, Scheme 2 shows the presumed catalytic mechanism of **1** rearrangement to 2,5-DHF.

Scheme 2

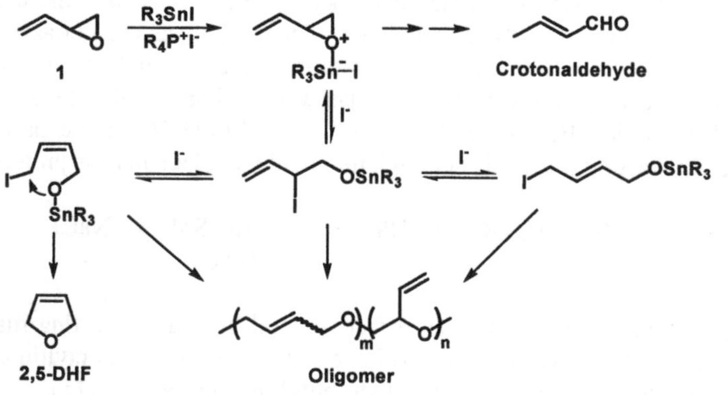

The catalyst system of **TOP18** and **TOT** is remarkably stable and long-lived if the reaction system is kept free of water and air. Contamination of the reaction mixture with water will lead to **TOT** decomposition (17). This reaction produces hydrogen iodide which reacts with epoxide **1**. Air is deleterious to both catalyst components. Introduction of air into the process can cause oxidation of iodide anion to iodine which will dealkylate **TOT**. These reactions lead to loss of iodine value from the catalyst system which is monitored by XRF during production.

$$2 \ (n\text{-Oct})_3SnI \ + \ H_2O \ \rightarrow \ [(n\text{-Oct})_3Sn]_2O \ + \ 2 \ HI$$

$$1 \ + \ HI \ \rightarrow \ \text{iodobutenols} \uparrow$$

$$(n\text{-Oct})_3SnI \ + \ I_2 \ \rightarrow \ (n\text{-Oct})_2SnI_2 \ + \ n\text{-OctI} \uparrow$$

Another mode for loss of catalyst is in the oligomer. Low levels of Sn, I and P were found in the oligomer waste (see below). Presumably some of this lost catalyst value is via covalent attachment (end groups) as well as unextracted active catalyst.

Process Development and Scale-up
Laboratory studies of the rearrangement process began with semi-continuous operation in a single, 200-mL, glass reactor, feeding **1** as a liquid and simultaneous distillation of 2,5-DHF, crotonaldehyde and unreacted **1**. Catalyst recovery was performed as needed in a separatory funnel with n-octane as the extraction solvent. Further laboratory development was performed with one or more 1000-mL continuous reactors in series and catalyst recovery used a laboratory-scale, reciprocating-plate, counter-current, continuous extractor (Karr extractor). Final scale-up was to a semiworks plant (capacity ca. 4500 kg/day) using three, stainless steel, continuous stirred tank reactors (CSTR).

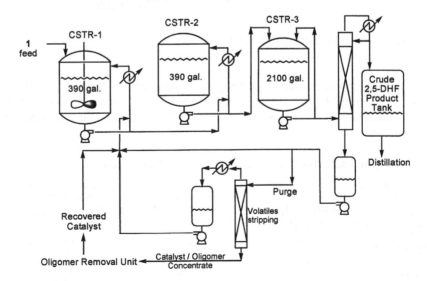

Figure 1. Semiworks-scale continuous process for 2,5-dihydrofuran.

The process is typically operated at 100-110°C and at, or slightly above, atmospheric pressure. The rearrangement reaction is exothermic so CSTR-1 and CSTR-2 require heat removal. CSTR-3 is operated adiabatically and its large size allows for increased overall residence time and high conversion of **1** to products. When the oligomer level has built up sufficiently, a small purge stream is removed and sent to the Oligomer Removal Unit (ORU) for catalyst recovery. This unit employs a 24-ft tall, 6-inch diameter Karr extractor and VM&P Naphtha extraction solvent (mixed heptanes, B.P. 118°C, specific gravity 0.76 g/mL). The alkane solvent is introduced at the bottom of the continuous extractor and the catalyst/oligomer concentrate at the top. Counter-current extraction occurs as the heavy catalyst and oligomer feed falls through the rising alkane while an axially-attached set of perforated plates agitates up and down (22). The oligomer (raffinate) is collected from the bottom of the column and is sent to the incinerator. Several extraction process variables were optimized including extractor rates, reciprocation rate, plate spacing, mass-transfer direction, and temperature profile (23).

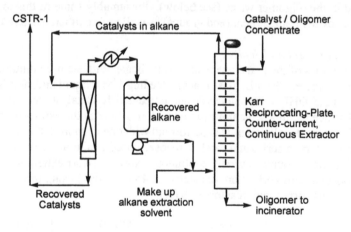

Figure 2. Oligomer Removal Unit (catalyst recovery).

Because the low molecular weight oligomers are soluble in the alkane extraction solvent, an unavoidably high concentration of oligomers is returned with recovered catalyst. Therefore during steady-state operation the reaction mixture is about one-third oligomer!

Table 3. Approximate steady-state process stream compositions.

Component	CSTR-3 Reaction Mixture	Karr Extractor Feed	Recovered Catalyst Stream
Volatiles	27%		
Oligomer	35%	48%	37%
TOT	17%	23%	27%
TOP18	21%	29%	36%

Normal operation gives about 99% conversion of **1** with selectivities of approximately 94% 2,5-DHF, 1% crotonaldehyde and 4.5% oligomer. The oligomer removed by the ORU is disposed of by incineration. Typically the catalyst level in oligomer is 3.5% **TOP18** and 0.5% **TOT**. In view of the high value of the product 2,5-DHF, this level of catalyst loss is fully acceptable. Optimized laboratory extraction conditions show 99.5% recovery of **TOP18** and 99.9% recovery of **TOT** from the oligomer. Second generation phosphonium iodide catalysts under development show promise for further improvement in catalyst recovery and cost savings.

Conclusions

A novel homogeneous process for the catalytic rearrangement of 3,4-epoxy-1-butene to 2,5-dihydrofuran has been successfully developed and scaled-up to production scale. A tri(n-alkyl)tin iodide and tetra-(n-alkyl)phosphonium iodide co-catalyst system was developed which met the many requirements for process operation. The production of a minor, non-volatile side product (oligomer) was the dominating factor in the design of catalysts. Liquid-liquid extraction provided the needed catalyst-oligomer separation process.

Acknowledgements

We gratefully acknowledge P. L. Dotson for catalyst screening experiments, M. K. Moore for catalyst extraction studies, T. R. Nolen for valuable engineering assistance, A. J. Robertson (Cytec Industries) for phosphonium iodide advice and samples, and the many members of the Eastman *EpB* Project Team.

References

1. Current address: Department of Chemical Engineering, University of South Carolina, Columbia, SC, 29208.
2. D. Denton, S. Falling, J. Monnier, J. Stavinoha, Jr. and W. Watkins, *Chimica Oggi*, **14**, 17 (1996).
3. S. N. Falling and B. L. Gustafson, US Pat. 4,962,210 to Eastman Kodak Company (1990).
4. J. R. Monnier, U.S. Pat. 5,536,851 to Eastman Chemical Company (1996); J. R. Monnier and C. S. Moorehouse, U.S. Pat. 5,670,672 to Eastman Chemical Company (1997); T. R. Nolen, S. N. Falling, D. M. Hitch, J. L. Miller and D. L. Terrill, US Pat. 5,681,969 to Eastman Chemical Company (1997); J. R. Monnier, J. W. Medlin and Y.-J. Kuo, *Appl. Catal. A*, **194-195**, 463 (2000).
5. S. N. Falling and G. W. Phillips, U.S. Pat. 5,254,701 to Eastman Kodak Company (1993).
6. W. A. Beavers, U.S. Pat. 5,945,549 to Eastman Chemical Company (1999).
7. S. Liang, T. Price, T. R. Nolen, D. B. Compton and D. Attride, U.S. Pat. 5,502,257 to Eastman Chemical Company (1996); S. Liang and T. Price, US Pat. 5,504,245 to Eastman Chemical Company (1996).

8. V. P. Kurkov, US Pat. 3,932,468 to Chevron Research Company (1976); R. G. Wall and V. P. Kurkov, US Pat. 3,996,248 to Chevron Research Company (1976).
9. M. Fischer, DE Offen. 3926147 to BASF AG (1989); J. R. Monnier, S. A. Godleski, H. M. Low, L. G. McCullough, L. W. McGarry, G. W. Phillips and S. N. Falling, US Pat. 5,082,956 to Eastman Kodak Company (1992).
10. J. R. Monnier and P. J. Muelbauer, US Pats. 4,897,498 and 4,950,773 to Eastman Kodak Company (1990); J. R. Monnier, J. W. Medlin and M. A. Barteau, *J. Catal.*, **203**, 362 (2001).
11. J. L. Stavinoha, Jr. and J. D. Tolleson, US Pat. 5,117,012 to Eastman Kodak Company (1992); J. L. Stavinoha, Jr., J. R. Monnier, D. M. Hitch, T. R. Nolen and G. L. Oltean, US Pat. 5,362,890 to Eastman Chemical Company (1994).
12. J. C. Matayabas, Jr. and S. N. Falling, US Pat. 5,434,314 to Eastman Chemical Company (1995); S. N. Falling, S. A. Godleski, P. Lopez-Maldonado, P. B. MacKenzie, L. G. McCullough and J. C. Matayabas, Jr., US Pat. 5,608,034 to Eastman Chemical Company (1997); T. Kuo, E. E. McEntire, S. N. Falling, Y.-C. Liu and W. A. Slegeir, US Pat. 6,451,926 to Eastman Chemical Company (2002).
13. A. Baba and T. Nozari, H. Matsuda, *Bull. Chem. Soc. Japan*, **60**, 1552 (1987).
14. D. B. Malpass, L. W. Fannin and J. J. Ligi, Organometallics-Sigma Bonded Alkyls and Aryls, *Kirk Othmer Encyclopedia of Chemical Technology*, Third Edition, 1981, Volume 16, pp. 573-579.
15. P. J. Smith, *Chemistry of Tin*, P. J. Smith, Ed., Blackie Academic & Professional, Thomson Science, London, Second Edition, 1998, pp. 431-435.
16. G. W. Phillips, S. N. Falling, S. A. Godleski and J. R. Monnier, US Pat. 5,315,019 to Eastman Chemical Company (1994).
17. S. N. Falling and P. Lopez, US Pat. 5,693,833 to Eastman Chemical Company (1997).
18. S. N. Falling, S. A. Godleski, L. W. McGarry and J. S. Kanel, US Pat. 5,238,889 to Eastman Kodak Company (1993).
19. M. Pereyre, J.-P. Quintard and A. Rahm, *Tin in Organic Synthesis*, Butterworths & Co. Ltd., London, 1987.
20. J.-C. Pommier, B. Delmond and J. Valade, *Tet. Lett.*, 5287 (1967).
21. B. Delmond, J.-C. Pommier and J. Valade, *J. Organomet. Chem.*, **50**, 121 (1973).
22. T. C. Lo and J. Prochazka, Reciprocating Plate Extraction Columns, *Handbook of Solvent Extraction*, T. C. Lo, M. H. I. Baird and C. Hanson, Eds., Krieger Publishing Company, Malabar, Florida, 1991, pp. 373-389.
23. D. Glatz and W. Parker, Enriching Liquid-Liquid Extraction, *Chem. Eng.*, November 2004, pp. 44-48.

38.　A New Economical and Environmentally Friendly Synthesis of 2,5-Dimethyl-2,4-Hexadiene

Alessandra O. Bianchi [a], Valerio Borzatta[a], Elisa Poluzzi [a] and Angelo Vaccari[b]

[a] Endura SpA, Viale Pietramellara 5, 40012 Bologna (Italy)
[b] Dipartimento di Chimica Industriale e dei Materiali, Alma Mater Studiorum Università di Bologna, Viale del Risorgimento 4, 40136 Bologna (Italy)

angelo.vaccari@unibo.it; +39 051 2093679

Abstract

A new vapour phase process for the synthesis of 2,5-dimethyl-2,4-hexadiene (or tetramethyl-butadiene or TMBD) by dehydration of 2,5-dimethyl-2,5-hexandiol on solid acid catalysts has been developed. Commercial acid catalysts (oxides, zeolites, clays and acid-treated clays) were preliminarily screened by using the reaction condition previously reported in literature. Subsequently, the study focused on a detailed set-up of the reaction conditions (temperature, LHSV value and feed concentration). In the best reaction conditions identified, yield values higher than 80% were obtained using some commercial and very inexpensive clay-based adsorbents (Mega Dry and Granosil 750 JF), with a stable activity of up to 400 h of time-on-stream. These results were significantly superior to those already published. Finally, the clay-based catalysts made it possible to recover high amounts of isomers by converting them to TMBD and may be partially regenerated.

Introduction

Pyrethrins and pyrethroids are probably the best known and safest classes of natural or synthetic insecticides, widely used in domestic and agricultural applications (1-7). Pyrethrins are natural insecticides derived from the *Chrysanthemum cineraria* flowers: the plant extract, called pyrethrum, is a mixture of six isomers (pyrethrin I and II, cinerin I and II, jasmolin I and II) which was first used in China in the 1st century AD, during the Chou Dinasty. The world pyrethrum market is worth half a billion US dollars [main producers are East Africa highlands (Kenia, Tanzania and Rwanda) and Australia]; however, its availability is subject to cyclical trends, due to rains and relations with farmers, who face high harvest costs also due to the fact that the flowers have to be harvested shortly after blooming (8). To by-pass this problem, the Australian variety has been genetically modified to ensure that all the flowers grow at the same speed, thus reducing the harvest costs. To overcame the availability problems, many chemical companies started to manufacture the synthetic equivalents (pyrethroids), which are esters of different acid and alcohol moieties (1-7,9). Pyrethroids may be classified on the basis of their biological and physical properties (knock down or killing effect, UV-resistant or non UV-resistant, for farming or domestic uses, respectively), which depend on chemical structures and stereoisomerisms.

The main intermediate for preparing pyrethroids is the chrysanthemic acid (CHA), which is currently obtained by the reaction of an appropriate diazo-acetic ester with 2,5-dimethyl-2,4-hexadiene (tetramethyl-butadiene or TMBD). The latter may be prepared in different ways (10-20), although the most relevant ones appear to be both the vapour phase reaction of i-butyralde hyde with i-butene (Prins reaction) on different oxides [14-16,18] and the liquid or vapour phase dehydration of 2,5-dimethyl-2,5-hexandiol (DMED) (Fig. 1) [11,12,19]. However, the Prins reaction shows low conversion values and the formation of high amounts of by-products, with a rapid catalyst deactivation [14,15], while in the vapour phase dehydration the surface acidity has to be finely tuned [19] in order to prevent the formation of huge amounts of useless isomers. In this study, a new economical and environmentally friendly synthesis of TMBD, by selective vapour phase dehydration of DMED on different commercial solid acids is reported. The results obtained were significantly better than those reported in literature [12], demonstrating the possibility to use cheap and commercially available raw clays or clay mixtures.

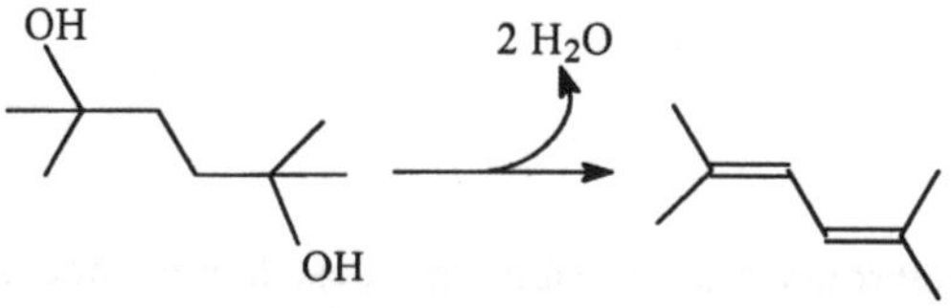

Figure 1. Dehydration of 2,5-dimethyl-2,5-hexandiol (DMED) to 2,5-dimethyl-2,4-hexadiene (TMBD).

Experimental

Various commercial heterogeneous catalysts have been investigated: 1) oxides = γAl_2O_3 (Sasol, D), SiO_2 (Engelhard, USA), ZrO_2 (Mel-Chemicals, UK); 2) zeolite = T4480 (Süd-Chemie, D); 3) clays = Bentonite de Cabanas and Esmectita EMV351 (TOLSA, E), K 306F (United Catalyst, USA) Granosil 750 JF, Mega Dry and 20/50 LVM (Süd-Chemie Inc., USA), Attagel 40 (Engelhard, USA); 4) acid treated clays = F20 (Engelhard, USA), KSF, Tonsil CO 610G (Süd-Chemie, D). 2,5-Dimethyl-2,5-hexandiol (DEMD), 2,5-dimethyl-2,4-hexadiene (TMBD), and CH_3OH had purity ≥ 99% and were used without any further purification. The vapour phase tests were carried out using 9 g of catalyst (30-40 mesh) in a stainless steel reactor (i.d. 1", length 50 cm), inserted in an electric oven controlled by a J thermocouple, operating at atmospheric pressure and in the 270-310 °C temperature range. The absence of axial profiles was determined by a J thermocouple sliding inside a 1/8" stainless steel tube inserted in the catalytic bed. The feed was a solution of 2,5-dimethyl-2,5-hexandiol (DMED) in CH_3OH (from 2.0 to 2.8 M) introduced by a Labflow 2000 HPLC pump (LHSV range = 0.46-1.15 h^{-1}) in a N_2 flow, regulated by a mass-flowmeter. The catalysts were first heated at 550°C for 4.5 h under a 200 mL/min flow of N_2/air mixture (62:38) and afterwards at 290°C *in-situ* for 18 h under a 180 mL/min flow of N_2. In each test, the products were collected for 4 h (after 1 h necessary to reach steady state activity) by condensing them in a trap at room temperature, followed by two traps cooled at 0°C and analysed off-line using a Hewlett-Packard gas chromatograph, equipped with FID and a wide bore HP1 column (length 30 m, i.d. 0.53 mm, and film width 1.25 μm). The catalyst-

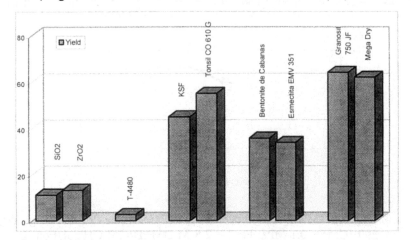

Figure 2. Yield in TMBD in the preliminary screening tests. (Cat. = 9 g; T = 270 °C; P = 0.1 MPa; Feed = 2.8 M DMED/CH_3OH solution in N_2 flow; LHSV = 0.46 h^{-1} [12].

stability tests were performed for 400 h of time-on-stream. The reaction products were identified using a Varian Saturn 2000 GC-MS apparatus, equipped with a wide bore HP1 column (length 30 m, i.d. 0.53 mm, film width 1.25 μm).

Results and Discussion

The preliminary screening tests on the different classes of catalysts were performed in the best conditions reported in a previous patent (12). These tests showed   interesting data for the vapour phase dehydration of DMED, using a commercial montmorillonite clay catalyst. All the investigated oxides and the T-4480 zeolite showed very low yield values in TMBD (with the following order of activity: ZrO_2 > SiO_2 > γAl_2O_3 > zeolite) and the formation of huge amounts of by-products, mainly C_8H_{14} isomers. On the other hand, acid and acid-treated clays, showed a very wide range of results, with increasing yield values in TMBD by decreasing the surface acidity. In fact, F20 showed a high activity, but practically formed only huge amounts of by-products, identified only in part, while very interesting data were obtained with the Tonsil CO 610G sample, a granulated material produced by acid activation of natural calcium montmorillonite (23). The latter is a very active acid clay used in a wide range of applications, having more than 200 m^2/g of surface area, a highly porous inner structure and a multitude of active sites, i.e produced by treatment in soft conditions (24, 25). However, the best performances were obtained with the non-acid-treated clays Granosil 750 JF and Mega Dry, simply produced by the thermal activation of natural attapulgite/montmorillonite clay mixtures. Attapulgite, also known as palygorkite, is a clay mineral with a needle-like shape, while montmorillonite is a layered 2:1 type clay (24,25). On the other hand, the role of the clay is confirmed by the two

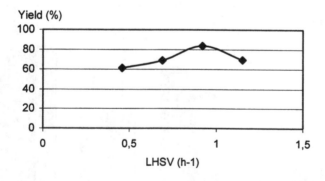

Figure 3. Yield in TMBD as a function of the LHSV value (9 g of Mega Dry clay; T = 290 °C; P = 0.1 MPa; Feed = 2.8 M DMED/ CH_3OH solution in N_2 flow).

Yield (%)

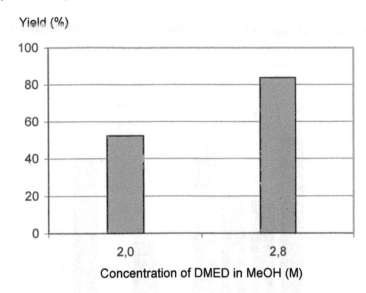

Concentration of DMED in MeOH (M)

Figure 4. Yield in TMBD as a function of the feed concentration. ((9 g of Mega Dry clay; T = 290 °C; P = 0.1 MPa; LHSV = 0.92 h^{-1}; Feed = DMED/CH_3OH solution in N_2 flow.

saponites (Bentonites de Cabanas and Esmectita EMV 351), which showed the significantly lowest catalytic performances. Saponite is also a 2:1 type clay, but it is characterised by the isomorphous substitution of Si by Al atoms in the tetrahedral layers and therefore has many and strong Lewis acid sites (24, 25).

In a following step, we attempted to optimise the reaction conditions, using Mega Dry clay as catalyst: the first increase in temperature (290 °C) produced an approximate 15% increase in the yield of TMBD, while a further increase to 310 °C worsened the catalytic performances. The rate of the liquid feed also played a significant role, with an increase in the yield of TMBD up to a LHSV value of about 0.92 h^{-1}, whereas for higher values the catalyst surface was partially stuffed (Fig. 3). Lastly, the DMED concentration also played an important role, since when the DMED concentration was decreased, the yield in TMBD almost halved (Fig. 4). However, according to literature (12), operating at high feed concentration is not simple: in fact special reactor arrangements, with heated feed, are required because of the limited solubility of DMED in CH_3OH.

The screening tests in optimised conditions confirmed the excellent performances of raw attapulgite/montmorillonite mixtures (Granosil 750 JF and Mega Dry) (Fig. 5). It is worth noting that these latter catalysts showed better

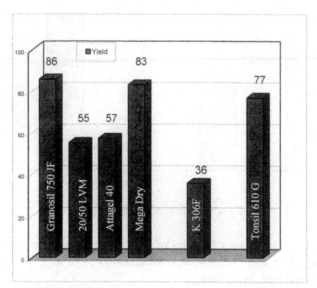

Figure 5. Catalytic activity in the optimized screening tests. (Cat. = 9 g of Mega Dry clay; T = 290 °C; P = 0.1 MPa; Feed = 2.8 M DMED/CH$_3$OH solution in N$_2$ flow; LHSV = 0.92 h^{-1}).

performances than both the previously patented montmorillonite clay catalyst (K 306 F) (12) and a pure attapulgite clay (Attagel 40). Compared to montmorillonite, attapulgite has a higher surface area and pore volume, but lower swelling capacity; therefore, it may be suggested that commercial mixtures of the two clays represent the best compromise for the reaction studied, because each acid-catalyzed reaction and substrate has specific requirements, as previously shown in both the acylation reactions with acid treated clay F20 (26) and the vapour-phase Fries rearrangement with T-4480 zeolite (27, 28). Both Granosil 750 JF and Mega Dry showed significantly better higher and more constant catalytic performances, up to 400 h of time-on-stream, than those of pure attapulgite (Attagel 40) (Fig. 6). Lastly, it must be noted that Granosil 750 JF is an ISO-2002 certified product (23), while the physical properties of Mega Dry, which is not certified, are kept constant during production. Thus, constant performances are warranted for both catalysts.

We investigated the possibility of regenerating the two most interesting catalysts - after an accelerated deactivation achieved feeding a DMED-rich solution at high LHSV value - by heating at 550 °C for 4.5 h, under a N$_2$/air flow.

Compared to the catalyst before regeneration, the regenerated catalyst showed a slightly higher yield value in TMBD, together with a higher formation of isomers (Fig. 7). Moreover, catalytic performances were significantly lower than those of the fresh catalyst (Fig. 5). Therefore, these catalysts may be only partially regenerated, but this is not a significant problem considering both the high stability with time-on-stream and the very low market price.

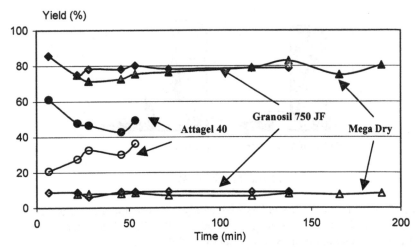

Figure 6. Yield in TMBD (full symbols) and in C_8H_{14} isomers (open symbols) with time-on-stream. (Cat. = 9 g; T = 290 °C; P = 0.1 MPa; Feed = 2.8 M DMED/CH$_3$OH solution in N$_2$ flow; LHSV = 0.92 h^{-1}).

As reported above, one of the main problems in the vapour phase dehydration of DMED to TMBD is the formation of significant amounts of by-products. Generally speaking, TMBD isomers may generally be classified in two broad categories: i) linear isomers (for example 5,5-dimethyl-hexa-1,3-diene, methyl-heptenes or methyl-heptadienes); ii) cyclic isomers (trimethyl-cyclopentenes or dimethyl-cylcohexene). While for the former compounds it is possible to hypothesize isomerization to TMBD, the latter must be considered lost in the carbon balance, since the ring opening reaction requires catalyst performances and reaction conditions which are not compatible with those required by the main reaction (29). In order to investigate the possibility of recovering the isomer fraction at least in part, a test was performed entailing the feeding of an isomer-rich solution, obtained as residue in the TMBD recovery by distillation, on Mega Dry and acid Tonsil CO 610G catalysts. The acid clay showed a higher conversion value, but a lower selectivity to TMBD, mainly triggering further isomerization reactions to cyclic isomers, while the Mega Dry clay, although less active, caused the formation of TMBD, thus making it possible to recover a sizable fraction of isomers and rendering the overall process even more economically appealing (Fig. 8).

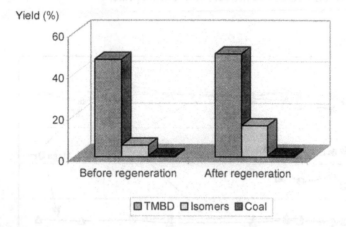

Figure 7. Activity of the Mega Dry clay catalyst, before and after regeneration at 550 °C for 4,5 h. (Cat. = 9 g; T = 290 °C; P = 0.1 MPa; Feed = 2.8 M DMED/CH$_3$OH solution in N$_2$ flow; LHSV = 0.92 h^{-1}).

Conclusions

The vapour phase dehydration of 2,5-dimethyl-2,5-hexanediol (DMED) to 2,5-dimethyl-2,4-hexadiene (TMBD) - key intermediate for the synthesis of chrysanthemic acid and, consequently, of pyrethroids (probably the safest and most used family of insecticides) - is an interesting example of an economical and environmentally-friendly process of industrial interest. In fact, it uses a non-toxic or non-hazardous raw material, commercially available at a reasonable price, which is thus supported by many companies. Furthermore, it functions in safe reaction conditions with heterogeneous catalysts, while showing high conversion and selectivity in TMBD values. The best results have been obtained by using as catalyst certain commercial raw clay mixtures, available at very low prices and being currently used as adsorbents. The best catalysts (Granosil 750 JF and Mega Dry) showed stable physical properties and very interesting catalytic performances, which remained constant with time-on-stream. Finally, they made it possible to recover a sizable fraction of isomers by their conversion to TMBD, thus making the overall process even more economically attractive.

Acknowledgments

Our thanks go to Engelhard (USA), Mel-Chemicals (UK); Sasol (D), Tolsa (E) and Süd-Chemie (D) for providing the commercial solid acid catalysts. The financial support by Endura and the Ministero dell' Istruzione, dell'Università e della Ricerca (MIUR, Roma) is gratefully acknowledged.

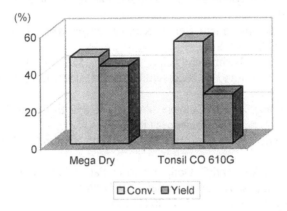

Figure 8. Catalytic activity with a TMBD:isomer: high molecular weight compounds (21.2:76.8: 2.0 mol.%) (Cat. = 9 g; T = 290°C; P = 0.1 MPa; LHSV = 0.92 h^{-1})

References

1. R.L. Metcalf, In *Ulmann's Encyclopedia of Industrial Chemistry*, B. Elvers, S. Hawkins, M. Ravenscroft and G. Schulz, eds., Vth Ed., Vol. A14, VCH, Weinheim, 1989, pp. 263-320.
2. R.L. Metcalf, In *Kirk-Othmer Encyclopedia of Chemical Technology*, J.I. Kroschwitz and M. Howe-Grant, eds., Vol. 14, Wiley, New York, 1995, pp. 524-602.
3. Agency for Toxic Substances and Disease Registry, *Toxicological Profile for Pyrethrins and Pyrethroids*, US Department of Health and Human Services, Atlanta GE, 2003.
4. C. Cox, *J. Pesticide Reform.* **22**, 14 (2002).
5. PAN Pesticides Dartabase-Chemicals, www.pesticideinfo.org
6. Cornell University, Ithaca NY, *Extension Toxicology Network*, http://extoxnet.orst.edu
7. National Pesticide Information Centre, http://npic.orst.edu

8. www.pyrethrum.org; www.new-agri.co.uk; www.dpiwe.tas.gov.au; www.zhongzhibiotech.com.

9. www.chemexper.com

10. D. Holland and D.J. Milner, *Chem. Ind. (London)* 706 (1979).

11. A. Prevedello, E. Platone and M. Morelli, US Pat. 4,409,419 to ANIC SpA (1983).

12. D.V. Petrocine and R. Harmetz, US Pat. 4,507,518 to Penick Co (1985).

13. K. Takahashi, T. Nakayama and Y. Too, Jpn Pat. 03,232,822 to Sumitomo Chemical Co (1991).

14. T. Yamaguchi, C. Nishimichi and A. Kubota, *Prepr. Am. Chem. Soc., Div. Petr. Chem.* **36**, 640 (1991).

16. T. Yamaguchi and C. Nishimichi, *Catal. Today* **16**, 555 (1993).

17. K. Takahashi, Y. Too and T. Nakayama, Jpn Pat. 06,074,218 to Sumitomo Chemical Co (1994).

18. X. Gao, L. Xu and X. Liu, Chinese Pat. 1,145,892 to Dalian Inst. Chem. Physics, Chinese Academy of Sciences (1997).

19. M. Yamamoto and Y. Too, Jpn Pat. 09,002,980 to Sumitomo Chemicals Co (1997).

20. X. Gao, L. Xu, X. Liu and C. Dong, Chinese Pat. 1,171,980 to Dalian Inst. Chem. Physics, Chinese Academy of Sciences (1998).

21. B. Notari, *Chim. Ind. (Milan)* **51**, 1200 (1969).

22. E. Poluzzi, A.Ò. Bianchi, D. Brancaleoni, V. Borzatta and A. Vaccari, Italian Pat. MIA000,247 to Endura SpA (2003).

23. Süd-Chemie Group, Technical Information Sheets.

24. A. Vaccari, *Catal. Today* **41**, 53 (1998).

25. M. Campanati and A. Vaccari, In *Fine Chemicals through Heterogeneous Catalysis,* R.A. Sheldon and H. van Bekkum, eds., Wiley-VCH, Weinheim, pp. 61-79. 2001

26. M. Campanati, F. Fazzini,, G. Fornasari, A. Tagliani and A. Vaccari, In *Catalysis of Organic Reactions,* F.E. Herpes, ed., Dekker, New York, pp. 307-318. 1998

27. V. Borzatta, E. Poluzzi and A. Vaccari, World Pat. 023,339A1 to Endura SpA (2001).

28. V. Borzatta, G. Busca, E. Poluzzi, V. Rossetti, M. Trombetta and A. Vaccari, *Appl. Catal.* **A257**, 85 (2004).

29. H. Dhu, C. Fairbridge, H. Yang and Z. Ring, *Appl. Catal.* **A294**, 1 (2005).

39. Supported Heteropoly Acid Catalysts for Friedel-Crafts Acylation

Kenneth G. Griffin[1], Peter Johnston[1], Roger Prétôt[2] and Paul A. van der Schaaf[2]

[1] Johnson Matthey plc, Catalyst Development, Royston, SG8 5HE, UK
[2] Ciba Specialty Chemicals Inc., CH-4002 Basel, Switzerland

johnsp@matthey.com

Abstract

The liquid phase Friedel-Crafts acylation of thioanisole with iso-butyric anhydride to produce 4-methyl thiobutyrophenone has been studied using supported silicotungstic acid catalysts. Reaction is rapid, giving the para-acylation product in high yield. Reactions have been performed in both batch slurry and trickle bed reactors. In both reactors catalyst deactivation due to strong adsorption of product was observed.

Introduction

The Friedel-Crafts acylation of aromatic compounds is an important synthesis route to aromatic ketones in the production of fine and specialty chemicals. Industrially this is performed by reaction of an aromatic compound with a carboxylic acid or derivative e.g. acid anhydride in the presence of an acid catalyst. Commonly, either Lewis acids e.g. $AlCl_3$, strong mineral acids or solid acids e.g. zeolites, clays are used as catalysts; however, in many cases this gives rise to substantial waste and corrosion difficulties. High reaction temperatures are often required which may lead to diminished product yields as a result of byproduct formation. Several studies detail the use of zeolites for this reaction (1).

In this paper we report the use of supported heteropoly acid (silicotungstic acid) and supported phosphoric acid catalysts for the acylation of industrially relevant aromatic feedstocks with acid anhydrides in the synthesis of aromatic ketones. In particular, we describe the acylation of thioanisole **1** with iso-butyric anhydride **2** to form 4-methyl thiobutyrophenone **3**. The acylation of thioanisole with acetic anhydride has been reported in which a series of zeolites were used as catalysts. Zeolite H-beta was reported to have the highest activity of the zeolites studied (41 mol % conversion, 150°C) (2).

Heteropoly acids are strong Brønsted acids containing a heteropoly anion and proton cations. In the case of silicotungstic acid (STA), used in this study, the anion consists of a heteroatom (silicon) surrounded by 12 metal atoms (tungsten) which are held together by oxygen bridges. The complex has the formula $H_4SiW_{12}O_{40}.nH_2O$ in a Keggin structure. Heteropoly acids, in either supported form or in bulk form have been widely used as acid- as well as oxidation catalysts (3-6). Previous studies describing the use of heteropoly acids in the acylation of anisole with acetic anhydride have been reported (7-9). Phosphotungstic acid was found to have slightly higher activity than silicotungstic acid under the conditions studied; this was ascribed to its stronger acid strength and resulted in the para- monoacylation product in high yield.

Results and Discussion

The activity and selectivity of silicotungstic acid as catalyst has been evaluated in the acylation of thioanisole with iso-butyric anhydride. Reactions were performed using a 3 fold molar excess of thioanisole. The performance of silicotungstic acid was evaluated in both supported (42%STA/silica) and bulk form. The results are shown in Table 1.

Table 1. Acylation of thioanisole with iso-butyric anhydride. Conditions: 60°C, 7.5wt% catalyst loading.

Catalyst	Reaction time min	Conv. 2 %	Product selectivity / %			
			o-ketone	p-ketone 3	Vinyl ester	Side products
42%STA/SiO$_2$	0.5	92.7	1.44	94.33	0	4.23
42%STA/SiO$_2$	1	94.1	1.40	94.30	0	4.30
42%STA/SiO$_2$	5	98.8	1.42	95.26	0	3.32
42%STA/SiO$_2$	7.5	99.4	1.29	94.95	0	3.76
42%STA/SiO$_2$	10	100	1.26	93.95	0.49	4.30
42%STA/SiO$_2$	20	100	1.28	94.77	0.43	3.52
Bulk STA	10	99.4	1.17	91.97	0.47	6.39
Bulk STA	20	99.7	1.18	92.16	0.51	6.15

As expected, the para-acylated ketone **3** is the major reaction product, with only minor yields of the ortho-acylated product formed. Reaction is very rapid and complete conversions are obtained. Catalyst TOF numbers are $>19000h^{-1}$ under these conditions. Enolisation and further reaction with the anhydride results in formation of the vinyl ester. Minor amounts of diacile and side products are formed. Similar performances are observed with both supported and non-supported STA (at equivalent STA loading in reactor). Use of non-supported STA resulted in agglomeration and deposition of the material onto the reactor wall, whereas the silica supported material could be readily removed by filtration.

As the loading of STA on the catalyst support is decreased, incomplete anhydride conversion is observed and significant hydrolysis of the anhydride to form iso-butyric acid is observed (Table 2). Use of silica supported phosphoric acid results in lower ketone yields and significant hydrolysis of the iso-butyric anhydride. Blank reactions (catalyst and anhydride, 90°C, 30 min) indicates that hydrolysis of anhydride is observed in the presence of these catalysts and may result from either dehydroxylation of the silica support or residual water in the catalyst. However this reaction is slow (42%STA/silica, 44% conversion and 70%H_3PO_4/silica, 86% conversion respectively).

Table 2. Acylation of thioanisole with iso-butyric anhydride. Conditions: 90°C, 10wt% catalyst loading.

Catalyst	Conv. 2 / %	Ketone 3 / %	Acid 4 / %	Side prods / %
42% STA/SiO$_2$	100	70.4	27.6	2.0
29% STA/SiO$_2$	100	69.8	26.9	3.3
17% STA/SiO$_2$	97.0	64.9	29.2	8.2
9% STA/SiO$_2$	84.5	56.5	41.6	1.9
70%H_3PO_4/SiO$_2$	91.3	52.6	44.7	2.7

Incorporation of alumina into the silica supports as silica-alumina mixed oxides or use of alumina as support results in materials exhibiting poor activity and poor selectivity to desired product, as shown in Table 3. Details of the support materials is given in Table 6. Increasing the alumina content of the silica-alumina supports results in an increase in Lewis acid sites as the alumina is largely located at the surface of the oxide. These sites appear to favour hydrolysis of the iso-butyric anhydride to acid over thioanisole acylation. The poor activity of STA/alumina may result from partial neutralisation of the STA at the basic sites on the support. Phosphotungstic acid, partially neutralised with Cs ($Cs_{2.5}H_{0.5}PW_{12}O_{40}$) supported on clay has been reported to show no activity in the acylation of thioanisole with acetic anhydride (10). Non-supported $Cs_{2.5}H_{0.5}PW_{12}O_{40}$ displayed high activity in the acylation of benzene with benzoic anhydride (11).

350 _Heteropoly Acids_

Table 3. Acylation of thioanisole with iso-butyric anhydride. Conditions: 90°C, 10wt% catalyst loading.

Catalyst [a]	Conv. **2** / %	Ketone **3** / %	Acid **4** / %	Side prods / %
29%STA/SiO$_2$	100	69.8	26.8	3.3
25%STA/SiO$_2$-Al$_2$O$_3$-1	92.0	65.0	25.2	6.3
38%STA/SiO$_2$-Al$_2$O$_3$-2	20.0	trace	98.4	1.6
29%STA/Al$_2$O$_3$	17.0	18.7	80.4	0.9

[a] analysis of the catalyst supports is given in Table 6.

Influence of Reaction Temperature

The variation in activity and selectivity of 42%STA/SiO$_2$ at different reaction temperatures (23°C, 60°C, 90°C, 127°C) is shown in Table 4.

Table 4. Variation in product selectivity with reaction temperature in the acylation of thioanisole with iso-butyric anhydride. Catalyst: 42%STA/silica, 10wt% catalyst loading.

Temperature /°C	Conv. **2** / %	Product selectivity /%			
		p-Ketone **3**	o-Ketone	Vinyl ester	Side prods
23	96.0	89.2	1.9	8.9	0
60	100	97.6	1.4	0.3	0.7
90	99.0	97.1	1.4	1.5	0
127	76.0	95.7	1.3	3.0	0

At room temperature (23°C) incomplete conversion of the anhydride is observed (96%) and the slower reaction leads to reduced selectivity to the desired p-ketone **3**. Highest conversions and selectivities are obtained when the reaction is performed in the range of 60°C to 90°C.

Catalyst Reuse

The possibility for reuse of the catalyst was investigated by filtration of the catalyst from the reaction mixture, washing twice with toluene and drying under vacuum. The dry catalyst was then reused for the acylation reaction with fresh reactants. Activity of the catalyst in the second cycle (first reuse) was 90% of the original activity. The catalyst displayed very poor activity upon a third cycle (5% of original activity). XRF analysis of the used, washed catalysts confirmed that no significant leaching of STA from the catalyst had occurred upon use. Deactivation is believed to result from strong adsorption of products onto the catalyst. This is consistent with previous reports describing strong product adsorption as the cause for catalyst deactivation in anisole acetylation over similar catalysts (6,7).

Fixed Bed Reaction

Reaction was also investigated in a fixed bed reactor under trickle flow conditions to investigate whether reaction under flow conditions would affect the rate of catalyst deactivation. Conversions and selectivities are shown in Table 5. Rapid deactivation of the catalyst was observed with concomitant progressive darkening of the catalyst bed indicative of strong adsorption (retention) of material on the catalyst and resulted in a lower catalyst productivity compared to equivalent reaction in a batch slurry reactor.

Table 5. Acylation of thioanisole with iso-butyric anhydride in a trickle bed reactor. Conditions: 60°C, 42%STA/silica, 4.0g, Liquid WHSV = 33h^{-1}.

Reaction TOL / min	Conversion 2 / %	Product selectivity / %			
		o-Ketone	p-Ketone 3	Vinyl ester	Side prods
11	99.9	1.3	92.2	0.3	3.8
23	94.1	1.2	91.3	0.4	7.1
34	92.7	1.2	90.5	0.4	7.9
45	91.0	1.4	91.0	0.6	7.0
Average	95.1	1.3	91.2	0.4	7.1

Other Acylation Reactions

The activity of 42%STA/silica catalysts for the acylation of related aromatic reactants with iso-butyric anhydride was investigated. In the presence of anisole and veratrole, 100% anhydride conversion was observed, leading to the expected para-acylation products. No reaction was observed in the presence of chlorobenzene and other deactivated aromatic systems.

Conclusions

The results reported in this paper show that supported STA catalysts are efficient catalysts for the acylation of thioanisole and related activated aromatic molecules in the presence of iso-butyric anhydride as the acylating agent. The para- substituted ketone isomer is the major acylation product. Optimal catalyst activity is in the range of 60°C to 90°C. Use of either lower STA concentrations or use of weaker acids eg phosphoric acid, decreases the reaction rate and selectivity; this results in greater hydrolysis of the anhydride. Use of supported STA catalysts is more efficient than bulk STA since the reaction medium is much cleaner and enables easier removal of the catalyst.

SiO₂ is the best catalyst support for this reaction. As Al₂O₃ is incorporated into the SiO₂ the catalyst activity and selectivity decreases significantly. In these mixed oxide support materials Brønsted acid sites favour acylation, whilst the presence of Lewis acid sites results in hydrolysis of the anhydride.

Supported catalysts could be reused once with little loss of activity; further reuse led to a significant drop in activity, as a result of strong absorption of products and by-products on the catalyst surface (indicated by colour change of the catalyst). More rapid catalyst deactivation was observed in a trickle bed reactor than in a batch slurry reactor.

Experimental Section

Catalysts were prepared by impregnation of the appropriate support material with aqueous solutions of silicotungstic acid using JM proprietary methods. The concentration of STA in the impregnation solution was such as to achieve the desired loading on the catalyst support. Thereafter the impregnated material was dried in air (LOD <5%). Catalysts were used as prepared. Details of the catalyst supports used in this study are given in Table 6.

Table 6. Analysis of catalyst supports used in this study.

Support	Silica content %	Alumina content %	Surface area m^2/g	Pore volume ml/g
SiO₂	100	0	320	1.14
SiO₂-Al₂O₃-1	98	2	356	1.37
SiO₂-Al₂O₃-2	86	14	573	0.76
Al₂O₃	0	100	155	0.45

Batch slurry reactions were carried out in liquid phase in a stirred glass vessel with condenser. Catalyst was added to a preheated solution containing aromatic reactant (35ml, Aldrich) and iso-butyric anhydride (16ml, Aldrich) in a 3:1 molar ratio. Samples of the reaction mixture were removed from the reaction mixture after various reaction times, filtered and analysed by gas chromatography (column: DB-5, 30m, He carrier gas, FID detector) to determine reaction progress. Product identification was made by comparison with appropriate reference materials.

Trickle bed reactions were carried out by passing a preheated solution of 70 ml thioanisole and 32 ml iso-butyric anhydride (3:1 molar ratio) through a fixed bed reactor containing 4.0g catalyst, heated to 60°C. Samples were collected from the reactor outlet at regular intervals during reaction and analysed as described above.

Acknowledgments

We thank R. Kugler and W. Wolf, Ciba Specialty Chemicals Inc., Coating Effects, Process Development, Schweizerhalle, Switzerland for assistance and helpful discussions. We thank Derek Atkinson, Grace Davison Specialty Catalysts for supply of the catalyst supports used in this study.

References

1. A. Corma and H. Garcia, *Chem. Rev.*, **103**, 4307 (2003).
2. D.P. Sawant and S.B. Halligudi, *Catal. Commun.*, **5**, 659 (2004).
3. M. Misono, *Catal. Rev. Sci. Eng.*, **29**, 269 (1987).
4. M. Misono, *Catal. Rev. Sci. Eng.*, **30**, 339 (1988).
5. I.V. Kozhevnikov, *Chem. Rev.*, **98**, 171, (1998).
6. F.T.T. Ng, T. Mure, M. Jiang, M. Sultan, J. Xie, P. Gayraud, W.J. Smith, M. Hague, R. Watt and S. Hodge, Chemical Industries (CRC Press), **104**, (*Catal. Org. React.*) 251-260 (2005).
7. J. Kaur, K.Griffin, B. Harrison and I.V. Kozhevnikov, *J. Catal.*, **208**, 448 (2002).
8. L.A.M. Cardoso, W. Alves Jr, A.R.E. Gonzaga, L.M.G. Aguiar, H.M.C. Andrade, *J.Mol. Catal. A: Chemical*, **209**, 189 (2004).
9. B. Bachiller-Baeza and J.A. Anderson, *J. Catal.*, **228**, 225 (2004).
10. G.D. Yadav and R.D. Bhagat, *J. Mol. Catal A: Chemical*, **235**, 98, (2005).
11. T. Tagawa, J. Amemiya and S Goto, *App. Catal. A: General*, **257**, 19 (2004).

40. Synthesis of Pseudoionones by Aldol Condensation of Citral with Acetone on Li-Modified MgO Catalysts

Verónica K. Díez, Juana I. Di Cosimo and Carlos R. Apesteguía

*Catalysis Science and Engineering Research Group (GICIC ,
Instituto de Investigaciones en Catálisis y Petroquímica (UNL-CONICET)
Santiago del Estero 2654, (3000) Santa Fe, Argentina*

capesteg@fiqus.unl.edu.ar

Abstract

The liquid-phase synthesis of pseudoionones by aldol condensation of citral with acetone was studied on MgO samples containing between 0.13 and 2.6 wt. % of Li^+. The Li^+ addition to MgO up to about 0.5% increased the density of strong O^{2-} basic sites, but increasing the Li^+ amount further diminished the sample basicity. Formation of pseudoionones increased with the sample basicity: the higher the density of strong basic sites the higher the pseudoionone yields. Thus, Li-MgO samples containing the highest density of O^{2-} sites were the most active catalysts for producing pseudoionones.

Introduction

Pseudoionones (PS) are valuable intermediate compounds for the synthesis of α and β-ionones, which are widely used as pharmaceuticals and fragrances. In particular, β-ionone is the preferred reactant for different synthesis processes leading to vitamin A (1). Pseudoionones are commercially produced via the aldol condensation of citral with acetone in a liquid-phase process involving the use of diluted bases, such as NaOH, $Ba(OH)_2$, LiOH, which pose problems of high toxicity, corrosion, and spent base disposal (2). The consecutive cyclization of pseudoionones to yield α and β-ionones is catalyzed by strong liquid acids. The two-step process for ionone synthesis is depicted in equation (1).

Increasing research work has been lately performed to find suitable and recyclable solid catalysts for efficiently promoting the liquid-phase aldol condensation of citral with acetone with the aim of developing an environmentally benign process for producing pseudoionones. Climent et al. (3) showed that the reaction is poorly catalyzed by employing acid zeolites or bifunctional acid-base aluminophosphates (ALPO). In contrast, the use of solid bases such as MgO or Mg-Al mixed oxides may selectively produce pseudoionones (3,4). Other authors (5,6,7) have noted that Mg-Al hydrotalcites of Mg:Al ratio near 3 also form preferentially pseudoionones, although the catalyst activity and selectivity are greatly affected by both the synthesis procedure and the activation steps. These previous works showed that solid bases are potential catalysts for the aldol condensation of citral with acetone but the exact requirements of basic site density and strength to efficiently promote the selective pseudoionone formation is lacking. In this paper, we have studied the citral/acetone reaction on Li-doped MgO, taking into account that we have previously observed that this type of catalysts is highly active for acetone condensation reactions (8). Our goal was to test new catalytic formulations and also to relate the structural properties and the base site density and nature of the solid with their ability to promote the pseudoionone formation.

Experimental

To obtain high-surface area MgO, commercial MgO (Carlo Erba, 99%) was hydrated with distilled water and the resulting $Mg(OH)_2$ was decomposed and stabilized at 773 K in a N_2 flow for 18 h. Li/MgO-x samples were prepared by incipient wetness impregnation from the same mother batch of MgO, using aqueous solutions of LiOH to obtain catalysts with 0.13, 0.3, 0.5, 1.2 and 2.6 wt. % of lithium. After incorporation of Li^+ cation, samples were dried at 353 K and then decomposed and stabilized at 773 K during 18 h in flowing N_2. The sample composition was determined by Atomic Absorption Spectrometry (AAS). BET surface areas (Sg) were measured by N_2 adsorption at 77 K in a Nova-1000 Quantachrom sorptometer. Structural properties of the samples were determined by X-Ray Diffraction (XRD) using a Shimadzu XD-D1 instrument. Crystallite sizes were evaluated using the Scherrer equation.

Catalyst base site densities (n_b) were measured by Temperature-Programmed Desorption (TPD) of CO_2 pre-adsorbed at room temperature. Samples were pretreated *in situ* in a N_2 flow at 773 K, cooled down to room temperature and then exposed to a flowing mixture of 3 % of CO_2 in N_2 until surface saturation. Weakly adsorbed CO_2 was removed by flushing in N_2. Finally, temperature was raised up to 773 K at 10 K/min. The desorbed CO_2 was converted to CH_4 over a methanation catalyst (Ni/Kieselghur) and then analyzed with a flame ionization detector.

The cross aldol condensation of citral (Millennium Chemicals, 40 % cis-isomer + 55 % trans-isomer) with acetone (Merck, PA) was carried out at 353 K in N_2 atmosphere under autogenous pressure ($\approx$ 250 kPa) in a batch PARR reactor, using an acetone/citral = 49 (molar ratio) and a catalyst/(citral+acetone) = 1 wt.% ratio. Catalysts were pre-treated ex-situ in flowing N_2 at 773 K for 2 h to remove adsorbed water and carbon dioxide and then quickly transferred to the reactor without exposing them to air. Reaction products were analyzed by gas chromatography. Selectivities (S_j, mol of product j/mol of citral reacted) were calculated as $S_j(\%) = C_j$ x 100/ ΣC_j where C_j is the concentration of product j. Product yields (η_j, mol of product j/mol of citral fed) were calculated as $\eta_j = S_j X_{Cit}$. Thirteen samples of the reaction mixture were extracted and analyzed during the 6-hour reaction. The main reaction product of citral conversion was pseudoionone, PS (cis- and trans- isomers).

Results and Discussion

The chemical composition, BET surface area, pore volume, and XRD analysis results of MgO and Li-promoted MgO samples are shown in Table 1.

Table 1. Physicochemical properties of the catalysts used in this work

Catalyst	Li loading (wt.%)	Li/Mg (molar ratio)	Sg (m²/g)	Vg (cm³/g)	Crystallite size (Å)
MgO	0	--	149	0.27	83
Li/MgO-1	0.13	0.008	138	--	107
Li/MgO-2	0.30	0.017	100	0.23	109
Li/MgO-3	0.50	0.029	87	0.25	114
Li/MgO-4	1.20	0.070	66	0.18	125
Li/MgO-5	2.60	0.151	56	0.13	139

Promotion of MgO with Li decreased the MgO surface area and pore volume. XRD analysis of the Li/MgO-x catalysts revealed that a single phase of MgO periclase is present at low Li loadings, but on sample Li/MgO-5 containing 2.6 wt. % Li an incipient Li_2CO_3 phase was also detected (Figure 1). MgO crystallite size increased with the Li loading which is in line with the observed concomitant drop of the sample surface area. These results are explained by particle agglomeration in the presence of the Li salt during MgO aqueous impregnation and further thermal decomposition of the resulting Li(OH)/Mg(OH)$_2$ which involves Li(OH) melting (9). From XRD diffractograms of Fig.1, we also calculated for Li/MgO-x samples the unit cell parameter a corresponding to the MgO structure with face-centered cubic symmetry. The obtained a values were almost constant indicating no detectable structural modification of the MgO lattice by Li doping. This result suggested that Li is located on the surface of Li/MgO-x samples rather than inside of the MgO matrix, probably forming small domains of amorphous Li_2O or crystalline Li_2CO_3. The surface base properties of the samples were investigated by TPD of CO_2 preadsorbed at room temperature. Figure 2A shows the CO_2 desorption rate as a function of

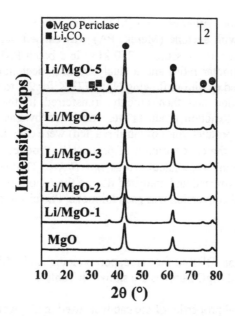

Figure 1. X-ray diffraction patterns of MgO and Li/MgO-x catalysts

desorption temperature for MgO and Li/MgO-x catalysts. The total amount of desorbed CO_2 (n_b) was measured by integration of TPD curves and the resulting values are reported in Table 2; n_b shows a maximum value of 11.3 µmol/m² for a Li loading of 0.5 wt. % and clearly decreases for higher Li loadings. The complex TPD profiles of Fig. 2A suggest that the surface of the Li/MgO-x oxides are nonuniform and contain several species formed by CO_2 adsorption; in other words, the solid surface contains oxygen atoms of different chemical nature that bind CO_2 with different coordination and binding energy. In previous work (10), we investigated the chemical nature of CO_2 adsorbed species on similar alkaline- promoted MgO catalysts by using FTIR and identified at least three different species: unidentate carbonate, bidentate carbonate, and bicarbonate. These three adsorption species are depicted in Fig. 2B. Unidentate carbonate formation requires isolated surface O^{2-} ions, i.e., low-coordination anions, such as those present in corners or edges and exhibits a symmetric O-C-O stretching at 1360-1400 cm⁻¹ and an asymmetric O-C-O stretching at 1510-1560 cm⁻¹. Bidentate carbonate forms on Lewis acid-Brönsted base pairs (M^{n+}-O^{2-} pair site, where M^{n+} is the metal cation Mg^{2+} or Li^+), and shows a symmetric O-C-O stretching at 1320-1340 cm⁻¹ and an asymmetric O-C-O stretching at 1610-1630 cm⁻¹. Bicarbonate species formation involves surface hydroxyl groups and shows a C-OH bending mode at 1220 cm⁻¹ as well as symmetric and asymmetric O-C-O stretching bands at 1480 cm⁻¹ and 1650 cm⁻¹, respectively. We also determined (10) the following base strength order: low coordination O^{2-} anions > oxygen in M^{n+}-O^{2-} pairs > OH groups. Based on these FTIR results, the TPD profiles of Fig. 2A were deconvoluted in three desorption bands in order to quantify the density of weak, medium and strong base sites. The

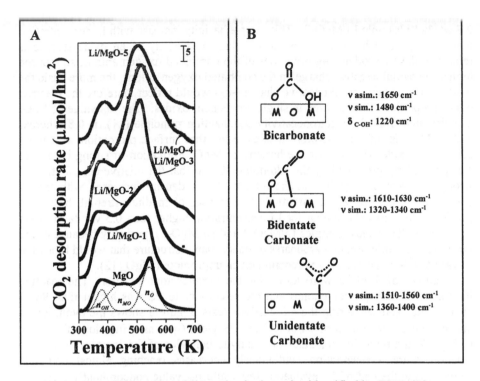

Figure 2. TPD of CO_2 (A) and CO_2 adsorbed species identified by FTIR (B)

low-temperature peak at 360-380 K was assigned to bicarbonates formed on weak Brönsted OH groups (n_{OH}); the middle-temperature peak at 450 K was attributed to bidentate carbonates desorbed from medium-strength metal–oxygen pairs (n_{MO}) and the high-temperature peak at 510-550 K was caused by the release of unidentate carbonates from low coordination oxygen anions (n_O). A fourth peak above 650 K was found in samples with Li loadings higher than 0.5 wt. % which can be assigned to decomposition of surface Li carbonates as determined by XRD. Thus, the area under this high-temperature peak was not taken into account for quantification of base site density. Results are shown in Table 2. The last column of Table 2 shows that the total density of base sites (n_b) increased with increasing the Li content up to

Table 2. Sample basicity determined by TPD of CO_2

Catalyst	Base Site Density ($\mu mol/m^2$)			
	Weak (n_{OH})	Medium (n_{MO})	Strong (n_O)	Total (n_b)
MgO	0.5	1.8	1.5	3.8
Li/MgO-1	0.8	2.4	2.3	5.6
Li/MgO-2	1.1	2.8	4.6	8.5
Li/MgO-3	1.8	3.0	6.4	11.3
Li/MgO-4	0.5	2.1	5.7	8.3
Li/MgO-5	0.7	0.9	5.5	7.1

0.5 wt. % Li (sample Li/MgO-3). This sample basicity increase with Li concentration may be explained by taking into account that the basicity of an oxide surface is related to the electrodonating properties of the combined oxygen anions, so that the higher the partial negative charge on the combined oxygen anions, the more basic the oxide. The oxygen partial negative charge ($-q_O$) would reflect therefore the electron donor properties of the oxygen in single-component oxides. The $-q_O$ value of Li_2O, as calculated from the electronegativity equalization principle (11), is 0.8 whereas that of MgO is 0.5 and therefore it is expected that surface promotion with more basic Li_2O oxide will increase the basicity of MgO. Formation of strong base sites was particularly promoted by the addition of Li, so that the relative contribution of strong base site density (n_O) to the total base site density increased with the Li content up to 0.5 wt. % Li, essentially at expenses of medium-strength base sites (n_{MO}). Table 2 also shows that for Li concentration higher than 0.5 wt. % the base site density diminishes (samples Li/MgO-4 and Li/MgO-5), probably because of the formation of stable lithium carbonates of bulky planar structure that would block the surface $Mg^{2+}-O^{2-}$ pairs that are predominant on unpromoted MgO (12).

Samples of Table 2 were tested for the citral/acetone reaction. The reaction formed essentially pseudoionones and the selectivity to PS was higher than 96 % over all the catalysts during the 6 h catalytic tests. Figure 3 shows the evolution of the relative concentration of citral and the PS yield (η_{PS}) as a function of time on Li/MgO-2 and typically illustrates the catalytic behavior of the samples during the reaction. Citral concentration continuously decreases reaching almost complete conversion at the end of the evaluation test while η_{PS} value concomitantly increases up to more than 90%.

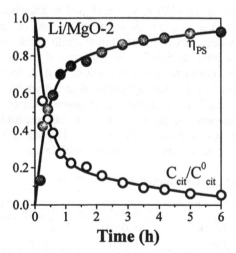

Figure 3. Evolution of relative citral concentration and PS yield on Li/MgO-2 sample

In Figure 4 we have represented citral conversions (X_{cit}) as a function of tW/n_{cit}^0 where W is the catalyst weight, t the reaction time, and n_{cit}^0 the initial moles

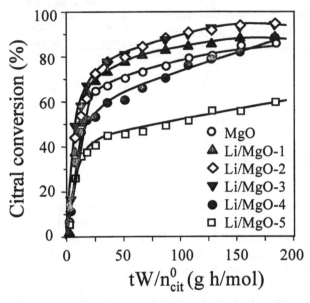

Figure 4. Citral conversion as a function of parameter tW/n_{cit}^0

of citral. The local slope of each curve in Fig. 4 gives the citral conversion rate at a specific value of citral conversion and reaction time. Thus, we determined the initial citral conversion rate on areal basis (r_{cit}^0, mol/hm^2) by calculating the initial slopes in Fig. 4 according to:

$$r_{Cit}^0 = \frac{1}{S_g}\left[\frac{dX_{Cit}}{d(tW/n_{Cit}^0)}\right]_{tW/n_{cit}^0=0}$$

Using a similar procedure, we determined the initial PS formation rate (r_{PS}^0, mol/hm^2) from η_{PS} vs tW/n_{cit}^0 curves (not shown here). The obtained r_{cit}^0 and r_{PS}^0 values for all the catalysts are shown in Table 3. It is observed that r_{cit}^0 and r_{PS}^0 increase when small amounts of Li are added, reaching a maximum at *ca.* 0.5 wt. % Li (sample Li/MgO-3); then, the initial catalyst activity decreases for higher Li concentrations. A similar trend with the amount of Li on the catalyst was verified for the pseudoionone yield. In fact, in Fig. 5 we have represented the evolution of η_{PS} values obtained at the end of the catalytic runs as a function of % Li and it is clearly observed that η_{PS} depend on the Li content, reaching a maximum of 93 % for about 0.5 wt. % Li.

Catalytic results of Table 3 and Fig. 4 and 5 show that the addition of small

Table 3. Catalytic activity data obtained on Li-promoted MgO samples [a]

Catalyst	Initial citral conversion rate r^0_{cit} (μmol/min m^2)	Initial PS formation rate r^0_{PS} (μmol/min m^2)	PS Yield [b], η_{PS} (%)	
			Cis-isomer	Trans-isomer
MgO	6.3	6.2	49.2	32.5
Li/MgO-1	7.8	7.6	48.5	36.7
Li/MgO-2	15.4	14.9	51.5	39.8
Li/MgO-3	21.2	20.5	51.1	41.9
Li/MgO-4	16.6	16.6	47.6	40.9
Li/MgO-5	15.2	15.2	32.7	25.5

[a] At T = 353 K, n^0_{DMK} = 0.8 moles, n^0_{cit} = 0.016 moles, $W_{Cat.}$ = 0.5 g

[b] At 6 h reaction time.

amounts of Li improve the MgO activity for the formation of pseudoionones; however, Li concentrations higher than about 0.5 wt. % are detrimental because cause the η_{PS} to decrease. In order to explain these results, and also in an attempt to relate the catalytic behavior of the samples with their surface base properties, we compared the catalyst activity data of Table 3 with the base site densities given in Table 2. We obtained a good correlation between the initial PS formation rate and the density of strong base sites (n_O) as it is shown in Fig. 6. In contrast, a poor correlation was found when r^0_{PS} was plotted against the density of weak (n_{OH}) or

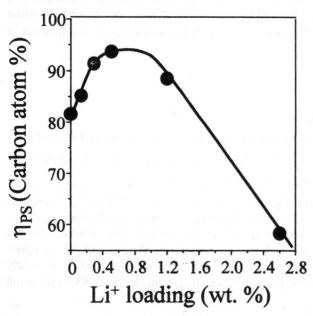

Figure 5. Effect of Li amount in Li/MgO samples on pseudoionone yield

medium (n_{MO}) strength basic sites. The observed proportionality between r_{PS}^0 and n_O in Fig. 6 indicates that the rate-determining step in the reaction mechanism is effectively

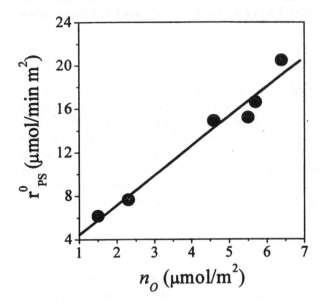

Figure 6. Initial PS formation rate as a function of the density of strong basic sites.

controlled by the surface base strength of the sample. The reaction pathway to produce pseudoionone from the aldol condensation of citral with acetone involves the initial abstraction of the α-proton from acetone forming a carbanion that consecutively attacks the carbonyl group of the contiguously adsorbed citral molecule. Then, the resulting unstable intermediate dehydrates forming pseudoionone and regenerating an active site on the catalyst surface. In this mechanism, the function of weak Lewis acid sites (Mg^{2+} and Li^+ cations) is to favor the adsorption of both reactants via their carbonyl groups. In general, the basis of any base-catalyzed aldol condensation is the abstraction of the α-proton, the most acidic site of the acetone molecule, with a pK_a of 20 (13). In this regard, it had to be noted that we have studied the liquid-phase self-condensation of acetone to produce diacetone alcohol using the catalysts of Table 2 (14), and observed that, similarly to the results reported in this work for acetone/citral reaction, the addition of small amounts of Li promotes the initial formation of diacetone alcohol. This result supports the assumption that the strong base sites of Li-doped MgO efficiently promote the abstraction of proton $H^α$ from the acetone molecule that would be the rate-limiting step for the synthesis of pseudoionones via the citral/acetone aldol condensation.

Finally, we remark that precisely because of its strong basicity, low coordination oxygen atoms O^{2-} are very reactive and may reconvert to surface OH^- in the presence of water formed during the pseudoionone synthesis (see reaction scheme 1). Recent reports confirm that the surface of metal oxides with basic character are often restructuring during the reaction (15). Thus, the formation of pseudoionones on Li-doped MgO may be accompanied by the gradual modification of the catalyst surface, by reconverting strong O^{2-} basic sites to weak OH^- basic groups. While it was not the intent of this work to ascertain the changes of active site nature during reaction, it is worthy of note that eventual changes on catalyst surface with reaction time would not affect the experimental results given in Table 3 and Fig. 6 that were obtained at initial reaction conditions.

Conclusions

The addition of Li to MgO up to about 0.5 wt. % increases the surface density of low coordination oxygen anions that are the strongest base sites on these samples. Higher Li loadings cause the formation of separate Li_2CO_3 phases that block the surface base sites and thereby diminish the sample basicity. Li-doped MgO are active and selective catalysts for the formation of pseudoionones from aldol condensation of citral with acetone in liquid phase. The initial pseudoionone formation rate increases linearly with the surface density of strong base sites, showing that the rate-determining step is preferentially promoted by low coordination oxygen anions.

Acknowledgements

We thank the Universidad Nacional del Litoral (UNL), Consejo Nacional de Investigaciones Científicas y Técnicas (CONICET), and Agencia Nacional Científica y Tecnológica (ANPCyT), Argentina, for the financial support.

References

1. H. Pommer, *Angew.Chem.*, **89**, 437 (1977).
2. P. Mitchell, US Pat 4,874,900, to Union Camp Corporation (1989).
3. M.J. Climent, A. Corma, S. Iborra, and A. Velty, *Catal. Lett.* **79**, 157 (2002).
4. C. Noda Perez, C.A. Henriques, O.A.C. Antunes, and J.L.F. Monteiro, *J. Mol. Catal. A: Chemical*, **233**, 83 (2005).
5. J.C. Roelofs, A.J. van Dillen, and K.P. de Jong, *Catal. Today*, **60**, 297 (2000).
6. M.J. Climent, A. Corma, S. Iborra, K. Epping, and A. Velty, *J. Catal.*, **225**, 316 (2004).
7. S. Abello, F. Medina, D. Tichit, J. Pérez-Ramírez, X. Rodríguez, J.E. Sueiras, P. Salagre, Y. Cesteros, *Appl. Catal. A: General*, **281**, 191 (2005)
8. J.I. Di Cosimo, V.K. Díez, and C.R. Apesteguía, *Appl. Catal.*, **137**, 149 (1996).
9. V. Perrichon and M.C. Durupty, *Appl. Catal.*, **42**, 217 (1988).
10. V.K. Díez, C.R. Apesteguía, and J.I. Di Cosimo, *Catal. Today*, **63**, 53 (2000).
11. R.T. Sanderson, in *Chemical Bonds and Bond Energy*, Academic Press, New York, 1976, pp. 75-94.
12. A. Grzechnik, P. Bouvier, and L. Farina, *J. Solid State Chemistry*, 173, 13 (2003)
13. J. Kijenski and S. Malinowski, *J. Chem. Soc. Faraday Trans. 1*, **74**, 250 (1978).
14. V.K. Díez, J.I. Di Cosimo, and C.R. Apesteguía, unpublished results.
15. H. Tsuji, A. Okamura-Yoshida, T. Shishido, and H. Hattori, *Langmuir*, **19**, 8793 (2003)

41.
The One-Step Synthesis of MIBK via Catalytic Distillation: A Preliminary Pilot Scale Study

William K. O'Keefe, Ming Jiang, Flora T. T. Ng and Garry L. Rempel

*The University of Waterloo, Department of Chemical Engineering,
200 University Avenue West, Waterloo, Ontario, Canada. N2L 3G1*

fttng@cape.uwaterloo.ca

Abstract

This paper reports the results of a preliminary pilot scale study in which MIBK was successfully produced in a single catalytic distillation reactor utilizing two separate catalytic reaction zones. The MIBK productivity was relatively low, primarily due to hydrogenation catalyst deactivation and the relatively low hydrogen partial pressures in this experiment. However, the experimental results are insightful and elucidate the immediate challenges that must be addressed in this ongoing process development. Specifically, improved catalyst performance is required. Catalyst characterization via DRIFT spectroscopy indicates that the ketone group of MIBK is strongly adsorbed on the commercial Pd/alumina hydrogenation catalyst used in this study.

Introduction

Methyl isobutyl ketone (MIBK) is undoubtedly the most valuable derivative of acetone. It is used primarily as an industrial solvent in the paint and coating industry as well as in metallurgical and solvent extraction processes. It is also used as a precursor in the production of specialty chemicals including rubber antioxidants, surfactants and pesticides. The synthesis of MIBK involves three major reaction steps outlined in Figure 1. First acetone is dimerized through an aldol condensation reaction to produce diacetone alcohol (DAA). Second, the dehydration of DAA gives mesityl oxide and water. Third, the olefin group of mesityl oxide is selectively hydrogenated to produce MIBK.

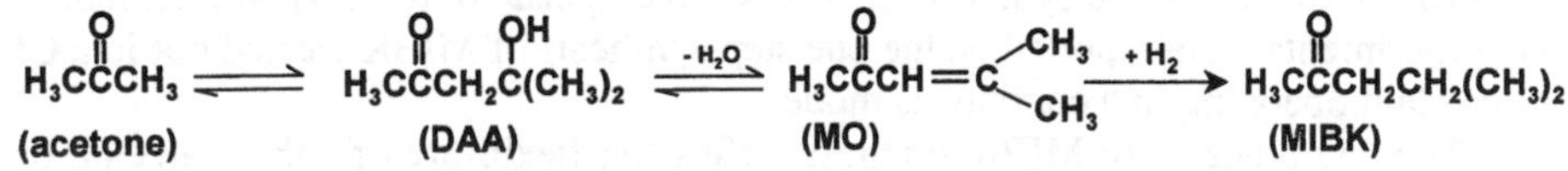

$$H_3CCCH_3 \rightleftharpoons H_3CCCH_2C(CH_3)_2 \xrightarrow{-H_2O} H_3CCCH=C\begin{smallmatrix}CH_3\\CH_3\end{smallmatrix} \xrightarrow{+H_2} H_3CCCH_2CH_2(CH_3)_2$$
(acetone)   (DAA)   (MO)   (MIBK)

Figure 1. Major reaction steps in the synthesis of MIBK from acetone

Due to its industrial importance, this organic synthesis has been studied extensively in both the liquid and gas phases. The acid catalyzed mechanism to produce mesityl oxide from acetone is well understood (1). Recently, we have proposed a novel mechanism for the hydrogenation of mesityl oxide (2). Since mesityl oxide is an α,β-unsaturated ketone with conjugated olefin and ketone groups, both functional groups may interact with the heterogeneous catalyst and as a consequence, mesityl oxide may coordinate with the catalyst in one of several possible adsorption modes. The adsorption mode of the substrate is of particular importance since it will not only determine the selectivity and activity of the hydrogenation reaction but may also constrain the possible reaction pathways and thus strongly influence the reaction mechanism (3). We have proposed that mesityl oxide will exist on a Pd/Al_2O_3 catalyst as an η_4 diadsorbed species if adjacent sites are available, with the olefin group forming an $\eta_2\pi(C,C)$ complex and the ketone group forming an $\eta_2\pi(C,O)$ complex (2). Mesityl oxide adsorbed at a single site will act as an inert as a consequence of the preferential adsorption via the ketone group, which interacts strongly with the catalyst (2). In this adsorption mode, the mesityl oxide is not hydrogenated since the chemisorption of the olefin group is required for its hydrogenation (3).

The need for an efficient one-step process for MIBK production is of heightened importance today in this era of rising energy costs and increased environmental regulation and presents an excellent opportunity for the application of catalytic distillation (CD) technology (4). The CD reactor combines catalytic reaction and separation in a single distillation column. As a consequence of this process intensification, a reduction in operating and capital expenditures may be realized (5). Within a CD reactor, heterogeneous catalyst is immobilized within one or more discrete reaction zones while liquid and vapour pass through the reactor in a counter current fashion. The continuous product removal from the reaction zone due to the distillation action shifts the reaction in favour of product formation in accordance with Le Chatelier's Principle resulting in product yields much greater than the theoretical equilibrium conversion would allow. Since the reaction occurs in a boiling medium, excellent temperature control is achieved which is often critical in organic synthesis, and mitigates hot spot formation (5). In addition, since heat transfer is maximized in a boiling medium, the energy evolved from the exothermic reaction is efficiently converted to drive the distillation process, which could lead to substantial energy savings and a reduction of CO_2 emissions from power production facilities (5). CD is a promising reactor technology for the MIBK synthesis since the first two reaction steps, particularly the aldol condensation of acetone, are known to be equilibrium limited and the overall synthesis is complex (4). Patents on the application of CD for the synthesis of MIBK have appeared (6,7). However, there is no experimental data reported on the one step synthesis of MIBK carried out in a CD pilot plant operating in a continuous mode.

The CD process for MIBK synthesis offers the flexibility of either carrying out the reaction in a single zone using a multifunctional catalyst or in separate optimized reaction zones (4). In the former configuration, it may be possible to avert undesirable parallel reactions for which mesityl oxide is a precursor (Figure 2) by rapidly hydrogenating mesityl oxide to produce MIBK. However, for this approach to succeed, hydrogen must not be the limiting reagent at the active sites.

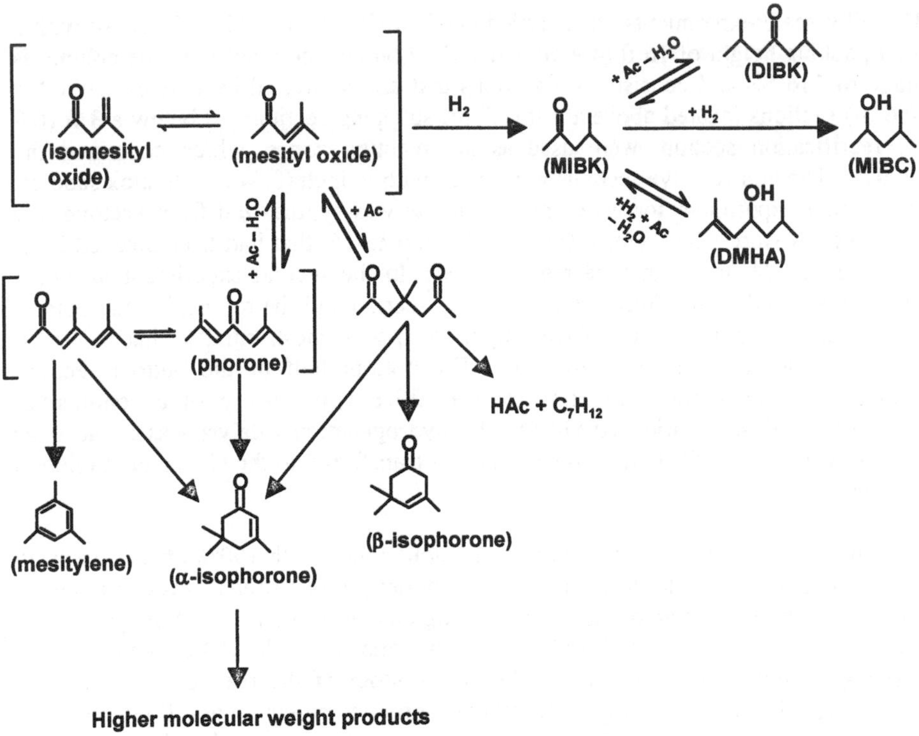

Figure 2. Undesirable consecutive reactions in the synthesis of MIBK

The latter approach affords greater flexibility allowing the catalyst and reaction zones to be designed and optimized independently for each reaction step. In this configuration, it is possible to design the hydrogenation catalyst in a manner to protect the active sites and facilitate hydrogen transport. Previously, we have developed a CD process for the production of DAA from acetone (8). A CD process for the one step synthesis of MIBK appears to be a simple extension of this DAA process. However, the introduction of hydrogen to this system opens up numerous possible reaction pathways. Most noteworthy is the hydrogenation of acetone to produce 2-propanol, which is a significant competing reaction for the expected CD process conditions (2,4). In this work, the results of a preliminary pilot scale investigation of the one step synthesis of MIBK in a CD pilot plant operating in a continuous mode utilizing two commercial catalysts in separate reaction zones is reported. The objective was to obtain a cursory assessment of the process by investigating the effects of the reaction temperature as well as the hydrogen and acetone feed rates on the MIBK yield and selectivity and to establish the technical feasibility of a CD process for MIBK synthesis utilizing two separate reaction zones.

Experimental Section

The pilot scale experiments were carried out in a CD column 23 ft (7 m) tall with a total packing height of 16 ft (4.9 m) and a 1" (2.54 cm) nominal I.D. The column is made of 316 SS and consists of 5 sections that are connected by flanges. Two 2 ft (0.6 m) sections located above a 9 ft (2.7 m) stripping section and below a 3 ft (0.9 m) rectification section, were used as the reaction zones, which contained the catalyst. The non-reactive sections were filled with ¼ inch (0.64 cm) Intalox saddles. In the first experiment for which mesityl oxide was synthesized from acetone, the two sections were filled with 130 mL of Amberlyst-15, that had been swelled in 2-propanol for 24 hours, in wire mesh bundles. In the second experiment in which MIBK was synthesized from acetone, the top section and the top half of the bottom section contained 135.0 mL of Amberlyst-15 in wire mesh bundles that had been swelled in acetone for over 24 hours. The bottom half of the bottom section, immediately below the Amberlyst 15, was filled with 50.1 g of a commercial Pd/Al$_2$O$_3$ catalyst (Aldrich 20,574-5). The hydrogenation catalyst was reduced *ex situ* in hydrogen at 350°C for 3 hours and was transferred to the CD column under a nitrogen blanket.

The CD column was operated in continuous mode with 100% of the overhead product being refluxed to the column and a reboiler product stream was continuously removed from the bottom of the column through a control valve. ACS reagent grade acetone (Aldrich 67-64-1, >99.5%) was fed continuously to the CD column at a feed port approximately 6 in. (15 cm) below the bottom of the reaction zone using a Milton-Roy LCD mini pump. The reboiler product stream mass flow rate was matched the acetone feed stream by maintaining a constant liquid level in the reboiler. In the first experiment, nitrogen was fed to the CD column to promote the convective transport of matter and energy up the column. In the second experiment, UHP Hydrogen was fed continuously to the bottom of the CD column using a Brooks 5850E mass flow controller (MFC) instead of nitrogen. Liquid samples were obtained from the reboiler product on an hourly basis and the average reboiler composition for the previous hour was ascertained by GC/FID using an Agilent Technologies 6890N gas chromatograph equipped a 7683 Series autosampler injector and a J&W Scientific DB-WAX column (30m X 0.53mm I.D. X 1mm film thickness). Certified 1-Propanol (Fisher A414-500 >0.998) was used as an internal standard. Data was acquired continuously from the CD column including the column temperature profile measured at 16 thermocouple locations along the column major axis, the column pressure and the pressure drop across the column in addition to the hourly reboiler product composition. Since the reaction zones are short relative to the total column height, the thermal profiles along the reaction zones were isothermal during steady state operation. Steady state was defined as the condition where the CD column temperature profile remained invariant to within ± 0.5 °C and the mass fraction of species in the reboiler product remained constant to within a maximum relative change of ± 5%. Once steady state had been achieved, the steady state condition was maintained from 5 to 10 hours depending on the process variability, during which time, sufficient observations were made under steady state conditions to provide narrow 95% confidence intervals and hence precise estimates of the reboiler product composition. For example, the reboiler composition corresponding to the second condition in Table 1 was based on 6 steady state observations. The resultant 95% confidence intervals for the mesityl oxide and DAA mass fractions for this entry correspond to relative errors of 1.44 and 1.73 % respectively.

The hydrogenation catalyst activity was tested before and after the CD experiment in a 300 mL Parr 4560 series microreactor with a Parr 4842 PID controller. Liquid samples were obtained periodically through a dip-tube with a custom made external heat exchanger and analyzed by GC/FID. The hydrogenation catalyst was characterized by Diffuse Reflectance Infrared Fourier Transform (DRIFT) spectroscopy. The adsorption properties of MIBK, mesityl oxide, acetone and carbon monoxide probe molecules on the Pd/Al_2O_3 catalyst were investigated using an Excalibur Bio-Rad FTIR spectrometer equipped with an MCT detector. The heat treatment and reduction of the catalyst as well as the adsorption and desorption of probe molecules were investigated *in situ* in a high temperature IR cell equipped with a ZnS dome while DRIFT spectra were recorded in 10s intervals.

Results and Discussion

Pilot Scale CD Experiments

In the first experiment, Amberlyst-15, a strongly acidic cation exchange resin, was used as a catalyst to synthesize mesityl oxide, the precursor of MIBK, from acetone without hydrogenation. The effects of acetone feed rate, reboiler duty and reaction temperature on the mesityl oxide productivity and product distribution were investigated. Preliminary results of this experiment are outlined in Table 1.

Table 1. The effects of reaction temperature, acetone feed rate and reboiler duty on the mesityl oxide productivity and product distribution for the first CD experiment

Acetone Feed Rate [mL/hr]	Reaction Temp [°C]	Reboiler Duty [Watts]	Acetone Conv. [%]	MO Productivity [g/hr*mL cat]	MO [wt%]	DAA [wt%]	Higher Mwt [%]
152	114	300	26.2	.1418	19.0	1.31	1.72
152	114	350	26.7	.1601	18.6	1.81	0.89
152	101	350	29.0	.1788	19.5	3.73	0.24
37	99	350	>99.9	.1413	78.2[a]	14.2	0.95

[a] Last condition is a preliminary result. Process had not yet achieved steady state after 24 hours. Two phases present, composition for organic phase is reported

It was found that the mesityl oxide productivity was a strong function of the reflux flow rate in the CD column, which suggests the reaction is controlled by the rate of external mass transfer. It is also evident from Table I that acetone conversions as high as essentially 100% can be achieved with the mesityl oxide concentration in the reboiler product reaching as high as 78 wt%. This is particularly noteworthy since the syntheses of DAA and mesityl oxide are strongly equilibrium limited reactions. The selectivity to mesityl oxide remained within a narrow range from 85 to 90%. Evidently, conditions that resulted in a higher conversion of DAA also resulted in an increased rate of production of undesirable higher molecular weight products including mesitlyene, phorone and isophorone. Similarly, the conditions for which these undesirable consecutive reactions were averted resulted in a greater amount of unreacted DAA remaining in the system. A major finding was that the undesirable consecutive reactions from mesityl oxide could be mitigated while simultaneously increasing mesityl oxide productivity by increasing the liquid reflux flow rate in the CD column. Note the effect of increasing the reboiler duty from 300 to 350 W at 114°C. This shows that CD technology allows improved selectivity to a desired intermediate by the rapid removal of the desired product from the reaction zone.

In the second CD experiment, MIBK was successfully produced from acetone in a single stage. However, the MIBK yield was relatively low in this experiment. For example, when hydrogen was introduced into the CD reactor at 60 L/hr (STP) with a reboiler duty of 350 W and a reaction temperature of 119°C, the MIBK productivity was 0.10 [g_{MIBK}/(hr*g_{cat})]. The MIBK productivity was calculated from an average MIBK concentration in the reboiler product of 3.98 ± 0.031 wt% based on 11 measurements over a 10 hour period of steady state operation. The mesityl oxide conversion was 15.1% and the hydrogen utilization was less than 2%. The hydrogenation of acetone to produce 2-propanol was the only significant competing hydrogenation reaction and the selectivity of the hydrogenation was 84.4% for the conditions described above. It is evident that the hydrogenation of mesityl oxide to produce MIBK is currently the limiting step of this CD process. It should also be noted that the locations and catalyst amounts for the reaction zones were not optimized for this preliminary experiment.

Although the MIBK productivity was comparable to the data reported by Lawson and Nkosi (6), there was evidence of significant hydrogenation catalyst deactivation. At the end of the experiment, the hydrogenation catalyst was recovered from the reactor under a nitrogen blanket and protected in solvent. Its activity for mesityl oxide hydrogenation was subsequently tested in an autoclave and was found to have an activity of 0.302 relative to the fresh catalyst. Spectroscopic data presented in the next section suggests that the strong adsorption of MIBK may have had a detrimental effect on the long-term performance of the catalyst in the CD reactor. The operating pressure for this experiment was constrained to less than 0.6 MPa due to the poor thermal stability of Amberlyst 15, which has a maximum operating temperature of 120°C. Consequently, a low MIBK yield was expected due to the relatively low hydrogen partial pressure in this experiment.

Catalyst Characterization via DRIFT Spectroscopy

The commercial Pd/Al_2O_3 catalyst used in this pilot scale study was characterized via DRIFT spectroscopy. *In situ* DRIFT spectra of carbon monoxide, MIBK, acetone and mesityl oxide probe molecules were obtained to ascertain their adsorption properties. The adsorption DRIFT spectra at ambient temperature of carbon monoxide probe molecules on the Pd/Al_2O_3 catalyst, which was pre-reduced in hydrogen for 1h at 120°C, are illustrated in Figure 3. The band at 2075 cm^{-1} corresponds to linearly adsorbed carbon monoxide. The bands at 1977 and 1916 cm^{-1} correspond to bridged and multibridged carbon monoxide, respectively. The doublet bands at 2173 cm^{-1} and 2119 cm^{-1} correspond to gaseous carbon monoxide over the catalyst. The existence of linear and bridged carbon monoxide not only confirms the existence of Pd^0 but is also indicative of the metal crystallite structure since the bridged species requires larger crystallite sizes while the linearly adsorbed carbon monoxide is characteristic of smaller more highly dispersed crystallites. Moreover, the prevalence of bridged and multibridged carbon monoxide in the specimen of Figure 3 indicates the existence of adjacent sites, which may facilitate the diadsorption of mesityl oxide, as we have previously proposed (2). Additional experiments in which the catalyst was reduced *in situ* over various temperatures prior to carbon monoxide adsorption revealed that the crystallites became increasingly aggregated with increasing reduction temperature up to 250°C, beyond which no morphological change was observed.

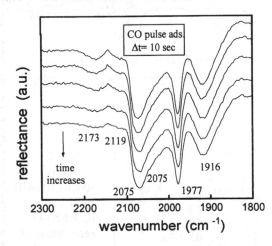

Figure 3 DRIFT spectra of carbon monoxide probe molecule pulse adsorbed on a commercial Pd/Al_2O_3 catalyst reduced in H_2 for 1h at 120°C (spectra were recorded in 10 sec interval after pulse).

The hydrogenation of mesityl oxide over the commercial Pd/Al_2O_3 catalyst was carried out in situ in the FTIR cell and was found to be facile proceeding readily at ambient temperature as illustrated in Figure 3. The spectra in the top half of Figure 4 correspond to mesityl oxide pulse adsorbed onto the catalyst. The bands at 1677 and 1610 cm^{-1} correspond to the carbonyl and olefin groups respectively. Note that as hydrogen is pulsed into the IR cell as illustrated in the bottom half of Figure 3, the band corresponding to the olefin group of mesityl oxide begins to disappear and the band corresponding to the ketone group begins to shift from 1677 to 1700 cm^{-1}, characteristic of the ketone group of MIBK, indicating that the C=C species in mesityl oxide is easy to hydrogenate. Note that MIBK and mesityl oxide are further distinguished by their characteristic CH_x stretching vibrations as shown in Figure 3.

In MIBK, the asymmetric and symmetric bands of ν(CH₃) at 2960 and 2875 cm⁻¹ become much stronger than those of the ν(CH₂) due to the saturation of the C=C bond.

Figure 4 (right) DRIFT adsorption spectra of mesityl oxide pulse adsorbed at 25°C on Pd/Al₂O₃ catalyst pre-reduced in H₂ for 1h at 120°C recorded in a time interval of 10 sec (bottom) *in situ* hydrogenation of mesityl oxide at 25°C as purged by H₂ recorded in a time interval of 10 sec.

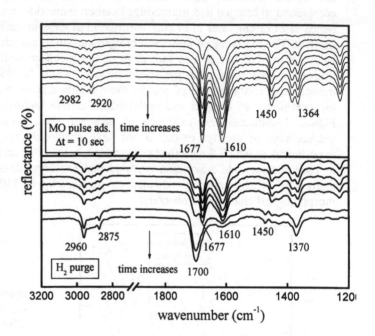

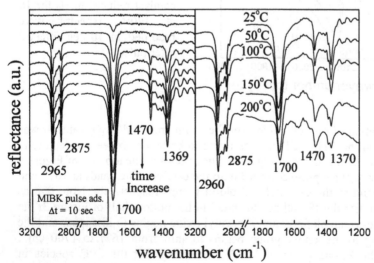

Figure 5
Left: DRIFT spectra of MIBK pulse adsorbed on Pd/Al₂O₃ recorded in 10 sec intervals
Right: DRIFT desorption spectra of MIBK as temperature is increased to 200°C.

The DRIFT spectra of MIBK are illustrated on the left side of Figure 5, as MIBK is pulse adsorbed onto the Pd/Al₂O₃ catalyst. The MIBK desorption DRIFT spectra are illustrated on the right side of Figure 5 as the temperature is increased from ambient temperature to 200°C in discrete increments. Note that even at 200°C, the band intensities characteristic of adsorbed MIBK are evident indicating a strong interaction between MIBK and the Pd/Al$_2$O$_3$ catalyst. Similar experiments were carried out for acetone and mesityl oxide pulse adsorbed on the Pd/Al$_2$O$_3$ catalyst. The desorption spectra for acetone on Pd/Al$_2$O$_3$ catalyst is given in Figure 6.

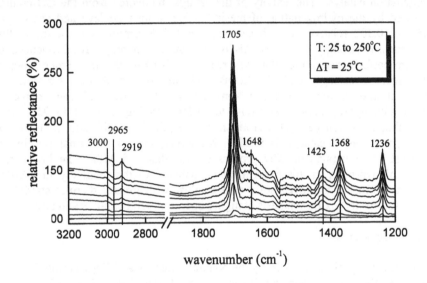

Figure 6 DRIFT desorption spectra of acetone on Pd/Al$_2$O$_3$ during heating from 25°C to 250°C at 25°C increments. The catalyst was first reduced in H$_2$ at 120°C for 1.5 h and purged in He for 1h at 25°C

The ketone group in acetone also interacts strongly with the catalysts. When mesityl oxide adsorbed on the catalysts was subjected to thermal desorption in helium, it can be converted to MIBK, most probably, due to the hydrogenation of C=C by the residual hydrogen from the pre-treatment of the catalysts with hydrogen. Therefore, mesityl oxide adsorbed on the catalysts is not stable and the strongly adsorbed species are ketone groups. Previous experiments in which HPLC grade MIBK was intentionally added to a mixture of mesityl oxide and acetone did not indicate product inhibition had occurred during the hydrogenations carried out in an autoclave (3). However, it is possible that the strong adsorption of MIBK may have affected the long-term catalyst performance in the CD reactor. DRIFT experiments with MIBK on alumina alone without Pd resulted in essentially the same spectra as that obtained with Pd/Al$_2$O$_3$. Therefore the DRIFT spectra could not provide an interpretation of the mechanisms of the reaction. However, the DRIFT data suggest a strong interaction of ketone group of MIBK with the Lewis acid sites of the alumina support.

Conclusions

MIBK has been produced successfully in a single CD reactor utilizing two different catalytic reaction zones. Although the MIBK productivity was comparable to the data presented by Lawson and Nkosi (6) utilizing a single catalytic reaction zone, there was experimental evidence of significant deactivation of the hydrogenation catalyst. The activity of the catalyst recovered from the CD column was tested by the hydrogenation of mesityl oxide in an autoclave and showed an activity of 0.302 relative to fresh catalyst. DRIFT desorption spectra shows that MIBK adsorbs strongly on the Pd/Al_2O_3 catalyst, which may have influenced its long-term performance in the pilot scale reactor. It is likely that the ketone group of MIBK interacts strongly with the Lewis acid sites of the alumina support. The hydrogenation of acetone to produce 2-propanol was the only significant competing reaction with hydrogenation selectivity to MIBK ranging from 84 to 95%. It is evident that improved catalyst performance is required. Specifically, a stable and active hydrogenation catalyst is needed. Moreover, the poor thermal stability of Amberlyst 15 constrains the CD process to less than 0.6 MPa. The low partial pressure of hydrogen in this experiment contributed to the relatively low MIBK yield. Therefore, an improved solid acid catalyst with greater thermal stability is required for mesityl oxide synthesis at higher operating temperatures and pressures.

Acknowledgements

Funding for this project provided by the Natural Sciences and Engineering Research Council (NSERC) of Canada and financial support provided to W.K O'Keefe from the Province of Ontario, Ministry of Training Colleges and Students, in the form of an Ontario Graduate Scholarship, is gratefully acknowledged.

References

L. Melo, P. Magnoux,, G. Giannetto, F. Alvarez, and M. Guisnet, *J. Mol. Catal. A.*, **124**, pp.155-161, (1997).
2. W. K. O'Keefe, M. Jiang, F.T.T. Ng and G. L. Rempel, *Chem. Eng. Sci.*, **60**, pp. 5131-4140, (2005).
3. F. Delbecq and P. Sautet, *J. Catal.* **152**, pp. 217-236, (1995).
4. W. K. O'Keefe, M. Jiang, F. T .T. Ng and G. L. Rempel, *(Cat. Org. React.)*, Ed. J. R. Sowa, (CRC Press, Taylor and Francis), pp. 261-266, (2005).
5. F.T.T. Ng and G.L. Rempel, "Catalytic Distillation", in the Encyclopedia of Catalysis, (John Wiley), (2003), pp. 477-509.
6. H. Lawson and B. Nkosi, U.S. Patent, 6,008,416 to Catalytic Distillation Technologies, Pasadena, TX., (1999).
7. N, Saayman, G.J. Lund and S. Kindemans, U.S. Patent, 6,518,462 to Catalytic Distillation Technologies, Pasadena, TX, (2003)
8. G. G. Podrebarac, F. T. T. Ng and G. L. Rempel, *Chem. Eng. Sci.,* **53(5)**, pp. 1067-1075 (1998).

VI. Symposium on "Green" Catalysis

42. Producing Polyurethane Foam from Natural Oil

Aaron Sanders, David Babb, Robbyn Prange, Mark Sonnenschein, Van Delk, Chris Derstine and Kurt Olson

The Dow Chemical Company, 2301 N Brazosport Blvd., Freeport, TX, 77541

awsanders@dow.com

Abstract

As part of the effort to reduce our dependence on fossil fuels, The Dow Chemical Company has been developing a seed oil based polyol to be used as a replacement to conventional petrochemical based polyether polyols in the production of flexible polyurethane foam. The general process for making natural oil polyols consists of four steps. In the first step, a vegetable oil (triglyceride) is transesterified with methanol, liberating glycerin, and forming fatty acid methyl esters or FAMEs. In the second step the FAMEs are hydroformylated giving a complex mixture of FAMEs that contain 0-3 formyl groups per chain. In the third step, the aldehydes and the remaining unsaturates are hydrogenated to yield a mixture of FAMEs that contain 0-3 hydroxymethyl groups. Finally, the poly(hydroxymethyl)fatty esters are transesterified onto a suitable initiator to produce the natural oil polyol.

Introduction

The preparation of polyester polyols from seed oils for the production of a variety of polyurethane products has been previously reported (1,2). The development of process and product technology that is sufficiently robust to compensate for inherent variability in products derived from natural resources is key to successful implementation. Product variability is primarily due to genetic variety in feedstocks and seasonal inconsistency, such as regional rainfall totals or pests and disease. Process technology that may be applied to a wide variety of potential feedstocks would be highly desirable.

Dow's process for producing natural oil polyols consists of four steps and is shown in Figure 1. In the first step, a vegetable oil (triglyceride) is transesterified with methanol, liberating glycerin and forming fatty acid methyl esters (FAMEs). In the second step the FAMEs are hydroformylated to create a complex mixture of FAMEs that contain 0-3 formyl groups per chain. In the third step, the aldehydes and the remaining unsaturation are hydrogenated to yield a mixture of FAMEs that contain 0-3 hydroxymethyl groups. In the final step the poly(hydroxymethyl)fatty esters are transesterified onto a suitable initiator to produce the natural oil polyol. This paper will review the process, emphasizing the catalysts employed in each step.

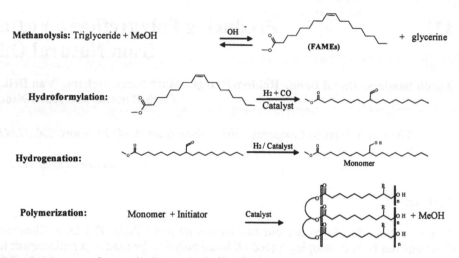

Figure 1. Process for producing natural oil polyols

Results and Discussion

The composition of seed oil triglycerides is well understood. Triglycerides are fatty acid esters of glycerin, and the composition depends on the source of the oil (Figure 2). The nomenclature used is standard in the fats and oils industry, with the number of carbons in the fatty acid indicated first, followed by the number of sites of unsaturation in parentheses.

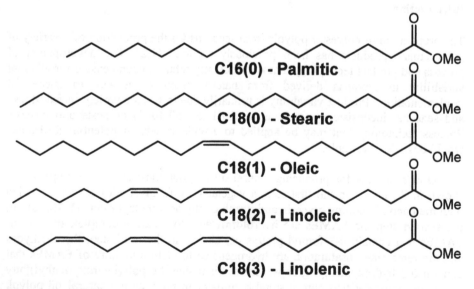

Figure 2. Methyl esters of common fatty acids found in vegetable oils

For the purposes of making polyols from these triglycerides, oils which contain a high level of unsaturation are desirable. Oils such as soy, canola, and sunflower are acceptable due to relatively low levels of saturated fatty acids, while feedstocks such as palm oil are considered unusable without further purification or refinement due to high levels of saturated fatty acids. Table 1 outlines the composition of several oils (3).

Table 1. Fatty acid content of selected vegetable oils in weight percent

	Soybean	Sunflower	Canola	Palm
Palmitic – C16(0)	11	6	4	44
Stearic - C18(0)	4	5	2	4
Oleic - C18(1)	22	20	56	40
Linoleic – C18(2)	53	69	26	10
Linolenic – C18(3)	8	0	10	0

Methanolysis

The transesterification of triglycerides with methanol is a simple and straightforward process. It is commercially practiced worldwide in the production of FAMEs, which have become popular as a replacement for diesel known as "biodiesel". The process consists of three separate equilibrium reactions that can be catalyzed by both acids and bases.(4) The overall process is described in Figure 3. Phase separation of the glycerin is the predominant driving force for this process.

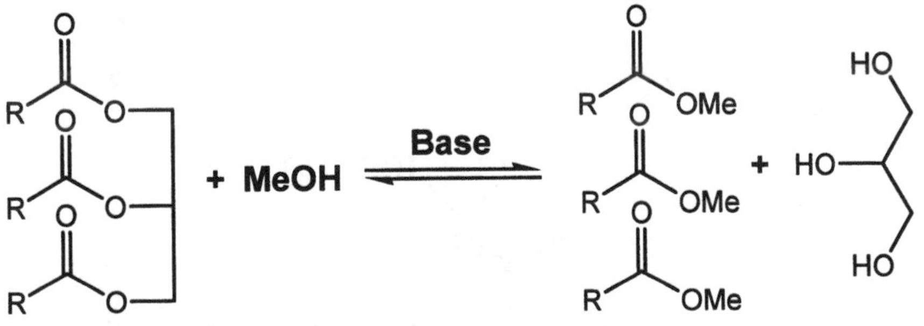

Figure 3. The methanolysis of triglycerides is an equilibrium reaction which is generally base catalyzed commercially

Industrial processes tend to favor base catalysis, since they have lower activation energies allowing the reactions to be carried out near or just above room temperature (5). Further, the carbonate or caustic bases are relatively inexpensive and easily separated with the glycerin product.

Hydroformylation

The important criteria for catalyst selection in the hydroformylation of FAMEs are activity, stability and catalyst-product separation. For this process, the feed is composed of a mixture of internal olefins. The differences in the placement of a hydroxymethyl group on the 9 vs 10 position has little impact on the polyol product.

The rhodium catalyst currently used for this step utilizes a monosulfonated phosphine ligand, dissolved N-methyl-2-pyrrolidinone (NMP).(6)   This catalyst system has shown adequate activity and stability.   More importantly, it enables product-catalyst phase separation.(7)

Each olefin component in the feed behaves somewhat differently during the course of hydroformylation. For the hydroformylation of methyl linoleate (shown in Figure 4) the reaction of olefin at the 9,10 or 12,13 positions occur at a similar rate (k_1). The hydroformylation rate (k_2) of the remaining olefin, however, is generally slower.

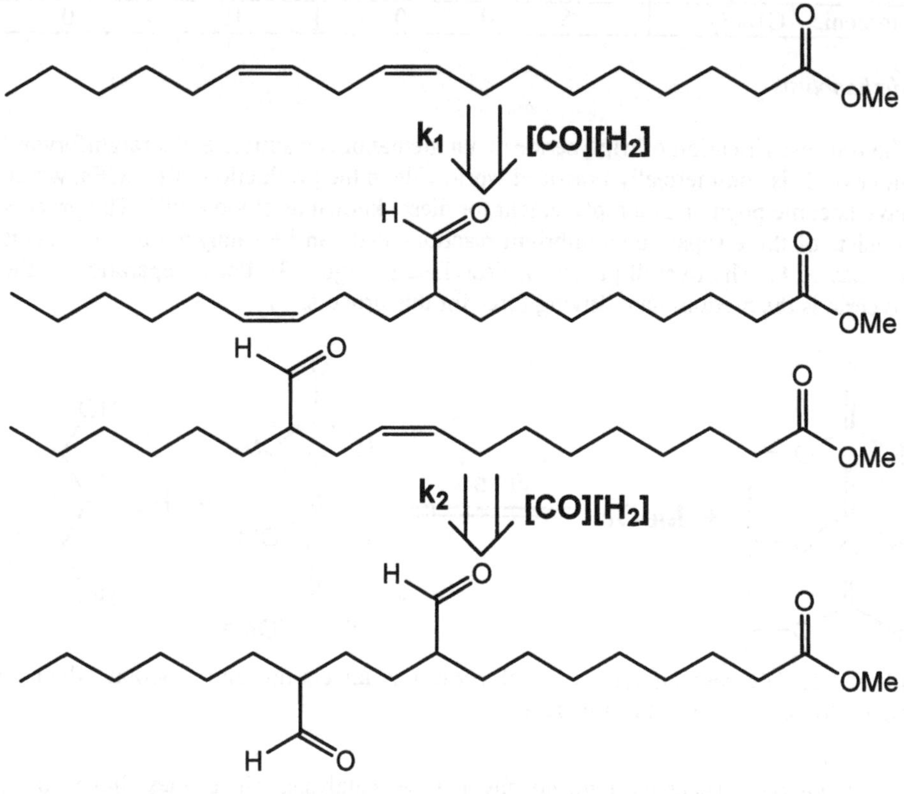

Figure 4. Hydroformylation products of methyl linoleate (excluding isomers)

Hydrogenation

After hydroformylation, the resulting mixture of saturates, unsaturates, and aldehydes, are hydrogenated over a fixed-bed commercial hydrogenation catalyst. Unreacted olefins are converted to saturates, and aldehydes are converted to the corresponding alcohol. Thus the hydroformylation product that contains a large number of components is converted to a simple mixture of saturated and hydroxy fatty esters. Since the apparent hydrogenation rate of the olefins is faster than the hydrogenation of the aldehydes, there is virtually no unreacted olefin in the final product.

Polymerization

The final step in the formation of the polyol is the reverse of the methanolysis step, however, the alcohol used is not limited to glycerin. Almost any alcohol can be used as the "initiator" in what might better be termed an oligomerization. Like the methanolysis, this reaction is an equilibrium and can be catalyzed by both acids and bases. The primary difference is that the methanolysis reaction is generally driven by the phase separation of the glycerin, while the polymerization is driven by removal of methanol (Figure 5).

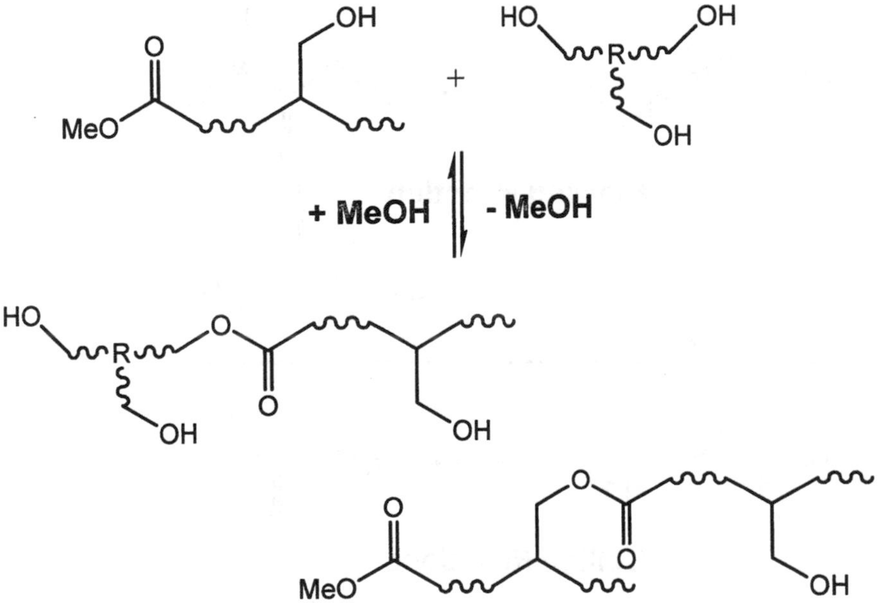

Figure 5. Transesterification of a hyroxymethylated fatty ester with a polyfunctional alcohol

The methanolysis catalyst is generally a base such as potassium carbonate, since the base catalyzed transesterification is generally lower in energy(5). For the transesterification of the hydroxymethylated fatty esters, however, a Lewis acid (stannous 2-ethylhexanoate) is employed. Although this catalyst requires higher temperatures to achieve rapid equilibrium, it has the benefit of not requiring removal from the final product. Stannous 2-ethylhexanoate is easily hydrolyzed with water, yielding an oxide that is inactive for both transesterifaction and urethane reactions (8). Although carbonate is less expensive and is a better catalyst for this reaction than the tin catalyst, the necessity of removing the residual salts from the product makes the carbonate process less attractive overall.

Polyurethane Foam

The polyurethane reaction that creates the foam is actually a balance of two separate reactions: blowing and gelling (Figure 6) (9). The blowing reaction takes place when water reacts with the isocyanate forming isocyanuric acid, which immediately decomposes to the amine and CO_2. The generation of CO_2 creates bubbles which continue to grow as more CO_2 is formed. Simultaneously, the gelling reaction is producing the polymer network that constitutes the foam cells and struts. In this reaction, the polyol or amine from the blowing reaction reacts with isocyanate to create a urethane or urea bond respectively.

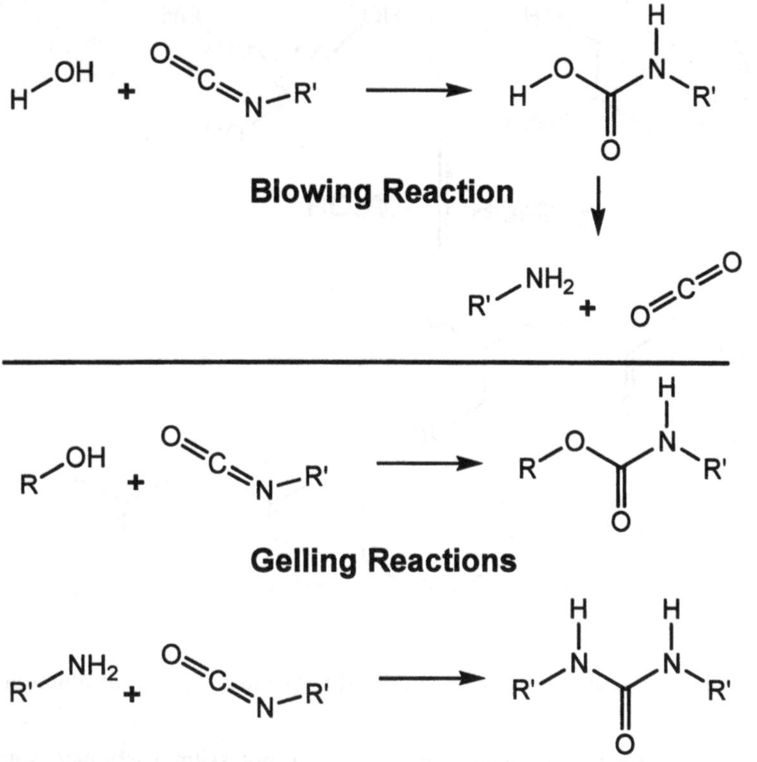

Figure 6. Polyurethane blowing and gelling reactions

The literature in this area is quite extensive, and summarizing it is beyond the scope of this discussion, however it must be pointed out that producing a usable polyurethane foam is not independent of the polyol. The gelling and blowing reactions must be balanced through the use of appropriate catalysts to achieve a "good" foam structure. The micrographs shown in Figure 7 illustrate the point. The foams were produced using the same natural oil polyol. The foam on the left is consistent with a good cellular structure, characterized by uniform well drained, opened cells. This type of cellular structure was obtained by balancing the blowing and gelling reaction to generate the optimal foam physical properties. Foams produced with imbalanced blowing and gelling reactions, shown on the right, form a cellular structure that consists of non-uniform, partially drained cells, and often possess undesirable physical properties.

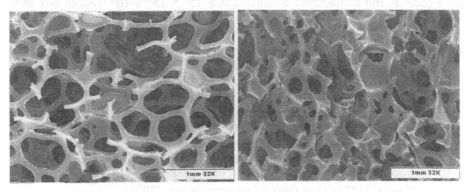

Figure 7. Foams prepared using natural oil polyols with well balanced catalysis (left) and unbalanced catalysis (right)

Experimental

Below are example or general procedures for each step of the process (patents pending see references 1f, 6 and 7).

Hydroformylation: A catalyst solution consisting of dicarbonylacetylacetonato rhodium (I) (0.063 g) and dicyclohexyl-(3-sulfonoylphenyl)phosphine mono sodium salt (1.10 g) in n-methyl pyrrolidinone (NMP) (16.0 g) was placed in a 100 mL stainless steel autoclave at 75 C under 200 psig synthesis gas. After 15 minutes soy methyl esters (34.05 g) were added and the synthesis gas pressure raised and maintained at 400 psig for 3 hrs resulting in the desired conversion of unsaturation.

Hydrogenation: To an up-flow tubular reactor packed with a supported Ni catalyst were fed a liquid stream comprised of 3.52 g/min of hydroformylated soy methyl ester and a recycle stream of 16.5 g/min (total liquid hourly space velocity of 3.65 hr^{-1}). Hydrogen was fed at 2000 sccm, at 159 °C and a reactor outlet pressure of 459, yielding the desired conversion of residual unsaturation and aldehydes.

Polymerization: To a 5 liter glass reactor was added hydroxymethylated fatty esters and 625 molecular weight poly(ethylene oxide) triol initiator in a ~6:1 molar ratio.

The reactor was purged with nitrogen and heated to 50 °C under 20 torr vacuum. The vacuum was broken and 500 ppm of stannous octoate catalyst obtained from City Chemical was added. The methanol was stripped under 20 torr vacuum at 195 °C for 4-6 hours yielding a viscous light yellow natural oil polyol with an equivalent weight of 660.

Polyurethane foam: Natural oil polyol, water, and surfactants were weighed into a 1 quart metal cup and premixed for 15 seconds at 1800 rpm using a pin type mixer. The catalyst was added and the mixture stirred an additional 15 seconds at 1800 rpm. The polyisocyanate was then added and the mixture stirred at 2400 rpm for 3 seconds and immediately transferred to a 15" x 15" x 10" wooden box lined with a polyethylene bag. The buns were allowed to cure overnight before testing.

References

1. List of leading references: a) J. John, M. Bhattacharya, R. Turner, *J. Appl. Polym. Sci.*, **86**, 3097 (2002). b) L. Mahlum, US Pat. Appl. 2001056196 to South Dakota Soybean Processors (2001). c) H. Kluth and A. Meffert, US Pat. 4,508,853 to Henkel K. (1984). d) S. Greenlee, US Pat. 3,454,539 to CIBA Ltd. (1969). e) T. Kurth, US Pat. 6,465,569 to Urethane Soy Systems Co. (2002). f) Z. Lysenko, A. Schrock, D. Babb, A. Sanders, J. Tsavalas, R. Jouett, L. Chambers, C. Keillor, J. Gilchrist, PCT Int. Appl. Pub. PCT/US 2004/012427 to Dow Global Technologies Inc. (2004).
2. P. Kandanarachichi, A. Guo, and Z. Petrovic, *J. Mol. Catal. A; Chem.*; **184**, 65 (2002). o) A. Guo, D. Demydov, W. Zhang, and Z. Petrovic *Polym. Matl. Sci. Eng.* **86**, 385 (2002).
3. Gunstone, Frank, *Fatty Acid and Lipid Chemistry*, Aspen Publishers Inc., Gathersberg, MD, 1999, p. .
4. a) H. Fukuda, A. Kondo, H. Noda, *J. of Biosc. and Bioeng,* **92**, 405 (2001). b) A. Srivastava and R. Prasad *Renewable and Sustainable Energy Reviews,* **4,** 111 (2000).
5. M. W. Formo, *J. Am. Oil Chem. Soc.*, **31**, 548 (1954).
6. Z. Lysenko, D. Morrison, D. Babb, D. Bunning, C. Derstine, J. Gilchrist, R. Jouett, J. Kanel, K. Olson, W. Peng, J. Phillips, B. Roesch, A. Sanders, A. Schrock, P. Thomas, PCT Int. Appl. Pub. PCT/US 2004/012246 to Dow Global Technologies Inc. (2004).
7. J. Kanel, J. Argyropoulos, A. Phillips, B. Roesch, J. Briggs, M. Lee, J. Maher, and D. Bryant, PCT Int. Appl. WO 2001068251 to Union Carbide Chemicals & Plastics Technology Corporation (2001).
8. L. Thiele and R. Becker; K. C. Frisch,, D. Klempner, eds. *Advances in Urethane Science and Technology*, 12, p. 59-85 1993
9. R. Herrington and K. Hock, *Flexible Polyurethane Foams*, 2nd Ed., The Dow Chemical Company, Midland, MI pp. 2.1-2.35. 1997

43. Carbonylation of Chloropinacolone: A Greener Path to Commercially Useful Methyl Pivaloylacetate

Joseph R. Zoeller and Theresa Barnette

Eastman Chemical Company, P.O. Box 1972, Kingsport, TN 37662

jzoeller@eastman.com

Abstract

Palladium catalyzed carbonylation of α-chloropinacolone (1) in the presence of methanol and tributyl amine provides a more efficient and environmentally sound process for the generation of methyl pivaloylacetate (2). After optimization, the preferred catalyst, [(cyclohexyl)₃P]₂PdCl₂, can be used to generate methyl pivaloylacetate under mild conditions (5-10 atm, 120-130°C) with extremely high turnover frequencies (>3400 mol MPA/mol Pd/h) and very high total turnover numbers (>10,000 mol MPA/mol Pd). The process requires very little excess amine or methanol and uses no extraneous reaction solvents. Further, extraction solvents are minimized or eliminated since the product spontaneously separates from the liquid tributylamine hydrochloride. The tributylamine can be readily recycled upon neutralization and azeotropic drying. A part of this study includes the first demonstrated use of a Pd-carbene complex as a carbonylation catalyst.

Introduction

Methyl pivaloylacetate (2) (methyl 4,4-dimethyl-3-oxo-pentanoate, MPA) is useful intermediate in the production of materials used in photographic and xerographic processes (1-6). However, current methods entail environmentally challenging methodology. For example, current best methodology produces methyl pivaloylacetate (MPA) in yields of 65-85% via a condensation of dimethyl carbonate (DMC) with a 3-5 fold excess of pinacolone in the presence of 1.5-2 fold excess of a strong base such as sodium methoxide or sodium hydride (1-3, 7-12). Unfortunately, in addition to using large excesses of reagents, these processes also utilize difficult to separate and environmentally challenging polar aprotic solvents such as a tertiary amides, sulfoxides, or hexaalkyl phosphoramides. The best alternative processes involve the addition of a dimethyl malonate or methyl aceetoacetate salt, normally as the Mg salt, to a solution of pivaloyl chloride and excess tertiary amine in ethereal or chlorinated solvents (4-6, 13-16). Unfortunately, these alternative processes (i) use more expensive reagents while demonstrating no improvements in yield, (ii) require excess reagents which must be removed and recycled, (iii) generate a co-product (dimethyl carbonate or methyl acetate) and (iv) generate large Mg and amine waste streams.

A process which offers the opportunity to reduce the waste and eliminate the use of hazardous solvents is the carbonylation of the commercially available α-chloropinacolone **(1)** (1-chloro-3,3-dimethyl-2-butanone). (See Reaction [1].) Unfortunately, earlier experience with the (Ph$_3$P)$_3$PdCl$_2$ catalyzed carbonylation of α-bromopinacolone was not encouraging. (17) Consistent with the previously reported carbonylation of α-bromoacetophenone (18), the carbonylation of α-bromopinacolone gave low yields (64%) while demonstrating limited turnover numbers (<100). However, recent reports on the carbonylation of chloroacetone and α-chloroacetophenones using (Ph$_3$P)PdCl$_2$ (19, 20) noted significant yield improvements over the bromo analogs and prompted us to re-examine the alternative carbonylation process.

[1]

1   **2**

Unfortunately, while these recent reports regarding carbonylation of chloroketones can provide helpful guidelines with respect to operating temperature, pressure, and base selection, the processes still had several serious drawbacks. Most importantly, the Pd turnover numbers were unacceptably low (<100 mol methyl aceotacetate/ mol Pd.) and the process used very large excesses of alcohol which translate into poor reactor productivities and high distillation costs. Before being considered for commercialization, a process would need to be developed that demonstrated Pd turnover numbers (TON) >5,000 mol product/mol Pd and which did so while markedly curtailing the quantities of methanol and amine so that acceptable reactor productivities and distillation costs could be achieved. As a benchmark, acceptable reactor productivities and distillation costs would likely be achieved if the product concentration was >10 wt.% methyl pivaloylacetate and was achieved in relatively short reaction times. These objectives would need to be met despite addressing a more sterically challenging α-chloropinacolone substrate.

This report will describe the realization of these goals by proper selection of the catalyst and operating conditions, which ultimately led to a process with exceptional turnover numbers (TON>10,000) and exceptional reactor productivities (product concentrations >25 wt% in 3 hrs). Further, the process ultimately requires little or no solvent while generating a benign NaCl waste stream.

Results and Discussion

Catalyst Screening

The first step in achieving the desired increase in turnover number and concentration was to optimize the catalyst choice. The earlier investigation of chloroacetone carbonylations was restricted to a description of $(Ph_3P)_2PdCl_2$ operating at 110°C and 10 atm (19, 20) under fairly dilute conditions. Based on analogy to the earlier literature, the only anticipated problem was loss of selectivity due to reductive dehalogenation of chloropinacolone to pinacolone (PA). A number of catalysts, including complexes of Ru, Rh, Co, Ir and Pd, were tested for activity in this study. However, only Pd demonstrated any significant activity. A summary of the results with Pd catalysts in the screening process is summarized in Table 1.

Table 1. Screening Pd Catalysts for the Carbonylation of α-Chloropinacolone.[a]

Entry	Catalyst	Conversion	Selectivity MPA[b]	Selectivity PA[b]	TON[b]
1	$(Ph_3P)_2PdCl_2$	59%	84%	8%	417
2	$(cy\text{-}hex_3P)_2PdCl_2$	82%	95%	2%	656
3	(2-pyridyl)PPh_2:$PdCl_2$ (2:1)	55%	81%	4%	378
4	$(o\text{-}toluyl)_3P$:$PdCl_2$ (2:1)	13%	0%	4%	0
5	$(Me_2Im)_2PdCl_2$[b]	61%	76%	11%	390
6	$(Ms_2Im)_2PdCl_2$[b,c]	71%	95%	3%	443
7	$Ph_2P(CH_2)_2PPh_2$:$PdCl_2$ (1:1)	18%	5%	10%	8
8	$Ph_2P(CH_2)_3PPh_2$:$PdCl_2$ (1:1)	23%	5%	16%	9
9	$Ph_2P(CH_2)_4PPh_2$:$PdCl_2$ (1:1)	28%	8%	17%	18
10	$Ph_2P(CH_2)_4PPh_2$:$PdCl_2$ (2:1)	47%	3%	26%	10
11	$Ph_2P(CH_2)_4PPh_2$:$PdCl_2$ (1:1)	19%	0%	22%	0
12	$(tert\text{-}Bu)_3P$:$PdCl_2$ (4:1)	18%	3%	41%	5
13	(Ph_3P):$(Ph_3P)_2PdCl_2$ (2:1) (P/Pd= 4)	57%	84%	3%	405
14	$(cy\text{-}hex_3P)$:$(cy\text{-}hex_3P)_2PdCl_2$ (2:1) (P/Pd = 4)	79%	90%	2%	591
15	$(Ph_3P)_2PdCl_2$[d]	49%	88%	7%	434
16	$(cy\text{-}hex_3P)_2PdCl_2$[d]	72%	91%	2%	650
17	$(Ph_3P)_2PdCl_2$[e]	75%	73%	9%	562
18	$(cy\text{-}hex_3P)_2PdCl_2$[e]	98%	91%	2%	884

[a] Conditions (unless otherwise noted): 11.0 mL (83.8 mmol) chloropinacolone, 30.0 mL (0.126 mol) n-Bu₃N, 110 mL MeOH, 0.1mmol Pd, 5.4 atm CO, 105°C, 3 h
[b] Abbreviations: MPA = methyl pivaloylacetate; PA = Pinacolone; TON = mol MPA produced/mol Pd used; Me₂Im = 1,3-dimethyl imidazoline-2-ylidene (dimethyl imidazolium carbene complex); Ms₂Im = 1,3-di-(2,4,6-trimethylphenyl) imidazoline-2-ylidene (dimesityl imidazolium carbene complex); cy-hex = cyclohexyl
[c] Used 0.127 mmol Pd catalyst
[d] Conditions: 13.1 mL (0.10 mol) chloropinacolone, 36 mL (0.151 mol) n-Bu₃N, 110 mL MeOH, 0.1mmol Pd, 10.0 atm CO, 105°C, 3 h
[e] Conditions: 11.0 mL (83.8 mmol) chloropinacolone, 30.0 mL (0.126 mol) n-Bu₃N (0.151 mol) , 110 mL MeOH, 0.1mmol Pd, 10.0 atm CO, 120°C, 3 h

As can be discerned from Table 1, processes using (cy-hex₃P)₂PdCl₂ (cy-hex = cyclohexyl) as the catalyst were superior to the earlier (Ph₃P)₂PdCl₂ over a range of conditions. (See entries 1,2, and 13-18.) One significant advantage in using (*cy*-hex₃P)₂PdCl₂ catalysts was that the reactions could be driven toward completion by increasing the temperature without deleterious impacts on selectivity. This was not true with (Ph₃P)₂PdCl₂ catalyzed processes where increasing temperatures led to measurable losses in selectivity. Further, the final concentration of methyl pivaloyl acetate in the best run (entry 18), was now in a useful concentration range (10.7 wt.%) and time frame (3 h) to give commercially viable reactor productivities.

Carbene complexes of Pd (entries 5 and 6) were also useful catalysts with (Ms₂Im)₂PdCl₂ demonstrating superior rates and selectivity when compared to the earlier (Ph₃P)₂PdCl₂ catalyst. The only prior example of a carbene complex being utilized in a carbonylation process entailed the carbonylation of iodobenzene in the presence of excess phosphine (21). Since the carbonylation was operated in the presence of excess phosphine, the nature of the complex was clearly in question. Therefore, these examples represent the first clear demonstration of a carbene complex being utilized in a carbonylation process. While noteworthy, the carbene complexes were still inferior to (*cy*-hex₃P)₂PdCl₂ with respect to rate.

A further drawback to the carbene complexes became apparent upon analyzing the returned solutions for Pd. Each of the successful carbonylation runs demonstrated some degree of Pd precipitation. However, reactions employing either (Ph₃P)₂PdCl₂ or (*cy*-hex₃P)₂PdCl₂ still retained ca. 70-85% of the Pd in solution. Unfortunately, in reactions using the carbene complexes >90% of the Pd precipitated and could not be readily recycled. Since there was no prospect of developing an advantageous Pd recycle with carbene complexes and the rates were slower than with (cy-hex₃P)₂PdCl₂, the carbene complexes were not examined further.

Since Pd-phosphine complexes normally demonstrate an optimum phosphine:Pd ratio, an attempt was made to determine the optimal tricyclohexyl phosphine:Pd ratio at this stage. (See Table 2.) However, under these conditions, any rate optimum is barely detectable although there appeared to be selectivity optimum. Large amounts of phosphine were deleterious to both selectivity and rate. There was little change in the levels of reduction to pinacolone as the ratio was altered.

Table 2. Optimization of Tricyclohexyl Phosphine Level in Screening Reactions

| Entry | cy-hex₃P /Pd ratio | Conversion | Selectivity | | TON |
			MPA	PA	
1	2	72%	91%	1.5%	650
2	3	74%	93%	1.4%	685
3	4	68%	93%	1.6%	629
4	5	67%	97%	1.5%	644
5	6	63%	91%	1.4%	577
6	7	64%	91%	1.6%	583
7	8	62%	91%	1.8%	560

*ᵃ*Conditions: 13.1 mL chloropinacolone, 36 mL nBu₃N, 110 mL MeOH, 0.1mmol Pd, 10.0 atm CO, 105°C, 3 h; tricyclohexyl phosphine as indicated..

Material accountability (the sum of recovered chloropinacolone, methyl pivaloyl acetate, and pinacolone) with the active Pd monophosphine and carbene complexes was in the range of 92-99% without accounting for impurities present in the starting chloropinacolone. A GC-MS examination of several product mixtures was undertaken to see if there were any additional, unanticipated by-products. The only additional material identified was α-methoxy pinacolone (1-methoxy-3,3-dimethyl-2-butanone). This compound was formed by methanolysis of the starting α-chloropinacolone and appears to be formed by a mixture of catalyzed and uncatalyzed processes Since this product was not anticipated, it was not quantified but represents the majority of, if not the only, remaining volatile product. No attempt was made to determine the presence of any quaternary ammonium salt formed by similar alkylation of the amine base by α-chloropinacolone.

Improving Catalyst Performance and Reactor Productivity.

The screening work demonstrated that with a $(cy\text{-hex}_3P)_2PdCl_2$ catalyst the targeted methyl pivaloylacetate concentrations and desired reactor residence times could be achieved. Unfortunately, the process still would not meet the targeted catalyst performance (TON >5,000 mol MPA/mol Pd) and required further development directed at improving catalyst and reactor productivity.

Carbonylation catalysts can demonstrate complex kinetics with variant rate determining steps and mechanisms. However, normally carbonylation reactions demonstrate first order behaviors in catalyst and organic halide and zero order dependence on alcohol. Kinetic behavior with respect to ligands and CO pressure are less predictable with inverse, zero, and first order behaviors as well as optima all being reported for these components. If this process follows the general trend toward first order behavior in the halide component and zero order in the methanol component, replacing a significant volume of the methanol with α-chloropinacolone should lead to an increase in the catalyst turnover frequency. Replacing methanol with reactive α-chloropinacolone would have the added benefit of increasing the concentration of methyl pivaloylacetate in the product solution.

As indicated in Table 3, reducing the excess methanol to only a 3 fold molar excess (rendering a nearly solvent free process) far exceeded expectations and allowed significant reductions in the catalyst levels. Under these conditions, catalyst turnover numbers exceeding 10,000 mol MPA/mol Pd were achieved with a turnover frequency of >3400 mol MPA/mol Pd/h. The reaction mixtures obtained from this process formed two liquid phases and the product spontaneously separated from the amine and amine hydrochloride. As a consequence of eliminating large methanol excesses, the methyl pivaloylacetate concentration in the product was raised to 26 wt. % without additional reaction time being required. This represents an additional ca. 2.5 fold improvement in reactor productivity. No attempt was made to reduce the methanol further.

Included in Table 3 (entries 5-9) is a survey of the effect of the tricyclohexyl phosphine level. Unlike the earlier screening runs, these highly concentrated, low catalyst level reactions demonstrated clearly discernible rate and selectivity effects with respect to the phosphine:Pd ratio with the optimal rate being observed at a ratio (*cy*-hex$_3$P):Pd ratio of ca. 7:1. Selectivity improvements increased consistently with rising phosphine levels. The presence of a rate optimum is consistent with original expectations for a carbonylation process requiring stabilizing ligands. In commercial practice, a slightly longer reaction time would be acceptable if it leads to improved selectivity since this approach conserves raw materials. As a consequence, the process would normally be operated at the higher phosphine levels. In the fastest operations (entries 3,4,7-9, Table 3), the material accountability remains high (92-96% without accounting for impurities in the chloropinacolone).

Table 3. Increasing Rate and Turnover Number using Reduced Methanol Concentrations to Attain High Turnover Process[a]

Entry	Pd (mmol)	Cy-hex$_3$P (mmol)[b]	P (atm)	Conversion	Selectivity		
					MPA	PA	TON
1	0.10	0.40	5.4	99%	86%	3.0%	2130
2	0.05	0.30	5.4	66%	91%	2.5%	3030
3	0.019	0.138	5.4	78%	91%	3.2%	9450
4	0.0095	0.069	5.4	47%	77%	2.6%	9620
5	0.019	0.038	8.5	73%	70%	2.0%	6760
6	0.019	0.088	8.5	84%	79%	3.0%	6410
7	0.019	0.138	8.5	89%	88%	3.7%	10280
8	0.019	0.188	8.5	77%	88%	2.8%	9020
9	0.019	0.238	8.5	74%	91%	2.4%	8970

[a]Conditions: 33.0 mL (0.251 mol) chloropinacolone, 90 mL (0.378 mol) Bu$_3$N, 31.0 mL (0.766 mol) MeOH, 120°C, 3 h; catalyst: (cy-hex$_3$P)$_2$PdCl$_2$ + cy-hex$_3$P; CO pressure as indicated.
[b]Includes tricyclohexyl phosphine contained in the (cy-hex$_3$P)$_2$PdCl$_2$ catalyst.

As indicated earlier, the effect of CO pressure is often unpredictable in carbonylations. To optimize this process, the effect of CO pressure was measured at 120°C and 130°C and the results appear in Table 4. With these highly active catalyst systems, there appeared to be an optimum CO pressure and excess CO pressures was deleterious to the reaction. While the presence of CO optima is not unknown in carbonylation chemistry, it is normally observed at significantly higher CO pressures. It is likely that the optimum observed in this study represented the transition from a mass transfer limited reaction to a chemically limited reaction. (The combination of a phosphine optima and rate reductions with increased CO likely indicate a rate determining dissociative process along the reaction pathway.)

Comparing data in Table 4 at 120°C (entries 1-4) with data at 130°C (entries 5-7) indicates that a decrease in selectivity occurred as the temperature was raised above 120°C. A closer look at the impact of temperature on selectivity is shown in Table 5 where the response to temperature was examined at (*i*) constant partial pressure of CO and (*ii*) at constant total pressure. Selectivity losses became increasingly severe with higher temperature and optimal performance was achieved at about 120-130°C. This loss of selectivity is similar to the decrease observed with increasing temperatures in the earlier screening runs using (Ph₃P)₂PdCl₂ as a catalyst. Fortunately, in the case of (cy-hex₃P)₂PdCl₂ catalysts this phenomenon is not observed until higher temperatures are attained. Access to higher temperature regimes allows (*cy*-hex₃P)₂PdCl₂ catalysts to be operated at substantially higher reaction rates without the significant selectivity losses observed when using the earlier (Ph₃P)₂PdCl₂ catalyst at higher temperatures.

Table 4. Effect of Pressure on Conversion and Selectivity [a]

		Gauge			Selectivity		
Entry	T (°C)	Pressure (atm)	P_{co}[b] (atm)	Conversion	MPA	PA	TON
1	120	5.4	1.4	78%	91%	3.2%	9450
2	120	8.5	4.4	89%	88%	3.7%	10280
3	120	17.0	12.9	68%	88%	2.6%	7960
4	120	34.0	29.9	57%	60%	2.1%	4570
5	130	8.5	2.7	92%	78%	3.3%	9470
6	130	17.0	11.2	78%	82%	3.3%	8420
7	130	34.0	28.2	84%	69%	2.4%	7600

[a] Conditions: 33.0 mL chloropinacolone, 90 mL Bu₃N, 31.0 mL MeOH; catalyst: 14.0 mg (0.019 mmol) (cy-hex₃P)₂PdCl₂ + 28.0 mg. (0.1 mmol) of cy-hex₃P, 3 h, CO pressure, and temperature as indicated.
[b] P_{co} (partial CO pressure) calculated by subtracting the partial pressure of MeOH and Bu₃N from the total pressure.

Table 5. Effect of Temperature at Constant Partial Pressure CO and Constant Gauge Pressure

		Gauge			Selectivity		
Entry	T (°C)	Pressure (atm)	P_{co}[b] (atm)	Conversion	MPA	PA	TON
1	110	7.1	4.4	38%	40%	2%	2000
2	120	8.5	4.4	89%	88%	4%	10280
3	130	10.2	4.4	89%	82%	3%	9740
4	140	12.2	4.4	97%	80%	6%	10220
5	150	13.6	4.4	97%	69%	11%	8870
6	120	17.0	12.9	68%	88%	3%	7960
7	130	17.0	11.2	78%	82%	3%	8420
8	140	17.0	9.2	97%	80%	9%	10340
9	150	17.0	6.7	99%	59%	24%	7690

*Conditions: 33.0 mL chloropinacolone, 90 mL Bu₃N, 31.0 mL MeOH; catalyst:
14.0 mg (0.019 mmol) (cy-hex₃P)₂PdCl₂ + 28.0 mg. (0.1 mmol) of cy-hex₃P, 3 h, CO
pressure, and temperature as indicated.
^b P_{co} (partial CO pressure) calculated by subtracting the partial pressure of MeOH
and Bu₃N from the total pressure.

The selectivity loss observed with either elevated temperature or elevated
pressure was likely due to competing formation of α-methoxy pinacolone via
methanolysis. Whereas the GC peak assigned to α-methoxy pinacolone was only a
trace component in reactions operated at low pressure and at 120-130°C, the peak
assigned to α-methoxy pinacolone became quite significant at conditions entailing
high temperature, high pressure, or both.

The earlier work on chloroacetone (18,19) already indicated that trialkyl
amines were superior to other bases for this reaction. Therefore the decision to use
trialkyl amines to scavenge HCl was already determined by the literature precedent.
However, when compared to the tributyl amine, smaller amines might be preferred
since they could boost the reactor productivity by reducing the volume.
Unfortunately, for unknown reasons, the process did not work as well with the
simpler amines. Both tripropyl amine and triethyl amine displayed both lower rates
and lower selectivity for methyl pivaloyl acetate. (See Table 6.)

Table 6. Effect of Using Simpler Amines ^a

Entry	Base	Pressure (atm)	Conversion	Selectivity MPA	PA	TON
1	Et₃N	9.5^b	88%	68%	2.3%	8020
2	Bu₃N	8.5^b	89%	88%	3.7%	10280
3	Pr₃N	5.4	65%	83%	2.4%	7110
4	Bu₃N	5.4	78%	91%	3.2%	9450

^a Conditions: 33.0 mL chloropinacolone; 378 mmol of amine; 31.0 mL MeOH;
catalyst: 14.0 mg (0.019 mmol) (cy-hex₃P)₂PdCl₂ + 28.0 mg. (0.1 mmol) of cy-
hex₃P; 120°C, 3 h, CO gauge pressure as indicated.
^b P_{co} (calc.) = 4.4 atm. (accounting for solvent vapor pressure.)

In a commercial process, product recovery in this newly developed process
would be simple since most of the product spontaneously separates into two liquid
layers and only a minor amount of product is retained in the amine hydrochloride
phase upon acidification of the remaining excess amine. Therefore, no additional
solvent is required although a small amount might be used to extract the minor
portion of product retained in the amine hydrochloride layer. (Any minor amount of
product remaining in the tributylamine hydrochloride layer may be recycled with the
amine and recovered in the subsequent batches.) Final product purification is readily
achieved by distillation and any unreacted α-chloropinacolone can be recycled to
subsequent batches. Neutralization of the Bu₃N·HCl layer with NaOH regenerates
tributyl amine with varying amounts of the Pd still being retained in the amine. The

tributyl amine can be dried by azeotropic distillation and reused in producing the next batch of methyl pivaloylacetate. (Bu_3N forms a minimum boiling azeotrope with water.) The resultant effluent from the reactor is a stream representing 1.5 mol of innocuous NaCl resulting from the neutralization.

In summary, compared with earlier processes which utilize hazardous solvents and generate large waste streams, the carbonylation of chloropinacolone offers an environmentally and operationally advantaged process since it requires (*i*) very little excess amine or methanol, (*ii*) no extraneous reaction solvents, (*iii*) little or no extraction solvent, and (*iv*) the tributylamine coreactant and methanol components are readily recycled. This was accomplished while also demonstrating similar yields, shorter reaction times, and higher product concentrations which result in a significant reduction in the number of batches and time required to produce significant quantities of methyl pivaloylacetate.

Experimental Section

Screening Reactions.
The following general procedure is typical for the screening reactions.
To a 300 mL Hastelloy-B autoclave was added 110 mL (87.0 g, 2.71 mol) of methanol, 30.0 mL (23.3 g, 126 mmol) of tributyl amine, 11.0 mL (11.3 g, 83.8 mmol) of 1-chloropinacolone, and 0.1 mmol of catalyst. The autoclave was sealed, flushed with carbon monoxide, and pressurized to 30 psi with CO. The autoclave was then heated to 105°C and the pressure was adjusted to 80 psi. The temperature and pressure were maintained using a continuous carbon monoxide feed for 3 h. The mixture was then cooled and analyzed by gas chromatography. Phosphine effects and additional comparison were conducted at 150 psi of CO as noted in the tables.

Catalyst Optimization.
The following procedure is typical for optimizing the catalyst utilizing low catalyst and methanol concentrations.
To a 300 mL Hastelloy-B autoclave was added 31.0 mL (24.5 g, 765 mmol) of methanol, 90.0 mL (70.0g, 378 mmol) of tributyl amine, 33.0 mL (33.8 g, 251 mmol) of 1-chloropinacolone, 14.0 mg (0.019 mmol) of $[(cyclohexyl)_3P)]_2PdCl_2$, and 28.0 mg (0.1 mmol) of tricyclohexylphosphine. The autoclave was sealed, flushed with carbon monoxide, and pressurized to 30 psi with CO. The autoclave was then heated to 120°C and the pressure was adjusted to 150 psi. The temperature and pressure were maintained using a continuous carbon monoxide feed for 3 h. The mixture was then cooled to yield a two phase reaction product. The entire product mixture (both layers) was diluted with 50.0 mL (39.55 g) of methanol to generate a homogeneous mixture and analyzed by gas chromatography. Variations in the procedure regarding temperature, pressure, catalyst levels (phosphine and Pd) are indicated in the tables and text.

Acknowledgements

Thanks to Eastman Chemical Company for permission to publish this work and Dr. Robert Maleski for helpful discussions in initiating this effort.

References

1. K. Harada, S. Yamada and M. Ogami, Jap. Pat. 09110793 (A2) to Ube Industries, Ltd (1997).
2. K. Harada, S. Yamada and M. Oogami, Jap. Pat. 07215915 to Ube Industries, Ltd. (1995).
3. G. Renner, I. Boie, and Q. Scheben, GB 1,491,606 to AGFA-Gavaert, A.G. (1977).
4. M. Eyer, U.S. Pat. 5,144,057 to Lonza, Ltd.(1992).
5. S. Yamada, Y. Oekda, T. Hagiwara, A. Tachikawa, K. Ishiguro, and S. Harada U.S. Pat. 6,570,035 B2 to Takasago International Corp. (2003).
6. R. A. Sheldon and H. J. Heijmen, U.S. Pat. No. 4,656,309 to Shell Oil Co. (1987).
7. Y. Sun and C. Yao, *Huanong Shikan*, **15**, 33 (2001).
8. N. W. Boaz and M. T. Coleman, U.S. Pat. No. 6,143,935 to Eastman Chemical Company (2001).
9. K. Harada and S. Ikezawa, Jap. Pat. 06279363 to Ube Industries, Ltd JP (1994).
10. K. Harada, S. Ikezawa and M. Oogami, Jap. Pat. 06279362 to Ube Industries, Ltd. (1994).
11. H. Iwasaki, H. Koichi, and T. Hosogai, Jap. Pat. 3371009 to Kuraray Co., Ltd. (2003).
12. G. Renner, I. Boie and Q. Scheben, US Pat. No. 4,031,130 to AGFA-Gavaert, A.G. (1977).
13. M. W. Rathke and P. J. Cowan, *J. Org. Chem.*, **50**, 2622 (1985).
14. M. Vlassa and A. Barabas, *J. fur Praktische Chemie (Liepzig)*, **322**, 821 (1980).
15. E. Sato and T. Furukawa, Jap. Pat. 10025269 A2 to Osaka Yuki Kagaku Kogyo Co., Ltd. (1998).
16. Y. Suenobe, N. Hanayama, T. Miura, and M. Kasagi, Jap. Pat. 63057416 to Yoshitomi Pharmaceutical Industries, Ltd. (1988).
17. P. N. Mercer, unpublished results.
18. J. K. Stille and P. K. Wong, *J. Org. Chem.*, **40**, 532 (1975).
19. A. L. Lapidus, O.L. Eliseev, T. N. Bondarenko, O.E. Sizan, and A. G. Ostapenko, Russian Chemical Bulletin, Int. Edit., **50**, 2239 (2001)
20. A. L. Lapidus, O.L. Eliseev, T. N. Bondarenko, O.E. Sizan, A. G. Ostapenko, and I. P. Beletskya, Synthesis, 317 (2002).
21. V. Calo, P. Giannoccaro, A. Nacci, and A. Momopoli, *J. Organomet. Chem.*, **645**, 152 (2002).

44. Recycling Homogeneous Catalysts for Sustainable Technology

Jason P. Hallett, Pamela Pollet, Charles A. Eckert and Charles L. Liotta

School of Chemistry and Biochemistry
School of Chemical & Biomolecular Engineering
Specialty Separations Center
Georgia Institute of Technology
Atlanta, GA 30332-0325

charles.liotta@carnegie.gatech.edu

Abstract

Homogeneous catalysts possess many advantages over heterogeneous catalysts, such as higher activities and selectivities. However, recovery of homogeneous catalysts is often complicated by difficulties in separating these complexes from the reaction products. The expense of these catalysts (particularly asymmetric catalysts) makes their recovery and re-use imperative.

We have developed several techniques using CO_2 as a "miscibility switch" to turn homogeneity "on" and "off". The goal is to create a medium for performing homogeneous reactions while maintaining the facile separation of a heterogeneous system. Our approach represents an interdisciplinary effort aimed at designing solvent and catalytic systems whereby a reversible stimulus induces a phase change enabling easy recover of homogeneous catalysts. The purpose is to preserve the high activity of homogeneous catalysts while taking advantage of simple separation techniques, such as filtering and extraction, normally applied to heterogeneous or biphasic catalytic systems. Specific examples include the application of gaseous CO_2 as a benign agent in gas-expanded liquids to induce organometallic catalytic recycle of water/organic, fluorous/organic biphasic systems. Additional applications involve the enhancement of solid-liquid phase transfer catalysis with supercritical solvents and improved recovery of phase transfer catalysts from biphasic liquid mixtures using gas-expanded liquids. Specific reaction systems include the hydroformylation of hydrophobic olefins using water-soluble catalysts, the hydrogenation of pro-chiral and achiral substrates using fluorous-modified catalysts captured in a fluorinated solvent or on a fluorous surface phase, the hydrolysis of hydrophobic esters using enzymatic biocatalysts in mixed aqueous/organic media, and nucleophilic substitutions using novel phase transfer catalyzed systems.

Introduction

Catalytic synthesis can be achieved by a variety of methods, including homogeneous and heterogeneous organometallic complexes, homogeneous enzymatic biocatalysts, phase transfer catalysts, and acid and base catalysts. However, each of these methods offers advantages and disadvantages that must be balanced cautiously. Homogeneously catalyzed reactions are highly efficient in terms of selectivity (i.e. regioselectivity, enantiomeric excesses) and reaction rates, due to their monomolecular nature. Unfortunately, catalyst recovery can be very difficult (due to the homogeneous nature of the solution) and product contamination by residual catalyst or metal species is a problem. In contrast, heterogeneously catalyzed reactions allow easy and efficient separation of high value products from the catalyst and metal derivatives. However, selectivity and rates are often limited by the multiphasic nature of this system and/or variations in active site distribution from the catalyst preparation.

Catalyst separation is crucial for industrial processes – to minimize the waste streams and to develop potential catalyst recycling strategies. Therefore, efforts have been made to improve the recovery of highly selective homogeneous catalysts by developing new multiphasic solvent systems. We have developed several techniques using CO_2-expanded liquids and supercritical fluids to create a medium for performing homogeneous reactions while maintaining the facile separation of heterogeneous systems.

Results and Discussion

One example of a recoverable homogeneous catalytic system involves the addition of CO_2 to fluorous biphasic systems (1,2). In fluorous biphasic systems, a fluorous solvent (perfluoroalkane, perfluoroether or perfluoroamine) is employed as an orthogonal phase, immiscible with most common organic solvents and water. An organometallic catalyst can be made preferentially soluble in a fluorous solvent by introduction of one or more fluorous side chains, or "ponytails" (3) with hydrocarbon spacers (4) to mitigate the electron-withdrawing effects of the fluorines. Usually, multiple ponytails are required to impart preferential solubility to most organometallic complexes (5). The mutual immiscibility of fluorous and organic solvents (6) provides an opportunity for facile separation of reaction components and the recycle of the expensive fluorous-derivatized homogeneous catalyst. However, mass transfer limitations in biphasic systems can limit overall reaction rate. In systems containing nonpolar solvents, such as toluene, heating the biphasic reaction mixture to around 90°C will induce miscibility (3). However for more polar or thermally labile substrates this is not a viable option as the consulate point is much higher than 100 °C (7,8). Thus, any polar reactants must be diluted into a nonpolar solvent, introducing an extra volatile organic compound into the process. Instead of heating, a homogenizing agent such as benzotrifluoride (BTF, 9) can be added to mixture. However, BTF is expensive and its recovery is not trivial.

Alternatively, we have shown that CO_2 can be used to induce miscibility of fluorocarbon-hydrocarbon mixtures (see Figure 1), even those involving polar compounds such as methanol (2). Fluorinated organometallic complexes have been well established to have significant solubility in supercritical CO_2, and their use as catalysts in this medium is well developed (10). This allows the homogeneously catalyzed reaction to be carried out in the CO_2-expanded homogeneous solution.

When the reaction is complete, depressurization induces a phase split, with the catalyst available for recycle with the fluorinated solvent, and the product ready for purification in the organic phase. We demonstrated that a wide variety of organic solvents are made miscible with fluorocarbon solvents by adding CO_2 pressure. This CO_2 "miscibility switch" was demonstrated on two model reactions: the hydrogenation of allyl alcohol and the epoxidation of cyclohexene. In each of these experiments, a fluorous-soluble catalyst was dissolved in a fluorous solvent and added to a system containing a neat organic reactant phase. In both cases, addition of enough CO_2 to merge the phases increased the average turnover frequency (TOF, mole product produced per mole catalyst per time) relative to the biphasic system by 50-70% (2).

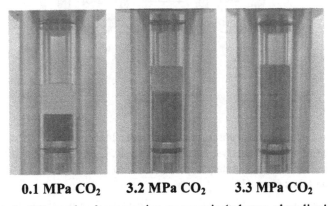

0.1 MPa CO_2   **3.2 MPa CO_2**   **3.3 MPa CO_2**

Figure 1. CO_2 used to homogenize an organic (toluene, clear liquid) and a fluorous (FC-75, colored liquid) phase. The fluorous phase is colored by a cobalt catalyst. Note the slight coloration of the organic phase (middle panel), indicating extensive mutual solubility just prior to miscibility (2).

Although these examples illustrate the effectiveness of the CO_2 switch for enhancing the catalytic activity of fluorous biphasic reactions, fluorinated solvents are undesirable because of high expense and environmental persistance. To alleviate these limitations, we tried using the CO_2 "miscibility switch" without any fluorinated solvent. We found that expansion of an organic solvent by the application of CO_2 pressure (creating a gas-expanded liquid or GXL) increases the fluorophilicity of the solvent to such an extent that the solvent is able to dissolve highly fluorinated complexes (11). This allowed us to recrystallize fluorinated catalyst complexes for purification or X-ray crystallography. The phenomenon could potentially be used as a miscibility trigger in solventless fluorous biphasic catalysis, analogous to the work of Gladysz et al. (12,13) using a temperature switch or PTFE support. Unfortunately, our efforts to crystallize a pure catalyst phase from depressurization proved unsatisfactory. Therefore, we have explored the use of fluorous silica as a solid supported for "capturing" the fluorinated catalysts. We covalently bonded fluorous "tails" of about 500 Daltons to silica beads to create a surface "phase" of highly fluorous character.

The fluorous silica concept involves the selective partitioning of a fluorous-modified catalyst between an organic liquid phase and the fluorinated surface phase. In the absence of CO_2, the fluorinated catalyst prefers the fluorous surface phase and remains partitioned onto the silica. When CO_2 pressure is added, the catalyst will partition off of the silica and into the GXL phase (containing reactants), where homogeneous reaction can take place. After the reaction is completed, the CO_2 is removed and the catalyst will partition back onto the fluorous silica surface, which can be easily recovered by filtration. A cartoon schematic is shown as Figure 2.

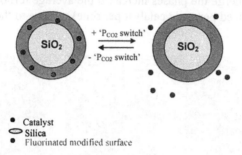

• Catalyst
⊂ Silica
• Fluorinated modified surface

Figure 2. Schematic representation of the fluorous silica concept. In the absence of CO_2 the catalyst partitions onto the fluorous silica surface. In the presence of CO_2 the catalyst partitions into the bulk liquid phase where reaction takes place (14).

We examined the extent of reversible partitioning of fluorinated compounds or complexes on and off of the fluorous silica support upon the expansion of the solvent with CO_2 (1). For a slightly fluorophilic molecule, bis(perfluoro-n-octyl) benzene, the partitioning in polar solvents, such as acetonitrile, was altered from 8:1 in favor of the silica surface to 45:1 in favor of the bulk fluid phase by adding modest CO_2 pressures (20-50 bar). For a more fluorophilic compound, a perfluoropolyether complex, the partitioning can be changed from 100:1 in favor of the silica to 99:1 in favor CO_2-expanded cyclohexane, a change of four orders of magnitude in partitioning (see Table 1).

Table 1. Reversible solubility of various fluorinated compounds in CO_2-expanded liquids at 25°C. PFPE = poly(hexafluoropropyleneoxide); Fl-Wilkinson's Catalyst = $RhCl(PR_3)_3$ with R = C_6H_4-p-$CH_2CH_2C_6F_{13}$ (1).

Compound	Solvent	Pressure \<bar\>	Partitioning w/o CO_2	Partitioning with CO_2
Co(O-PFPE)$_2$	C_6H_{12}	68.3	0.010	99.06
Bis(perfluorooctyl)benzene	CH_3OH	31.7	0.132	45.97
Bis(perfluorooctyl)benzene	CH_3CN	31.1	0.123	45.15
Fl-Wilkinson's Catalyst	C_6H_{12}	28.6	0.024	40.84
Fl-Wilkinson's Catalyst	CH_3OH	28.6	0.012	68.07

Our fluorous silica technology was also tested (1) on the catalytic hydrogenation of styrene. The fluorous silica phase contained a fluorinated version of Wilkinson's catalyst (Figure 3) deposited onto the surface of the fluorous silica. The organic phase consisted of styrene dissolved in cyclohexane. No fluorous solvent was used.

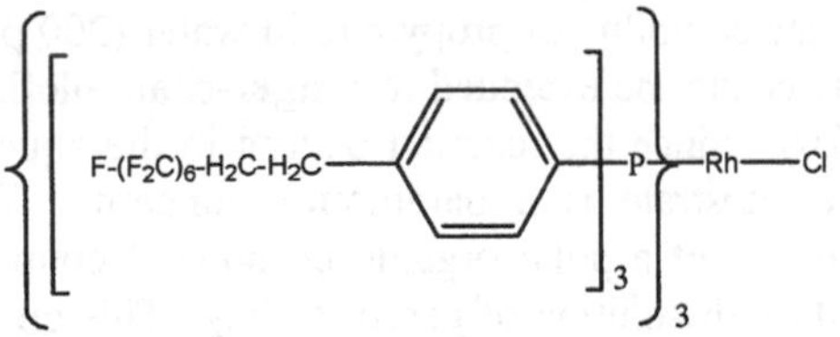

Figure 3. Fluorinated Wilkinson's catalyst

H_2 and then CO_2 pressure were applied, forming a GXL. The fluorinated catalyst then partitioned off of the fluorinated silica support and into the CO_2-expanded organic phase. The reaction was assumed to occur in the expanded liquid phase in which reactants (styrene, hydrogen) and catalyst (fluorinated Wilkinson's catalyst) are homogeneously present. After the reaction was completed, the pressure was released and the catalyst then partitioned back onto the silica surface.

The recyclability of the fluorinated catalyst was investigated. Five consecutive runs were carried out successfully with the same initial fluorinated catalyst/silica. The styrene hydrogenation activity proved to be relatively consistent (TOF from 250-400 h^{-1}) for each of the five runs, indicating minimal loss of catalytic activity.

Another way to omit the fluorous solvent would be to utilize a catalyst immobilization solvent that is not fluorinated, such as water. We demonstrated the application of a phase change after reaction permits facile recycle of hydrophilic catalysts. This method is called OATS (Organic-Aqueous Tunable Solvent) (15).

Changes in composition or temperature give relatively incomplete separations – the addition of CO_2 has a far more profound effect on system phase behavior. CO_2 is miscible with most organics but virtually immiscible with water. Addition of CO_2 can result in a phase separation of a miscible organic/water mixture (16), or drastically change distribution coefficients in a two-phase organic/water system (17). In addition, it provides for benign recycle of hydrophilic organometallic catalysts (14). For example, with the traditional aqueous biphasic catalysis technique, popularized by the Ruhrchemie/Rhône-Poulenc process (18), catalyst recovery can be achieved most easily by maintaining an aqueous catalyst-rich phase separate from the substrate-containing (nonpolar) organic phase. The catalyst is immobilized in an aqueous phase by modifying the catalyst ligands with one or more polar functionalities, such as sulfonate or carboxylate salts (19). This renders the catalyst completely insoluble in the substrate- (and later product-) containing organic phase, so that decantation of the organic phase results in no loss of catalyst. The aqueous layer can be recycled many times, yielding high catalyst turnover with little metallic contamination of the product. This "heterogenized" homogeneous system can improve the lifetime of organometallic catalysts by orders of magnitude.

The Ruhrchemie/Rhône-Poulenc process is performed annually on a 600,000 metric ton scale (18). In this process, propylene is hydroformylated to form butyraldehyde. While the solubility of propylene in water (200 ppm) is sufficient for catalysis, the technique cannot be extended to longer-chain olefins, such as 1-octene (<3 ppm solubility) (20). Since the reaction occurs in the aqueous phase (21), the hydrophobicity of the substrate is a paramount concern. We overcame these limitations via the addition of a polar organic co-solvent coupled with subsequent phase splitting induced by dissolution of gaseous CO_2. This creates the opportunity to run homogeneous reactions with extremely hydrophobic substrates in an organic/aqueous mixture with a water-soluble catalyst. After CO_2-induced phase separation, the catalyst-rich aqueous phase and the product-rich organic phase can be easily decanted and the aqueous catalyst recycled.

The OATS concept was tested on the catalytic hydroformylation of 1-octene, a hydrophobic substrate. This reaction was selected because it has previously been shown to be inactive for traditional aqueous biphasic systems (18). The catalyst used was a Rh/TPPTS complex, an industrial water soluble catalyst (22). The application of the OATS concept increased catalytic efficiency by a factor of 65 (TOF increased from 5 h^{-1} for biphasic to 325 h^{-1} for monophasic).

In this example, we started with a single homogeneous phase of tetrahydrofuran and water, which has the advantage of dissolving both hydrophobic and hydrophilic species. After the reaction occurs in this single homogeneous phase, addition of just one percent CO_2, which requires less than 10 bars pressure, causes the split of into one phase rich in water and another rich in tetrahydrofuran. Again the aqueous phase, with catalyst, is readily decanted and subsequent depressurization returns the organic phase with the product free of catalyst. Since CO_2 and water form two distinct phases, we would expect the partitioning of solute to depend on the nature of the solute, with polar and hydrogen-bonding molecules and salts to favor the water phase, and non-polar molecules to favor the carbon dioxide phase. What we were not expecting was the magnitude of partitioning that we found. To show the enormous change in distribution coefficient possible with this method, we show Figure 4, where a THF/water mixture at 1 bar contains a colored hydrophilic catalyst surrogate. At only 30 bars, the separation observed is better than a factor of 10^6!

There have been other modified catalyst systems used in conjunction with CO_2 for recycling purposes. Prominent examples include the use of scCO$_2$ with ionic liquids (23) in biphasic systems and scCO$_2$ with poly(ethylene glycol) in biphasic (24) and reversible monophasic (25) systems.

Figure 4. Left side: Miscible Solution of THF/water at 1 bar. Right side: phase separation at 30 bar CO_2. The partition coefficient of the water-soluble dye is greater than 10^6! (14)

Supercritical fluids are benign alternatives to conventional organic solvents that may offer improvements in reaction rate, product selectivity, and product separation. We reported the first use of SCFs for phase-transfer catalysis (PTC), where these benign alternatives also offer greatly improved transport, product separation, catalyst recycle, and facile solvent removal (26-29).

PTC is a powerful and widely used technique for conducting reactions between two or more reactants in two or more immiscible phases (30,31). Polar aprotic solvents have traditionally been used to dissolve the reactants into a single phase, however these solvents are frequently expensive, environmentally undesirable, and greatly complicate product purification. PTC eliminates the need for these solvents by employing a phase-transfer agent to transfer one of the reacting species from one phase into a second phase where reaction can occur. Generally, PTC involves the transfer of an ionic reactant into an organic phase across a phase boundary. Once the reaction is complete, the catalyst transports the ionic product back to the aqueous or solid phase to start a new catalytic cycle. The most PTCs are quaternary ammonium salts (31), which are inexpensive and separated from the reaction products by water extraction. In both liquid-liquid and solid-liquid PTC, a three-phase PTC system can occur if the phase-transfer catalyst has limited solubility in both phases.

The utility of SCFs for PTC was demonstrated for several model organic reactions – the nucleophilic displacement of benzyl chloride with bromide ion (26) and cyanide ion (27), which were chosen as model reversible and irreversible S_N2 reactions. The next two reactions reported were the alkylation and cycloalkylation of phenylacetonitrile (28,29). Catalyst solubility in the SCF was very limited, yet the rate of reaction increased linearly with the amount of catalyst present. Figure 5 shows data for the cyanide displacement of benzyl bromide, and the data followed pseudo-first order, irreversible kinetics. The catalyst amounts ranged from 0.06 (solubility limit) to 10% of the limiting reactant, benzyl chloride.

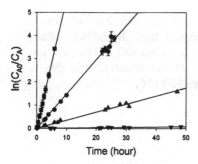

Figure 5. Pseudo-first-order kinetics at 75°C and 138 bar with 5 mol % acetone: (▼) 8.8×10^{-6} mol catalyst (solubility limit); (▲) 1.6×10^{-4} mol catalyst; (●) 4.0×10^{-4} mol catalyst; (■) 1.6×10^{-3} mol catalyst (27).

These results suggested that the reaction mechanism involved the formation of a catalyst-rich third phase in which the reaction actually occurred. The data seem to be consistent with the concept of the reaction occurring in an interfacial phase called "omega phase," which has previously been found to be important in other PTC reactions (32). Figure 6 shows this phase at the interface between the solid PTC and the SCF phase, where the SCF transports substrate in and product out. The reaction occurs in the catalyst-rich omega phase, which is sufficiently organic to solubilize the reactant, but also sufficiently polar for ions. Since the reaction is generally limited by contacting of the reactants and catalyst, the SCF offers diffusivity about two orders of magnitude better than that of a liquid. While this reaction could be run homogeneously in a polar organic such as DMSO, solvent removal is a tremendous problem; with CO_2, this issue is trivial.

A final application of CO_2 for environmentally benign and efficient recovery of homogeneous catalysts is in the recovery of phase transfer catalysts with aqueous extraction. A PTC is by definition a compound that partitions between an aqueous phase and an organic phase. In current practice, the recovery is achieved by repeated washings of the organic phase, followed by an expensive reconcentration process.

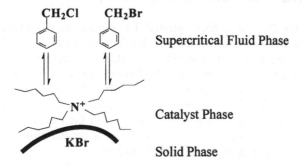

Figure 6. Suggested PTC reaction mechanism in an omega phase (16).

Using CO_2 can alter the distribution of phase transfer catalysts so dramatically that even in dilute solutions they can be separated selectively from an organic reaction mixture with only a small fraction of the water required in a traditional aqueous extraction. Addition of CO_2 can change the distribution coefficient of the PTC by two orders of magnitude (17). The aqueous phase with the recovered catalyst can then be decanted and the catalyst recycled. Depressurization returns the organic phase with the product free of catalyst. This method is useful for reducing the amount of washwater used in liquid-liquid extraction processes typically used to recover PTCs from industrial processes by 95%. The efficiency of this method toward dilute solutions creates a distinct advantage over the gas anti-solvent (GAS) crystallization method, which requires larger amounts of CO_2 to induce supersaturation of the PTC-containing organic solution.

An example of CO_2 enhanced aqueous extraction is shown in Figure 7 for tetrabutylammonium picrate (TBAP). Three common solvents for PTC: butyl acetate, methyl isobutyl ketone (MIBK) and methylene chloride were used as the organic phase.

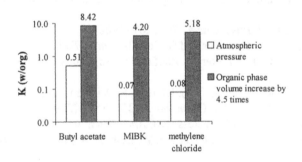

Figure 7. TBAP distribution coefficient as a function of CO_2 pressure at room temperature (23-25 °C). Organic solvents used: butyl acetate, methyl isobutyl ketone and methylene chloride. 20 mL 8.770 x 10^{-5} M TBAP aqueous solution, 20 mL organic solvent (17).

References

1. C. D. Ablan, J. P. Hallett, K. N. West, R. S. Jones, C. A. Eckert, C. L. Liotta and P. G. Jessop, *Chem. Commun.*, 2972 (2003).
2. K. N. West, J. P. Hallett, R. S. Jones, D. Bush, C. L. Liotta and C. A. Eckert, *Ind. Eng. Chem. Res.*, **43**, 4827 (2004).
3. I. T. Horváth and J. Rábai, *Science*, **266**, 72 (1994).
4. L. J. Alvey, R. Meier, T. Soos, P. Bernatis and J. A. Gladysz, *Eur. J. Inorg. Chem.*, 1975 (2000).
5. P. Bhattacharyya, D. Gudmunsen, E. G. Hope, R. D. W. Kemmitt, D. R. Paige and A. M. Stuart, *J. Chem. Soc., Perkin Trans. 1*, 3609 (1997).
6. R. L. Scott, *J. Phys. Chem.*, **62**, 136 (1958).
7. J. H. Hildebrand and D. R. F. Cochran, *J. Am. Chem. Soc.*, **71**, 22 (1949).
8. J. H. Hildebrand, *J. Am. Chem. Soc.*, **72**, 4348 (1950).
9. A. Ogawa and D. P. Curran, *J. Org. Chem.*, **62**, 450 (1997).
10. P. G. Jessop, T. Ikariya, and R. Noyori, *Chem. Rev.*, **99**, 479 (1999).

11. P. G. Jessop, M. M. Olmstead, C. D. Ablan, M. Grabenauer, D. Sheppard, C. A. Eckert and C. L. Liotta, *Inorg. Chem.*, **41**, 3463 (2000).
12. M. Wende, R. Meier and J. A. Gladysz, *J. Am. Chem. Soc.*, **123**, 11490 (2001).
13. M. Wende and J. A. Gladysz, *J. Am. Chem. Soc.*, **125**, 5861 (2003).
14. C. A. Eckert, C. L. Liotta, D. Bush, J. S. Brown and J. P. Hallett, *J. Phys. Chem. B*, **108**, 18108 (2004).
15. J. Lu, M. J. Lazzaroni, J. P. Hallett, A. S. Bommarius, C. L. Liotta and C. A. Eckert, *Ind. Eng. Chem. Res.*, **43**, 1586 (2004).
16. M. J. Lazzaroni, D. Bush, R. S. Jones, J. P. Hallett, C. L. Liotta and C. A. Eckert, *Fluid Phase Equilib.*, **224**, 143 (2004).
17. X. Xie, J. S. Brown, J. P. Jayachandran, C. L. Liotta and C. A. Eckert, *Chem. Commun.*, 1156 (2002).
18. C. W. Kohlpaintner, R. W. Fischer and B. Cornils, *Appl. Cat. A*, **221**, 219 (2001).
19. W. A. Herrmann and C. W. Kohlpaintner, *Angew. Chem. Int. Ed.*, **32**, 1524 (1993).
20. C. McAuliffe, *J. Phys. Chem.*, **70**, 1267 (1966).
21. O. Watchsen, K. Himmler and B. Cornils, *Catal. Today*, **42**, 373 (1998).
22. M. Beller, J. G. E. Krauter, A. Zapf and S. Bogdanovic, *Catal. Today*, 279 (1999).
23. P. B. Webb, M. F. Sellin, T. E. Kunene, S. Williamson, A. M. Z. Slawin and D. J. Cole-Hamilton, *J. Am. Chem. Soc.*, **125**, 15577 (2003).
24. D. J. Heldebrant and P. G. Jessop, *J. Am. Chem. Soc.*, **125**, 5600 (2003).
25. M. Solinas, J. Jiang, O. Stelzer and W. Leitner, *Angew. Chem. Int. Ed.*, **44**, 2291 (2005).
26. A. K. Dillow, S. L. J. Yun, D. Suleiman, D. L. Boatright, C. L. Liotta and C. A. Eckert, *Ind. Eng. Chem. Res.*, **35**, 1801 (1996).
27. K. Chandler, C. W. Culp, D. R. Lamb, C. L. Liotta and C. A. Eckert, *Ind. Eng. Chem. Res.*, **37**, 3252 (1998).
28. C. Wheeler, D. R. Lamb, J. P. Jayachandran, J. P. Hallett, C. L. Liotta and C. A. Eckert, *Ind. Eng. Chem. Res.*, **41**, 1763 (2002).
29. J. P. Jayachandran, C. Wheeler, B. C. Eason, C. L. Liotta and C. A. Eckert, *J. Supercritical Fluids*, **27**, 179 (2003).
30. C. M. Starks and C. L. Liotta, *Phase Transfer Catalysis: Principles and Techniques*, Academic Press, New York, (1978).
31. C. M. Starks, C. L. Liotta and M. Halpern, *Phase-Transfer Catalysis : Fundamentals, Applications, and Industrial Perspectives*, Chapman & Hall, New York, (1994).
32. N. C. Pradhan and M. M. Sharma, *Ind. Eng. Chem. Res.*, **29**, 1103 (1990).

45. "Green" Catalysts for Enhanced Biodiesel Technology

Anton A. Kiss, Gadi Rothenberg and Alexandre C. Dimian

*University of Amsterdam, van 't Hoff Institute for Molecular Sciences,
Nieuwe Achtergracht 166, 1018 WV Amsterdam, The Netherlands*

alexd@science.uva.nl

Abstract

The benefits of using biodiesel as renewable fuel and the difficulties associated with its manufacturing are outlined. The synthesis via fatty acid esterification using solid acid catalysts is investigated. The major challenge is finding a suitable catalyst that is active, selective, water-tolerant and stable under the process conditions. The most promising candidates are sulfated metal oxides that can be used to develop a sustainable esterification process based on continuous catalytic reactive distillation.

Introduction

Developing sustainable energy sources is a key scientific challenge of the 21st century. Automotive applications consume a large fraction of the available global energy, making the implementation of sustainable fuels a critical issue worldwide. Biodiesel is an environmentally-friendly renewable fuel (Figure 1) that can be manufactured from vegetable oils, animal fats or even recycled greases from the food industry (1). It is also a viable alternative to petroleum diesel, especially following the tighter legislation on vehicle emissions. Remarkably, biodiesel is the only fuel with an overall positive life cycle energy balance. It is a *green fuel*, with many advantages over conventional diesel (1-4): it is safe, non-toxic, biodegradable and a good lubricant. Moreover, it has a higher cetane number and an eco-friendly life cycle. It also emits ~70% less gas pollutants and ~50% less soot particles. Moreover, it is easily blended with petroleum diesel, giving a mixture that can power regular vehicles without any changes to the engine or the fuel distribution infrastructure.

Biodiesel is a mixture of fatty acid alkyl esters, derived typically from short chain alcohols. Methanol is more suitable for biodiesel manufacturing, but other alcohols can in principle also be used, depending on the feedstock available. Fatty acid methyl esters (FAME) are currently manufactured mainly by trans-esterification using a homogeneous base catalyst (NaOH). This catalyst is corrosive to the equipment and must be neutralized afterwards. Moreover, due to the presence of free fatty acids (FFA) it reacts to form soap as unwanted by-product, requiring expensive separation. In this process the biodiesel composition depends heavily on the types of fatty acid groups building the triglycerides from the raw material.

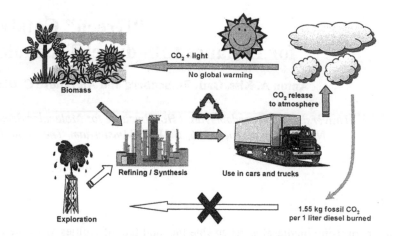

Figure 1. The life cycle of biodiesel compared to that of petroleum diesel.

Another method to produce fatty esters is the by batch esterification of fatty acids, catalysed by H_2SO_4. The problem is that the batch operation mode again involves costly neutralization and separation of the homogeneous catalyst (5). Hence, the current biodiesel manufacturing is an energy intensive process that consumes large amounts of energy, primarily from fossil sources. Production costs and pollution produced by the present process may outweigh advantages of using biodiesel. Thus, biodiesel remains an attractive but still costly alternative fuel.

During the last decade many industrial processes shifted towards using solid acid catalysts (6). In contrast to liquid acids that possess well defined acid properties, solid acids contain a variety of acid sites (7). Solid acids are easily separated from the biodiesel product; they need less equipment maintenance and form no polluting by-products. Therefore, to solve the problems associated with liquid catalysts, we propose their replacement with solid acids and develop a sustainable esterification process based on catalytic reactive distillation (8). The alternative of using solid acid catalysts in a reactive distillation process reduces the energy consumption and manufacturing pollution (*i.e.* less separation steps, no waste/salt streams).

Here we present the catalyst screening, highlight the pros and cons of several solid catalyst types, and discuss the possible applications of this novel process.

Results and Discussion

In large scale processes, a good esterification catalyst must fulfil several conditions that may not seem so important in the laboratory. The catalyst must be very active and selective (by-products formed in secondary reactions are likely to render the process uneconomical), water-tolerant and stable at relatively high temperatures. In addition, it should be an inexpensive material that is readily available on an industrial scale. Considering these conditions, we searched for a strong Brønsted acid with increased hydrophobicity, and high thermal stability (up to 200–250 °C). Hydrophobic surfaces are preferable to avoid the covering with water of the solid acid surface and prevent the adsorption of organic materials.

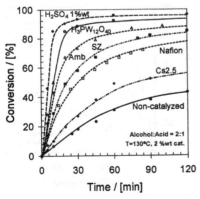

$$C_{10}H_{23} \overset{O}{\underset{OH}{\bigvee}} + HO-\overset{}{\underset{}{\bigvee}}-n\text{-}Bu \overset{160\,°C}{\underset{120\,min}{\rightleftharpoons}} C_{10}H_{23}\overset{O}{\underset{O}{\bigvee}}\overset{}{\underset{}{\bigvee}}n\text{-}Bu + H_2O \quad (1)$$

The following experimental results are presented on the use of solid catalysts in esterification of dodecanoic acid with 2-ethylhexanol and methanol. In the next figures, conversion is defined as: $X\,[\%] = 100 \cdot (1 - [Acid]_{final} / [Acid]_{initial})$, and the amount of catalyst used is normalized: $W_{cat}\,[\%] = M_{cat} / (M_{acid} + M_{alcohol})$.

In our experiments we screened zeolites, ion-exchange resins, heteropoly compounds and mixed metal oxides. Several alcohols were used to show the range of applicability. The selectivity was assessed by testing the formation of side products in a suspension of catalyst in alcohol (e.g. SZ in 2-ethylhexanol) under reflux for 24 hours. Under the reaction conditions, no by-products were detected by GC analysis.

Three types of zeolites were investigated: H-ZSM-5, Y and Beta. Zeolites showed only an insignificant increase in acid conversion compared to the non-catalysed reaction. This agrees with previous findings suggesting that the reaction is limited by the diffusion of the bulky reactant into the zeolite pores (9). The Si/Al ratio can be used to control the acid strength and hydrophobicity of zeolites. While acid strength increases at lower Si/Al ratio, hydrophobicity increases at higher Si/Al ratios. Thus, a balance is necessary for optimal performance (9). In our experiments, we observed only minor activity differences even for large variations of Si/Al ratio.

We tested then two ion-exchange organic resins: Amberlyst-15 and Nafion-NR50 (10). In spite of the high activity (Figure 2, left), Amberlyst-15 is not stable at temperatures higher than 150 °C, making it unsuitable for industrial RD applications. As for Nafion, its activity was lower than our prepared sulfated metal oxides.

The tungstophosphoric acid shows high activity, close to H_2SO_4 used as a benchmark. Regrettably, this acid is soluble in water and hence not usable as a solid catalyst. However, the corresponding cesium salt (Cs2.5) is also super acidic and its mesoporous structure has no limitations on the diffusion of the reactants. Cs2.5 exhibits low activity per weight, hence it is not suitable for industrial applications.

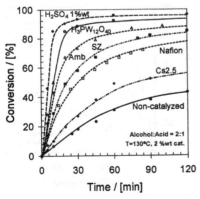
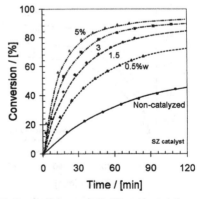

Figure 2. Esterification of dodecanoic acid with 2-ethylhexanol (left); non-catalyzed and catalyzed (0.5-5 wt% SZ catalyst) reaction profiles (right).

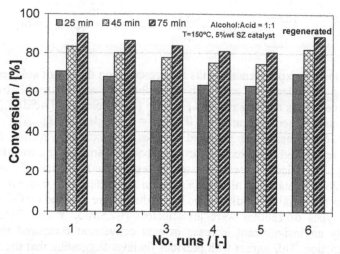

Figure 3. Stability of the sulfated zirconia catalyst in the esterification reaction of dodecanoic acid with 2-ethylhexanol – conversions given at 25, 45 and 75 minutes.

Sulfated titania and tin oxide performed slightly better than the sulfated zirconia (SZ) catalyst. However, SZ is cheaper and readily available on an industrial scale. It is already applied in several processes (7,8). Zirconia can be modified with sulfate ions to form a superacidic catalyst, depending on the treatment conditions (11-16). The calcination temperature strongly affects its activity, the optimal interval being 600–700 °C. However, the concentration of H_2SO_4 used did not affect the catalytic activity. In our experiments, SZ showed high activity and selectivity for the esterification of fatty acids with a variety of alcohols, from 2-ethylhexanol to methanol. Increasing the amount of catalyst leads to higher conversions (Figure 2, right). This makes SZ suitable for applications where high activity is required over short time spans. SZ is also regenerable and thermally stable.

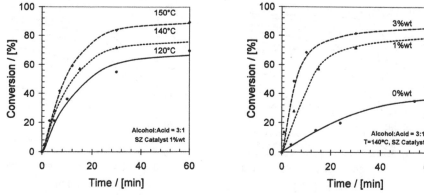

Figure 4. Esterification of dodecanoic acid with methanol, using an alcohol:acid ratio of 3:1 and sulfated zirconia (SZ) as catalyst.

Table 1. Advantages and disadvantages of tested catalysts.

Catalyst	Advatanges	Disadvantages
H_2SO_4	Highest activity	Liquid catalyst
Ion-exchange resins	Very high activity	Low thermal stability
$H_3PW_{12}O_{40}$	Very high activity	Soluble in water
$Cs_{2.5}H_{0.5}PW_{12}O_{40}$	Super acid	Low activity per weight
Zeolites	Controlable acidity and	Small pore size
(H-ZSM-5, Y and Beta)	hydrophobicity	Low activity
Sulfated metal oxides	High activity	Deactivates in water
(zirconia, titania, tin oxide)	Thermally stable	(but not in organic phase)

In a separate set of experiments, we tested the catalyst reusability and robustness. In five consecutive runs, the activity dropped to ~90% of the original value, and remained constant thereafter. Re-calcination of the used catalyst restored the original activity (Figure 3). Considering the promising results with 2-ethylhexanol, we tested the applicability of SZ also for esterification with methanol (Figure 4). High conversions can be reached even at 140 °C providing that an increased amount of catalyst is used. Note that the esterification with methanol takes place at higher rates compared to the one with 2-ethylhexanol. This can be explained by the alcohols' relative sizes.

For a convenient comparison, Table 1 summarizes the pros and cons for each catalyst tested. Clearly, the sulfated metal oxides are the best choice.

Although the reaction mechanism for the heterogeneous acid-catalysed esterification was reported to be similar to the homogeneously catalysed one (17), there is a major difference concerning the relationship between the surface hydrophobicity and the catalyst's activity (Figure 5, left). *Reaction pockets* are created inside a hydrophobic environment, where the fatty acid molecules can be absorbed and react further. Water molecules are unlikely to be absorbed on sites enclosed in hydrophobic areas. SZ has large pores, thus not limiting the diffusion of the fatty acid molecules. It does not leach under the reaction conditions nor give rise to side reactions. This makes SZ a good candidate for catalytic biodiesel production.

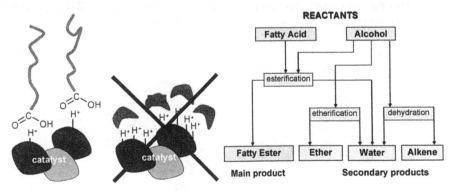

Figure 5. Surface influence on activity (left). Reaction pathways (right).

Reactive Distillation Design. In this part we present the design for fatty acids esterification using SZ as 'green solid acid catalyst'. Several secondary reactions are possible (Figure 5, right). These can be avoided by using a selective solid catalyst such as SZ. The following results were accomplished by rigorous simulations (using AspenONE Aspen Plus) that integrate the experimental findings. The problem is highly complex, as it involves chemical and phase equilibria, vapour-liquid equilibria and vapour-liquid-liquid equilibria, catalyst activity and kinetics, mass transfer in gas-liquid and liquid-solid, adsorption on the catalyst and desorption of products (8, 18-20). The catalyst development was integrated in the process design at an early stage, by data mining and embedding of reaction kinetics in the process simulation. The analysis of physico-chemical properties shows very high boiling points for dodecanoic acid and esters (Table 2).

Table 2. Normal boiling points of chemical species involved in the process.

Chemical name	Chemical formula	M_w (g/mol)	T_b (K)	T_b (°C)
Dodecanoic (lauric) acid	$C_{12}H_{24}O_2$	200	571	298
Methanol	CH_4O	32	338	65
2 Ehyl hexanol	$C_8H_{18}O$	130	459	186
Methyl dodecanoate	$C_{13}H_{26}O_2$	214	540	267
2 ethylhexyl dodecanoate	$C_{20}H_{40}O_2$	312	607	334
Water	H_2O	18	373	100

Hence, the ester will be separated in the bottom of the RD column and water by-product is removed as top product. By removing water by-product the equilibrium is shifted towards ester formation. The fatty acid feed must be fed in the top of reactive zone to reduce the contamination of the final product. Figure 6 (left) presents the flowsheet. High purity final products are feasible. An additional evaporator is used for further ester purification. Alternatively, both reactions (trans-esterification & esterification) can be combined with separation, shifting the equilibrium towards products by continuous removal of water and glycerol (Figure 6, right). In the first case, the RD column has 10 reactive stages and uses an acid reflux. For the optimal reflux ratio the maximum reaction rate is located in the centre of RDC, providing complete conversion of reactants at the ends of the column.

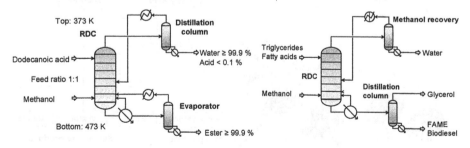

Figure 6. Flowsheet of FAME production by esterification (left) or combined esterification + trans-esterification process (right).

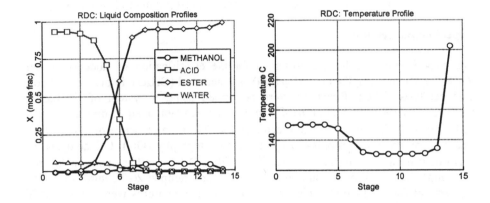

Figure 7. Liquid composition profiles (left) and temperature profile (right) in RDC.

The composition and temperature profiles in the RDC are shown in Figure 7. The ester product with traces of methanol is the bottom product, whereas a mixture of water and fatty acid is the top product. This mixture is then separated in the additional distillation column and the acid is refluxed back to the RDC. The ester is further purified in a small evaporator and methanol is recycled back to the RDC.

In conclusion, the hydrophobicity of the catalyst surface and the density of the acid sites are of paramount importance in determining the activity and selectivity. The systematic study of reaction rates under controlled process conditions (temperature, pressure, reactants ratio) is indeed a suitable method for screening catalyst candidates for fatty acids esterification (21). Catalysts with small pores, such as zeolites, are not suitable for making biodiesel because of diffusion limitations of the large fatty acid molecules. Ion-exchange resins are active strong acids, but have low thermal stability, which is problematic as relatively high temperatures are needed for increased reaction rates. Heteropolyacids are super-acidic compounds but due to the high molecular weight, their activity per weight of catalyst is not sufficient for industrial applications. However, of the mixed metal oxides family, sulfated zirconia was found as a good candidate. It is active, selective, and stable under the process conditions hence suitable for industrial reactive distillation applications.

Biodiesel can be produced by a sustainable continuous process based on catalytic reactive distillation. The integrated design ensures the removal of water by-product that shifts the chemical equilibrium to completion and preserves the catalyst activity. The novel alternative proposed here replaces the liquid catalysts with solid acids, thus dramatically improving the economics of current biodiesel synthesis and reducing the number of downstream steps. The key benefits of this approach are:

1. High unit productivity, up to ~6-10 times higher than of the current process
2. Lower excess alcohol requirements, with stoichiometric ratio at reactor inlet
3. Reduced capital and operating costs, due to less units and lower energy use

4. Sulfur-free fuel, since SAC do not leach into the product
5. No waste streams because no salts are produced (no neutralization step)

Experimental Section

The experimental results are presented for the esterification of dodecanoic acid ($C_{12}H_{24}O_2$) with 2-ethylhexanol ($C_8H_{18}O$) and methanol (CH_4O), in presence of solid acid catalysts (SAC). Reactions were performed using a system of six parallel reactors (Omni-Reacto Station 6100). In a typical reaction 1 eq of dodecanoic acid and 1 eq of 2-ethylhexanol were reacted at 160 °C in the presence of 1 wt% SAC. Reaction progress was monitored by gas chromatography (GC). GC analysis was performed using an InterScience GC-8000 with a DB-1 capillary column (30 m × 0.21 mm). GC conditions: isotherm at 40 °C (2 min), ramp at 20 °C min^{-1} to 200 °C, isotherm at 200 °C (4 min). Injector and detector temperatures were set at 240 °C. Reaction profiles were measured for both non-catalysed and catalysed reactions.

Chemicals and catalysts.
Double distilled water was used in all experiments. Unless otherwise noted, chemicals were purchased from commercial companies and were used as received. Dodecanoic acid 98 wt% (GC), methanol, propanol and 2-ethylhexanol 99+ wt% were supplied by Aldrich, zirconil chloride octahydrate 98+ wt% by Acros Organics, 25 wt% NH_3 solution and H_2SO_4 97% from Merck. Zeolites beta, Y and H-ZSM-5 were provided by Zeolyst, and ion-exchange resins by Alfa.

Preparation of mixed metal oxides.
The sulfated metal oxides (zirconia, titania and tin oxide) were synthesized using a two steps method. The first step is the hydroxylation of metal complexes. The second step is the sulfonation with H_2SO_4 followed by calcination in air at various temperatures, for 4 h, in a West 2050 oven, at the temperature rate of 240 °C h^{-1}. **Sulfated zirconia:** $ZrOCl_2.8H_2O$ (50 g) was dissolved in water (500 ml), followed by precipitation of $Zr(OH)_4$ at pH = 9 using a 25 wt% NH_3 soln. The precipitate was washed with water (3×500 ml) to remove the chloride salts (Cl⁻ ions were determined with 0.5 N $AgNO_3$). $Zr(OH)_4$ was dried (16 h at 140 °C, impregnated with 1N H_2SO_4 (15 ml H_2SO_4 per 1 g $Zr(OH)_4$), calcined at 650 °C. **Sulfated titania:** HNO_3 (35 ml) was added to an aqueous solution of $Ti[OCH(CH_3)_2]_4$ (Acros, >98%, 42 ml in 500 ml H_2O). Then 25% aqueous ammonia was added until the pH was raised to 8. The precipitate was filtered, washed and dried (16 h at 140°C) .The product was impregnated with 1N H_2SO_4 (15 ml H_2SO_4 per 1 g $Ti(OH)_4$). The precipitate was filtered, washed, dried and then calcined. **Sulfated tin oxide:** $Sn(OH)_4$ was prepared by adding a 25% aqueous NH_3 solution to an aq. sol. of $SnCl_4$ (Aldrich, >99%, 50 g in 500 ml) until pH 9-10. The precipitate was filtered, washed, suspended in a 100 ml aq. sol. of 4% CH_3COONH_4, filtered and washed again, then dried for 16 h at 140°C. Next, 1N H_2SO_4 (15 ml H_2SO_4 per 1 g $Sn(OH)_4$) was added and the precipitate was filtered, washed, dried and calcined.

Preparation of Cs2.5 catalyst [Cs$_{2.5}$H$_{0.5}$PW$_{12}$O$_{40}$].
Cs_2CO_3 (1.54 g, 10 ml, 0.47 M) aqueous solution were added dropwise to $H_3PW_{12}O_{40}$ (5 ml, 10.8 g, 0.75 M aq. sol.). Reaction was performed at room temperature and normal pressure while stirring. The white precipitate was filtered and aged in water for 60 hours. After aging, the water was evaporated in an oven at 120 °C. White solid glass-like particles of $Cs_{2.5}H_{0.5}PW_{12}O_{40}$ (9.0375 g, 2.82 mmol) were obtained.

Table 3. Catalyst characterization.

Catalyst sample	Surface area	Pore volume	Pore diameter max./mean/calc.	Sulfur content
$Cs_{2.5}H_{0.5}PW_{12}O_{40}$	163 m^2/g	0.135 cm^3/g	2 / 5.5 / 3 nm	N/A
ZrO_2/SO_4^{2-} / 650°C	118 m^2/g	0.098 cm^3/g	4.8 / 7.8 / 7.5 nm	2.3 %
TiO_2/SO_4^{2-} / 550°C	129 m^2/g	0.134 cm^3/g	4.1 / 4.3 / 4.2 nm	2.1 %
SnO_2/SO_4^{2-} / 650°C	100 m^2/g	0.102 cm^3/g	3.8 / 4.1 / 4.1 nm	2.6 %

Catalyst characterization.
Characterization of mixed metal oxides was performed by atomic emission spectroscopy with inductively coupled plasma atomisation (ICP-AES) on a CE Instruments Sorptomatic 1990. NH$_3$-TPD was used for the characterization of acid site distribution. SZ (0.3 g) was heated up to 600 °C using He (30 ml min^{-1}) to remove adsorbed components. Then, the sample was cooled at room temperature and saturated for 2 h with 100 ml min^{-1} of 8200 ppm NH$_3$ in He as carrier gas. Subsequently, the system was flushed with He at a flowrate of 30 ml min^{-1} for 2 h. The temperature was ramped up to 600 °C at a rate of 10 °C min^{-1}. A TCD was used to measure the NH$_3$ desorption profile. Textural properties were established from the N$_2$ adsorption isotherm. Surface area was calculated using the BET equation and the pore size was calculated using the BJH method. The results given in Table 3 are in good agreement with various literature data. Indeed, stronger acid sites lead to higher catalytic activity for esterification.

Catalyst leaching. The mixture may segregate leading to possible leaching of sulfate groups. The leaching of catalyst was studied in organic and in aqueous phase. First, a sample of fresh SZ catalyst (0.33 g) was stirred with water (50 ml) while measuring the pH development in time. After 24 h, the acidity was measured by titration with KOH. The suspension was then filtered and treated with a BaCl$_2$ solution to test for SO$_4^{2-}$ ions. In a second experiment, the catalyst was added to an equimolar mixture of reactants. After 3 h at 140 °C, the catalyst was recovered from the mixture, dried at 120 °C and finally stirred in 50 ml water. The pH was measured and the suspension titrated with a solution of KOH. SO$_4^{2-}$ ions in the suspension were determined qualitatively with BaCl$_2$. In a third experiment, the procedure was repeated at 100 °C when the mixture segregates and a separate aqueous phase is formed. From the leaching tests it can be concluded that SZ is not deactivated by leaching of sulfate groups when little water is present in the organic phase but it is easily deactivated in water or aqueous phase. There are several methods to prevent aqueous phase formation and leaching of acid sites: 1) use an excess of one reactant,

2) work at low conversions, and 3) increasing the temperature exceeding the boiling point of water to preserve the catalyst activity and drive reaction to completion.

Selectivity and side reactions. Typically, the alcohol-to-acid ratio inside an RD unit may vary over several orders of magnitude. Especially for stages with an excess of alcohol, the use of a SAC may lead to side reactions. Selectivity was assessed by testing the formation of side products in a suspension of SZ in pure alcohol under reflux for 24 h. No ethers or dehydration products were detected by GC analysis.

Acknowledgements

We thank M.C. Mittelmejer-Hazeleger and J. Beckers for technical support, and the Dutch Technology Foundation STW (NWO/CW Project Nr. 700.54.653) and companies Cognis, Oleon, Sulzer, Uniquema, Engelhard for financial support.

Acknowledgment is made to the Donors of The American Chemical Society Petroleum Research Fund, for the partial support of this activity.

References

1. F. Maa and M. A. Hanna, *Bioresource Technol.*, **70**, 1 (1999).
2. B. Buczek and L. Czepirski, *Inform*, **15**, 186 (2004).
3. A. Demirbas, *Energy Exploration & Exploitation*, **21**, 475 (2003).
4. W. Körbitz, *Renewable Energy*, **16**, 1078 (1999).
5. M. A. Harmer, W. E. Farneth and Q. Sun, *Adv. Mater.*, **10**, 1255 (1998).
6. J. H. Clark, *Acc. Chem. Res.*, **35**, 791 (2002).
7. K. Wilson, D. J. Adams, G. Rothenberg and J. H. Clark, *J. Mol. Catal. A: Chem.*, **159**, 309 (2000).
8. F. Omota, A. C. Dimian and A. Bliek, *Chem. Eng. Sci.*, **58**, 3175 & 3159 (2003).
9. T. Okuhara, *Chem. Rev.*, **102**, 3641 (2002).
10. M. A. Harmer and V. Sun, *Appl. Catal. A: Gen.*, **221**, 45 (2001).
11. S. Ardizzone, C. L. Bianchi, V. Ragaini and B. Vercelli, *Catal. Lett.*, **62**, 59 (1999).
12. H. Matsuda and T. Okuhara, *Catal. Lett.*, **56**, 241 (1998).
13. M. A. Harmer, Q. Sun, A. J. Vega, W. E. Farneth, A. Heidekum and W. F. Hoelderich, *Green Chem.*, **2**, 7 (2000).
14. G. D. Yadav and J. J. Nair, *Micropor. Mesopor. Mater.*, **33**, 1 (1999).
15. M. A. Ecormier, K. Wilson and A. F. Lee, *J. Catal.*, **215**, 57 (2003).
16. Y. Kamiya, S. Sakata, Y. Yoshinaga, R. Ohnishi and T. Okuhara, *Catal. Lett.*, **94**, 45 (2004).
17. R. Koster, B. van der Linden, E. Poels and A. Bliek, *J. Catal.*, **204**, 333 (2001).
18. H. G. Schoenmakers and B. Bessling, *Chem. Eng. Prog.*, **42**, 145 (2003).
19. R. Taylor and R. Krishna, *Chem. Eng. Sci.*, **55**, 5183 (2000).
20. H. Subawalla and J. R. Fair, *Ind. Eng. Chem. Res.*, **38**, 3696 (1999).
21. A. A. Kiss, A. C. Dimian, G. Rothenberg, *Adv. Synth. Catal.*, **348**, 75 (2006).

46. Continuous Deoxygenation of Ethyl Stearate: A Model Reaction for Production of Diesel Fuel Hydrocarbons

Mathias Snåre, Iva Kubičková, Päivi Mäki-Arvela, Kari Eränen and Dmitry Yu. Murzin

Laboratory of Industrial Chemistry, Process Chemistry Centre, Åbo Akademi University, Biskopsgatan 8, FIN-20500 Turku/Åbo, Finland

dmitry.murzin@abo.fi

Abstract

In this paper the continuous deoxygenation of biorenewable feeds for diesel fuel production is addressed. The model reactant, ethyl stearate (stearic acid ethyl ester, $C_{20}H_{40}O_2$) is deoxygenated by removal of a carboxyl group via decarboxylation and/or decarbonylation reactions yielding an oxygen-free diesel fuel compound, heptadecane. The reaction was carried out in a fixed-bed tubular reactor over a heterogeneous catalyst under elevated temperatures and pressures. The effect of catalyst pretreatment and catalyst mass as well as influence of reaction temperature were investigated.

Introduction

Currently significant research effort is devoted to developing environmentally friendly liquid fuels from renewable sources. Natural oils and fats are produced via extraction or pressing of renewable materials such as vegetable and animal feeds. The natural oil and fats consist primarily of triglycerides (98%), which are made up of a glycerol moiety and three fatty acid moieties[1]. Typically fatty acids in vegetable oils and animal fats have a straight hydrocarbon chain with carbons varying from C_6 to C_{24}, the chain can be saturated, monounsaturated or polyunsaturated, Table 1 [2,3]. Diesel fuel compounds derived from crude oil generally have 10-20 carbon atoms , thus removal of the carboxyl group from the fatty acid molecule would result in a paraffinic hydrocarbon similar to fossil diesel compounds. The paraffinic fuel compounds, converted from fatty acids and their derivates, would have a superior cetane number compared to both fossil diesel compounds and the conventional biodiesel compounds, FAME, produced via the transesterification method [4]. Furthermore, fatty acid derived fuels are very ecologically benign, e.g. they exhibit a low sulphur and aromatic content and are biodegradable [5].

Table 1. Common fatty acids in vegetable oil and animal fat triglycerides [2,3]

	Trivial name	Systematic name	Carbon atoms
saturated	Butyric acid	butanoic acid	4
saturated	Caproic Acid	hexanoic acid	6
saturated	Caprylic Acid	octanoic acid	8
saturated	Capric Acid	decanoic acid	10
saturated	Lauric Acid	dodecanoic acid	12
saturated	Myristic Acid	tetradecanoic acid	14
saturated	Palmitic Acid	hexadecanoic acid	16
saturated	Stearic Acid	octadecanoic acid	18
saturated	Arachidic Acid	eicosanoic acid	20
saturated	Behenic acid	docosanoic acid	22
saturated	-	tetracosanoic acid	24
monounsaturated	Palmitoleic Acid	9-hexadecenoic acid	16
monounsaturated	Oleic Acid	9-octadecenoic acid	18
monounsaturated	Vaccenic Acid	11-octadecenoic acid	18
monounsaturated	Gadoleic Acid	9-eicosenoic acid	20
monounsaturated	Erucic acid	13-docosenoic acid	22
polyunsaturated	Linoleic Acid	9,12-octadecadienoic acid	18
polyunsaturated	α-Linolenic Acid	9,12,15-octadecatrienoic acid	18
polyunsaturated	γ-Linolenic Acid	6,9,12-octadecatrienoic acid	18
polyunsaturated	Arachidonic Acid	5,8,11,14-eicosatetraenoic acid	20
polyunsaturated	EPA	5,8,11,14,17-eicosapentaenoic acid	20

The plausible deoxygenation routes for production of diesel like hydrocarbons from fatty acids and their derivates are decarboxylation, decarbonylation, hydrogenation and decarbonylation/hydrogenation. The main focus in this study is put on liquid phase decarboxylation and decarbonylation reactions, as depicted in Figure 1. Decarboxylation is carried out via direct removal of the carboxyl group yielding carbon dioxide and a linear paraffinic hydrocarbon, while the decarbonylation reaction yields carbon monoxide, water and a linear olefinic hydrocarbon.

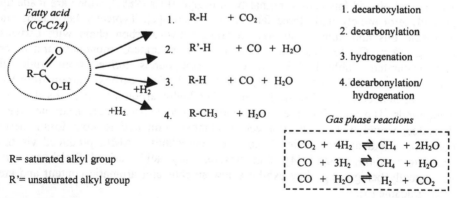

Figure 1. Schematic deoxygenation path of fatty acids for biodiesel production.

The production of hydrocarbons for diesel fuel via decarboxylation of vegetable based feeds has been verified in our previous work [6]. The fatty acid, stearic acid, the fatty acid ester, ethyl stearate, and a triglyceride of stearic acid, tristearin, were successfully decarboxylated resulting in the paraffin, n-heptadecane, originating from the stearic acid alkyl chain. The experiments were carried out in a semi-batch reactor over a commercial palladium catalyst. The results showed that ethyl stearate deoxygenation proceeds over the corresponding fatty acid intermediate which is subsequently deoxygenated to heptadecane and carbon dioxide. In the present study the continuous deoxygenation of ethyl stearate was investigated in a tubular reactor in order to evaluate the catalyst stability and the industrial applicability.

Experimental Section

The fresh and spent catalysts were characterized with the physisorption/chemisorption instrument Sorptometer 1900 (Carlo Erba instruments) in order to detect loss of surface area and pore volume. The specific surface area was calculated based on Dubinin-Radushkevich equation. Furthermore thermogravimetric analysis (TGA) of the fresh and used catalysts were performed with a Mettler Toledo TGA/SDTA 851e instrument in synthetic air. The mean particle size and the metal dispersion was measured with a Malvern 2600 particle size analyzer and Autochem 2910 apparatus (by a CO chemisorption technique) , respectively.

The catalytic deoxygenation experiments were carried out in a tubular fixed bed reactor. The reactor length and the inner diameter were 175 mm and 4.4 mm, respectively. The catalyst powder was placed between a layer of quartz sand and quartz wool. The experiments were typically carried out in an upward two-phase system and volumetric flow 0.1 ml/min using a high performance liquid chromatography pump (HPLC). The system was pressurized by adding 10 ml/min pressurizing media (Ar and N_2) using a mass controller and a pressure controller, which were placed downstream. The experimental setup is illustrated in Figure 2.

Decarboxylation of ethyl stearate in the continuous system is a challenge, due to its high melting point and low solubility in inert solvents. The solvent caused some additional challenges since the reaction must be carried out in the liquid phase and the reaction temperature should be sufficiently high for the reaction to occur. The reaction mixture containing 5 mol% ethyl stearate (Aldrich, >97%) in hexadecane (Fluka >98%) was continuously fed through a fine catalyst powder bed (particle size<50 μm), typically 0.2-0.6 g of catalyst. The composition of a reaction mixture as a function of time was monitored by continuous sampling of the liquid outlet and analyzed by gas chromatography. The first sample was taken when the first drop was observed except for the pressurized system. In this case the zero sample was withdrawn after 40 minutes of the liquid mixture to the reactor pumping (it took approximately 30 minutes for the first drop to appear). Samples were derivatised by BSTFA (N,O-bis-(trimethylsilyl)trifluoroacetamide) in order to detect the fatty acid compounds in the gas chromatography. Additionally a gas chromatography coupled to a mass spectrometer was utilized for complementary product identification. Prior to reaction, *in situ* catalyst pretreatment was performed with constant flow of hydrogen at 2 bar and 200°C for 2 h.

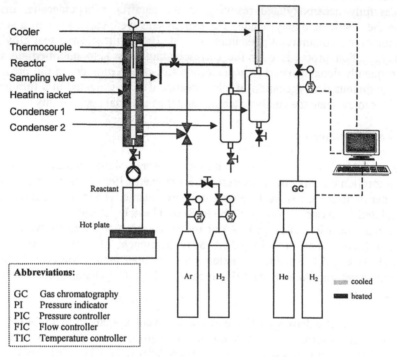

Figure 2. Experimental setup for the deoxygenation of renewable feedstocks.

Results and Discussion

Textural properties of the commercial powder catalyst (e.g. surface area and pore size information) were obtained by a physisorption method. Additionally the mean particle size of the fresh catalyst was measured by light scattering apparatus. The surface area of the fresh catalyst was 1214 m^2/g, while all the surface area measurements of spent catalyst showed a substantial decrease of surface area, indicating coke formation. Similarly the micropore volume decreased for all the spent catalysts compared to the fresh catalyst. The TGA analysis showed that non-pretreated lost 11 wt% of total amount of combustibles at 450 °C, while pretreated and fresh catalyst lost 22 wt% and 37 wt%, respectively, of total combustibles (in the temperature range 270 - 800 °C) indicating that more severe coking was present on the non-pretreated catalyst than on the pretreated catalyst. The highest rate of weight loss was noticed at 600°C and 560°C for non-pretreated and pretreated catalyst, respectively implying that the coke molecule is in average bigger on the non-pretreated than on the pretreated catalyst. The mean particle size of the fresh catalyst was 15 μm and the majority of particles (97%) were below 50μm. The characterization results as well as reaction conditions for the fresh and spent catalyst are summarized in Tables 2 and 3, respectively.

Table 2. Characterization results of the fresh catalyst: surface area (Dubinin), micropore volume, mean particle size and metal dispersion

Metal loading (wt%)	Mean particle size (μm)	Surface area (m^2/g)	Specific pore volume (ml/g)	Dispersion (%)
5	15.0	1214	0.431	18

Table 3. Characterization results of spent catalysts

Run nr.	Catalyst mass (g)	Temperature (˚C)	Pressure (bar)	Surface area (m^2/g)	Specific pore volume (ml/g)
I*	0.2	270	1	-	-
II	0.2	270	1	349	0.123
III	0.4	270	1	302	0.107
IV	0.6	270	1	280	0.100
V**	0.6	270	1	-	-
VI***	0.4	270	1	302	0.107
VII	0.4	300	3	290	0.102
VIII	0.4	330	5	287	0.102
IX	0.4	360	7	305	0.110
X	0.6	300	3	-	-

*non-pretreated
**volumetric flow, V'=0.07 ml/min
***repeated experiment of run III (characterization results obtained from run III)

The deoxygenation reaction of ethyl stearate was carried out for 4-6 h at reaction temperatures between 270-360°C and under reactor pressure of 0-7 bar. The effect of the catalyst pretreatment and the catalyst mass as well as the influence of reaction temperature were studied. The observed products in the ethyl stearate reaction are the deoxygenation products (desired), intermediate product (fatty acid), hydrogenation products (unsaturated Et-SA) and by-products (Figure 3).

Initially the effect of catalyst pretreatment by hydrogen was investigated (Figure 4). The pretreated catalyst converted 17 % of ethyl stearate (Et-SA) while the non-pretreated catalyst converted 66% at reaction temperature 270°C. The high initial conversion over a non-pretreated catalyst was due to dehydrogenation of ethyl stearate to monounsaturated fatty acid ethyl esters (unsaturated Et-SA) , such as ethyl oleate, indicating that the deficit of hydrogen on the catalyst surface initially favors the dehydrogenation reaction rather than the desired deoxygenation routes.

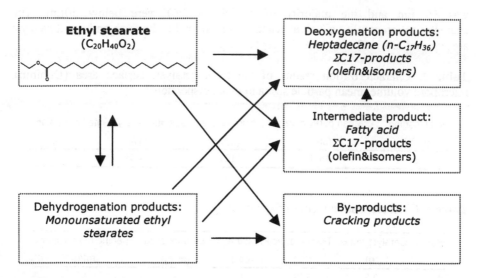

Figure 3. Reaction products detected during the deoxygenation of ethyl stearate.

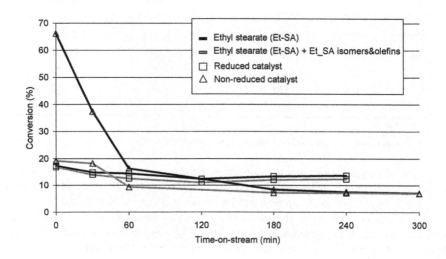

Figure 4. The effect of catalyst reduction in deoxygenation of ethyl stearate. Reaction conditions: 5 mol% ethyl stearate in hexadecane, T = 270 °C, p = 1 bar, m_{cat} = 0.4 g and liquid flow rate (V') = 0.1 ml/min.

The dehydrogenation of ethyl stearate over the non-pretreated catalyst subsequently led to a rapid catalyst deactivation (from 66% to 16% of conversion within 1h of reaction) whereas the pretreated catalyst deactivated only slightly (17%

to 13% of conversion), thus leading to a conclusion that the unsaturated ethyl stearates caused the rapid deactivation. Similar dehydrogenation phenomena and catalyst deactivation in the transformation of cyclohexene were reported over a palladium supported catalyst [7]. Calculating total conversion based on ethyl stearate and dehydrogenated ethyl stearates, one gets similar initial conversion (~17%) for both cases, nevertheless the non-pretreated catalyst is deactivating faster.

The pretreated catalyst maintained a more stable performance than the non-pretreated catalyst. Product distribution as a function of conversion was fairly constant for the former catalyst. Selectivity towards desired product, heptadecane (n-C17), was 80% at low conversions and it increased with increasing conversion (Figure 5). Minor formation of ethyl oleate as well as C17 isomers and olefins (ΣC17-products) were detected. Generally ΣC17-products comprises of 1-heptadecene, 3-heptadecene, 8-heptadecene and 1-methyldecylbenzene. Over the non-pretreated catalyst the formation of dehydrogenation products of ethyl stearate increased as a function of conversion, while formation of heptadecane and stearic decreased as a function of conversion. Furthermore the by-product (cracking) formation seemed to be more profound over the non-pretreated catalyst (Figure 5 and 6).

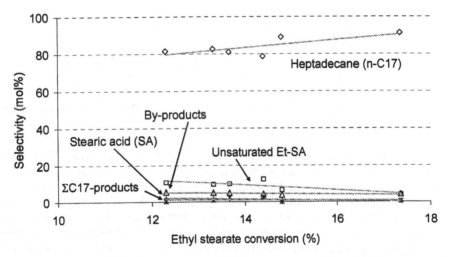

Figure 5. Product selectivity versus ethyl stearate conversion for the pretreated catalyst in the deoxygenation of ethyl stearate. Reaction conditions: 5 mol% ethyl stearate in hexadecane, T=270 °C, p=1 bar, m_{cat}=0.4 g and V'=0.1 ml/min.

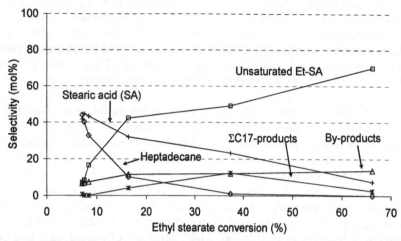

Figure 6. Product selectivity versus ethyl stearate conversion for the non-pretreated catalyst in the deoxygenation of ethyl stearate. Reaction conditions: 5 mol% ethyl stearate in hexadecane, T=270 °C, p=1 bar, m_{cat}=0.4 g and V'=0.1 ml/min

The mass of catalyst was varied in order to achieve higher conversions and investigate diffusion limitations (Figure 7). The experiments over 0.2 g of catalyst at 270°C (residence time: 4 min) gave an initial conversion of 16% while doubling the mass of catalyst (residence time: 8.2 min) the initial conversion increased to 23%. A three time increased catalyst mass (0.6 g, residence time: 12 min) converted 58% of ethyl stearate, however the catalyst was rapidly deactivated within 10 minutes from the initial conversion of 58% to 32% of conversion. The non-linear relationship between the catalyst mass and the initial conversion gave additional indication that the catalyst bed was partially deactivated, thus concluding that mechanistic studies of the initial rates should take into account the very fast deactivation. Experiments carried out with different catalyst masses at 300°C showed similar tendencies, the higher catalyst mass, 0.6 g, deactivated more rapidly then 0.4 g of catalyst. The deactivation could be caused by coke formation, the concept which is also strengthened by the observed loss of initial surface area, however, the similar decrease of surface area for the three catalyst mass does not explain the non-linear trend appearing. Thus, concluding that the deactivation is additionally affected by catalyst poisoning and/or sintering. Furthermore reproducibility was proven to be good by repeating the experiment over 0.4 g of catalyst at 270°C. The complete conversion of ethyl stearate was achieved over 0.6 g of catalyst with the volumetric flow of 0.07 ml/min (residence time: 17.2 min) at 270°C and at 300°C over 0.6 g of catalyst. The effect of catalyst mass on conversion and selectivity is shown in Figure 7.

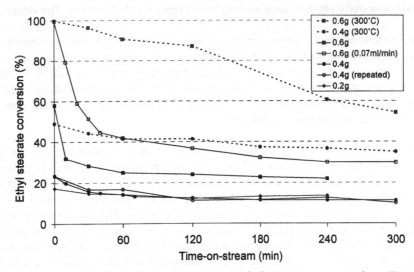

Figure 7. The effect of catalyst mass on ethyl stearate conversion. Reaction conditions: 5 mol% ethyl stearate in hexadecane, T=270 °C, p= 1 bar and V'= 0.1ml/min.

Table 4. Experimental results in the deoxygenation of ethyl stearate

Run nr.	Catalyst mass	Temperature	Residence time	Initial conversion	Selectivity (mol%)	
	(g)	(°C)	(min)	(%)	$S_{n\text{-}C17}$	$S_{\Sigma C17}$
I**	0.2	270	4.0	67	-	2.6
II	0.2	270	4.0	17	95.0	0.4
III	0.4	270	8.2	23	95.5	-
IV	0.6	270	12.0	59	93.8	1.7
V***	0.6	270	17.2	100	99.5	0.5
VI	0.4	270	8.2	20*	93.0	1.1
VII	0.4	300	8.2	48*	91.0	4.3
VIII	0.4	330	8.2	89*	94.2	4.9
IX	0.4	360	8.2	100*	95.4	3.7
X	0.6	300	12.0	100*	99.5	0.5

*sample withdrawn after 40 min. of pumping
**non-pretreated
***volumetric flow, V=0.07 ml/min

In order to elucidate the effect of liquid products on the catalyst deactivation, selectivity versus conversion was investigated. As indicated above particularly interesting was the rapid decrease of conversion over 0.6 g compared to 0.4 g and 0.2 g of catalyst, however a comparison of product selectivities as a function of conversion gave no direct explanation for this rapid deactivation. Very similar product selectivity patterns were noticed for all three catalyst masses. The selectivity to the desired product, heptadecane, was above 75% and in all three cases minor concentrations of ethyl stearate dehydrogenation products were present. Complexity of relationship between liquid phase composition and catalyst deactivation calls for, further investigations of the impact of the gas-phase reactions, which result in products, strongly influencing catalyst deactivation. . The initial conversion and the initial selectivities towards the desired C17-products in the deoxygenation experiments are summarized in Table 4.

The influence of the reaction temperature on the decarboxylation rate was studied in the temperature range of 270-360°C. The pressure was adjusted depending on the vapor pressure of the solvent, hexadecane, at specific temperatures in order to keep the reaction in the liquid phase. As expected the conversion of ethyl stearate increased with increasing reaction temperature (Figure 8). The complete conversion of ethyl stearate was achieved at 360°C. Furthermore the catalyst was very stable at these temperatures, hence loosing not more than 10% of ethyl stearate conversion during 5 h of time-on-stream. The surface areas of the spent catalysts at different temperatures were approximately 300 m^2/g for all four temperatures, which is equal to a 75% decrease of the initial surface area and moreover is an indication of a heavy coke formation on the catalyst surface. Based on the Arrhenius equation, the apparent activation energy was calculated to 69 kJ/mole in a temperature range of 270-330°C, applying zero order kinetics.

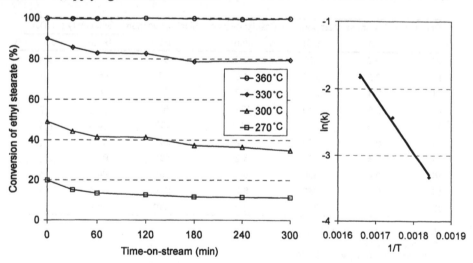

Figure 8. The effect of reaction temperature on ethyl stearate conversion as a function of time-on-stream and Arrhenius plot (based on zero order kinetics) Reaction conditions: 5 mol% ethyl stearate in hexadecane, m_{cat}=0.4 g, $p_{270°C}$=1 bar, $p_{300°C}$=3 bar, $p_{330°C}$=5 bar, $p_{360°C}$=7 bar and V'=0.1 ml/min.

The effect that temperature had on selectivity is minor, thus formation of heptadecane was maintained stable over 90% as a function of temperature, while an inoroaoo in the formation of C17 Isomers and olefins was observed above 300°C (Table 4).

Conclusion

A novel method for production of paraffinic hydrocarbons, suitable as diesel fuel, from renewable resources was illustrated. The fatty acid ethyl ester, ethyl stearate, was successfully converted with high catalyst activity and high selectivity towards formation of the desired product, heptadecane. Investigation of the impact of catalyst reduction showed that the reduction pretreatment had a beneficial effect on the formation of desired diesel compound. The non-pretreated catalyst dehydrogenated ethyl stearate to ethyl oleate. The experiments at different reaction temperatures, depicted that conversion of ethyl stearate was strongly dependent on reaction temperature with E_{act}=69 kJ/mole, while product selectivities were almost constant. Complete conversion of ethyl stearate and very high selectivity towards desired product (95%) were achieved at 360°C.

Acknowledgements

This work is part of the activities at the Åbo Akademi Process Chemistry Centre within the Finnish Centre of Excellence Programme (2000-2011) by the Academy of Finland.

References

1. B.K. Barnwal and M. P. Sharma, *Renewable and Sustainable Energy Reviews*, **9**, 363-378 (2005)
2. M. Fangrui and A.H. Milford, *Bioresource Technology*, **70**, 1-15 (1999)
3. E. H. Pryde, *Fatty Acids*, American Oil Chemists' Society, 2-14. (1979)
4. J. V. Gerpen, *Fuel Processing Technology*, **86** 1097-1107 (2005)
5. N. Martini and S. Schell, *Plant Oils as Fuels: Present State of Science and Future Developments*, Proceedings of the Symposium, Potsdam, Germany, 16-18. February, (1997), Berlin: Springer, 6. (1998)
6. I. Kubičková, M. Snåre, P. Mäki-Arvela, K. Eränen and D.Yu. Murzin, *Catalysis Today*, **106** 197-200 (2005).
7. M.A. Aramendía, V. Boráo, Im. García, C. Jiménez, A. Marinas, J.M. Marinas, and F.J. Urbano, *J. Mol. Catal. A: Chem.* **151**, 261-269. (2000)

47. Glycerol Hydrogenolysis to Propylene Glycol under Heterogeneous Conditions

Simona Marincean, Lars Peereboom, Yaoyan Xi, Dennis J. Miller and James E. Jackson

Department of Chemistry, Department of Chemical Engineering and Materials Science, Michigan State University, East Lansing, MI 48824-1322

Abstract

In aqueous media, glycerol (GO), a low-cost renewables-based feedstock, can be catalytically converted under mild conditions to the commodity products propylene glycol (PG), lactic acid (LA) and ethylene glycol (EG). This report chronicles a comparative study of catalysts, solvents, and reaction conditions aimed at optimizing selectivity towards PG and at the same time augmenting our understanding of catalyst-substrate interactions. Two catalysts were evaluated: Ni/Re on carbon was more active and selective than Ru on carbon for GO hydrogenolysis to PG. Partial replacement of water with other hydroxylic solvents—the simple alcohols—increased the selectivity and conversion in the order ethanol/water < water < iso-propanol/water ~ tert-butanol/water, while unmixed solvents yield the following trend: water < ethanol < iso-propanol < tert-butanol, with a three-fold increase in PG yield in tert-butanol relative to water. We interpret these results as evidence of facilitated GO access to the catalyst surface in the presence of larger solvent molecules with a hydrophobic carbon backbone. The reaction is also pH limited because of formation of acids such as lactic and formic, which neutralize the base promoter. We discuss the relationship between solvent structure and PG yield in terms of interactions at the catalyst surface.

Introduction

Depletion of fossil oil resources and concern about environmental pollution, especially heavy metals and greenhouse gases, have brought biomass into focus as a renewable source of raw materials for large-scale chemicals and energy production (1). The "biomass refinery" of the future will require a powerful toolbox of processes for converting complex plant matter into useful commodity and specialty products.

Since biomass consists mainly of carbohydrates, sugars are the major renewable feedstock from which economically feasible organic compounds are to be developed to replace the ones traditionally obtained from petrochemical sources. Glucose, fructose and other monosaccharides as well as disaccharides are inexpensive, accessible in large amounts, and suited for transformations into low molecular weight organic commodities. Research endeavors to transform carbohydrates into useful commodity and specialty products need to develop strategies to selectively decrease the oxygen content, introduce unsaturation in the form of C=C and C=O, and cleave C-C bonds. The technology used to convert fossil resources is fundamentally different since the raw materials are hydrocarbons, oxygen free, hydrophobic, volatile and thermally tough.

A byproduct of vegetable oil transesterification to make biodiesel fuel, glycerol (GO) has captured our attention for several reasons. In aqueous medium, GO itself can be converted to value-added commodity products such as propylene glycol (PG), lactic acid (LA) and ethylene glycol (EG) in the presence of a metal catalyst at mild conditions (2-7). An array of metals deposited on various supports have been examined as catalysts for the above reaction (6, 7).

But GO is also an archetype for more complex polyols. Catalytic hydrogenolysis of polyols leads to C-C and C-O bond cleavage as well as activation via dehydrogenation of HCOH to C=O sites (8, 9). With its three vicinal hydroxyl groups, and its two types of carbon atoms (all bearing hydroxyl functionalities) GO can serve as a simple model in which to study the competition between these processes. We aim to gain insight into the individual steps and apply that information to increase selectivity toward desired products.

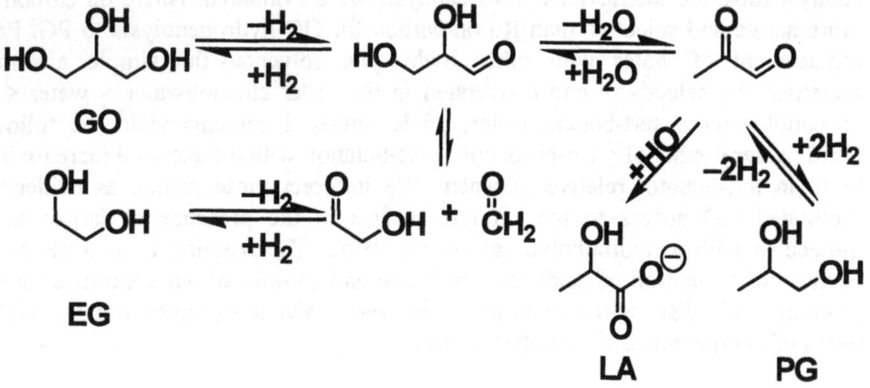

Scheme 1

Recent isotopic labeling studies of GO hydrogenolysis over Ru/C catalyst at basic pH found the first step to be C-1 dehydrogenation. The glyceraldehyde analog formed then faces two possible paths (Scheme 1): enolization, water loss, and hydrogenation or base addition and to pyruvaldehyde intermediate isomerization to form PG and LA; or retro aldol cleavage to give, after hydrogenation, EG and formaldehyde (10). In this report we examine GO hydrogenolysis over metal catalysts at different base concentrations and different solvents. HPLC is used to monitor the catalyst activity and selectivity toward the products.

Experimental Section

The Ru/C catalyst was provided by PMC, Inc. and consists of 5.0 wt% Ru on activated carbon support. The Ni/Re/C catalyst comprises 2.5 wt% Ni and 2.5 wt% Re on activated carbon. Both catalysts were characterized via N_2 physisorption and H_2 chemisorption using a Micromeritics ASAP 2010 apparatus; the results are presented in **Table 1**.

Table 1. Catalyst Properties

Catalyst	BET[a] (m^2/g)	Pore Diam.[a] (Å)	Total Pore Vol.[a] (cm^3/g)	Micropore Vol.[a] (cm^3/g)	Metal Dispersion[b] (%)	Metal Surface Area[b] (m^2/g)
Ru/C	670	34	0.56	0.17	8.8	1.6
Ni/Re	1500	20	0.76	0.35	5.6	1.2

[a]Determined by N_2 physisorption at 78K [b]Determined by H_2 chemisorption at 308K

All reactions were run in a Parr stirred autoclave (Model 4561) at 1000 psi H_2 and 200°C for 6h. A weighed quantity (0.5 g dry basis) of the catalyst, Ru/C or Ni/Re/C, was introduced into the reactor and reduced at 250°C and 200 psi H_2 (Ru/C) or at 280°C and 500 psi H_2 (Ni/Re/C) for 13 hours. After cooling, 100ml of solution (1.0M GO and 0.1-1.0M KOH) was added to the closed reactor. For reactions in solvent mixtures, entries 1-4 in **Table 2**, the water/solvent ratio was 1/9 (v/v). When the solvent was either t-BuOH or 1,4-dioxane, 1.5 g of water was added to the solution to facilitate dissolution of KOH, because of its low solubility these solvents. Once steady state was achieved following heatup, samples were taken at 30 minute intervals for the first hour, and then hourly, and analyzed via HPLC. The HPLC column was a BIORAD Aminex HPX-87H run at 65°C with 5mM H_2SO_4 as the mobile phase at a flow rate of 0.6ml/min, using both UV (210 nm) and refractive index (RI) detection.

Final quantitative evaluation of feed conversion and product distribution, given in **Tables 2** and **3**, were based on HPLC analyses after 6h reaction time for most reactions. In several cases evaluation is based on the samples taken at 5h. Selectivity is defined as mol product formed/mol glycerol converted; yield is mol product formed/mol initial glycerol. The carbon balance is defined as:

C Balance(%)= {[GO]+ [PG] + [EG] + [LA]}$_{final}$ / [GO]$_{initial}$ x 100

We did not explicitly account for the number of carbon atoms because all compounds have the same number of carbons except EG. In the case of EG (Scheme 1), for each molecule of EG formed there is a corresponding one-carbon compound formed as well.

In reaction, time t = 0 is defined as the point when reaction conditions are reached (200°C) and the reaction vessel is pressurized with H_2. Hence, for the solvents i-PrOH or t-BuOH there is PG in the reaction mixture at t=0 because of reaction during the 15 minute heatup. However, it is clear that base is necessary for reaction as a control experiment in which base was added to the reaction mixture only after heating to 200°C showed no PG at t=0.

Results and Discussion

The distribution of GO hydrogenolysis products, PG, LA, and EG, is influenced by the catalyst used and the amount of base present in the reaction mixture (Table 2). When compared to Ru, Ni/Re appears to be more selective towards PG at the expense of EG.

Table 2. GO Hydrogenolysis in Aqueous Medium[a]

Exp.	Catalyst	KOH(M)	Conv(%)	Selectivity(%) PG	LA	EG	C Balance(%)
1	Ru	1.00	92.0	38.8	47.2	8.4	94.9
2	Ru	0.50	70.5	38.7	45.3	11.3	96.7
3	Ru	0.25	52.1	51.0	33.7	15.0	99.8
4	Ru	0.10	47.1	46.9	13.7	16.0	89.0
5	Ni/Re	1.00	99.1	63.8	20.6	6.3	90.4
6	Ni/Re	0.50	74.4	44.1	35.9	8.6	91.5
7	Ni/Re	0.25	43.9	45.7	30.7	6.9	92.8
8	Ni/Re	0.10	22.9	60.9	26.6	8.1	99.0

[a]Reactions carried out on 100ml samples of 1M GO for 6h in H_2O at 200°C and 1000psig H_2 using 0.5g catalyst.

Figure 1. GO conversion and solution pH[a]

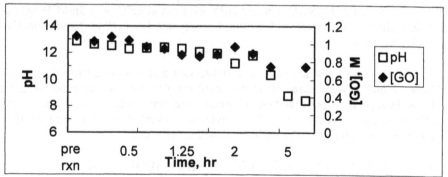

[a]Reaction carried out on 100ml solution of 1M GO and 0.25M KOH for 6h in H_2O at 200°C and 1000psig H_2 using 0.5g Ni/Re catalyst.

The C-C cleavage pathway that leads to EG takes place to a smaller extent in the presence of Ni/Re, when keto-enol tautomerization is favored as is shown by the higher sensitivity of the reaction to the amount of base. GO consumption rate closely follows the solution pH, slowing greatly when the reaction medium becomes close to neutral (Figure 1). The acids identified in the reaction mixture, formic and lactic, are not present in high enough quantities to account entirely for the pH decrease (e.g. to neutralize all base present). Formic acid (as formate) is produced via Cannizzaro reaction of formaldehyde, a byproduct in the retro-aldol cleavage pathway, with itself or with other aldehydes, leading to a maximum formic acid:EG molar product ratio of 0.5:1. The actual detected ratio of formic acid:EG was 0.13. Control experiments with 1.0M formic acid in 0.1 M KOH solution over Ni/Re/C showed that base present is neutralized, but the free acid is degraded completely over 6h, without detection of other one-carbon compounds in the reaction solution. We suspect that CO_2 is formed via degradation of formic acid and is present in solution as carbonic acid or bicarbonate, which contributes to the pH decrease. However, we did not detect carbonate via HPLC even though standard carbonate solutions do show a peak. We believe that our preparation of HPLC samples, involving neutralization in the acidic mobile phase (5 mM H_2SO_4), leads to acidification of the solution and thus loss of carbonic acid via CO_2 evolution.

In order to understand the mechanism of GO hydrogenolysis in heterogeneous conditions, one must gain insight into the interactions that take place between the catalyst and the reactants on one hand and catalyst and the solvent on the other hand. Recent studies in our group showed that under our typical reaction conditions, the Ru/C catalyst is capable of exchanging hydrogen between D_2O and molecular H_2, presumably via exchange of H and OD sites bound to the surface of the catalyst[10]. Being an avid hydrogen bonder, water may also agglomerate at the surface of the catalyst around the metal centers. Other OH-bearing compounds should similarly be able to participate in such an interaction as long as the OH has access to the surface of the catalyst, i.e. is not sterically hindered. So water and GO may be competitors for catalytically relevant sites. At the same time, the catalyst support carbon is hydrophobic and GO and reaction products, with their carbon backbones, should be favored over water to adsorb in the carbon pore structure. Thus species concentrations within the carbon pores may be significantly different, particularly in terms of water, than in the bulk solution, which may strongly affect reaction rates.

To further investigate the effects of hydrogen bonding and hydrophobicity on the reaction, water was partially (water/solvent ratio 1/9 (v/v)) replaced with compounds similar to GO in that they possessed OH groups and a carbon backbone. The GO conversion (Figure 2) increased in the order PG < water < EtOH < i-PrOH ~ t-BuOH. The initial GO conversion rate is several times larger in i-PrOH/water and t-BuOH/water than in water alone. The PG yield (Figure 3) shows the same trend. Interestingly, i-PrOH/water and t-BuOH/water mixtures as solvents have very similar effects on the GO conversion and PG yield, which leads us to believe that the reaction is more sensitive to the presence of the -OH than to the increase in the carbon backbone size, given that both mixtures included the same amount of water. t-BuOH has a carbon backbone similar in size to that of GO, but its hydroxyl site is much more shielded by the large, hydrophobic t-butyl group. The low GO reactivity in PG/water is a strong indicator of PG inhibition, presumably via direct competition for surface sites.

As expected, the product distribution is affected by the solvent environment. While PG is obtained as hydrogenation of pyruvaldehyde takes place, LA (as lactate) is formed via a hydride transfer in pyruvaldehyde[11]. PG should be favored at the expense of LA as the polarity of the reaction medium decreases, and indeed their respective selectivities vary in opposite directions (**Table 3**).

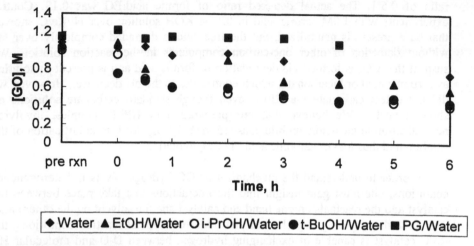

Figure 2. GO Conversion in Solvent/Water Mixtures[a]
[a]Reactions carried out on 100ml solution of 1M GO and 0.1M KOH for 4-6h in H_2O at 200°C and 1000psig H_2 using 0.5g Ni/Re catalyst.

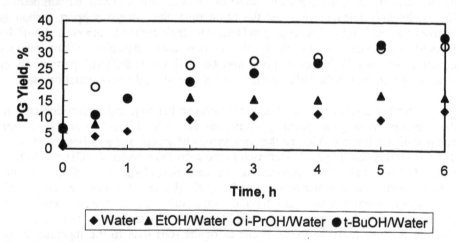

Figure 3. PG Yield in Solvent/Water Mixtures[a]
[a]Reactions carried out on 100ml solution of 1M GO and 0.1M KOH for 6h at 200°C and 1000psig H_2 using 0.5g Ni/Re catalyst.

The lower GO conversion rate in water could arise via several conditions. If GO is competing for adsorption sites on the catalyst surface with solvents, it may do so more successfully against i-PrOH and t-BuOH than against water because the -OH groups in GO are all less shielded than the ones in i-PrOH and t-BuOH. This is further supported by the low lactic acid yields of reactions run in these alcohols, suggesting that water (a strong competitor) is not present in the vicinity of the catalyst in alcohol solvent. Second, the presence of PG clearly reduces reaction rate - water may be ineffective in removing product PG from the reaction environment, whereas alcohols may preferentially displace PG from the carbon pore structure. Finally, water is a poor solvent for hydrogen, as hydrogen solubility in water is only 20 to 30% that in simple alcohols (12).

Full replacement of water with organic solvents leads to further increases in the GO conversion and PG selectivity in the order water < EtOH < i-PrOH < t-BuOH (**Figure 4**). As one can see from **Table 3**, GO conversion almost doubles going from EtOH to t-BuOH, and yield of PG (**Figure 5**) increases with conversion. Yet PG selectivity only increases modestly, while at the same time the LA and EG selectivities decrease. Taken together these experimental observations suggest that while the pathway for LA formation becomes less important when little water is present, presumably aldol reactions remain accessible. Glyceraldehyde, the first intermediate of GO hydrogenolysis, is known to be very reactive towards aldol condensation in basic media (13). The lower carbon balances may be due to the to the reactivity of the carbonyl compounds formed as intermediates toward condensation reactions, resulting in higher molecular weight unsaturated compounds, consistent with the appearance of color in the reaction samples.

Table 3. GO Hydrogenolysis in Different Solvents[a]

			Selectivity (%)			
Exp	Solvent	Conv(%)	PG	LA	EG	C Balance (%)
1	Water	22.9	60.9	26.6	8.1	99.0
2	EtOH/Water	34.4	57.1	15.5	10.4	94.1
3	i-PrOH/Water	59.8	54.9	7.5	9.2	83.1
4	t-BuOH/Water	57.6	61.5	7.6	8.0	86.8
5.	PG/Water	10.8	n/a	46.2	9.2	95.2
6.	EtOH	50.8	45.0	5.5	8.1	79.0
7.	i-PrOH	66.7	69.3	3.1	8.9	87.5
8.	t-BuOH[b]	89.9	58.2	1.2	5.6	68.6
9.	1,4-Dioxane[b]	54.5	22.8	0.0	14.4	65.8

[a]Reactions carried out for 4-6h at 200°C and 1000psig H_2 using 0.5g Ni/Re catalyst, 0.1M KOH, and 1.0M GO. [b]1.5 g water was added to the reaction solution to enhance KOH solubility.

The experiment with 1,4-dioxane as solvent was characterized by poor solubility of GO and KOH into the solvent, hence water was added to the initial mixture. The absence of the hydroxyl functionality in the solvent significantly decreased the system's reactivity. The overall conversion and reactivity in this medium was intermediate, but interesting is that (a) no LA was formed and (b) the selectivity toward EG was substantially higher than in any other solvents.

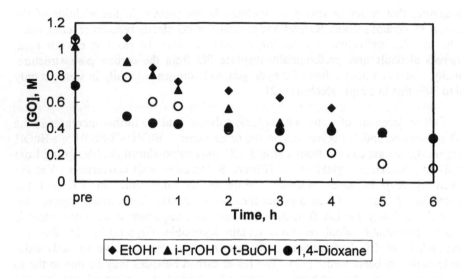

Figure 4. GO Hydrogenolysis in Organic Solvents[a]
[a]Reactions carried out on 100ml solution of 1M GO and 0.1M KOH for 4-6h at 200°C and 1000psi H_2 using 0.5g Ni/Re catalyst.

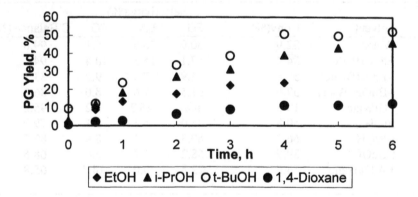

Figure 5. PG Yield in Organic Solvents[a]
[a]Reactions carried out on 100ml solution of 1M GO and 0.1M KOH for 4-6h at 200°C and 1000psi H_2 using 0.5g Ni/Re catalyst.

Pyruvaldehyde conversion to LA is base-catalyzed while hydrogenation of pyruvaldehyde to PG is not (Scheme 1). Thus, the influence of solvent on basicity of solution must be considered as an additional factor in the observed differences in reactivity. From control experiments, we know that without base GO hydrogenolysis in water does not take place. In Figure 6, it is seen that increasing KOH concentration from 0.1 to 0.25M in water resulted in a PG yield increasing from 13% to 25%, while similar experiments in i-PrOH/water solvent with KOH concentration increasing from 0.05 to 0.1M led to PG yields of 16% and 33% - a similar increase. In contrast, for 0.1 M KOH the increase in PG yield from water to i-PrOH/water to i-PrOH is 13% to 32% to 45%. Thus, while we cannot exclude the possible role of base strength in different solvents, the solvent structure appears to have a larger effect.

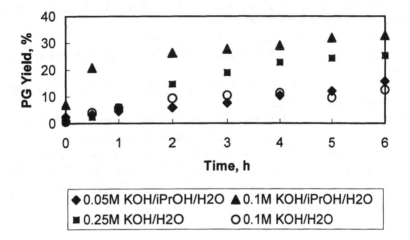

Figure 6. PG Yield at Different Base Concentrations[a]
[a]Reactions carried out on 100ml solution of 1M GO, 0.1 M KOH for 6h at 200°C and 1000psig H_2 using 0.5g Ni/Re catalyst.

Finally, we note that carbon balance closures are generally poorer in the alcohols than in water. A control experiment in which the entire reaction was carried out without sample collection, and another in which reactor and contents were carefully weighed at each stage of reaction, offered no hint as to the fate of lost GO or products. We measured gas formation in the reactor headspace and found < 1% of initial carbon present as gaseous products, primarily methane. We suspect that glycerol and alcohols are forming ethers at the elevated reaction temperatures, and that these ethers are not detected in HPLC. We are continuing efforts to better understand interactions of the solvents with substrates and reaction products.

Conclusions

GO hydrogenolysis to PG in presence of a Ni/Re catalyst is favored by a solvent that possesses both a hydroxyl group and a bulky carbon backbone. We attribute this phenomenon to facilitated access of GO to the surface of the catalyst. We are conducting absorption experiments to elucidate in more detail the phenomena that take place at the surface of the catalyst.

References

1. F. Lichtenthaler *Acc. Chem. Res.* 35, 728 (2003).
2. B. G. Casale and A. M. Gomez U.S. Pat. 5,214,219, to Novamont (1993).
3. B. G. Casale and A. M. Gomez U.S. Pat. 5,276,181, to Novamont (1994).
4. T. N. Haas, A. Neher, D. Arntz, H. Klenkand W. Girke, U.S. Pat. 5,426,249, to Degussa Akteingesellschaft (1995).
5. A. D. Mohanprasad, P.-P. Kiatsimkul, W. R. Sutterlin, G. J. Suppes, *Applied Cataly. A:* General **281**, 225 (2005).
6. D. G. Lahr, B. H. Shanks, *J. Catal.,* **232**, 386 (2005).
7. J. Chaminand, L. Djakovitch, P. Gallezot, P. Marion, C. Pinel, C. Rosier, *Green Chem.,* **6**, 359 (2004).
8. T. A. Werpy, J. G. Frye, A. H. Zacher, D. J. Miller, US Pat. 6,841, 085, to Battelle Memorial Institute (2005).
9. M. Dubeck, G. G. Knapp U.S. Pat. 4,476,331 to Ethyl Corporation (1984).
10. D. G. Kovacs, D. J. Miller and J. E. Jackson, unpublished results.
11. M. K. J. Fedoronko, *Coll. Czechoslovak Chem. Comm.,* **36**, 3424 (1971).
12. M. S. A. Wainwright, T. Ahn, D. L. Trimm, *J. Chem. Eng. Data,* **32**, 22 (1987).
13. R. W. Nagorski and J. P. Richards, *J. Am. Chem. Soc.,* **123**, 794 (2001).

48. Developing Sustainable Process Technology

Joseph A. Kocal

UOP LLC, 25 East Algonquin Road, Des Plaines, IL 60017-5017

Abstract

The key to sustainable growth is innovative technology that addresses performance elements for economic growth, environmental protection, and social progress. Environmental regulations, the push for social progress, and profitability demands will drive refining, chemical, and other industries to meet this innovation challenge. Near-term efforts focus on process optimization, energy utilization, and emissions reduction using incremental technology improvements. In the future, hazard reduction, feedstock substitution, and new energy sources will become more critical. This paper describes approaches to developing sustainable technology and provides pertinent examples in areas such as direct natural gas utilization, desulfurization, and elimination of hazardous liquid acids as catalysts.

Introduction

There are many definitions for sustainable development in use today. The World Resources Institute definition: "Growth that meets the economic, social, and environmental needs without compromising any of them" encompasses most concepts. The chemical, petrochemical, and petroleum refining process industries have slowly evolved toward maintaining sustainable development. Although the progress toward sustainable development has been significant, it has largely been directed at increasing efficiency, minimizing waste, and eliminating harmful emissions for existing processes. Much of this has been driven by the savings resulting from more efficient operations and government legislation. For example, Dow Chemical has utilized new recovery systems and more efficient processing equipment to reduce emissions by 43% and waste by 30%. This resulted in a 180% return on investment at the Midland, Michigan facility (1). Improvements in technology such as this will extend our resources, but in the long run are not sustainable.

Although there will continue to be advances in more efficient and cleaner operation of existing processes, there is clearly a need to be proactive and develop technology based on the tenet of strategic development. A long-term strategic plan is necessary to successfully move forward to the development of sustainable technologies. The goal of creating the most efficient processes and better products, with minimal use of resources and generation of waste and pollution, is common sense and good business. Identifying what is truly sustainable will be a moving target as complexities in life cycle analysis are better understood. As a result, the strategies to achieve sustainable development will be modified. UOP does not have the answers to these complex issues. At this time we are developing a vision for sustainable development and plan to address the critical issues in a logical sequence.

The challenges to creating sustainable technology are numerous and the magnitude daunting. Hart has estimated it will be necessary to grow the economy tenfold to sustain a doubling of the population (2). In order to maintain the current environmental standard, he estimates it will require approximately a twenty-fold improvement in technology.

Innovation is the enabler to succeed in developing technology that not only maintains, but exceeds current quality of life standards and dramatically improves our environment. Throughout the history of the process industries, innovation has been the key to success. Breakthrough innovations have enabled step changes in performance in the past. The response to fuel quality challenges such as the elimination of lead and the development of ultra-low sulfur fuels are examples. The identification of a solid catalyst in the Detal^{TM} process for production of linear alkylbenzene used in preparation of a biodegradable surfactant allowed for the elimination of hazardous hydrofluoric acid in a cost-effective manner. No new HF-based processes for this technology have been built since the introduction of Detal. These discoveries improved the environmental situation and local safety in an economical way. However, these new technologies are clearly not sustainable as they consume irreplaceable resources. Two examples of sustainable technology based on renewable resources are the commercialization of polylactic acid (PLA) from corn starch by Cargill Dow LLC and 1,3-propanediol (PDO) from glucose by Dupont (3, 4). Although these two processes are based on renewable resources, it is not clear that they will be sustainable. Only time will tell. UOP believes breakthroughs such as these as well as process innovations utilizing renewable feedstocks will make sustainability objectives achievable.

Results and Discussion

UOP is in the process of developing a vision for sustainable growth. We have developed a preliminary road map (Figure 1) to reach this goal. This map is an initial proposal and indicates areas we feel may be important in developing sustainable technology. UOP will be clarifying its vision, developing an improved road map and plan to achieve sustainable development as time progresses and will work with its customers to ensure that social, environmental, and economic needs of all involved are satisfied.

There are progressive stages in the development of sustainable growth that span from optimization efforts to the ultimate goal of a society that utilizes renewable resources. The Y-axis on the figure indicates the degree of technical difficulty that is increasing with time. The exact contribution to sustainable development made by each of the topics outlined in the figure needs to be determined. Additional areas will be identified and others deemphasized or dropped as we learn more. We are beginning to put together information related to these topics to determine their relevance to sustainable growth. As expected, there is currently substantially more effort in the early stages along this journey, but increasing realization that we must focus our research and development efforts on the latter stages. A detailed description of the four stages to sustainable development with specific examples follows.

Figure 1. Sustainable Technology Road Map

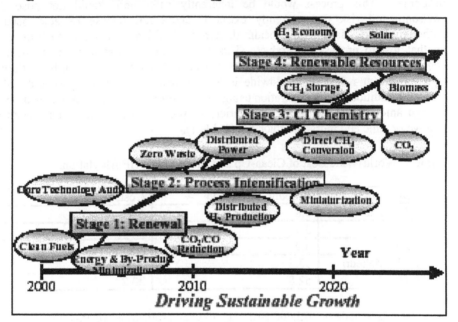

Stage 1 of this process is renewal. Past and continuing efforts have primarily been focused in this stage of development. Activities include improving feedstock utilization, driving down energy intensity of existing processes, developing clean fuels, minimizing undesirable by-products, and reducing environmental impact of existing processes while maintaining profitability. This is the first step toward sustainable development.

UOP and others have numerous efforts in this stage. Some of the programs include desulfurization technology from hydroprocessing to polishing off-spec product to meet "zero" sulfur specifications. Other programs involve upgrade of LCO feedstock to be utilized as a clean fuel or as a petrochemical feedstock. Additionally, UOP is continually looking at ways of managing its catalyst products for the various processes. This includes improved catalyst production yields and catalyst formulations that are environmentally compatible from cradle to grave.

An example of UOP technology in this stage is the Alkylene™ process for the production of motor fuel alkylate gasoline. Currently, alkylate gasoline is produced via the reaction of isobutane with C_3-C_5 olefins in the presence of either hydrofluoric or sulfuric liquid acids. The alkylate gasoline prepared by this route is almost completely paraffinic and is an excellent blending stock because it has essentially no sulfur, nitrogen, aromatics, olefins, or volatile organic compounds. In addition, it has acceptable vapor pressure, boiling range, and high octane number. The

combination of these properties makes alkylate gasoline a preferred component for
our transportation fuel. It would be desirable to increase the amount of alkylate in
our gasoline pool and minimize or eliminate the undesirable components.
Development of a solid acid catalyst to replace these hazardous liquid acids is highly
desirable.   This process would be inherently safer and would not pose the
environmental risk that currently exists if accidental release of acid occurs.
Although this risk is generally localized, there is significant transport of HF to make
up that lost to process by-product. This heavy fluoride-containing by-product is
generally burned resulting in unwanted emissions. The total amount of hydrofluoric
acid on site at refineries worldwide is approximately 3.3 MM Kg or 0.87 MM
gallons. Additionally, the operation using hydrofluoric acid results in the creation of
a large amount of solid waste as a variety of process streams must be treated to
remove organic and inorganic fluorides (Table 1).

Table 1. World-Wide Cost of Clean-Up and Make-Up in HF Alkylation.

Material	Amount, MM lbs/yr	Cost, $MM/yr
Alumina	9.8	5.9
KOH	3.9	2.0
Lime	5.9	1.9
HF Makeup	33.2	23.4
TOTAL	52.8	33.2

Note: Estimate made on 975,000 BBL/day alkylate production world-wide

Over time the operating efficiency of hydrofluoric acid alkylation units has
improved in order to minimize waste streams as well as minimize the actual
hydrofluoric acid inventory. However, the only way to eliminate these issues
completely is to develop a process based on a heterogeneous acid catalyst. Such a
process would eliminate the hydrofluoric acid transportation and inventory, result in
no waste polymer product, and would not require fluoride scrubbing and
concomitant disposal of fluoride-containing solids.

Technologically, the Alkylene process is a break-through. Several significant
inventions were required to make it technically feasible. The development of a
unique solid-acid catalyst and transport reactor by UOP allows for the potential
elimination of hazardous liquid acid processes. About 1 MM gallons of hydrofluoric
acid inventory could be eliminated, transport of 33 MM lbs (4 MM gallons) of
hydrofluoric acid per year would be stopped, and ca. 20 MM lbs per year of other
fluoride containing solids would not have to be land filled.

The costs for the new and sustainable technology are found to be similar to those
of the existing technology (for an 8,000 BBL/day plant, an EEC of $MM 43.5 for the
Alkylene process and $MM 40.5 for hydrofluoric acid alkylation). The increased
costs for the novel Alkylene reactor are offset by costs for HF mitigation capital and

metallurgy in parts of the unit. With maturity and experience, the costs for this new technology are expected to decrease. UOP is currently developing an improved catalyst and process that will further reduce the cost of the Alkylene process thereby making it more economically competitive with hydrofluoric acid technology. Commercial operation of this technology is expected in 2008.

Stage 2 of UOP's roadmap entails process intensification which involves substantial simplification to minimize process steps, reduce capital costs, and improve product yields. Early efforts in process intensification included combining process steps such as reaction and separation. Reactive distillation has been effectively demonstrated in commercial environments for methylacetate and MTBE production, respectively (5, 6). In these cases separation during reaction also allows high conversion to be achieved by eliminating the equilibrium constraints at process conditions. This process intensification via combining process steps usually leads to lower capital and operating costs and many time improved yields, more efficient use of feedstock, minimization of by-products, and lower energy requirements. Again it is a step in the right direction, but not sustainable.

Process intensification has recently involved miniaturization of the reactor in order to greatly reduce heat and mass transfer resistances. Dramatic changes in technology can occur with the advent of order of magnitude increases in the rates of heat and mass transfer. Distributed production of power and chemicals would be realized as a direct result of effective miniaturization. Process intensification would allow precise control of reaction conditions that could dramatically impact product yields and purity. These improvements lead to reduced use of raw materials, decreased cost of purification and waste handling, and lower energy consumption. Because total volume is dramatically reduced, difficult to produce or hazardous chemicals can be produced as needed and in a distributed way at point of use. This minimizes the inherent handling problems or environmental and safety issues resulting from transportation of these materials. Distributed production of hydrogen is a prime example of this miniaturization for distributed production. In the Hythane^{TM} process, mini-reformers are used to convert methane to hydrogen or a hydrogen/natural gas blend for fuel use in cities around the world, including Chicago, Los Angeles, and Palm Springs (7).

Miniaturization might also be used in the distributed production of chemicals. Two examples include the safe production of hydrogen peroxide from hydrogen and oxygen or the production of chemicals or fuels from synthesis gas. More detailed study into the potential improvements in product yields, waste minimization, and costs are required to fully understand the impact of miniaturization. At this time it offers the potential to produce hazardous materials or conduct potentially dangerous oxidation reactions on a small scale that is inherently safer.

Stage 3 shifts the source of hydrocarbon for transportation and chemicals to methane. Methane currently is reformed at elevated temperatures and pressures to synthesis gas. This mixture of hydrogen and carbon monoxide can then be converted via the well-known technologies of methanol synthesis and Fischer-Tropsch synthesis to eventually produce a variety of chemicals and fuels. In this stage, focus will be given to direct methane conversion technologies that will substantially reduce dependence on petroleum and coal. It will also provide substantial reductions in energy and emissions resulting from the high temperature indirect routes to power, chemicals, and fuels involving synthesis gas. Storage of methane for transport or use on demand is also a prime target for development in the future. A much more detailed analysis of the potential for direct conversion approaches to replace indirect and established routes for production and utilization of synthesis gas must be completed. A hypothetical process is considered below for discussion. At this time UOP is considering its opportunities in the area of direct conversion of methane to chemicals or fuels.

The direct catalytic conversion of methane has been actively pursued for many years. Much of the emphasis has been on the direct production of methanol via selective partial oxidation (8), coupling of methane to ethylene (9), or methane aromatization (10). At this time none of these technologies has been demonstrated commercially due to low yields of desired products due to combustion by-products or low equilibrium conversion at reasonable process temperatures and pressures. The potential benefits of a hypothetical process for the direct partial oxidation of methane to methanol (11) are presented as an example.

According to data from 2001, methanol is consumed at a rate of about 67 billion pounds per year and is the eighth ranked chemical in production value in the United States. Current energy usage for production is 0.015 MMBTU/lb and nearly 950 trillion BTU energy per year consumed globally for methanol production. About 80 percent of the methanol is produced from natural gas. Therefore, a more efficient and direct process should result in dramatic energy savings and eliminate substantial amounts of CO_2 emissions. Table 2 lists the benefits of a hypothetical process for direct methanol production. Assuming a modest methanol yield of 9.5% (10%

Table 2. A Cost Comparison between of Methanol Produced by Conventional Technology in the Mid/Far-East and a Proposed Process

	Conventional Process (cents/lb)	Hypothetical Methane to Methanol Process (cents/lb)
Raw Material Cost	0.74	0.61
Catalyst and Chemicals	0.06	0.08
Operating Cost	0.52	0.26
Tax/overhead	0.50	0.25
Depreciation	1.25	0.63
Cash Cost	1.82	1.20
Total Cost	3.07	1.83

conversion and 95% selectivity), it was calculated that raw material costs decrease from 0.74 to 0.61 cents/lb, operating costs reduce by 50% (0.52 to 0.26 cents/lb.), and total cost is decreased from 3.06 to 1.83 cents/lb. Additionally, it is estimated that capital investment could be reduced by as much as 50%. Most importantly, this new process could result in a net 17 trillion BTU/yr savings. Development of a process with further improved yields maintaining high selectivity will produce greater energy and economic savings. The validity of these estimates will need to be checked as process development progresses.

Table 3 lists the environmental benefits. Substantial reductions in oxide and VOC emissions will be realized upon commercializing such a process. Furthermore, development of a process with improved yields maintaining high selectivity will produce greater energy and economic savings and further minimize pollutants. These benefits are a result of milder operating conditions and increased overall selectivity to methanol per unit methane converted (including utilities). It is important to note this hypothetical process is still a long way from commercialization and several technological breakthroughs in catalyst and process design will be required to realize these benefits. UOP is currently working on developing a novel process for direct conversion of methane to methanol and has been awarded a $5MM contract over 3 years by ATP/NIST (Award 70NANB4H3041) to conduct the research.

Table 3. Methane to Methanol: Calculated Reduction in Pollutants   (assuming technology is commercialized in 2012)

Pollutant Reductions (lbs)	2005	2010	2015	2020
Carbon (MMTCE/yr)	N/A	N/A	0.026	0.082
Nitrogen Oxides (NOX)	N/A	N/A	405,463	1,269,365
Sulfur Oxides (SOX)	N/A	N/A	360,238	1,127,782
Carbon Monoxide (CO)	N/A	N/A	79,533	248,991
Volatile Organic Compounds (VOCS)	N/A	N/A	8,811	27,584
Particulates	N/A	N/A	6,628	20,749
Other (million lbs)	N/A	N/A	130	407

Stage 4 moves to the utilization of renewable resources.   As technology progresses into this stage, the world's energy will come from primarily renewable resources, such as solar, biomass, and thermo chemical. The desired fuel would be hydrogen which is ultimately derived from water and not a hydrocarbon. Efficient hydrogen storage will also have to be developed for use by this time.

Considerable research effort is being placed into the development of renewable resources. Although steady progress is being made in areas such as photovoltaics, there are numerous discoveries yet to be made. At this time, the cost of producing electricity via electrolysis of water using photovoltaics is about $0.12/kWh. In order to be competitive, this cost must be reduced substantially via improved efficiency of the overall system and reduced cost resulting from mass production.

444

The utilization of biomass feedstock is in most cases not economically competitive today. One process that is commercially viable is the production of polylactic acid (PLA) from corn starch.[3] Although lactic acid production from the fermentation of carbohydrates has been known for over 150 years, processing the lactic acid into useful polymers was a difficult and expensive operation. Cargill's primary technical innovation was the development of melt-phase technology for the ring opening polymerization of the lactide dimer. Upon demonstration of the technology at the pilot scale, Cargill teamed with Dow Chemical to provide leverage into the polymer market. Cargill Dow LLC was formed in 1997 to commercialize the PLA technology. Its PLA polymer products are 100% derived from renewable resources and are produced with the cost and performance necessary to compete with petroleum-based packaging materials and traditional fibers. Detailed life-cycle assessments of PLA compared with petrochemical derived polymers have been performed and published; net fossil fuel reduction of 20-50%, with 15-60% reduction in the generation of greenhouse gases is claimed.

For economic viability of biomass-derived processes to be achieved, innovations that produce high added-value chemicals will likely be required. For use as a source of fuels or bulk chemicals, the cost of collecting and processing the biomass, as well as improved conversion technology, will need to be developed. As the availability of oil and gas eventually decreases and the cost increases, technology involving use of biomass should become more important without the need of government subsidies. At this time UOP is considering where it can apply its core competencies in the area of renewable resources. During this study period we are trying to identify which potential technologies will truly lead to sustainable development. This is a difficult analysis requiring detailed life cycle studies.

Conclusion

Sustainable growth is a major issue facing the chemical, petrochemical, and refining industries in the foreseeable future. The challenge to meet environmental, societal, and economic performance simultaneously will require both breakthrough and incremental innovations. The industry has met such challenges successfully in the past. At UOP, we are convinced that by delivering innovation in process technology and equipment, catalysts and advanced materials, as well as solutions and services we will contribute to developing a sustainable future. This effort will be conducted with the input of our customers to ensure we are meeting their needs.

References

1. M.D. Parker, *Pressing on with Our Journey - The Imperative of Sustainable Development*, CAEC, 2001.
2. S.L. Hart., Strategies for a Sustainable World, (*Harvard Business Review*), (1997).
3. Drumright, Gruber, and Henton, "Polylactic Acid Technology", **12**(23), (*Adv. Mat.*) 1841, (2000).
4. a) US Patent 5,686,276, (1995) "Bioconversion of a fermentable carbon source to 1,3-propanediol by a single microorganism" b) US Patent 6,514,733, (2003) "Process for the biological production of 1,3-propanediol with high titer".

5. J.J. Siirola, An Industrial Perspective on Process Synthesis., **92**(304), *AIChE Symp. Ser.*, 222, (1995).
6. S. Davis., "UOP Ethermax Process for MTBE, ETBE, and TAME Production", *Handbook of Petroleum Refining Processes - second edition*, 13.9-13.12, (1997).
7. U.S. Patent 5,139,002, "Special purpose blends of hydrogen and natural gas". (1992).
8. Q. Zhang, D. He, Q. Zhu, Recent Progress in Direct Partial Oxidation of Methane to Methanol, **12**, *Journal of Natural Gas Chemistry*, 81-89, (2003).
9. J.H. Lunsford., The Catalytic Oxidative Coupling of Methane, **34**, *Angew. Chem. Ed. Engl.*, 970, (1995).
10. G.V. Echevsky, E.G. Kodenev, O.V. Kikhtyanin, V.N. Parmon,, Direct Insertion of Methane into C3-C4 Paraffins Over Zeolite Catalysts: a Start to the Development of New One-Step Catalytic Processes for Gas to Liquid Transformation, **258**, *Applied Catalysis A: General*, 159-171, (2004).
11. J.M. Maher, Development of Novel Catalysts for the Oxidation of Methane to Methanol with High Selectivity and Throughput, Proposal for the Chemicals and Forest Products Industries of the Future, Solicitation DE-PS36-03GO93015, (2004).

49. Selective Oxidation of Propylene to Propylene Oxide in CO_2 Expanded Liquid System

Hyun-Jin Lee, Tie-Pan Shi, Bala Subramaniam and Daryle H. Busch

Center for Environmentally Beneficial Catalysis, University of Kansas, Lawrence, KS 66047

bsubramaniam@ku.edu, and busch@ku.edu

Abstract

Chemists have long recognized the need to replace the Chlorohydrin and Halcon industrial processes for propylene oxide production because they pose environmental pollution concerns and also yield large amounts of undesirable by-products. Although O_2 and H_2O_2 have been suggested as clean oxidants, their use in propylene oxide production is still confronted by many problems. In this work, "green" methods for the epoxidation of propylene were explored by studying systems based on CO_2 expanded liquid solvents (CXLs). The phase behavior of several CXLs involving C_3H_6, CH_3CN, H_2O_2 and H_2O were evaluated to systematically apply these media to the epoxidation of propylene under mild reaction conditions in the presence of a variety of catalysts. These catalyst systems give good conversion and selectivity of propylene oxide at moderate operating temperatures and pressures, paving the way for novel routes for the production of propylene oxide and propylene glycol using environmentally benign solvents such as CO_2 and H_2O.

Introduction

Propylene oxide is mainly used for the production of polyesters, polyols, and propylene glycol; the latter is the starting material for polyurethane and unsaturated resins, and other products. Industrially over six million tons of propylene oxide (PO) is produced each year using either the Chlorohydrin or the Halcon process. In the Chlorohydrin process, propylene reacts with Cl_2 and H_2O to produce 1-chloro-2-propanol and HCl, which is then treated with base to generate propylene oxide and salt. There are two pounds of salt waste for each pound of PO produced. Both present methods consume large amounts of chlorine and lime which are finally converted to useless and environmentally polluting waste (1). Minimizing waste in the selective oxidation of propylene to PO has long been an important objective of industrial chemistry. Direct oxidation of propylene to PO with O_2 would be highly desirable, but the presence of propylene's highly combustible allylic hydrogens renders this quite difficult.

In pursuit of a more environmentally friendly process, CO_2 has been investigated as a medium (2). CO_2 is considered environmentally acceptable, non-toxic, relatively cheap (~5 cents/lb), non-flammable, inert toward oxidation and readily available (3).

Supercritical media, in general, have the potential to increase reaction rates, to enhance the selectivity of chemical reactions and to facilitate relatively easy separations of reactants, products, and catalysts after reaction (3). However reactions involving CO_2 and water are typically conducted as biphasic processes, with the organic substrate dissolved mostly in the CO_2-rich phase and the water-soluble catalysts and/or oxidant dissolved in the aqueous phase. Such systems suffer from inter-phase mass-transfer limitations (4).

In this project, a 'green" method for the epoxidation of propylene is developed by employing CO_2 expanded liquid solvents (CXLs) in the absence of a catalyst. Ternary CXLs are homogeneous mixtures of dense CO_2, an organic solvent such as acetonitrile or methanol, and water. The oxidant is formed *in situ* through the reaction of carbon dioxide and hydrogen peroxide to produce peroxycarbonic acid as shown in Figure 1. A base will be used to control the acidity of the solution during reaction between the peroxycarbonic acid and substrate propylene. By tuning the solvent properties of the CXL containing $CH_3CN/H_2O_2/H_2O$ with dense-phase CO_2 both substrate and oxidant will be present in a homogeneous reaction mixture inside the *view cell* reactor.

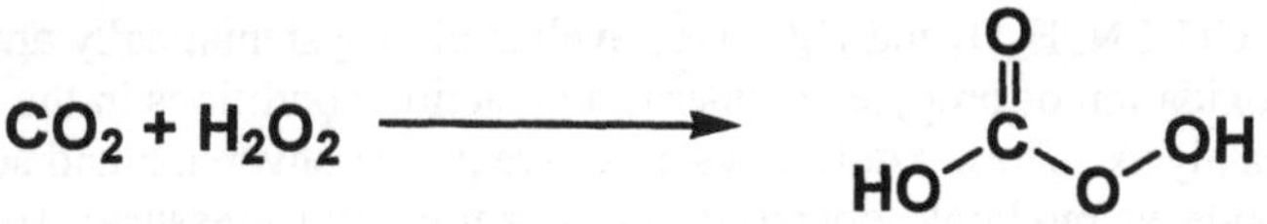

Peroxycarbonic Acid

Figure 1. *In situ* formation of oxidant for the epoxidation of propylene in CXLs.

Richardson and co-workers (5). showed that epoxidation of alkene can achieved using the percarbonate ion which is formed by reaction of H_2O_2 and sodium carbonate under basic conditions. Beckman and his group (2, 6). also reported the low conversion of propylene to propylene oxide using percarbonate formed via reaction of CO_2, H_2O and H_2O_2. Presumably, the reaction between H_2O_2 and CO_2 yields a peroxycarbonic acid species, i.e. a compound in which the peroxy group of hydrogen peroxide binds to the carbon of carbon dioxide (7). According to this view, the peroxycarbonic acid transfers oxygen to the olefin.

Building on earlier work in these laboratories (8) we have overcome the typical mass transfer limitations of phase transfer catalysis for propylene oxidation by the use of 3-component liquid phases based on CO_2 expanded liquids (CXLs). For the application to oxidations by aqueous H_2O_2, the organic component of the CXL is chosen because it is miscible with both dense CO_2 and water. In this way, homogeneous systems are produced which decrease mass-transfer limitations and intensify chemical reactions. Previous reports using CXL systems have shown that they enhance the oxidation of the substrate and improve the selectivity at moderate reaction temperatures and pressures (3, 8, 9).

Results and Discussion

In order to provide clear operating conditions, it is necessary to understand the phase relationships in systems where a single three-component liquid phase is desired for conducting chemical reactions. In preliminary experiments, we examined acetonitrile and methanol as organic solvents to form ternary CXL systems with homogeneous regions that may be exploited for performing the reactions in single liquid phases. Therefore phase behavior studies were carried out with acetonitrile $(AN)/H_2O/CO_2$ and $MeOH/H_2O/CO_2$ ternary systems using a Thar Variable Volume Phase Equilibrium Analyzer. In this manner, the bifurcation between single phase and biphasic operation in P-T space was identified for fixed compositions of the ternary system (Figure 2). The epoxidation reactions (at 40°C) showed that reactions in systems based on acetonitrile are more productive and more selective that those in methanol-based systems. Based on such measurements, the system with molar composition 0.343/0.453/0.093 $(X_{CO_2}/X_{AN}/X_{H_2O_2})$, the remaining being water, was applied to maintain a single liquid phase during the reaction.

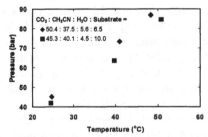

Figure 2. Single liquid phase condition for $CO_2/AN/H_2O/Propylene$ system

Three control experiments were performed to identify critical operating conditions of the reaction system, including low temperature limit, necessity for a base and essential nature of the terminal oxidant. First, 0.120 ml of pyridine, 0.360 ml of H_2O_2 and 3 ml of CH_3CN were loaded in the *view cell* reactor (Figure 3), which was then charged with propylene and liquid CO_2 at 545 psi. The reaction was carried out at 21°C for 24 h. The second control experiment was conducted without base at 40°C. The third such experiment was planned to learn the role of pyridine N-oxide in the reaction. The experiment was performed with pyridine N-oxide replacing both pyridine and hydrogen peroxide. In view of performance observed in other systems (unpublished work), pyridine N-oxide might function as either or both the oxidant and the base. All three control experiments yielded no detectable products. The reactor contents were analyzed for products by GC and GC/MS and the results are summarized in Table 1. From these preliminary experiments operating conditions were partially defined. Both the oxidant and the base were found to be essential components of the reaction system and operating condition substantially below 40°C were shown to be nonproductive.

Table 1. Control experiments

Run	Reaction condition	H_2O_2/H_2O	Base (mmoles)	Products detected
1	21°C for 24 h	0.36 ml	Pyridine (1.48)	None
2	40°C for 48 h	0.36 ml	None	None
3	40°C for 12 h	None	Pyridine N-oxide (1.43)	None

Some consideration was given to appropriate bases for this oxidation system. Pyridine and sodium bicarbonate had been shown to be useful earlier[8] and were included. Additional bases considered were a proton sponge (1,8-Bis(dimethylamino)naphthalene) and, as stated above, pyridine N-oxide. Because of the low solubilities in CXL systems of $NaHCO_3$ and the proton-sponge, pyridine was selected for these exploratory studies.

Propylene oxidation was carried out in the absence of catalysts, but in the presence of the base pyridine, at 40-70°C for 3, 6, and 12 hr time periods. The results are summarized in Table 2. The control experiments were designed to include H_2O_2 and a proper base because earlier studies[4, 8] had shown that both are required for the oxidation of other olefins, specifically cyclohexene. Water is only slightly soluble in supercritical carbon dioxide, and propylene is miscible with CO_2.[10] However, the addition of propylene in the system will not enhance the solubility of water or H_2O_2. At 40°C, the yield of propylene oxide was about 7% with pyridine in 6 hrs. The PO yield was decreased and trace amounts of propylene glycol and unknown products were produced at 60 and 70°C. As expected, the PO yield was increased by longer reaction times. When the amount of propylene was decreased by about half, the PO yield was also reduced by roughly the same fraction. From these results, we concluded that a single phase reaction has been achieved and that the single phase reaction does enhance the yield of PO as well as the selectivity of this CXL system. This and related systems are subject to on-going studies.

Table 2. Propylene oxidation at various temperatures

Rxn. time (Hrs)	Rxn. Temp. (°C)	Solvent	H_2O_2 (mmoles)	Pyridine (mmoles)	Yield of PO (%)
3	40	CO_2/AN	10.41	1.49	< 3
6	40	CO_2/AN	10.41	1.49	7.07
6	60	CO_2/AN	10.41	1.49	1.66
6	80	CO_2/AN	10.41	1.49	1.73
12	40	CO_2/AN	10.41	1.49	10.7

Experimental Section

The view-cell reactor is made of titanium and has two sapphire windows, a gas inlet valve and an outlet valve, as shown in Figure 3. The view cell is interfaced with a pressure transducer, a thermocouple, and a pressure relief valve. The pressure and temperature are computer-monitored during the reaction. 0.6 ml of 50 wt% H_2O_2/H_2O (10.41 mmoles), 0.20ml of pyridine (2.47 mmoles), or some other base, was dissolved in 5 ml of acetonitrile or methanol, and was added to the reactor. 2.2 ml of supercritical CO_2 was charged after 100mg of propylene (2.38 mmoles) had been added to the reactor. The reactor was heated with a band heater at 40 – 70°C for 3, 6, 12, and 24 hr reaction periods. Following a batch conversion experiment, the amounts of products formed were determined by GC and GC/MS.

 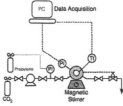

Figure 3. The view-cell reactor and diagram of system

Acknowledgements

This work was supported by the National Science Foundation Grant CHE-0328185 and, in part, by the facilities created under NSF/ERC Grant EEC-0310689.

References

1. E. J. Beckman, *Environ. Sci. Technol.* **37**, 5289-5296 (2003).
2. E. J. Beckman, *Green Chemistry* **5**, 332-336 (2003).
3. B. Subramaniam; D. H. Busch, *ACS Symp. Ser.* **809**, 364-386 (2002).
4. S. A. Nole, J. Lu, J. S. Brown, P. Pollet, B. C. Eason, K. N. Griffith, R. Glaeser, D. Bush, D. R. Lamb, C. L. Liotta, C. A. Eckert, G. F. Thiele and K. A. Bartels, *Ind. Eng. Chem. Res.* **41**, 316-323 (2002).
5. H. Yao and D. E. Richardson, *J. Am. Chem. Soc.* **122**, 3220-3221 (2000).
6. D. Hancu, J. Green and E. J. Beckman, *Acc. Chem. Res.* **35**, 757-764 (2002).
7. G. Thiele, S. A. Nolan, J. S. Brown, J. Lu, B. C. Eason, C. A. Eckert and C. L. Liotta, U.S. Pat 6100412, (2000).
8. B. Rajagopalan, M. Wei, G. T. Musie, B. Subramaniam and D. H. Busch, *Ind. Eng. Chem. Res.* **42**, 6505-6510 (2003).
9. M. Wei, G. T. Musie, D. H. Busch and B. Subramaniam, *J. Am. Chem. Soc.* **124**, 2513-2517 (2002).
10. S. J. Macnaughton and N. R. Foster, *Ind. Eng. Chem. Res.* **33**, 2757-63 (1994).

VII. Symposium on Other Topics in Catalysis

50.   cat*A*Sium® M : A New Family of Chiral Bisphospholanes and their Application in Enantioselective Hydrogenations

Thomas H. Riermeier,[a] **Axel Monsees,**[a] **Jens Holz,**[b] **Armin Börner**[b] **and John Tarabocchia**[c]

[a]*Degussa AG, Degussa Homogeneous Catalysts, Rodenbacher Chaussee 4, 63457 Hanau-Wolfgang, Germany*
Thomas.Riermeier@degussa.com
[b] *Leibniz-Institut für Organische Katalyse an der Universität Rostock e.V., Buchbinderstr. 5/6, 18055 Rostock, Germany*
[c]*Degussa Corporation, 379 Interpace Parkway, Parsippany, NJ07054*
john.tarabocchia@degussa.com

Abstract

The development of bisphospholanes and their application in enantioselective hydrogenation have influenced the academic as well as the industrial research during the last years. The excellent results that were obtained with bisphospholanes made them part of so called privileged ligands. We will shortly summarize the recent developments in this field and focus on a new type of bisphospholanes: catASium[□] M. The structure of catASium[□] M is shown in Figure 1. The tuneable backbone of these ligands opens a wide range of bite angles. We have developed a toolbox of different ligands that combines the high asymmetric induction of the phospholanes with tuneable bite angle modulated by the backbone to solve a given synthetic challenge.

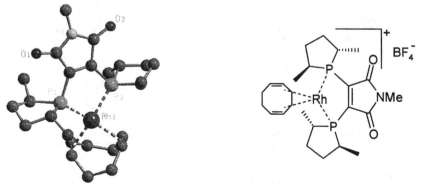

Figure 1. Molecular structure of catASium[□] M (see Figure 5 for other variants)

Introduction

Today, several pharmaceuticals, agrochemicals, flavors and fragrances as well as functional materials are used as enantiomerically pure material with an increasing importance. This tendency has made the enantioselective preparation of chiral compounds in an industrial scale an important topic (1). Among various strategies employed, homogeneous asymmetric hydrogenations using catalysts based on ⎾soft⏋ transition metals like Rh(I), Ru(II) or Ir(I) has become one of the most versatile strategies for the preparation of such compounds in large scale (2). In the overwhelming case as chiral ancillary ligands bidentate - recently also monodentate - trivalent phosphorus compounds have been used (3). Up to now, more than two thousand of such P-ligands have been synthesized and tested in the enantioselective hydrogenation of functionalized and non-functionalized olefins, ketones, imines and other prochiral substrates. Landmarks in this development were Kagan⎾s DIOP (4), Knowles⎾s DIPAMP (5), Noyori⎾s BINAP (6), and Burk⎾s DuPHOS (7). The overall success and popularity of these privileged ligands is anchored in the fact that they give good or excellent high enantiomeric excesses for a variety of substrate classes in hydrogenation.

In particular noteworthy is the high reactivity and intrinsic enantioselectivity of Rh-complexes based on the C_2-symmetric bisphospholane ligand DuPHOS, including its ethylene bridged analogue, BPE, for the hydrogenation of a broad range of substrates. These marked features stimulated the development of related bisphopholanes, such as PennPHOS (8), RoPHOS (9), BASPHOS (10) ButiPHANE (11), BeePHOS (12) and ligands of the bisphosphetane series, namely CnrPHOS (13) (Figure 2). The driving forces of these efforts were to elude the claims of the DuPHOS patents and to find easier synthetic access to related and more efficient ligands. Several of these diphosphines have been synthesized in collaboration with chemical companies and small samples are available now.

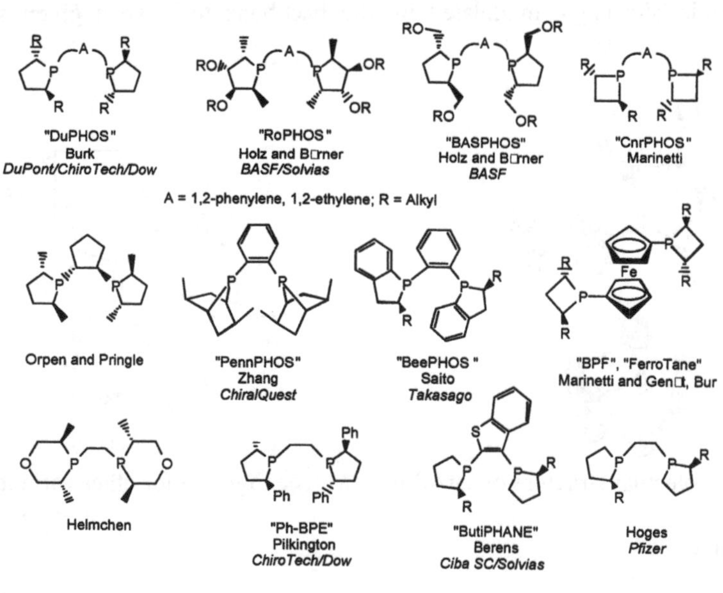

Figure 2. Selected chiral bisphospholanes and related cyclic diphosphines

It is interesting to note, that in these diphosphines, with the exception of ButiPHANE, both chiral units are linked by an 1,2-phenylene and 1,2-ethylene bridge, respectively, or in case of BPF/FerroTane by an 1,1'-ferrocenyl moiety. Based on the original procedure of Burk such ligands are prepared by the double nucleophilic substitution of chiral cyclic sulfates with corresponding primary diphosphines in the presence of strong bases. This tedious approach is limited to the synthesis of bisphospholanes (bisphosphetanes) with those bridging units where the corresponding primary phosphines are synthetically accessible. Due to this obstacle to date DuPHOS-type Rh-catalysts with definite P-Rh-P bite angles cannot be constructed. Hence, the ligand synthesis does not allow flexibility and the performance can not be tuned for a given challenging substrate.

Results and Discussion

As a part of the collaboration between Degussa AG and the Leibniz-Institut of Organic Catalysis in Rostock we envisaged the modular synthesis of diverse arrays of chiral vicinal bisphospholanes. These ligands, now commercially available in multi-kg scale under the trademark cat*A*Sium$^\square$ M, can be characterized by varying P-C=C angles and electronic properties in the bridge (Figure 3). For the construction of cat*A*Sium$^\square$ M ligands, we searched for a general structural principle, which allows in the corresponding Rh-catalysts the stepwise variation of the bite angle (α) over a broad range. We anticipated that due to their large variety 1,2-functionalized heterocycles derived from maleic acid could be a promising scaffold for this goal. In particular, we targeted larger bite angles (14) than usually found in DuPHOS-Rh-complexes. We speculated that an increase of the bite angle could reinforce the diastereo-discriminating interactions between catalyst and coordinated prochiral substrate. Furthermore, small changes, eg. O vs. NMe, have incremental effects and should allow for fine tuning of these new ligands.

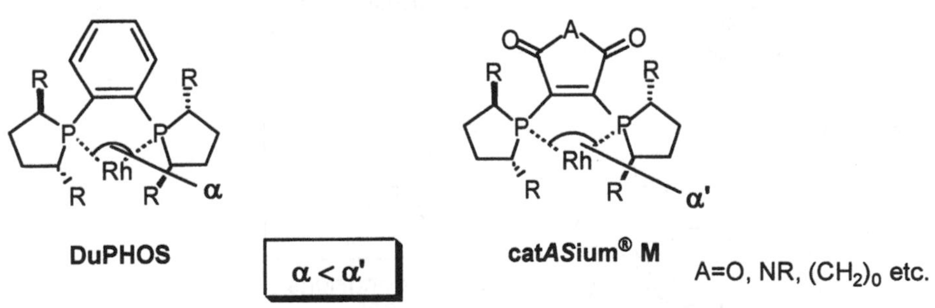

DuPHOS $\alpha < \alpha'$ cat*A*Sium® M   A=O, NR, (CH$_2$)$_0$ etc.

Figure 3. Influence of the shape of the backbone on the bite angle α

The synthesis of the new ligands should be based on a convergent methodology, that is easy to perform and convenient to scale up.

As mentioned above, for several reasons the commonly used method for the preparation of bisphospholanes and bisphosphetanes tracing back to the seminal work of Burk does not match the requirements for the synthesis of cat*A*Sium$^\square$ M ligands.

First, the preparation of required primary phosphines is rather tedious and expensive. Since for each ⎕bridge⎕ the individual synthesis of relevant primary phosphines is necessary, the preparation of a family of primary diphosphines is - even if chemically feasible ⎕ not practicable. Secondly, most heterocycles bearing primary phosphines are not stable due to the attack of the phosphine group on the functionalities of the heterocycle. Third, the nucleophilic substitution of cyclic sulfates by the metal phosphide in the presence of an excess of strong bases may seriously interfere with the heterocyclic ring or other functional groups. Therefore a new and convergent synthetic strategy was necessary which allows the coupling of the bridging scaffold with a phospholane moiety under neutral conditions without using primary phosphines or phosphides as the building block.

As a first approach we found that a suitable 2,5-dimethyl substituted P-silylated phospholane can be easily derived from the corresponding chiral bis-mesylate by reaction with dichlorophenylphosphine (Figure 4) (15). This TMS-phospholane proved to be a remarkablely stable reagent. It could be distilled under vacuum and stored under argon for a long time without any tendency of decomposition or epimerization. When this building block was reacted with 2,3-dichloromaleic anhydride in etheral solution at 0⎕C, in the absence of any base, a smooth and clean coupling reaction took place. The optically pure ligand **1a** could be precipitated in 53 % yield as dark-red crystals from the reaction solution. The same methodology could also be successfully employed for the synthesis of related ligands such as **1b** and **1c** based on maleimide and squaric acid, respectively. As another possibility of variation, a silylphospholane bearing ethyl groups in 2,5-positions was employed in the coupling procedure. Due to this uniform methodology a broad variety of ligands are accessible in a technically feasible and practicable manner.

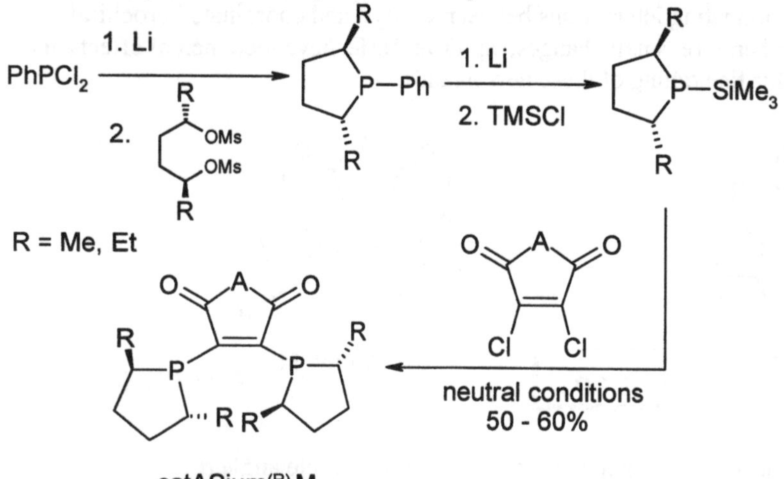

Figure 4. Modular synthesis of cat*A*Sium⎕ M ligands

Reaction of the diphosphines with [Rh(COD)₂]BF₄ afforded precatalysts of the type [Rh(COD)(ligand)]BF₄. DFT calculations and X-ray structural analyses revealed that the P-Rh-P bite angles of complexes based on **1a-c** are significantly larger than that of the corresponding DuPHOS complex. Compared to the benzene backbone in DuPHOS, the five-membered and four-membered backbone, respectively of cat*A*S*ium*□ M ligands cause a larger *exo* PC=C angle. This in turn widens the bite angle in the rhodium complexes.

Corresponding rhodium precatalysts were tested in the enantioselective hydrogenation. In the first trials, as a benchmark test standard substrates like methyl (*Z*)-*N*-acetamido cinnamate or itaconates were employed. In general, catalysts based on new ligands afforded good or excellent enantioselectivities. In some cases a decline in the enantioselectivity was noted by application of the ligand with the squaric acid backbone **1c**. The corresponding Rh-complex has the largest bite angle among the catalyst investigated.

Obviously, the enantioselective hydrogenation has an optimum for each ligand and solvent used. Figure 5 exemplifies a typical fine tuning approach employing four related bisphospholane ligands. After three levels of testing, 99 % ee could be achieved in the hydrogenation.

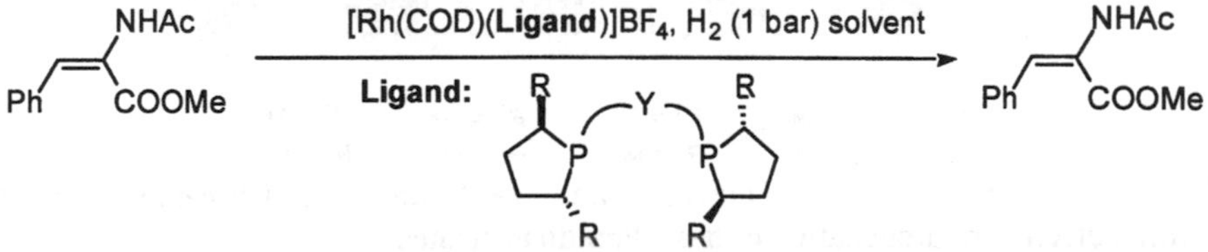

	1.Level		2.Level		3.Level	
	R: Me solvent: MeOH		R: Me solvent: THF		R: Et solvent: THF	
Y		%ee		%ee		%ee
(benzene)		97.6		97.4		-
(N-Me maleimide)		96.9		97.3		-
(maleic anhydride)		94.4		98.6		99.0
(squaric)		83.9		82.3		-

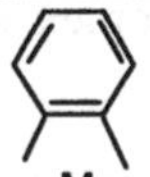

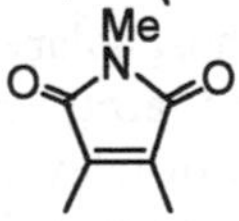

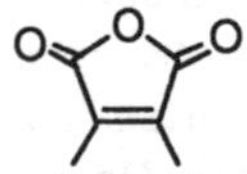

Figure 5. Fine tuning of enantioselective hydrogenation (only the best ligand after level 2 was taken to level 3 of the optimization)

Interesting results were also observed when the hydrogenation of α-acylamido acrylates being important intermediates for the synthesis of enantiopure α-amino acids was investigated.

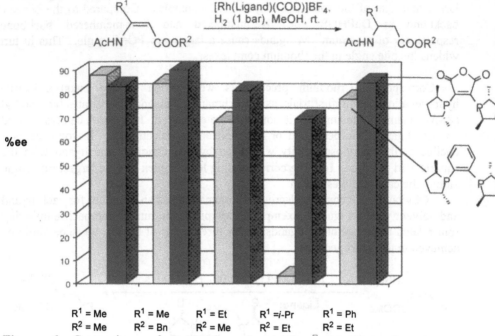

Figure 6. Comparison of DuPHOS and cat*A*S*ium*□ M Rh-catalyst in the enantioselective hydrogenation of α-acylamido acrylates

In particular for the hydrogenation of the important *Z*-configured substrates bearing bulkier substituents in 3-position (Et, *i*-Pr, Ph) a cat*A*S*ium*□ M Rh-catalyst bearing the maleic anhydride backbone (**1a**) gave significantly higher enantioselectivities than the related DuPHOS-complex (Figure 6). Differences by up to 75 % ee were observed. A similar superiority of the cat*A*S*ium*□ M complex was noted in THF as solvent.

In summary, the modular synthesis of a new type of vicinal bisphospholanes called cat*A*S*ium*□ M have been discovered, which affords a broad array of diphosphine ligands with small and definite differences in the bite angle of relevant metal complexes. Besides new applications, the new ligands can be advantageously employed for the fine tuning of those enantioselective hydrogenations where DuPHOS-type ligands need optimization. Due to the modular synthesis each diphosphine ligand can be easily provided in kg-quantities by Degussa AG, *Degussa Homogeneous Catalysts*.

References

1. I. Ojima, *Catalytic Asymmetric Synthesis*, Wiley-VCH: New York, 2000. H. U. Blaser and E Schmidt, *Asymmetric Catalysis on Industrial Scale*, Wiley-VCH: Weinheim 2003.
2. H.-U. Blaser, F. Spindler and M. Studer, *Appl. Catal.: General*, **221**, 119 (2001). H.-U. Blaser, *Chem. Commun.*, 293 (2003).
3. W. Tang and X. Zhang, *Chem. Rev.* **103**, 3029 (2003).
4. T. P. Dang and H. B. Kagan, *J. Am. Chem. Soc,.* **94**, 6429 (1972).
5. B. D. Vineyard, W. S. Knowles, M. J. Sabacky, G. L. Bachman and D. J. Weinkauff, *J. Am. Chem. Soc.* **99**, 5946 (1977).
6. R. Noyori, M. Ohta, Y. Hsiao, M. Kitamura, T. Ohta and H. Takaya, *J. Am. Chem. Soc.,* **108**, 7117 (1986).
7. M. J. Burk, J. E. Feaster, W. A. Nugent and R. L. Harlow, *J. Am. Chem. Soc.*, **115**, 10125 (1993).
8. Q. Jiang, Y. Jiang, D. Xiao, P. Cao and X. Zhang, *Angew. Chem. Int. Engl.*, **37**, 1100 (1998).
9. J. Holz, M. Quirmbach, U. Schmidt, D. Heller, R. St□rmer and A. B□rner, *J. Org. Chem.*, **63**, 8031 (1998); W. Li, Z. Zhang, D. Xiao and X. Zhang, *J. Org. Chem.*, **65**, 3489 (2000).
10. J. Holz, R. St□rmer, U. Schmidt, H.-J. Drexler, D. Heller, H.-P. Krimmer and A. B□rner, *Eur. J. Org. Chem.*, 4615 (2001).
11. Compare H.-U. Blaser, C. Malan, B. Pugin, F. Spindler, H. Steiner and M. Studer, *Adv. Synth. Catal.*, **345**, 103 (2003), Ref. 26b.
12. H. Shimizu, T. Saito and H. Kumobayashi, *Adv. Synth. Catal.*, **345**, 185 (2003).
13. A. Marinetti and D. Carmichael, *Chem. Rev.*, **102**, 201 (2002).
14. P.W.N.M. van Leeuwen, C.P. Casey and G.T. Whiteker, *Rhodium Catalyzed Hydroformylation*, P.W.N.M. van Leeuwen, C. Claver, Eds., Kluwer Academic Publishers, Dordrecht, The Netherlands, p.63-96 (2002)
15. T. Riermeier, A. Monsees, J. Almena Perea, R. Kadyrov, B. Gotov, W. Zeiss, I. Nagl, A. Boerner, J. Holz, K. Drauz, and W. Meichelboeck, Wilfried, WO 2005049629 to Degussa AG (2005).

51.   Synthesis of Chiral 2-Amino-1-Phenylethanol

Robert Augustine, Setrak Tanielyan, Norman Marin, Gabriela Alvez

Center for Applied Catalysis, Seton Hall University, South Orange, NJ 07079

Abstract

Two methods for the synthesis of chiral 2-amino-1-phenylethanol have been developed. The first utilizes a chiral oxaborolidine catalyzed borane reduction of phenacyl chloride to give the chiral chloro-alcohol in very good yield with an ee in the 93%-97% range. Reaction with dilute ammonium hydroxide produced the amino-alcohol in very good yield with a high ee. The second approach involved first conversion of the phenacyl chloride to the succinimide which was then hydrogenated using a chiral ruthenium complex in conjuction with a base and an optically active amine (Noyori procedure) to give the optically active succinimido alcohol in very good yield with an ee of 98%. Hydrolysis with dilute base produced the optically active amino alcohol in very good yield and excellent enantioselectivity.

Introduction

Chiral β-amino alcohols such as (R) or (S) 2-amino-1-phenylethanol (**1**) are important intermediates in the synthesis of a variety of pharmaceutically important compounds. While there were some early procedures reported for the synthesis of **1**,(1,2) several years ago an evaluation of the more promising of these approaches led to the conclusion that the most practical method for preparing **1** in larger quantities was by the resolution of the racemic material using di-O-toluoyltartaric acid (3) or dehydroabietic acid (4). In these resolutions, though, the chiral **1** was produced with 99% ee but was obtained in only about a 25% yield. In recent years, however, some efficient methods for the enantioselective reduction of ketones have been developed and it was considered that application of some of these newer reactions could lead to the efficient and enantioselective production of β-amino alcohols such as **1**. The two procedures selected for investigation were the chiral oxazaborolidine catalyzed reduction of phenacyl chloride (**2**)(5-7) followed by amination of the chloro alcohol product (8) (Eqn. 1) and the Noyori hydrogenation (9) of an appropriately blocked β-aminoacetophenone followed by unblocking of the amine group (Eqn. 2).

Results and Discussion

The chiral reduction of phenacyl chloride (**2**) was run using either the methyl- or methoxy- oxazaborolidine (**3**) as the catalyst. After optimization of the reaction parameters the reduction was run at 65°C using a BH_3 : ketone : catalyst ratio of 0.6 : 1 : 0.003. Reduction of a 0.9 M solution of **2** in THF under these conditions gave the chiral chloroalcohol (**4**) with 95-96% ee at 100% conversion. The aminoalcohol (**1**) was produced by treating a methanol solution of **4** with a large excess of 30% NH_4OH at room temperature for 2-3 days. After evaporation of the methanol, **1** was isolated as a crude product having an ee of 95% in 85% yield.

Even though **2** was successfully converted to **4** by a chiral transfer hydrogenation (10), attempts at running a classic Noyori hydrogenation of **2** were unsuccessful. Trying to hydrogenate a primary aminoketone would only lead to problems arising from self condensation. A blocked amine group was needed. The phthalimido ketone (**5**) could easily be prepared by reaction of **2** with potassium phthalimide. It has been reported that **5** could be hydrogenated enantioselectively using a chiral rhodium catalyst but the reaction only took place at high pressures (11). Another problem with the use of **5** as a precursor in the preparation of **1** was the very low solubility of **5** in methanol or *i*-propanol which are the solvents we used in the Noyori hydrogenations. However, the analogous succinimide, **6**, was sufficiently soluble under the hydrogenation conditions and was, therefore, used in the screening of the various ligands, diamines and bases used in the Noyori hydrogenation (Eqn. 2).

The ligands tested were: tol-binap, dipamp, P-phos, xyl-P-phos, binam, phanephos and xyl-phanephos with the results lised in Table 1. Table 2 shows the effect which

Table 1. Effect of ligand and solvent on the hydrogenation of 1 mmole of **6** using 10 µmole of Ru(ligand)(S,S)-DPEN at 50°C and 60 psig with 0.25 mmole of KO-*t*Bu.

Ligand	*MeOH/i-PrOH*	ee (%)	r (mmol/min)	Conv. (%)
(S)-Tol-Binap	30/70	73	0.068	87
(R)-Tol-Binap	30/70	-59	0.072	100
(R)-P-Phos	30/70	6	0.006	100
(R)-Xylyl P-Phos	30/70	7	0.020	100
(R)-Binam	60/40	55	0.030	100
(R)-Binam	100/0	68	0.020	100
(S,S)-Dipamp	30/70	93	0.005	95
(R)-Phanephos	30/70	64	0.070	100
(R)-Xylyl Phanephos	30/70	86	0.020	96
(R)-Xylyl Phanephos	60/40	>99	0.021	100
(R)-Xylyl Phanephos	100/0	>99	0.140	100
(R)-Xylyl Phanephos	EtOH	>99	0.039	22

different bases used in conjuction with the chiral diamine, DPEN, had on the reaction rate and product ee in the hydrogenations run using the xyl-phanephos ligand. The optimized reaction conditions were found to be the use of Ru(Xyl-Phanephos) modified with DPEN as the catalyst and running the reaction in a methanol solution at 30°C and 60 psig of hydrogen in the presence of potassium *t*-butoxide. A substrate:catalyst ratio of 2500 was routinely used giving the succinimido alcohol, **7**, in >98% yield with an ee of > 99%. Increasing the hydrogen pressure to 200 psig significantly reduced the reaction time while still maintaining these same ee and conversion values. Hydrogenation of **6** using dichloro [(R)-Xylyl-phanephos][1S,2S-DPEN]ruthenium (II) gave the S enantiomer of **7** with 99% ee at 100% conversion. Using (S)-xylyl phanephos and (1R,2R)-DPEN resulted in the formation of the R enantiomer of **7**, again with 99% ee at 100% conversion.

Table 2. Effect of the base used on the hydrogenation of 2 mmole of **6** using 2 µmole of Ru((R)-Xylyl Phanephos) (S,S)-DPEN at 30°C and 60 psig H_2.

Base (0.25 mmole)	ee (%)	r (mmol/min)	Conv. (%)
None	>99	0.0004	2
KO-*t*-Bu	>99	0.090	100
LiO-*t*-Bu	>99	0.069	100
K_2CO_3	-	0	0
K_2CO_3/12-Cro-4(1:1)[a]	>99	0.016	52
K_2CO_3/15-Cro-5(1:1)[a]	>99	0.010	68
K_2CO_3/18-Cro-6(1:1)[a]	>99	0.011	68
Li_2CO_3/12-Cro-4(1:1)[a]	>99	0.007	50
Li_2CO_3/15-Cro-5(1:1)[a]	>99	0.006	70
Li_2CO_3/18-Cro-6(1:1)[a]	>99	0.013	64
DABCO	0	0.004	0

[a] Carbonates were used in conjunction with crown ethers.

It was originally thought that one should be able to remove the succinic acid group by treatment of **7** with hydrazine in the same way one is able to produce a primary amine by treating a phthalimide with hydrazine in the classical Gabriel synthesis (12). This was not the case, though, since **7** did not react with hydrazine. However, it was found that treatment of **7** with dilute sodium hydroxide readily hydrolyzed the succinimide to produce the amino alcohol, **1**, in 90% yield and having a 98 - 99% ee.

Experimental Section

All reagents and ligands were obtained from Aldrich, Acros or Strem and were used without further purification. The methanol, *i*-propanol and DMF were distilled over calcium hydride under an argon atmosphere and stored in sealed flasks under argon. The THF was purified by distilled over sodium ketyl under an argon atmosphere and stored in a sealed flask under argon. The (S)-5,5-diphenyl-2-methyl-3,4-propano-1,3,2-oxazaborolidine (**3**) was purchased from Aldrich as a 1M solution in toluene [(S)-2-methyl-CBS-oxazaborolidine]. All gases used were zero grade. The HPLC analyses were accomplished using a 250 x 4.6mm Chiracel OJ column with appropriate internal standards.

Oxazaborolidine reductions:

A 100 mL jacketed reaction flask equipped with an addition port, reflux condenser and a magnetic stir bar was first purged with argon for 10 minutes and then charged with a solution of 0.125 mmole of catalyst **3** in 11 mL of THF and 34 mL of a BH_3-THF solution containing 24 mmole BH_3, added portionwize over 10 min. The reaction flask was heated to 65°C and a solution of 6.19g (40 mmole) of phenacyl chloride (**2**) in 7.6 mL of THF was added via syringe pump over three hours at a rate of 0.08 mL/min. After addition was complete, the reaction mixture was stirred for 60 min, cooled to ambient and the remaining hydride was decomposed with 5 mL of dry methanol. The reaction mixture was passed through a short bed of SiO_2 and the solvent evaporated giving the crude chloro alcohol, **4**, in 95% yield with ee = 95%.

Five and a half grams of this material was dissolved in 30 mL of methanol followed by the addition of 100 mL of 30% aqueous ammonium hydroxide. The resulting mixture was stirred at room temperature for 2-3 days after which time the light precipitate which formed was removed by filtration and the methanol only removed by rot-evaporation. To the cold water phase was added 15 g of sodium chloride and 30% aqueous ammonium hydroxide to pH 12.5. This was extracted twice with 50 cc portions of ether. The organic phase was dried over magnesium sulfate and evaporated giving 3.4 g (83%) of the crude amino alcohol, **1**, ee = 95%.

Catalytic hydrogenations

Succinimido Ketone, 6: To 17.22 g (0.17 mole) of succinimide in a 1000 mL round bottomed flask was added 150 mL of THF and the suspension stirred at 65°C until the succinimide was dissolved. In a separate 250 mL flask 20.4 g (0.18 mole) potassium *t*-butoxide was suspended in 120 mL of THF and the suspension sonicated for 10 min to give a cloudy solution which was added dropwise to the stirred succinimid solution at such a rate that the internal temperature remained below 20°C. After the addition was completed the mixture was sonicated for an additional one hour. To this solution (suspension) was added dropwise over a period of one hour with stirring a solution of 27.84g (0.18 mole) of phenacyl chloride (2)in 120 mL of DMF. After addition the reaction mixture was stirred overnight during which time the initial red solution changed color to orange. Water (2.5 L) was added dropwise to the stirred suspension and the pale yellow precipitate was filtered and dried in a vacuum oven at 45°C/3-4 mm Hg to give 30 g of the crude succinimido ketone (79% yield). This material was recrystallized twice from ethanol (700 mL) to produce a white solid which after drying in a vacuum oven at 40°C gave 20 g (67%) of the succinimido ketone, **6**, of sufficient purity for the hydrogenation step.

Catalyst preparation: Five milligrams (0.02 mmol Ru) of {[RuCl$_2$(benzene)]$_2$} and 14 mg of CTH-(R)-3,5-xylylphanephos (0.02 mmol) were placed in a 50 mL Schlenk flask. After completely replacing all of the air with argon, 2 mL of DMF was added to the flask via cannula and the mixture heated to 100°C for 10 min with stirring to give a reddish brown solution. After cooling to ambient, 4 mg S,S-DPEN (0.19 mmol) dissolved in 2 mL of degassed DMF was added and the mixture stirred for 3 hours. The DMF was removed under 1 mm Hg pressure at 25°C and then at 50°C to give a light yellow solid containing 20 μmole Ru. This was dissolved in 10 mL of degassed methanol to give a stock catalyst solution containing 2 μmol Ru/mL. This solution was kept in a completely de-aerated air tight septum sealed flask until use.

A second stock methanol solution containing 0.25 mmol / mL of potassium *t*-butoxide was prepared using 700 mg (6.25 mmol) of potassium *t*-butoxide dissolved in 25 mL of degassed methanol. This was also kept in a deaerated air tight septum sealed flask until use.

Asymmetric hydrogenation: The succinimido ketone, **6**, (4.34 g, 20 mmole) was placed in a 250 mL jacked glass reactor vessel (13) and the air in the reactor completely replaced by argon using fill-release cycles. Degassed methanol (250 mL) was added to the reactor via cannula and the mixture stirred at 30°C to dissolve the ketone. Four milliliters of the catalyst solution (8 μmol Ru) and 12 mL of the K-O-*t*Bu solution (3 mmole) were injected into the reactor using gas tight syringes. The argon in the reactor was replaced with hydrogen (fill/release cycles) and the reactor pressurized to 60 psig with hydrogen. The reaction mixture was stirred at 1000 rpm overnight at 30°C with the hydrogen uptake recorded as described previously (13). The reaction mixture was passed through a short alumina column and the solvent removed to give the (S) succinimido alcohol, **7**, with an ee of 99% at 100% conversion.

Amino alcohol, 1: The succinimido alcohol, **7**, (1.74g) was dissolved in 60 mL of 95% ethanol and 36 mL of 20% aqueous sodium hydroxide added to the solution. The solution was the refluxed for 18 hours and cooled to ambient which resulted in the separation of two layers. The top, organic, layer was separated and evaporated to dryness to give a solid material which was refluxed with 30 mL of MTBE and 10 mL of methylene chloride for 30-40 minutes to extract the amino alcohol from the solid sodium succinate. After cooling to ambient the solid was removed by filtration through a Celite pad and the filtrate decolored by stirring with Norit. Filtration gave a clear solution which, after evaporation, produced 980 mg of the white amorphous amino alcohol, 1 (90% yield, e = 98+%). The NMR and IR spectra and the HPLC traces of 1 were identical to those of an authentic sample of (S)-(+)-2-amino-1-phenylethanol obtained from Alfa Aesar

Acknowledgements

We would like to acknowledge financial support for this project from Sapphire Therapeutics.

References

1. R.K. Atkins, J. Frazier, L. Moore, and L.O. Weigel, *Tetrahedron Lett.*, **27**, 2451, 1986.
2. A.I. Meyers and J. Slade, *J. Org. Chem.*, **45**, 2785 (1980)
3. O. Lohse and C. Spondlin, *Org. Process Res. Dev.*, **1**, 247 (1997)
4. Z. Guangyou, L. Yuquing, W. Zhaohui, H. Nohira and T. Hirose, *Tetrahedron Asymmetry*, **14**, 3297 (2003).
5. E.J. Cory, R.K. Bakshi and S. Shibata, *J. Amer. Chem. Soc.,* **109**, 5551 (1987).
6. E.J. Corey and C. J. Helal, *Angew. Chem. Int. Ed.*, **37**, 1986 (1998).
7. R. Hett, C.H. Senanayake and S.A. Wald, *Tetrahedron Lett.*, **39**, 1705 (1998).
8. J. Duquette, M. Zhang, L. Zhu and R.S. Reeves, *Org. Process Res. Dev.*, **7**, 285 (2003).
9. T. Ohkuma, D. Ishii, H. Takeno, R. Noyori, *J. Am. Chem. Soc.*, **122**, 6510 (2000).
10. T. Hamada, T. Torii, K. Izawa, R. Noyori and T. Ikariya, *Org. Lett.*, **4**, 4373 (2002).
11. A. Lei, S. Wu, M. He and X. Zhang, *J. Am. Chem. Soc.*, **126**, 1626 (2004).
12. M.S. Gibson and R. W. Bradshaw, *Angew. Chem. Int. Ed.*, **7**, 919 (1968).
13. R.L. Augustine and S.K. Tanielyan, *Chem. Ind. (Dekker)*, **89** (Catal. Org. React.), 73 (2003).

52.   Selective *Tertiary*-Butanol
Dehydration to Isobutylene via Reactive
Distillation and Solid Acid Catalysis

John F. Knifton and John R. Sanderson

P.O. Box 200333, Austin, TX 78720

Abstract

Selective *tertiary*-butanol (tBA) dehydration to isobutylene has been demonstrated using a pressurized reactive distillation unit under mild conditions, wherein the reactive distillation section includes a bed of formed solid acid catalyst. Quantitative tBA conversion levels (>99%) have been achieved at significantly lower temperatures (50-120°C) than are normally necessary using vapor-phase, fixed-bed, reactors (*ca.* 300°C) or CSTR configurations. Substantially anhydrous isobutylene is thereby separated from the aqueous co-product, as a light distillation fraction. Even when employing crude tBA feedstocks, the isobutylene product is recovered in ca. 94% purity and 95 mole% selectivity.

Introduction

There has been an enormous technological interest in *tertiary*-butanol (tBA) dehydration during the past thirty years, first as a primary route to methyl *tert*-butyl ether (MTBE) (1) and more recently for the production of isooctane and polyisobutylene (2). A number of commercializable processes have been developed for isobutylene manufacture (eq 1) in both the USA and Japan (3,4). These processes typically involve either vapor-phase tBA dehydration over a silica-alumina catalyst at 260–370°C, or liquid-phase processing utilizing either homogenous (sulfonic acid), or solid acid catalysis (e.g. acidic cationic resins). More recently, tBA dehydration has been examined using silica-supported heteropoly acids (5), montmorillonite clays (6), titanosilicates (7), as well as the use of compressed liquid water (8).

$$(CH_3)_3C-OH \quad \rightarrow \quad (CH_3)_2C=CH_2 \quad + \quad H_2O \qquad (1)$$

In this research initiative, we have examined the potential of reactive distillation (9) for *tertiary*-butanol dehydration to isobutylene using solid acid catalysis. Advantages to employing reactive distillation for reaction (1) include: a) the mild operating conditions required (<120°C), b) quantitative tBA conversions per pass, and c) the option to use lower purity/lower cost, tBA feedstocks.

Results and Discussion

The dehydration of aliphatic alcohols to olefins is known to be catalyzed by acidic materials, wherein the reactant alcohol interacts with Bronsted acid sites and dehydrates to the corresponding alkene via a carbenium ion mechanism (7). *Tertiary*-butanol dehydration to isobutylene is an endothermic reaction (38 kJ/mol) (10) and the forward reaction (eq 1) is generally favored by high operating temperatures and low pressure (3). However, too high a temperature (>340°C) can lead to the formation of polyisobutylene by-products (3). The use of reactive distillation techniques, where the co-product water is immediately separated from the isobutylene as it is formed, allows the equilibrium of eq 1 to be shifted far to the right and high tBA conversions achieved under relatively mild operating conditions. Here we demonstrate quantitative tBA conversions to isobutylene using a pressurized reactive distillation unit (illustrated below, in Figure 1), in combination with four classes of highly-active, inorganic solid acid catalysts (9), namely:

- Beta zeolites
- HF-treated β–zeolites
- Montmorillonite clays
- Fluoride-treated clays

Typical data for crude (94.9%) tBA feedstocks are illustrated in Table 1. In the synthesis of ex. 1, 350 cc of zeolite Beta, having a silica-to-alumina ratio of 24 and a surface area of ca. 630 m²/g (comprising 80% Beta and 20% alumina binder), is initially charged to the unit as 1/16″ diameter extrudates. Under steady state conditions, with the reboiler temperature set at ca. 110°C, the column temperature 64-99°C, and the crude tBA feed rate 100 cc/hr, the overhead product fraction comprises ca. 94% isobutylene. The corresponding bottoms product is 97% water, and includes just 0.2% unreacted *tertiary*-butanol. The estimated tBA conversion is then >99%, and the isobutylene selectivity 95 mole%. Ex. 3-5 illustrate somewhat similar data for a second sample of zeolite Beta, as well as an HF-treated montmorillonite clay and an HF-treated Beta-zeolite catalyst. The normal column operating temperature range is generally 50-120°C under equilibrium conditions (10,11).

Impurities in the crude tBA feed, most notably water, methanol, acetone, and "heavies", do not appear to significantly inhibit the dehydration process (eq 1), although the presence of methanol clearly leads to the formation of additional MTBE (either through etherification of the tBA feedstock, or the isobutylene co-product).

Extended life for the zeolite Beta catalyst has been demonstrated in this work using the same, or similar, crude tBA feedstocks to those employed in Table 1. Isobutylene generation has been monitored over ca. 500 to 1000 hours of service, under steady state reactive distillation conditions, without significant losses in activity or changes in product compositions.

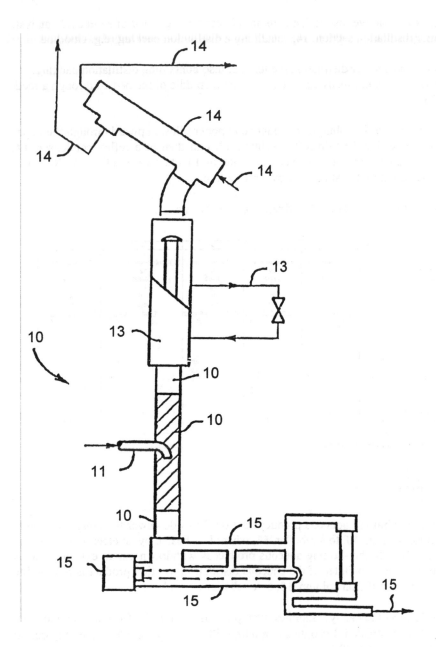

Figure 1. Reactive distillation unit design.

Experimental Section

The *tertiary*-butanol dehydration experiments described herein were conducted in a pilot unit reactive distillation unit of the type shown in Figure 1. The unit comprises a reactive distillation column, **10**, containing a bed of solid acid catalyst, an upper distillation section, **14**, containing a distillation packing (e.g. Goodloe

packing) and a lower distillation section, **13**, also containing distillation packing. The *tertiary*-butanol feedstock is charged to the middle of the bed **10** through a feed line, **11**.

The more volatile isobutylene product component flows upward through the upper distillation section **10** to a reflux splitter, **13**, and then to a reflux condenser, **14**, where it is cooled by room temperature water and withdrawn via line **14**. Reflux is recirculated within the splitter **13** by a reflux line.

Table 1. *Tertiary*–butanol dehydration to isobutylene

Ex.	Feed	1		2		3		4		5	
Catalyst		Beta-zeolite (a)		Beta-zeolite (a)		Beta-zeolite (b)		HF/Montmorillonite clay (c)		HF/Beta-zeolite (d)	
Reboiler Temp. (C)		110		99		92		127		112	
Column Temp. (C)		64-99		72-97		67-83		50-105		87-96	
Reflux Temp. (C)		50		51		46		(e)		(e)	
Reactor Pressure (kPa) (f)		80		60		60		140		80	
Feed Rate (g/hr)		100		145		103		100		100	
		Overhead	Bottoms	Overhead	Bottoms	Overhead	Bottoms	Overhead	Bottoms	Overhead	Bottoms
Water	1.4	0.1	97.2		77.8		56.8	0.2	96.5	1.3	95.3
Methanol	0.4	0.5		0.04				0.1		0.5	
Isobutylene		94.4		93.3		76.9	0.2	80.7		59.9	
Acetone	1.4	1.6		1.7				0.9		10.7	
Isopropanol	0.3			0.04	0.5	8.4	0.9				
tBA	94.9		0.2	3.2	18.7	6.1	40			12.7	0.7
MTBE	0.05	2.2		1.3	0.2	6.3	0.2	5.1	1.9	10.3	1.8
Methyl ethyl ketone	0.2	0.05			1.1		0.2				
Diisobutylene	0.06	0.2	0.6	0.08	0.9	0.3	0.5	12.4	0.7	1.2	1.2
Unknowns	1.4	0.9	2	0.3	0.7	1.9	1.2	0.6	0.6	3.4	1
tBA Conversion (%)		>99		95.6		89.1		(e)		91.7	
Isobutylene Sel. (mole%)		94.5		88.5		67.2		(e)		64.9	

(a) From PQ Corp., 1/16 ins diameter extrudates
(b) From UOP, 1/16 ins diameter extrudates
(c) 0.6% HF on montmorillonite clay
(d) 5.7% HF treated Beta zeolite
(e) Not determined
(f) Reactor pressure above atmospheric (101 kPa)

The higher boiling aqueous product fraction flows downwards through the lower distillation section, **10**, to a reboiler, **15**, where it is heated by an electrical heater. A portion of this higher-boiling aqueous product is withdrawn via an exit line, **15**, as shown, and the remainder of the aqueous distillation reaction product is returned to the reactive distillation column, **10**, by a reboiler return line.

The Beta-zeolite catalyst samples were purchased from PQ Corporation and from UOP. The HF-treated β-zeolite and montmorillonite clay samples were prepared as described previously (9,12).

References

1.   P. M. Morse, *C&E News* , 26 (April 12 1999).
2.   *C&E News*, 9 (November 8 1999).

3. O. C. Abraham and G. F. Prescott, *Hydrocarbon Processing*, 51 (February 1992).
4. See for example: US Pats 3,665,048 and 5,436,382, to Arco Chemical Technology (1972 and 1995), US Pats 4,155,945 and 4,165,343, to Cities Service Company (1979), and US Pat. 4,873,391, to Mitsubishi Rayon Company (1989).
5. R. Ohtsuka, Y. Morioka, and J. Kobayashi, *Bull. Chem. Soc. Jpn*, **62**, 3195 (1989).
6. M. L. Kantam, P. L. Santhi, and M. F. Siddiqui, *Tetrahedron Lett.*, **34**, 1185 (1993).
7. A. Philippou, M. Naderi, J. Rocha, and M. W. Anderson, *Catal. Lett.*, **53**, 221 (1998).
8. X. Xu and M. J. Antal, *Am. Inst. Chem. Eng. J.*, **40**, 1524 (1994).
9. J. F. Knifton, P. R. Anantaneni, P. E. Dai, and M. E. Stockton, US Pats 5,770,782, 5,777,187, and 5,847,254, to Huntsman Petrochemical Corporation (1998).
10. J. F. Knifton, J. R. Sanderson, and M. E. Stockton, US Pat. 5,811,620, to Huntsman Specialty Chemicals Corporation (1998).
11. J. F. Knifton, J. R. Sanderson, and M. E. Stockton, *Catal. Lett.*, **73**, 55 (2001).
12. J. F. Knifton and J. C. Edwards, *Appl. Catal. A* **183**, 1 (1999).

53. Leaching Resistance of Precious Metal Powder Catalysts – Part 2

Tim Pohlmann, Kimberly Humphries, Jaime Morrow, Tracy Dunn, Marisa Cruz, Konrad Möbus, Baoshu Chen

Degussa Corporation, 5150 Gilbertsville Highway, Calvert City, KY 42029

tim.pohlmann@degussa.com

Abstract

Carbon supported powdered palladium catalysts have been widely used in the chemical industry. In addition to activity and selectivity of those catalysts, the recovery rate of the incorporated precious metal has a major impact on the economic performance of the catalyst. In this study, the effects of catalyst age, oxidation state of the incorporated metal and temperature treatment on the palladium leaching resistance as well as on activity and dispersion of carbon supported palladium catalysts were investigated.

Introduction

Palladium-based precious metal powder catalysts are used for a wide variety of industrial hydrogenation reactions. In general, these catalysts are refined after use to capture the value of the incorporated metal. Due to attrition and metal leaching during the hydrogenation reaction, a significant part of the precious metal can be lost. Even though the majority of the leached metal can be recovered using metal scavengers like Degussa's Deloxan®, the metal loss can have an important influence on the overall economics of the catalytic process. Aim of catalyst research is thus to minimize the amount of metal losses. As a continuation of a previous study, this work investigates the influence of the catalyst age, degree of reduction, wetness of the catalyst and type of metal deposition (egg shell and uniform) on the leaching resistance as well as on hydrogenation activity and dispersion of powdered carbon supported palladium catalysts. (1) The leaching resistance of the catalysts was investigated by stirring the catalysts in a solution of ammonium chloride, which is known to be a model system to mimic the precious metal leaching in hydrogenation reactions. (2) The resulting mixtures were filtered and the mother liquor was analyzed for Pd by inductively coupled plasma (ICP).

Another part of our investigation deals with the effect of heat treatment on the leaching behavior of palladium on activated carbon catalysts. Heat treatment is a known technique to increase the performance of catalysts. (3) Therefore, standard carbon supported palladium catalysts were exposed to different temperatures ranging from 100 to 400 °C under nitrogen. The catalysts were characterized by metal leaching, hydrogenation activity and CO-chemisorption.

Results and Discussion

The influence of the catalyst age on the leaching resistance of precious metal powder catalysts was investigated. Eggshell and uniform, reduced and unreduced, dry and wet type palladium on activated carbon catalysts were prepared and characterized by their hydrogenation activity, metal dispersion and palladium leaching. The tested catalysts show a relatively high leaching in the first few days after preparation. This value drops remarkably in the first weeks after preparation. The hydrogenation activity and palladium dispersion did not change significantly during the same time period. The properties of the investigated catalysts of this study as well as their hydrogenation activities and palladium dispersions are summarized in table 1 and table 2.

Table 1. Properties of investigated catalysts

ERD	ERW	END	ENW	URD	URW	UND	UNW
eggshell	eggshell	eggshell	eggshell	uniform	uniform	uniform	uniform
reduced	reduced	non-reduced	non-reduced	reduced	reduced	non-reduced	non-reduced
dry	wet	dry	wet	dry	wet	dry	wet

The reduced catalysts show a significantly higher metal leaching compared to the corresponding non-reduced catalysts. The amount of Pd detected when leaching the catalysts four weeks after preparation was between 100 and 150 ppm for the investigated reduced catalysts and between 20 and 40 ppm for the non-reduced catalysts. No significant differences between the egg-shell and the uniform type catalysts were observed in this test.

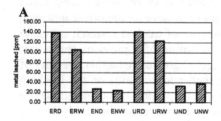 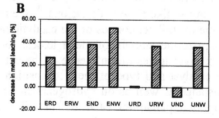

Figure 1A, B. Leaching results one month after catalyst preparation and relative decrease in Pd leaching compared to the leaching of a fresh catalyst

Similar trends were found for the palladium leaching values when leaching fresh catalysts. However, the overall amount of palladium leached is higher in this case. The results of the leaching tests performed one day after the catalyst preparation show values between 140 and 240 ppm for the reduced and of 30 to 60 ppm for the non-reduced catalysts. A comparison of the decrease of metal leaching over time shows different aging effects for the dry catalysts compared to the wet catalysts of this study. In the case of reduced catalysts, the catalysts ERW and URW show a relatively high drop in palladium leaching after four weeks compared to the test performed the day after the catalyst preparation. In contrast to these results, the dry analogues of these catalysts (ERD and URD) showed a much smaller decrease in palladium leaching (Figure 1).

Table 2. Relative change of activity and dispersion after four weeks

catalyst	ERD	ERW	END	ENW	URD	URW	UND	UNW
rel. activity	581	702	1053	925	395	403	689	568
rel. activity after one month	611	697	1074	1010	394	393	695	693
dispersion [%]	15.8	16.3	19.5	15.6	19.8	18.5	18.1	15.34
dispersion – after one month [%]	16.7	15.3	19.9	19.3	19.9	17.0	17.1	16.5

The hydrogenation activity and dispersion of all catalysts remains relatively constant during the evaluated period of time. The aging effect thus cannot be explained by sintering of the palladium crystallites, since this would also reduce the overall activity and dispersion of the catalyst. TPR experiments of the fresh and aged catalysts were carried out to investigate the influence of the oxidation state of the metal on the observed aging effect. A comparison of the TPR results of the fresh catalysts with the TPR results after two months shows that the reduced catalysts were slowly oxidized over time. The total extent of reduction in the TPR experiment denoted by the amount of H_2 consumed in the TPR to reduce the catalyst increased over a period of two months for the wet reduced catalysts from 50 ml/g of Pd to 120 ml/g of Pd of a 5 % H_2 in Ar mixture for the catalyst ERW and from 32 ml/g of Pd to 156 ml/g of Pd for the catalyst UNW. The corresponding results for the dry catalysts ERD (45 ml/g of Pd after two months) and URD (64 ml/g of Pd after two months) did not increase as strongly indicating that the dry catalysts are less prone to oxidation over time. This could also explain the smaller tendency of dry catalysts to show a reduced amount of leaching when storing the catalyst compared to the wet catalysts. As reported above the non reduced samples show a higher resistance against metal leaching compared to the reduced samples. These results also indicate that it is beneficial to use dried catalysts for applications that require a high degree of reduction of the catalyst, since dry catalysts show a higher resistance against oxidation during storage.

In a second part of this study, the effect of heat treatment under nitrogen of the reduced palladium catalysts A, B and C with an egg-shell type metal distribution on the metal leaching was investigated. The reduced catalysts were tested for metal leaching after they underwent a heat treatment at temperatures of 100 to 400 °C. The metal leaching of the investigated catalysts decreased after the heat treatment of

478 Palladium Leaching

temperatures higher than 200 °C. Heat treatment at lower temperatures showed only minor changes of the metal leaching (Figure 2).

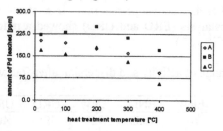

Figure 2 Pd leaching of heat treated catalysts

To identify the cause for this reduced amount of metal leaching after heat treatments the hydrogenation activities and dispersions of all investigated catalysts were determined. As expected, lower activities and dispersions were observed for the heat treated catalysts compared to the non-treated ones (Figure 3).

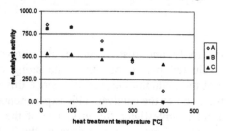

Figure 3 Relative hydrogenation activity of heat treated catalysts

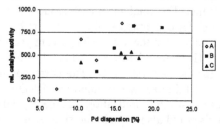

Figure 4 Correlation of relative hydrogenation activity with Pd-dispersion

This detected drop of activity and the increased resistance against palladium leaching when exposing the catalysts to an increased temperature can be explained by a sintering of the metal particles at elevated temperatures, which reduces the metal surface. The reduced palladium surface would cause lower hydrogenation activities and palladium dispersions and a reduced metal leaching (Figure 4). The good correlation of the relative hydrogenation activity with the palladium dispersion of the tested catalysts supports this theory.

Experimental Section

All catalysts were prepared using activated powder carbons using slurry methods. After precipitation of the metal salt, some of the catalysts were reduced. Some catalysts were dried after preparation and subsequently stored in a dry environment. Liquid phase hydrogenation of cinnamic acid or nitrobenzene at low pressures was performed to investigate the hydrogenation activity of each catalyst. The leaching resistance of the catalysts was investigated by stirring the catalysts in a solution of ammonium chloride. The resulting mixtures were filtered and the mother liquor was analyzed for Pd by inductively coupled plasma (ICP). Precious metal dispersions experiments by CO chemisorption at 25 °C on the catalysts and temperature programmed reduction (TPR) experiments were carried out using a Micromeritics Autochem 2910 unit by heating the sample from –35 °C to 700 °C at a rate of 30 °C/min in a 5% H_2/95% Ar mixture. The effluent gases were analyzed using a TCD to monitor the changes in the composition of the reducing gas as a function of the temperature.

Conclusions

Several carbon supported palladium catalysts were tested for hydrogenation activity, metal dispersion and metal leaching. These tests were repeated over a period of eight weeks. While the amount of metal leached reduces over time, activity and metal dispersion of the catalysts remains relatively constant. This trend was observed for eggshell and uniform type catalysts. Reduced catalysts showed a higher amount of palladium leached. The amount of palladium leaching was reduced over the period of the investigation. This reduced leaching effect was stronger for the catalysts that were stored in a wet form. TPR experiments of the reduced catalysts showed that the wet catalysts were slowly oxidized during the storage time, while the degree of oxidation of the dry catalysts was minimal. This oxidation effect is thus suspected to be the cause for the lower amount of palladium leaching of reduced catalysts after a storage time of several weeks. In a second part of this investigation, it was shown that palladium on activated carbon catalysts show a stronger resistance against metal leaching when heat-treated at temperatures higher than 200 °C compared to the non heat-treated analogues. Since the hydrogenation activity at low pressures and the metal dispersion was also reduced for the heat treated samples, it is probable that this effect is caused by sintering of the palladium crystallites of the catalyst.

References

1. The previous study was presented as a poster at the 19th North American Catalysis Society Meeting in 2005 (P-182)
2. A. J Bird, D. T. Thompson, *Catalysis in Organic Syntheses*, **91**, 61-106 (1980).
3. T. J. McCarthy, B. Chen, M. L. Ernstberger, F. P. Daly, Chemical Industries (Dekker), **82**, (*Catalysis of Organic Reactions*), 63-74 (2001).

54. Optimization of Reductive Alkylation Catalysts by Experimental Design

Venu Arunajatesan, Marisa Cruz, Konrad Möbus and Baoshu Chen

Degussa Corporation, 5150 Gilbertsville Hwy, Calvert City, KY 42029

Venu.arunajatesan@degussa.com

Abstract

Reductive alkylation is an efficient method to synthesize secondary amines from primary amines. The aim of this study is to optimize sulfur-promoted platinum catalysts for the reductive alkylation of p-aminodiphenylamine (ADPA) with methyl isobutyl ketone (MIBK) to improve the productivity of N-(1,3-dimethylbutyl)-N-phenyl-p-phenylenediamine (6-PPD). In this study, we focus on Pt loading, the amount of sulfur, and the pH as the variables. The reaction was conducted in the liquid phase under kinetically limited conditions in a continuously stirred tank reactor at a constant hydrogen pressure. Use of the two-factorial design minimized the number of experiments needed to arrive at the optimal solution. The activity and selectivity of the reaction was followed using the hydrogen-uptake and chromatographic analysis of products. The most optimal catalyst was identified to be 1%Pt-0.1%S/C prepared at a pH of 6.

Introduction

The synthesis of an N-alkylarylamine by the reductive alkylation of an aromatic primary amine with a ketone is used in the preparation of antioxidants for polymers and rubber. The alkylation of an amine with a ketone is typically carried out in the liquid phase using heterogeneous catalysts such as Pd, Pt, Rh, or Ru supported on carbon (1,2). The reaction of ADPA with MIBK yields an imine, which then is hydrogenated over a Pt or sulfur promoted-Pt catalyst to yield 6-PPD.

Optimization of a process or catalyst by experimental design such as two-factorial design can lead to significant reduction in time required to achieve the goal. In their excellent work, Mylroie *et al.* (3) reduced the time required for the optimization of the reductive alkylation process conditions by a factor of 10. Here, we turn our attention to the optimization of the catalyst rather than the process.

Since the preparation of a catalyst can be a complex process involving a number of variables, a thorough examination of all these variables would involve preparation of hundreds of catalysts. Platinum loading was a natural choice as one of the variables. It is well known that acids catalyze the formation of imine (2,4). It is also known that treatment of carbon with acids lead to the formation of acidic surface groups, hence pH during catalyst preparation was chosen as another variable (5). In the case of reductive alkylation, Thakur *et. al.* (6) showed that the reaction rate increases with sulfur loading therefore, sulfur loading was chosen as the third variable. Table 1 shows the range of values for each of the chosen parameters.

Experimental Section

Reactions were carried out in liquid phase in a well-stirred (1000 rpm) high-pressure reactor (Parr Instruments, 300 mL) at 30 bar and 150°C. The reaction mixture consisted of 61 g of ADPA (Acros Chemicals), 53 g MIBK (Acros Chemicals) and 370 mg of catalyst. The test procedures used here is similar to that described earlier by Bartels et al. (7). The reactor was operated at a constant pressure with the liquid phase in batch mode and the hydrogen fed in at a rate proportional to its consumption. The reaction was monitored by hydrogen uptake and the product yield was determined from gas chromatographic (Agilent Technologies, 6890N) analysis.

The catalysts were commercial catalysts from Degussa with Pt-loading from 1%-5%, S-loading varied between 0.1 and 0.5%. The pH of the catalyst during preparation was varied from 2-6. The dispersion of Pt was 52+/-5% for all the catalysts tested.

Results and Discussion

The reactions were conducted according to a two factorial design with three variables, which contains experimental points at the edges and the center of a face-centered cube leading to 9 different experiments. Typically, the experiment at the center point is conducted at least 3 times to add degrees of freedom that allow the estimation of experimental error. Hence a total of 11 experiments are needed to predict the reaction rate within the parameter space. The parameter space for the catalysts to be prepared is shown in columns 2-4 in Table 1.

The reactions were conducted in the liquid phase at conditions described in the experimental section. Test reactions were conducted to establish that the reactions were kinetically limited. In cases where the rate of reaction was >5 mmol/(g*min), the selectivity to 6-PPD was >97% and the yield of 6-PPD was >96%. Hence, the rate of hydrogen uptake was taken to be directly proportional to the formation of 6-PPD. This rate calculated at constant temperature and conversion was normalized to the amount of catalyst used and is shown in column 6 of Table1. The two cases where Pt/S ratio was high (Run 3 & 4), hydrogenation of the ketone (MIBK) to the alcohol, methyl isobutyl carbinol (MIBC), was observed. In cases where the Pt/S ratio was low (Run 5 & 6), significant amounts of the imine was detected in the GC.

It is apparent from Table 1 that the pH of catalyst preparation has only a small effect on the rate of reaction (compare run 1 & 2, 3 & 4, 7 & 8). But, it is not evident how either Pt-loading or S-loading affects the catalyst performance. In some cases (Run 1 & 3) it appears that lowering the S loading leads to lower activity while in other cases, the contrary appears to be true (Run 5 & 7). However, it is clear that at a constant molar ratio of Pt/S of 1.64, the activity of the catalyst remains consistently high (Run 1, 2, 7, 8 & 9). Significantly high (Run 3 & 4) or low (Run 5 & 6) Pt/S ratio appears to be detrimental to the catalyst activity. The vertical line in the standardized Pareto chart (Figure 1) delineates the parameters that are significant at 5% level indicating that the interaction variable AB (Pt-S) is the only variable that affects the catalyst activity. That is, based on our data, the probability that any variable other than AB (Pt-S) has an effect on the catalyst activity is <5%.

Table 1. The rate of consumption of hydrogen for various catalysts

Run	Pt (wt%)	S (wt%)	pH	Pt/S (mol/mol)	Rate (mmol/(g*min))
1	5	0.5	6	1.64	14.5
2	5	0.5	2	1.64	14.4
3	5	0.1	6	8.21	5.0
4	5	0.1	2	8.21	3.3
5	1	0.5	6	0.33	0.0
6	1	0.5	2	0.33	0.0
7	1	0.1	6	1.64	18.6
8	1	0.1	2	1.64	11.0
9	3	0.3	4	1.64	13.7
10	3	0.3	4	1.64	14.1
11	3	0.3	4	1.64	10.9

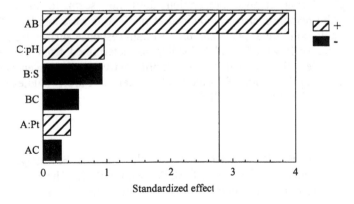

Figure 1. Standardized Pareto chart for rate of reaction

Table 2. Regression coefficients for the rate of reductive alkylation

Parameter	Coefficient	Parameter	Coefficient
Constant	16.0	-	-
Pt	-3.51	Pt*S	15.69
S	-41.19	Pt*pH	-0.18
pH	1.99	PH*S	-2.88

The data from Table 1 was fitted to a multiple regression model and the regression coefficients determined from the model are listed in Table 2. Thus the rate of reaction at any given value of Pt-loading, S-loading, and pH, within the parameter space, can be determined from the regression equation shown below:

Rate = 16.0 - 3.51*Pt - 41.19*S + 1.99*pH + 15.69*Pt*S - 0.18*Pt*pH - 2.88*S*pH

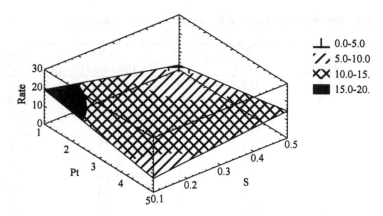

Figure 2 Response surface for the rate of reaction for catalysts prepared at pH=6

Since pH during catalyst preparation does not seem to affect the rate significantly, the response surfaces obtained at pH 2 to 6 were similar. Thus, this variable was held constant at pH=6 and the resultant response surface (reaction rate) is shown in Figure 2 with Pt and S loading as determined from the regression equation. It is clear that one can obtain a highly active catalyst even at a low Pt loading by choosing the optimal S loading in the catalyst. Further studies need to be conducted to establish this optimal Pt/S loading.

Conclusions

The current work clearly establishes that the sulfided Pt catalyst is highly effective for the reductive alkylation of ADPA with MIBK. The most significant parameter that determines the reductive alkylation rate is the Pt/S ratio.

References

1. H. Greenfield, Chemical Industries (Dekker), **53**, (*Catal. Org. React.*), 265-277 (1994).
2. P. Rylander, *Catalytic Hydrogenation in Organic Synthesis* (Academic Press), 165-174 (1979).
3. V. Mylroie, L. Valente, L. Fiorella, and M. A. Osypian, Chemical Industries (Dekker), **68**, (*Catal. Org. React.*), 301-312 (1996).
4. G. M. Loudon, *Organic Chemistry*, Oxford University Press, New York, 2002, p. 875,
5. Suh, D. J., Park, T. -J., and Kim, S. –K., Carbon, **31**, 427 (1993).
6. D. S. Thakur, B. D. Roberts, T. J. Sullivan, G. T. White, N. Brungard, E. Waterman, and M. Lobur, Chemical Industries (Dekker), **53**, (*Catal. Org. React.*), 561-568 (1994).
7. K. Bartels, K. Deller, B. Despeyroux, and J. Simon, Pat. DE 4,319,648, to Degussa AG (1994).

55. Accelerated Identification of the Preferred Precious Metal Powder Catalysts for Selective Hydrogenation of Multi-functional Substrates

Dorit Wolf, Steffen Seebald and Thomas Tacke

Degussa AG, Exclusive Synthesis & Catalysts EC-KA-RD-WG
Rodenbacher Chaussee 4, 63457 Hanau (Wolfgang)

dorit.wolf@degussa.com

Abstract

A heuristic method (so called "profiling analysis") was developed. The method is based on a large library of precious metal powder catalysts with different metal loading, metal dispersion, degree of reduction and functionalities of support materials and a data base comprising activity data from hydrogenation of mono-functional substrates. The data base allows the fast identification of potential catalysts for hydrogenation of multifunctional substrates from the library. In the presentation the usage of the profiling method is demonstrated for identification of a highly selective catalyst for hydroxyl-olefin hydrogenation to the hydroxy-alkane.

Introduction

The development of heterogeneous catalysts is related to the challenge that solid properties determining the catalytic properties are not easily accessible. This is especially true for fine tuning of selectivity and long-term catalytic stability where gradual changes by 1 % are already of importance. Regardless, the fact that catalysts do not show obvious differences with respect to solid properties (e. g. metal particle size, particle dispersion or solid phase and oxidation state of the active metal) they often reveal differences in their catalytic behaviour. For industrial application of catalysts in fine chemistry these circumstances are serious obstacles for a straightforward rational development and the identification of suitable catalysts for conversion of certain substrates [1].

Against this background, a method (so called "catalytic performance profiling") was developed with which catalytic characteristics of heterogeneous catalysts for fine chemical application can be efficiently elucidated. The catalytic performance profiling method introduced in the presentation is a heuristic method which takes into account for the complexity of the relationship between catalyst preparation method and catalytic performance in a large diversity of classes of reactions of (multi-functional) substrates. For this purpose catalytic tests with a variety of reactions are performed. The particular performance values of the catalytic reactions are summarized by catalytic performance profiles which are unique fingerprint for individual catalysts allowing a fast identification of strength and weaknesses of a catalyst.

Based on this approach, libraries of heterogeneous catalysts can be built up which cover a wide range of fine-chemical application of solid catalysts from which suitable catalysts can be chosen rapidly.

In this paper, principles and potential of the catalytic performance profiling is introduced and illustrated with respect to the identification of a suitable Pd/C catalyst for selective hydrogenation of a hydroxy-olefin from a library of diverse palladium catalysts.

Reaction scheme 1 (R1, R2 = alkyl)

Results and Discussion

Methodology: In Figure 1 a) and b) the principles of catalytic profiling analysis are explained: Catalytic profiling analysis includes a set of test reactions which are very sensitive with respect to catalyst properties and/or the recipes of preparation. From activity and selectivity values measured for a set of test reactions (Fig. 1a) corresponding performance profiles (Fig. 1b)) can be derived which can be understood as catalytic fingerprints for individual catalysts. Thus, performance profiles allow a statistical analysis of similarities (Fig. 1b).

Example for demonstration: The example shall demonstrate that a data base comprising activity data from hydrogenation of mono-functional substrates allows a pre-selection of potential catalysts for hydrogenation of multifunctional substrates. Based on this pre-selection concept the process of identifying the optimal precious metal powder catalysts is accelerated.

For proof of principle, a catalyst which leads to selective formation of saturated alcohol according to reaction scheme in Figure 2 was identified among a group of sixteen different Pd-catalysts prepared by different methods and showing different metal particle size, metal dispersion and oxidation state of Pd. For pre-selection of promising catalysts, profiling data concerning C=C double bond hydrogenation and hydrogenolysis are of interest. Activity data for hydrogenation of cinnamic acid which represents C=C double bond hydrogenation and debenzylation of dibenzylether, which represents hydrogenolysis, were chosen as pre-selection criteria and visualized by performance profiles (Fig. 3). The profiles refer to the following activity values:

Hydrogenation of cinnamic acid
     (1) Activity of hydrogen conversion at reaction time t = 0

Hydrogenation of dibenzylether
     (2) Activity for hydrogen conversion at reaction time t = 0
     (3) Activity for hydrogen conversion at reaction t = 80 min

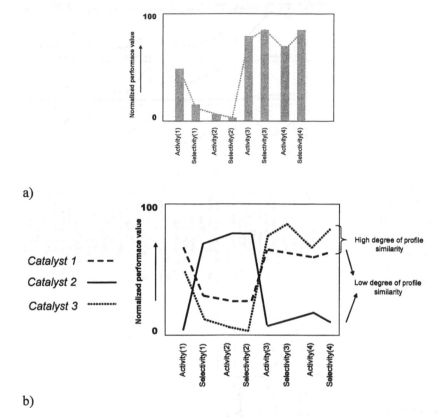

a)

b)

Figure 1: Illustration of the concept for catalyst profiling based on a set of sensitive test reactions:
a) Determination of particular performance data for an individual catalyst (Activity and selectivity values in different test reactions)
b) Visualization of performance profiles from the particular performance values as fingerprints for individual catalysts as a basis of similarity analysis

Catalysts which are expected to be highly selective in the hydrogenation of hydroxy-olefin should reveal high activity in C=C-double bond hydrogenation but low activity in hydrogenolysis. Accordingly, a hypothetical performance profile can be drawn which reflects an ideal catalyst revealing highest activity in C=C-double bond hydrogenation and zero activity in the hydrogenolysis as shown in Figure 3. Now, the profiles of the real catalysts shown in Fig. 2 can be compared with the hypothetical ideal profile based on statistical similarity analysis. Those Pd catalysts showing most similar profiles with respect to the hypothetical profile should be the preferable catalysts for selective hydrogenation of the hydroxy-olefin.

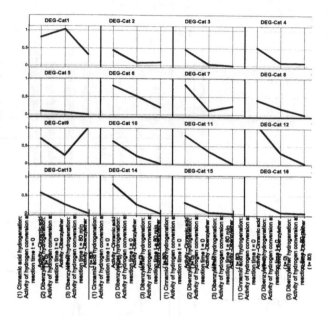

Figure 2.
Selected performance profiles focussing on catalyst activity in C=C-double bond hydrogenation and hydrogenolysis for various Pd/C catalysts

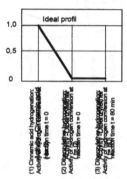

Figure 3.
Hypothetical ideal performance profile for selective hydrogenation of hydroxyl-olefin to hydroxyl-alkane

The statistical similarity analysis was performed based on determination of Euclidean distance between hypothetical and catalyst profile according to the following formula:

$$similarity = \sum \frac{(real\ perfomance\ value - ideal\ performance\ value)_i^2}{(ideal\ performance\ value)_i^2}$$

Since relative distances are considered in this formula, small positive deviations from zero-activity for the hydrogenolysis have strong impact on the similarity value.

Figure 4 indicates the ranking of similarity of the 16 Pd catalysts with respect to the hypothetical ideal profile shown in Figure 3. Accordingly, catalyst DEG-3 appears to be the preferable catalyst for the selective hydrogenation of hydroxy-olefin. DEG-16, DEG-14 and DEG-12 are also expected to give high yield of the saturated alcohol.

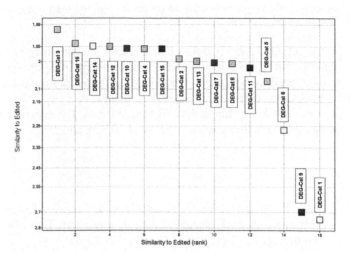

Figure 4.
Ranking of
similarities of the
Pd catalysts with
respect to the hy-
pothetical ideal
performance pro-
file shown in
Figure 3

The proof that this pre-selection meets indeed the most selective catalysts for the hydrogenation of the hydroxy-olefin is derived from Figure 5. Here, the yield of saturated alcohol obtained by conversion of the hydroxy-olefin is plotted vs. the similarity values derived from the Euclidean distance between performance profiles of real catalysts (Figure 2) and the ideal profile (Figure 3). The correlation between yield and similarity is significant. Therefore, the catalytic performance profiling appears to be a fast and unerring method for pre-selection of catalysts for hydrogenation of multi-functional substrates.

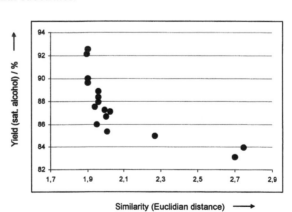

Figure 5.
Correlation
between yield in
hydroxy-alkane
and similarity
value derived
from profiling
analysis

Experimental section

For the activity tests an 8-fold batch reactor system (reactor volume 20 ml) with magnetic stirring which allows the measurement of hydrogen uptake at constant hydrogen pressure was used. Analysis of substrates and products was performed off-line by GC for determining selectivity values. Activity values were derived from hydrogen up-take within a defined time interval. Hydrogenation of both cinnamic acid and dibenzylether were carried out at 10 bars and 25°C.

References

1. Chen, B., Dingerdissen, U., Krauter, J. G. E., Lansink Rotgerink, H. G. J.,
    Möbus, K., Ostgard, D. J., Panster , P., Riermeier, T. H., Seebald, S., Tacke, T.
    Trauthwein, H., *Appl. Catal. A* **280** *17-46* (2005)

56. Transition Metal Removal from Organic Media by Deloxan® Metal Scavengers

Jaime B. Woods[1], Robin Spears[1], Jürgen Krauter[2], Tim McCarthy[3], Micheal Murphy[3], Lee Hord[4], Peter Doorley[4], Baoshu Chen[1]

[1] Degussa Corporation, Business Line Catalysts, 5150 Gilbertsville Highway, Calvert City, KY 42029, USA
[2] Degussa AG, Business Line Catalysts, Rodenbacher Chaussee 4, D-63457, Hanau, Germany
[3] Degussa Corporation, Business Line Catalysts, 379 Interpace Parkway, Parsippany, NJ, 07054, USA
[4] Mobile Process Technology, 2070 Airways Boulevard, Memphis, TN 38114, USA

jaime.woods@degussa.com

Abstract

Catalysis is a valuable and indispensable tool in organic synthesis. Transition metals, most preferably precious metals, are often used. However, potential residual metal contamination from these heterogeneously or homogeneously catalyzed reactions may be detrimental to product quality or, as in the case of active pharmaceutical ingredients (API's), the metal concentration in the final product may be regulated. Degussa's Deloxan® Metal Scavengers recover valuable precious metals from reaction mixtures and reduce the metal concentration in process solutions to an acceptable level (<5 ppm).

In this study, two Deloxan® Metal Scavengers were investigated. The first, THP II, is a thiourea functionalized polysiloxane while the second, MP, is mercapto functionalized. Both resins have been tested in solutions containing $20 - 100$ ppm Pd(II), Pd(0) or Ru(II). In addition to different metals and oxidation states, the effects of solvent (polar vs. nonpolar), temperature ($25 - 80\ ^\circ$C) and mode (fixed bed vs. batch) were explored. These resins were found to reduce precious metal concentrations in process solutions to levels at or below the target concentration of 5 ppm, even at room temperature in the case of Pd(II) and Pd(0). The results of this study will be discussed.

Introduction

Transition metal catalyzed reactions are becoming commonplace in synthetic chemistry. Heterogeneous or homogeneous catalysts containing valuable metals such as palladium, platinum, rhodium and ruthenium are frequently used in the manufacture of active pharmaceutical ingredients (API).[1] The use of such catalysts can lead to metal contamination of the product. The amount of precious metal residue permitted to remain in pharmaceutical products is strictly regulated and for Pd and other platinoid metals this level should be below 5 ppm.[2]

Several methods of metal removal from process solutions exist. Some of these approaches include crystallization, distillation and extraction. However, these methods can be very time-consuming and often result in loss of some valuable product. To overcome these shortcomings, Degussa has developed Deloxan® Metal Scavengers as an attractive option to solve problems associated with transition metal contamination of process solutions.[3, 4, 5] Two resins are commercially available. Deloxan® THP II (**1**) is a thiourea functionalized, macroporous, organofunctional polysiloxane while Deloxan® MP (**2**) is mercapto functionalized.

These Deloxan® beads have an inert siloxane matrix which is advantageous over other adsorbents such as activated carbon or organic polymer-based resins and fibers because (1) valuable API product is not adsorbed and lost as is the case when activated carbons are used and (2) they are chemically resistant to most solvents and stable over a wide pH range (0–12).

$$
\begin{array}{cc}
\text{—O—Si—CH}_2\text{-CH}_2\text{-CH}_2\text{-NH} & \text{—O—Si—CH}_2\text{-CH}_2\text{-CH}_2\text{-SH} \\
\text{C=S} & \\
\text{—O—Si—CH}_2\text{-CH}_2\text{-CH}_2\text{-NH} & \text{—O—Si—CH}_2\text{-CH}_2\text{-CH}_2\text{-SH}
\end{array}
$$

1   **2**

In addition to these, Deloxan® THP II and MP possess many other characteristics that are essential for good metal scavenging. First, they have good to excellent scavenging properties for most transition metals. They are structurally stable showing no shrinking or swelling and compatible with all solvents. These are highly attrition resistant and thermally stable. Deloxan® Metal Scavengers have a well-defined pore structure and due to the bead size (0.4–1.4 mm) and shape they can be applied in either fixed bed or batch mode. They can be refined allowing for recovery of precious metals. Importantly, Deloxan® THP II and MP are commercially available in large quantities.

Results and Discussion

Due to the particle size of Deloxan® THP II and MP, these resins can be used in either batch or fixed bed mode to scavenge transition metals from contaminated process solutions. In this work both modes of operation were investigated. In addition to mode of use, many other variables can influence the effectiveness of a metal scavenger. Oxidation state of the metal, solvent characteristics (polar or nonpolar) and nature of the metal complex (ligands) are just a few of these variables and these were chosen for investigation in this body of work.

Effect of oxidation state and solvent

Several factors can influence the effectiveness of a metal scavenger, one of which is the oxidation state of the metal. In these studies the removal of Pd(II) as trans-di(μ-acetato)bis[o-(di-o-tolylphosphino)benzyl]dipalladium(II) (**3**) and Pd(0) as tetrakis-(triphenylphosphine)palladium(0) (**4**) was investigated.

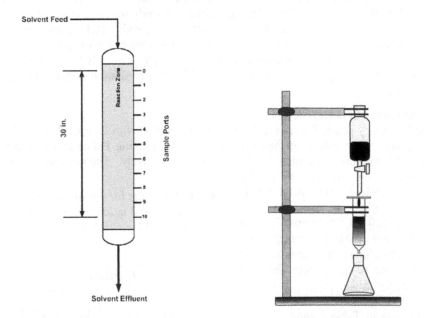

3   **4**

In pilot scale studies, a 30 in column was packed with 400 mL of Deloxan® THP II or Deloxan® MP (Figure 1a). A dimethylformamide (DMF) solution containing either 62.5 ppm or 71.0 ppm of both Pd(II) and Pd(0) as **3** and **4** was passed through the column at 4 EBV's (empty bed volumes) per hour at room temperature.

Figure 1 (a) Illustration of pilot column used in this study; (b) Illustration of lab-scale setup

Deloxan® THP II and MP were found to be effective at removing Pd from DMF solutions containing both Pd(II) and Pd(0) in the fixed bed mode. As Table 1 shows by the time the process solution had reached the second sample port the level of Pd remaining in the effluent was reduced from 62.5 ppm or 71.0 ppm to <1 ppm at room temperature.

Table 1 Results for removal of Pd(II) and Pd(0) in DMF by Deloxan® THP II and Deloxan® MP at room temperature.

Sample Port Number[a]	Deloxan® THP II ppm Pd	Deloxan® MP ppm Pd
Feed	62.5	71.0
0	3.78	1.83
2	0.03	0.57
10	0.03	N/D
Final	0.03	0.23

[a] 3.5 in distance between ports

Table 2 Results for removal of Pd(II) and Pd(0) in toluene by Deloxan® THP II and Deloxan® MP at room temperature.

Sample Port Number[a]	Deloxan® THP II ppm Pd	Deloxan® MP ppm Pd
Feed	62.5	71.0
0	2.30	1.03
2	0.04	0.15
10	0.04	N/D
Final	0.04	0.00

[a] 3.5 in distance between ports

The same concentration of **3** and **4** in toluene yielded similar results to those obtained for DMF solutions (Table 2). Therefore, this study illustrates two important concepts: (1) whether in polar or nonpolar solvent Deloxan® THP II and MP are capable of removing Pd and (2) regardless of oxidation state of Pd, even if Pd(0) and Pd(II) species are present in the same process solution, Deloxan® THP II and MP are effective at removing Pd to levels well within the regulation limits for API manufacture.

Effect of nature of metal complex

On the laboratory scale, 8.5 cm x 2.1 cm columns packed with Deloxan® THP II or MP (9 g wet) were tested for their effectiveness at removing Pd(II) as either **3** or H_2PdCl_4 (**5**) from their toluene or DMF solutions, respectively. The purpose of these experiments was to see if having different ligands on the metal would affect adsorption of the metal by Deloxan®. 100 mL of approx. 20 ppm solutions was passed through the conditioned columns at a flow rate of 5 mL/min. The setup is given above in Figure 1b.

The data in Table 3 shows that Pd is removed from process solutions by both Deloxan® THP II and MP independent of solvent or nature of Pd complex. All effluent Pd levels were

Table 3 Results for removal of Pd(II) as **5** in DMF and **3** in toluene by Deloxan® THP II and Deloxan® MP at room temperature.

Effluent	Deloxan® THP II ppm Pd	Deloxan® MP ppm Pd
Feed (**5**)	17.3	17.3
Final (**5**)	0.01	0.12
Feed (**3**)	20.5	20.5
Final (**3**)	0.11	0.15

<1 ppm. These values are well within the platinoid group metal specification limits for API applications.

Effect of temperature
The effect of temperature on the scavenging abilities of Deloxan® THP II and MP was explored for Cl₂Ru(PPh)₃ in toluene or DMF. Since Deloxan® THP II and MP metal scavengers have been shown to be excellent at removing Pd(0) and Pd(II) even at room temperature, the effect of temperature on their removal was not investigated. Typically, at higher temperatures the rate of adsorption increases. This effect is primarily due to an increase in the rate of diffusion of the metal contaminant through the solution to the scavenger. If solubility of the metal contaminant is an issue, increasing the temperature increases the solubility thereby increasing the diffusion rate.

As can be seen in Figure 2, after passing 100 mL of 20 ppm Cl₂Ru(PPh)₃ in toluene through an 8.5 cm x 2.1 cm column filled with Deloxan® THP II or MP (9 g wet weight) at a flow rate of 5 mL/min the residual Ru concentration in the solutions was reduced to ≤2 ppm at room temperature. Ru was undetectable by the time the temperature of the process solution was 60 °C. Under the same conditions, the Cl₂Ru(PPh)₃ /DMF system gave acceptable results as all final concentrations were <10 ppm. Besides the 7 ppm Ru observed in the DMF/MP system at 60 °C, it is clear that increasing the temperature of the process solution increases the effectiveness of Deloxan® Metal Scavengers at removing Ru(II).
Although it was shown in a previous section that, in general Deloxan® scavenging abilities are not influenced by the nature of the solvent, from this study it was observed that Deloxan® MP seems to be somewhat less effective in removing Ru from DMF solutions.

Figure 2
Temperature effect on the scavenging abilities of Deloxan® THP II and MP for Ru(II) in toluene or DMF.

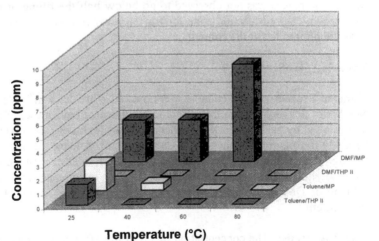

Effect of mode of use
So far, all the studies discussed have involved Deloxan® used in the fixed bed mode. Its use in batch mode was also investigated. When 9 g wet Deloxan® was added to a beaker containing 100 mL of 20 ppm Cl₂Ru(PPh)₃ in either toluene or DMF at 50 °C and stirred with an overhead stirrer for 3 h, the concentration of Ru in the filtrate was

reduced to single-digit ppm levels. Table 4 summarizes these results. The concentration levels obtained in this experiment are comparable to those obtained when similar conditions were used for studying adsorption of Pd in the fixed bed mode (see section "Effect of temperature"). However, in the fixed bed mode the experiment time was 20 min vs. 3 h in batch mode. In addition, an extra step for filtration of the scavenger from the process solution was required when batch mode was employed.

Table 4 Results from batch study of Deloxan® scavenging of Ru(II).[a]

Deloxan®	Toluene ppm Ru[b]	DMF ppm Ru[b]
MP	3	6
THP II	1	6.5

[a] Cl$_2$Ru(PPh$_3$)$_3$
[b] initial concentration 20 ppm Ru

Each mode of use has its own advantages and disadvantages. In batch mode, the entire amount of process solution and the entire amount of scavenger are in contact with each other from the beginning. Metal adsorption starts off quick because there are a large number of free sites available. However, over time the most easily accessible available sites become occupied leading to a decrease in the rate of adsorption of the metal contaminant. In fixed bed mode, the concentration of the metal contaminant decreases as the process solution moves down the column because, if the column is long enough, the process solution is constantly exposed to fresh adsorbent. If the metal contaminant forms a colored complex with the adsorbent, the zone of reaction of the metal with the scavenger is easily observable. Usually the reaction zone is much darker colored at the inlet than the outlet of the column. In the present work, the reaction zone for all Pd complexes was very narrow compared to the reaction zone for Ru(II). The orange-colored reaction zone for Pd complexes was not observed to go below half the filling of metal scavenger. The brown-colored reaction zone for Ru(II), however, was observed to go down to near the bottom frit of the column.

Experimental Section

General Procedure: All columns were packed with Deloxan® THP II or Deloxan® MP Metal Scavenger from Degussa and conditioned before use. The columns were conditioned by passing MeOH (methanol, Acros, extra dry) followed by either DMF (dimethylformamide, Fisher ACS certified) or toluene (Fisher ACS certified) through the column. 20 ppm solutions of H$_2$PdCl$_4$ in DMF, trans-di(μ-acetato)bis[o-(di-o-tolylphosphino)benzyl]dipalladium(II) (3) in toluene and Cl$_2$Ru(PPh$_3$)$_3$ in toluene or DMF were prepared by standard procedures. 3 and 4 were purchased from Strem.

Metal Analysis: The concentration of Pd in the effluent was determined by ICP or AA and the concentration of Ru was extrapolated from a UV-Vis calibration curve. A Leeman Labs, Inc. Direct Reading Eschelle ICP-AES was used for Pd concentration determination in the Deloxan® MP studies while a Varian Spectra AA 55B flame atomic absorption spectrophotometer was used to determine the Pd concentration in the Deloxan® THP II pilot studies. Once the organic solvent was removed the samples were prepared by first reducing the metal with 85 % hydrazine followed by dissolution of the metal in boiling aqua regia. The dissolved metal in aqua regia was diluted to 50 mL in deionized water. A Thermo Electron Corp. GENESYS 10S UV-Vis Spectrophotometer was used for Ru analysis.

Concluding Remarks

Deloxan® THP II and MP Metal Scavengers are effective for removal of metal contamination from process solutions, particularly in the development of APIs. Several factors such as temperature, solvent, oxidation state and nature of the metal, mode of operation as well as number of equivalents of scavenger among other things influence the effectiveness of these scavengers at reducing metal concentrations. Thus, these factors need to be considered when optimizing operating conditions for metal removal. Degussa provides the expertise and support in assessing technical feasibility and scale-up to commercial operations.

Acknowledgements

A special thanks to Mobile Process Technology for pilot scale testing of Deloxan® THP II and MP metal scavengers.

References

1. C. E Garrett, and K. Prasad, *Adv. Synth. Catal.346*, 889 (2004).
2. *Note for Guidance on Specification Limits for Residues of Metal Catalysts*, The European Agency for the Evaluation of Medicinal Products: London, 17 December **2002**; http://www.emea.eu.int
3. S. Wieland, E. Auer, A. Freund, H. L. Rotgerink, P. Panster, Chemical Industries (Dekker), **68**, (*Catal. Org. React.*) 277-286 (1996).
4. a) P. Panster, S. Wieland, EP 0415079 A1 to Degussa AG (1990); b) P. Panster, S. Wieland, EP 0416272 A1 to Degussa AG (1990).
5. G. Cote, F. M. Chen, D. Bauer, *Solvent Extraction and Ion Exchange*, 9(2), 289–308 (1991).

57. Electroreductive Pd-Catalyzed Ullmann Reactions in Ionic Liquids

Laura Durán Pachón and Gadi Rothenberg*

Van 't Hoff Institute for Molecular Sciences, University of Amsterdam,
Nieuwe Achtergracht 166, 1018 WV Amsterdam, The Netherlands.

*gadi@science.uva.nl

Abstract

A catalytic alternative to the Ullmann reaction is presented, based on reductive homocoupling catalyzed by palladium nanoparticles. The particles are generated *in situ* in an electrochemical cell, and electrons are used to close the catalytic cycle and provide the motivating force for the reaction. This system gives good yields using iodo- and bromoaryls, and requires only electricity and water. Using an ionic liquid solvent combines the advantages of excellent conductivity and cluster stabilising.

Introduction

Symmetrical biaryls are important intermediates for synthesising agrochemicals, pharmaceuticals and natural products (1). One of the simplest protocols to make them is the Ullmann reaction (2), the thermal homocoupling of aryl chlorides in the presence of copper iodide. This reaction, though over a century old, it still used today. It has two main disadvantages, however: First, it uses stoichiometric amounts of copper and generates stoichiometric amounts of CuI_2 waste (Figure 1, left). Second, it only works with aryl iodides. This is a problem because chemicals react by their molarity, but are quantified by their mass. One tonne of iodobenzene, for example, contains 620 kg of 'iodo' and only 380 kg of 'benzene'.

In the past five years, we showed that heterogeneous Pd/C can catalyse Ullmann-type reactions of aryl iodides, bromides, and chlorides. Two reaction pathways are possible: Reductive coupling, where Pd^{2+} is generated and reduced back to Pd^0, and oxidative coupling, which starts with Pd^{2+} and needs an oxidising agent. Different reagents can be used for closing the reductive coupling cycle, including HCO_2^- (3), H_2 gas (4), Zn/H_2O (5), and alcohols (6). The two pathways can even be joined, giving a tandem system that converges on one product (7). All of these examples, however, require an extra chemical reagent. In this short communication, we present a different approach, using electrochemistry to close the catalytic cycle (Figure 1, right).

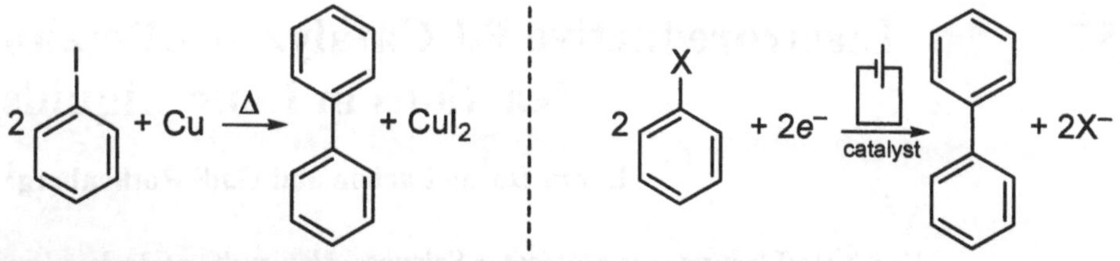

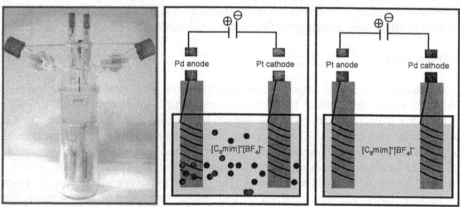

Figure 1 Ullmann reaction (left) and electrochemical catalytic alternative (right).

Results and Discussion

In a typical experiment, the aryl halide was stirred in a specially constructed electrochemical cell (Figure 2, left) containing Pd and Pt electrodes, using the ionic liquid [octylmethylimidazolium]$^+$[BF$_4$]$^-$ as a solvent (8). This solvent combines two important advantages: It is an excellent conductor and it can stabilise metal nanoparticles *via* an ion bilayer mechanism (9). The electrolysis was done using a constant current intensity of 10 mA at 1.6 V (*cf.* $E^0 = -0.83$ V for Pd$^0 \rightarrow$ Pd^{2+} + 2e^-).

Figure 2 Photo of the electrochemical cell (left) and schematics showing the generation of Pd clusters using a Pd anode (middle) and the reverse case (right).

Using PhI as the substrate, the reaction mixture turned from a light yellow solution to a dark brown suspension after 20 min. However, no conversion was observed by GC analysis. We assumed that Pd^{2+} ions, oxidised from the anode, were in turn reduced to adatoms at the Pt cathode and formed Pd^0 nanoparticles, *ca.* 11 nm in diameter (10). After 8 h, the PhI was totally consumed, giving 80% biphenyl and 20% benzene. Weighing the electrodes before and after the reaction showed difference of ~ 2.5 mg in the Pd anode, equivalent to 0.1 mol% of the aryl halide substrate. This corresponds to a TON of 1000 at least (assuming that all the 'missing' Pd participates in the catalysis).

To further investigate the role of palladium nanoparticles in this system, we switched the current between the two electrodes, so that now the Pd electrode was the cathode (Figure 2, right). The rationale behind this experiment was that in theory, the coupling reaction could occur on the cathode surface. Electron transfer from a Pd^0 atom on the cathode surface to PhI would give a $[PhI]^{-\bullet}$ radical anion, that would then dissociate to $Ph^\bullet$ and I^- (11). The constant supply of electrons to the cathode would ensure the electron transfer. However, we did not observe any reaction in this case (nor were any Pd nanoparticles formed). Thus, we conclude that palladium nanoparticles are necessary for catalysing the homocoupling of aryl halides.

Table 1 Ullmann homocoupling of various haloaryls.[a]

Entry	Aryl halide	Biaryl product	Conversion (%)[b]	Yield (%)[b]	TON[c]	Time (h)
1			99	80	816	8
2			99	82	811	8
3			99	75	701	14
4			74	61	540	20
5			80	65	550	24
6			5	4	32	24

[a] Reaction conditions: 20 mmol aryl halide, 50 mL $[omim]^+[BF_4]^-$, 25 °C.
[b] Based on GC analysis, corrected for the presence of an internal standard.
[c] Based on the difference in weight in the Pd anode before and after the reaction.

We then tested several other iodo-, bromo- and chloroaryl substrates. Table 1 shows the conversions, yields and corresponding turnover number (TON). PhI and PhBr gave biphenyl in good yields, and the *p*-nitrophenyls were also active. The corresponding *p*-nitrophenylchloride was much less active.

An interesting question is what closes here the catalytic cycle? Although we do not have a full mechanistic picture at this stage, we think that the complementary half-reaction of the oxidation of the aryl halide is the oxidation of water, *i.e.* $2H_2O \rightarrow O_2 + 4H^+ + 4e^-$ ($E^0 = -1.229$ V). Ionic liquids are notoriously hygroscopic, and a small water impurity is enough to close the cycle. Indeed, control experiments in the presence of 1 molar equivalent of water gave a faster reaction (complete conversion after 6 h, *cf.* with 8 h for the 'dry' system). No difference was found when excess water was added.

In summary, palladium nanoparticles generated *in situ* catalyse Ullmann-type reaction. Using electrochemistry is a simple and efficient way to perform this catalytic reductive homocoupling, and the reaction gives good yields with aryl bromides and iodides. To the best of our knowledge, this is the first electro-reductive palladium-catalysed demonstration of a Ullmann reaction (12). Moreover, this reaction proceeds smoothly at room temperature. Further work in our laboratory will include kinetic and mechanistic studies to gain further understanding into this interesting system.

Experimental Section

Materials and instrumentation. Experiments were performed using a special home-made cell coupled to a dual current supply with a maximum output of 10 V/40 mA. A detailed technical description of this system is published elsewhere (10). ^{1}H NMR spectra were recorded on a Varian Mercury vx300 instrument at 25 °C. GC analysis was performed on an Interscience GC-8000 gas chromatograph with a 100% dimethylpolysiloxane capillary column (DB-1, 30 m × 0.325 mm). GC conditions: isotherm at 105 °C (2 min); ramp at 30 °C min^{-1} to 280 °C; isotherm at 280 °C (5 min). Pentadecane was used as internal standard. The ionic liquid [omim]$^+$[BF$_4$]$^-$ was prepared following a published procedure and dried prior to use (8).All other chemicals were purchased from commercial sources (> 98% pure).

Procedure for Pd clusters-catalysed Ullmann homocoupling. *Example:* biphenyl from PhI. The electrochemical cell was charged with PhI (4.09 gr, 20.0 mmol) and 50 mL [omim][BF$_4$]. After stirring 5 min, a constant current (10 mA, 1.6 V) was applied and the mixture was further stirred for 8 h at 25 °C. Reaction progress was monitored by GC. After 8 h, the product was extracted with ether (3 × 50 mL). The ether phases were combined and evaporated under vacuum to give 1.12 g (75 mol% based on PhI) as a colourless crystalline solid. The solvent can be recycled by washing with aqueous NaBF$_4$. δ_H (ppm, Me$_4$Si): 7.36–7.42 (m, 2H), 7.45–7.51 (m, 4H), 7.62–7.68 (m, 4H). Good agreement was found with the literature values (13).

Acknowledgements

We thank B. van Groen and P. F. Collignon for excellent technical assistance and Dr. F. Hartl for discussions.

References

1. G. Bringmann, R. Walter and R. Weirich, *Angew. Chem. Int. Ed.*, **29**, 977 (1990).
2. F. Ullmann, *Ber. dtsch. chem. Ges.*, **36**, 2359 (1903).
3. S. Mukhopadhyay, G. Rothenberg, D. Gitis, H. Wiener and Y. Sasson, *J. Chem. Soc., Perkin Trans. 2*, 2481 (1999).
4. S. Mukhopadhyay, G. Rothenberg, H. Wiener and Y. Sasson, *Tetrahedron*, **55**, 14763 (1999).
5. S. Mukhopadhyay, G. Rothenberg, D. Gitis and Y. Sasson, *Org. Lett.*, **2**, 211 (2000).
6. D. Gitis, S. Mukhopadhyay, G. Rothenberg and Y. Sasson, *Org. Process Res. Dev.*, **7**, 109 (2003).
7. S. Mukhopadhyay, G. Rothenberg, D. Gitis and Y. Sasson, *J. Org. Chem.*, **65**, 3107 (2000).
8. J. D. Holbrey and K. R. Seddon, *J. Chem. Soc. Dalton*, 2133 (1999).
9. M. J. Earle and K. R. Seddon, *Pure Appl. Chem.*, **72**, 1391 (2000).
10. L. Durán Pachón and G. Rothenberg, *Phys. Chem. Chem. Phys.*, **8**, 151 (2006).
11. T. T. Tsou and J. K. Kochi, *J. Am. Chem. Soc.*, **101**, 6319 (1979).
12. D. Kweon, Y. Jang and H. Kim, *Bull. Kor. Chem. Soc.*, **24**, 1049 (2003).
13. V. Calo, A. Nacci, A. Monopoli and F. Montingelli, *J. Org. Chem.*, **70**, 6040 (2005).

Author Index

Keyword Index